Handbook of Human Factors for Automated, Connected, and Intelligent Vehicles

Handbook of Human Factors for Automated, Connected, and Intelligent Vehicles

Edited by
Donald L. Fisher, William J. Horrey,
John D. Lee, and Michael A. Regan

CRC Press
Taylor & Francis Group
Boca Raton London New York

CRC Press is an imprint of the
Taylor & Francis Group, an **informa** business

First edition published 2020
by CRC Press
6000 Broken Sound Parkway NW, Suite 300, Boca Raton, FL 33487-2742

and by CRC Press
2 Park Square, Milton Park, Abingdon, Oxon, OX14 4RN

© 2020 Taylor & Francis Group, LLC

CRC Press is an imprint of Taylor & Francis Group, LLC

Library of Congress Cataloging-in-Publication Data
Names: Fisher, Donald L., editor. | Horrey, William J., editor. | Lee, John D.,
1965– editor. | Regan, Michael A., editor.
Title: Handbook of human factors for automated, connected, and intelligent
vehicles / edited by Donald L. Fisher, William J. Horrey, John D. Lee,
and Michael A. Regan.
Description: Boca Raton, FL : CRC Press, 2020. | Includes bibliographical
references and index.
Identifiers: LCCN 2019057860 (print) | LCCN 2019057861 (ebook) |
ISBN 9781138035027 (hardback) | ISBN 9781315269689 (ebook)
Subjects: LCSH: Automobiles—Automatic control. | Driver assistance
systems. | Automobile driving—Human factors. |
Traffic safety—Technological innovations.
Classification: LCC TL152.8 .H357 2020 (print) | LCC TL152.8 (ebook) |
DDC 629.2/046—dc23
LC record available at https://lccn.loc.gov/2019057860
LC ebook record available at https://lccn.loc.gov/2019057861

ISBN: 978-1-138-03502-7 (hbk)
ISBN: 978-1-3152-6968-9 (ebk)

Typeset in Times
by codeMantra

Contents

Preface ix

Editors xi

Contributors xiii

Chapter 1   Introduction 1

                    Donald L. Fisher, William J. Horrey, John D. Lee, and
                    Michael A. Regan

Chapter 2   Automated Driving: Decades of Research and Development
                    Leading to Today's Commercial Systems 23

                    Richard Bishop

Chapter 3   Driver's Mental Model of Vehicle Automation 55

                    Bobbie Seppelt and Trent Victor

Chapter 4   Driver Trust in Automated, Connected, and Intelligent Vehicles 67

                    John D. Lee

Chapter 5   Public Opinion About Automated and Self-Driving Vehicles:
                    An International Review 95

                    Mitchell L. Cunningham and Michael A. Regan

Chapter 6   Workload, Distraction, and Automation 107

                      John D. Lee, Michael A. Regan, and William J. Horrey

Chapter 7   Situation Awareness in Driving 127

                    Mica Endsley

Chapter 8   Allocation of Function to Humans and Automation and the
                    Transfer of Control 153

                      Natasha Merat and Tyron Louw

Chapter 9   Driver Fitness in the Resumption of Control 173

Dina Kanaan, Birsen Donmez, Tara Kelley-Baker,
Stephen Popkin, Andy Lehrer, and Donald L. Fisher

Chapter 10 Driver Capabilities in the Resumption of Control 217

Sherrilene Classen and Liliana Alvarez

Chapter 11 Driver State Monitoring for Decreased Fitness to Drive 247

Michael G. Lenné, Trey Roady, and Jonny Kuo

Chapter 12 Behavioral Adaptation... 263

John M. Sullivan

Chapter 13 Distributed Situation Awareness and Vehicle Automation:
Case Study Analysis and Design Implications................................ 293

Paul M. Salmon, Neville A. Stanton, and Guy H. Walker

Chapter 14 Human Factors Considerations in Preparing Policy and
Regulation for Automated Vehicles 319

Marcus Burke

Chapter 15 HMI Design for Automated, Connected, and Intelligent
Vehicles 337

John L. Campbell, Vindhya Venkatraman, Liberty Hoekstra-
Atwood, Joonbum Lee, and Christian Richard

Chapter 16 Human–Machine Interface Design for Fitness-Impaired
Populations 359

John G. Gaspar

Chapter 17 Automated Vehicle Design for People with Disabilities.................. 377

Rebecca A. Grier

Chapter 18 Importance of Training for Automated, Connected, and
Intelligent Vehicle Systems 395

Alexandria M. Noble, Sheila Garness Klauer,
Sahar Ghanipoor Machiani, and Michael P. Manser

Chapter 19 Connected Vehicles in a Connected World: A Sociotechnical
Systems Perspective 421

Ian Y. Noy

Chapter 20 Congestion and Carbon Emissions.. 441

Konstantinos V. Katsikopoulos and Ana Paula Bortoleto

Chapter 21 Automation Lessons from Other Domains 455

Christopher D. Wickens

Chapter 22 HF Considerations When Testing and Evaluating ACIVs 473

Sheldon Russell and Kevin Grove

Chapter 23 Techniques for Making Sense of Behavior in Complex Datasets 497

Linda Ng Boyle

Chapter 24 Future Research Needs and Conclusions 519

*Donald L. Fisher, William J. Horrey, John D. Lee, and
Michael A. Regan*

Preface

Automobile crashes are the seventh leading cause of death worldwide, resulting in over 1.25 million deaths annually. Death lingers forever in the lives of those who have been forced suddenly to say goodbye. In America's most congested city, Washington, DC, the average commuter spends almost two weeks (216 hours) stuck in traffic each year. Congestion wastes the precious time we have to give to others, to meet our own growth needs, and to contribute to overall workforce productivity. A total of 31% of the greenhouse gas emissions are due to the transportation sector, of which 84% are produced by vehicles. Carbon emissions leave future generations at the mercy of our generations' self-indulgence. Health challenges leave millions of individuals confined to their homes without access to jobs and much that life has to offer. To have at our fingertips a new technology like automated, connected, and intelligent vehicles— estimated to reduce crashes, congestion, and carbon emissions significantly and increase access to transportation for those with health issues—is extraordinary at this point in the history of the planet. In spite of this strong potential, we, the editors, believe that the promise of automated, connected, and intelligent vehicles will only be achieved if humans (all road users) are fully considered.

The transition to a completely automated fleet will take decades. In the meantime, the human will be in the loop and therefore a critical part of the system. Researchers, automobile manufacturers and related OEMs, fleet operators, insurance industries, driving instructors, regulatory agencies, policy makers, and the driving public them- selves each have overlapping interests in these systems and in the link between drivers' behavior, the behavior of the systems, and the decrease in crashes, con- gestion, carbon emissions, and confinement to home or institution that the systems were designed to produce. The human factors knowledge that is central to address- ing these various interests is spread widely across disciplines and often difficult to track down. This handbook provides the diverse range of users with the human fac- tors knowledge that they need to make and the advances in understanding that are required if automated, connected, and intelligent vehicles are going to realize their full potential.

Researchers turning to the investigation of automated vehicles will need to under- stand issues like trust in automation, especially as it relates to safety; route choice and fuel efficiency; driver–vehicle interfaces; warning and control systems; transfer of control; measures of driver state; situation awareness, and its converse—cognitive and visual distraction; training; special populations; vulnerable road users; systems and smart cities; and measurement and evaluation. Automobile manufacturers will need to understand both how to design driver–vehicle interfaces that do not over- load the driver at lower levels of automation, bring the driver back into the loop as quickly as possible at the higher levels of automation, and are designed for drivers with a wide range of impairments at all levels of automation. Insurance companies will want to identify those features that drivers can elect to purchase that are most likely to lead to reductions in claims. Driving instructors and other stakeholders will have to generate new methods for training drivers of automated, connected, and intelligent vehicles how to respond when the automated driving suite reaches the boundary of its operating envelope. Regulatory agencies and policy makers will presumably want to understand how best to license drivers to operate automated vehicles and what guidelines original equipment manufacturers of driver–vehicle interfaces could use in order for an interface not to be considered a potentially lethal distraction. Internationally, low- and middle-income countries will need to make hard decisions about what features of automated, connected, and intelligent vehicles will save the most lives and still remain affordable.

In summary, the Handbook combines in one volume the information needed by a diverse community of stakeholders, including human factors researchers, transportation engineers, regulatory agencies, automobile manufacturers, fleet operators, driving instructors, vulnerable road users, and special populations. The Handbook will be particularly important for the engineers developing the new technologies. Technological advances often outpace related, yet fundamental research efforts into the safety and broader implications of newly developed systems. Technology developers are just now beginning to realize the importance of human factors in this burgeoning space, making the proposed Handbook a timely, valuable, and unique resource for the various individuals who will need to design their systems with people in mind and then evaluate those systems in ways that capture the impact of these systems on safety.

The Handbook is an integrated volume, the chapters clearly referencing one another. We hope that this helps readers navigate the Handbook. The Handbook contains chapters written by some of the world's leading experts that provide individuals with the human factors knowledge that will determine in large measure whether the potential of automated, connected, and intelligent vehicles goes realized or unrealized, not just in developed countries, but in low- and middle-income countries as well. The Handbook addresses four major transportation challenges—crashes, congestion, carbon emissions, and confinement—from a human factors perspective in the context of automated, connected, and intelligent vehicles. We hope that Handbook will help set the stage for future research in automated, connected, and intelligent vehicles by defining the space of unexplored questions. Researchers will not only have a place to turn for answers to questions but also a place to search for the next generation of questions.

A final comment on the cover of the Handbook is in order. The history of automated vehicles has its ups and downs, like a road with many twists and turns. But these diversions are not a function of change in the hope we and many involved in this history have for such vehicles, vehicles which can literally address many of the major challenges facing the world today: the economic, psychological or personal cost of automobile crashes, congestion, carbon emissions, and lack of accessibility to transportation. We hope, like the picture of the winding roads (our interpretation of the cover art), that the progress is ever upward over time, towards a future where the problems, like the roads, disappear.

Editors

Donald L. Fisher is a principal technical advisor at the Volpe National Transportation Systems Center in Cambridge, MA, a professor emeritus and research professor in the Department of Mechanical and Industrial Engineering at the University of Massachusetts (UMass) Amherst, and the director of the Arbella Insurance Human Performance Laboratory in the College of Engineering. He has published over 250 technical papers, including recent ones in the major journals in transportation, human factors, and psychology. While at the Volpe Center he has worked extensively across the modes on research designed to identify the unintended consequences of automation and, when such are identified, to develop and evaluate countermeasures. Additionally, he has developed a broad, interdisciplinary approach to understanding and remediating functional impairments in transportation, including distraction, fatigue, and alcohol. While at UMass Amherst, he served as a principal or co-principal investigator on over 30 million dollars of research and training grants, including awards from the National Science Foundation, the National Institutes of Health, the National Highway Traffic Safety Administration, MassDOT, the Arbella Insurance Group Charitable Foundation, the State Farm Mutual Automobile Insurance Company, and the New England University Transportation Center. He is a former associate editor of *Human Factors* and editor of both the recently published *Handbook of Driving Simulation for Engineering, Medicine and Psychology* (2011) and the *Handbook of Teen and Novice Drivers* (2016). He currently is a co-chair of the Transportation Research Board (TRB) Committee on Simulation and Measurement of Vehicle and Operator Performance. He has chaired or co-chaired a number of TRB workshops and served as a member of the National Academy of Sciences Human Factors Committee, the TRB Younger Driver Subcommittee, the joint National Research Council and Institute of Medicine Committee on the Contributions from the Behavioral and Social Sciences in Reducing and Preventing Teen Motor Crashes, and the State Farm® Mutual Automobile Insurance Company and Children's Hospital of Philadelphia Youthful Driver Initiative. Over the past 25 years, Dr Fisher has made fundamental contributions to the understanding of driving, including the identification of those factors that: determine how most safely to transfer control from an automated vehicle to the driver; increase the crash risk of novice and older drivers; impact the effectiveness of signs, signals, and pavement markings; improve the interface to in-vehicle equipment, such as forward collision warning systems, back over collision warning systems, and music retrieval systems; and influence drivers' understanding of advanced parking management systems, advanced traveler information systems, and dynamic message signs. In addition, he has pioneered the development of both PC-based hazard anticipation training and PC-based attention maintenance training programs, showing that novice drivers so trained actually anticipate hazards more often and maintain attention better on the open road and in a driving simulator. This program of research has been made possible by the acquisition in 1994 of more than half a million dollars of equipment, supported in part by a grant from the National Science Foundation. He has often spoken about his results, including

participating in a congressional science briefing on the novice driver research spon-
sored several years previously. The Human Performance Laboratory was recognized
by the Ergonomics Society, receiving the best paper award for articles that appeared in
the journal *Ergonomics* throughout 2009. The paper described the work in the Human
Performance Laboratory on hazard anticipation. Most recently, he published with his
Volpe colleagues a review of human factors issues critical to the development of intelligent vehicles in the inaugural volume of *IEEE Transactions on Intelligent Vehicles*.
Dr Fisher received an Bachelor of Arts (AB) from Bowdoin College in 1971 (philosophy),
a Master of Education (EdM) from Harvard University in 1973 (human development),
and a PhD from the University of Michigan in 1982 (mathematical psychology).

William J. Horrey, PhD, is the traffic research group leader at the AAA
Foundation for Traffic Safety. Previously, he was a principal research scientist in the
Center for Behavioral Sciences at the Liberty Mutual Research Institute for Safety.
He earned his PhD in engineering psychology from the University of Illinois at
Urbana–Champaign in 2005. He has published over 50 papers on numerous topics
including visual (selective) and divided attention, automation, driver behavior, and
distractions from in-vehicle devices. He chairs the Transportation Research Board
Standing Committee on Vehicle User Characteristics (AND10) and the Publications
Division at the Human Factors and Ergonomics Society. He is an associate editor
of the *Human Factors* Journal and has served on several national and international
committees related to transportation safety and human factors.

John D. Lee, PhD, is the Emerson Electric Professor in the Department of Industrial
and Systems Engineering at the University of Wisconsin, Madison and director of
the Cognitive Systems Laboratory. Dr Lee's research seeks to better integrate people
and technology in complex systems, such as cars, semi-autonomous systems, and
telemedicine. His research has led to over 400 publications and presentations, including 13 books. He helped to edit *The Oxford Handbook of Cognitive Engineering*, the
Handbook of Driving Simulation for Engineering, Medicine, and Psychology, and
two books on distraction *Driver Distraction: Theory, Effects, and Mitigation* and
Driver Distraction and Inattention. He is also the lead author of a popular textbook
Designing for People: An Introduction to Human Factors Engineering.

Michael A. Regan is a professor of human factors with the Research Centre for
Integrated Transport Innovation at the University of New South Wales in Sydney,
Australia. He has BSc (Hons) and PhD degrees in engineering psychology from the
Australian National University and has designed and led more than 200 research
projects in transportation human factors and safety—spanning aircraft, motorcycles,
cars, trucks, buses, and trains. Mike is the author/co-author of around 200 peer-
reviewed publications, including three books on driver distraction and inattention,
and driver acceptance of new technologies. He was the 25th president of the Human
Factors and Ergonomics Society of Australia and is a Fellow of the Australasian
College of Road Safety.

Contributors

Liliana Alvarez
Occupational Therapy
University of Western Ontario
London, ON, Canada

Richard Bishop
Bishop Consulting
Highland, MD, USA

Ana Paula Bortoleto
Sanitation and Environment
University of Campinas, School of Civil
    Engineering, Architecture and Urban
    Planning
Campinas, Brazil

Linda Ng Boyle
Industrial and Systems Engineering
University of Washington
Seattle, WA, USA

Marcus Burke
National Transport Commission
    Australia
Melbourne, Australia

John L. Campbell
Exponent, Inc.
Bellevue, WA, USA

Sherrilene Classen
Occupational Therapy
University of Florida
Gainesville, FL, USA

Mitchell L. Cunningham
School of Psychology
The University of Sydney
Sydney, Australia

Birsen Donmez
Mechanical and Industrial
    Engineering
University of Toronto
Toronto, ON, Canada

Mica Endsley
SA Technologies
Gold Canyon, AZ, USA

Donald L. Fisher
Transportation Human Factors (V314)
Volpe National Transportation
    Systems Center
Cambridge, MA, USA

John G. Gaspar
National Advanced Driving Simulator
University of Iowa
Iowa City, IA, USA

Rebecca A. Grier
Independent
Dearborn, MI, USA

Kevin Grove
Center for Truck and Bus
    Safety and Automated Vehicle
    Systems
Virginia Tech Transportation Institute
Blacksburg, VA, USA

Liberty Hoekstra-Atwood
Battelle
Seattle, WA, USA

William J. Horrey
AAA Foundation for Traffic Safety
Washington, DC, USA

Dina Kanaan
Mechanical and Industrial
  Engineering
University of Toronto
Toronto, ON, Canada

Konstantinos V. Katsikopoulos
Decision Analytics and Risk
University of Southampton
Southampton, UK

Tara Kelley-Baker
AAA Foundation for Traffic Safety
Washington, DC, USA

Sheila Garness Klauer
Center for Vulnerable Road
  User Safety
Virginia Tech Transportation
  Institute
Blacksburg, VA, USA

Jonny Kuo
Seeing Machines Ltd
Fyshwick, Australia

John D. Lee
Department of Industrial and Systems
  Engineering
University of Wisconsin-Madison
Madison, WI, USA

Joonbum Lee
Battelle
Seattle, WA, USA

Andy Lehrer
DIGITALiBiz, Inc.
Cambridge, MA, USA

Michael G. Lenné
Seeing Machines Ltd.
Fyshwick, Australia

Tyron Louw
Institute for Transport Studies
University of Leeds
Leeds, UK

Sahar Ghanipoor Machiani
Department of Civil, Construction, and
  Environmental Engineering
San Diego State University
San Diego, CA, USA

Michael P. Manser
Texas A&M Transportation Institute
Austin, TX, USA

Natasha Merat
Institute for Transport Studies
University of Leeds
Leeds, UK

Alexandria M. Noble
Virginia Tech Transportation Institute
Blacksburg, VA, USA

Ian Y. Noy
Independent
Fort Myers, FL, USA

Stephen Popkin
Volpe National Transportation
  Systems Center
Cambridge, MA, USA

Michael A. Regan
School of Civil and Environmental
  Engineering and Research Centre
  for Integrated Transport Innovation
University of New South Wales
Sydney, Australia

Christian Richard
Battelle
Seattle, WA, USA

Trey Roady
Seeing Machines Ltd
Fyshwick, Australia

Sheldon Russell
Center for Automated Vehicle Systems
Virginia Tech Transportation Institute
Blacksburg, VA, USA

Paul M. Salmon
Centre for Human Factors and
  Sociotechnical Systems
University of the Sunshine Coast
Maroochydore, Australia

Bobbie Seppelt
AgeLab
Massachusetts Institute of Technology
Cambridge, MA, USA

Neville A. Stanton
Transportation Research Group
University of Southampton
Southampton, UK

John M. Sullivan
University of Michigan Transportation
  Research Institute
Ann Arbor, MI, USA

Vindhya Venkatraman
Battelle
Seattle, WA, USA

Trent Victor
Volvo Cars Safety Centre
Chalmers University of Technology
Göteborg, Sweden

Guy H. Walker
School of the Built Environment
Heriot-Watt University
Edinburgh, UK

Christopher D. Wickens
Department of Psychology
Colorado State University
Fort Collins, CO, USA

1 Introduction

Donald L. Fisher
Volpe National Transportation Systems Center

William J. Horrey
AAA Foundation for Traffic Safety

John D. Lee
University of Wisconsin Madison

Michael A. Regan
University of New South Wales

CONTENTS

Key Points 2
1.1 Background 2
1.2 Definitions 5
    1.2.1 Levels of Automation and Active Safety Systems 5
        1.2.1.1 Levels of Automation 5
        1.2.1.2 Active Safety Systems 7
    1.2.2 Automated, Connected, and Intelligent Vehicles 8
        1.2.2.1 Automated Vehicles 8
        1.2.2.2 Connected Vehicles 8
        1.2.2.3 Intelligent Vehicles 8
    1.2.3 Operational Design Domain 9
1.3 The Handbook: A Quick Guide 10
    1.3.1 The State of the Art: ACIVs (Chapter 2) 11
    1.3.2 Issues in the Deployment of ACIVs (Problems) 11
        1.3.2.1 Driver's Mental Model of Vehicle Automation
                (Chapter 3) 11
        1.3.2.2 Driver Trust in ACIVs (Chapter 4) 12
        1.3.2.3 Public Opinion about ACIVs (Chapter 5) 12
        1.3.2.4 Workload, Distraction, and Automation (Chapter 6) 12
        1.3.2.5 Situation Awareness in Driving (Chapter 7) 13
        1.3.2.6 Allocation of Function to Humans and Automation and
                the Transfer of Control (Chapter 8) 13
        1.3.2.7 Driver Fitness in the Resumption of Control (Chapter 9) 13
        1.3.2.8 Driver Capabilities in the Resumption of Control
                (Chapter 10) 14

          1.3.2.9   Driver State Monitoring for Decreased Fitness to Drive
                    (Chapter 11).. 14
          1.3.2.10 Behavioral Adaptation (Chapter 12) 14
          1.3.2.11 Distributed Situation Awareness (Chapter 13)...................... 14
          1.3.2.12 Human Factors Issues in the Regulation of Deployment
                    (Chapter 14).. 15
     1.3.3 Human-Centered Design of ACIVs (Solutions).................................. 15
          1.3.3.1   HMI Design for ACIVs (Chapter 15).................................... 15
          1.3.3.2   HMI Design for Fitness Impaired Populations (Chapter 16)... 16
          1.3.3.3   Automated Vehicle Design for People with Disabilities
                    (Chapter 17).. 16
          1.3.3.4   Importance of Training for ACIVs (Chapter 18) 16
     1.3.4 Special Topics 17
          1.3.4.1   Connected Vehicles in a Connected World: A
                    Sociotechnical Systems Perspective (Chapter 19) 17
          1.3.4.2   Congestion and Carbon Emissions (Chapter 20) 17
          1.3.4.3   Automation Lessons from Other Domains (Chapter 21) 18
     1.3.5 Evaluation of ACIVs 18
          1.3.5.1   Human Factors Considerations in Testing and
                    Evaluating ACIVs (Chapter 22) 18
          1.3.5.2   Techniques for Making Sense of Behavior in Complex
                    Datasets (Chapter 23).. 18
1.4 Conclusion 19
Acknowledgments.. 19
References.. 20

KEY POINTS

- Automated, connected, and intelligent vehicles hold great promise—increasing safety for all and mobility for underserved populations while decreasing congestion and carbon emissions.
- There may be unintended consequences of advances in automated technologies that affect the benefit that drivers can derive from these technologies, potentially slowing the development of the technologies themselves.
- Many of these unintended consequences center around human factors issues, issues between the driver and the vehicle, other road users, and the larger transportation system.
- Human factors research can be used to identify and seek to explain the unintended consequences, to develop and evaluate countermeasures, and to decrease, if not entirely avoid, any delay in the deployment of these technologies.

1.1 BACKGROUND

We as humans cannot help but wonder what the future will hold and how it will unfold in time. When it comes to the effect of advanced technologies on our behaviors and on the behavior of the vehicles that we drive, the public speculation has been especially intense over the last ten years, starting in 2009 when Google[1] began its self-driving car program, now called Waymo. Such vehicles have the potential to substantially reduce the number of crashes, the level of carbon emissions, the congestion in our road systems, and the spread of wealth inequality, while at the same time increasing substantially opportunities for those who are mobility impaired (National Highway Traffic Safety Administration, 2017; Department of Transportation, 2019; Chang, 2015). Although some individuals are skeptical about early presumptions regarding the benefits of automated, connected, and intelligent vehicles (ACIVs) (Noy, Shinar, & Horrey, 2017; Bonnefon, Shariff, & Rahwan, 2016), the introduction of vehicles with advanced features continues to increase exponentially. As with the advent of the smartphone, anticipating the long-term positive and negative consequences of new technology is nearly impossible (e.g., Islam & Want, 2014; Twenge, Martin, & Campbell, 2018). It may be some time before we actually know the real benefits of such vehicles and features.

However, it is possible to take the bumps out of the road to full automation even without knowing the long-term consequences. This Handbook will focus specifically on the changes that will be wrought and the corresponding human factors challenges that need to be addressed by advances in the autonomy, connectivity, and intelligence of the vehicles that are being introduced into the fleet today and are likely to be introduced over the next several years. For readers relatively new to the discussion of why human factors concerns might be relevant to advanced technologies in the automobile, a simple example from one of the editors' and authors' long list of examples might help. This particular editor was driving 60 mph on a highway with two travel lanes in each direction and for a brief second or two fell asleep (had what is technically referred to as a "microsleep"). He drifted into the adjacent lane, woke up, and returned to his own lane. Had there been a large truck overtaking him in the adjacent lane, he might not be here to tell the story. Others' lives may have been destroyed as well. But, fortunately, there was no truck and all was well. This speaks directly to the lifesaving potential of technologies which, in this case, could have kept the car in the lane and maintained speed adaptively. But it also points out just how beguiling these technologies can be.

Most vehicles on the road today that keep the car centered and adjust the speed require the driver to constantly monitor the driving environment (SAE International, 2018). Why? If we consider just automatic steering, there are many situations in which it may unexpectedly deactivate. The driver really does need to be in the loop. But, we also know that, perversely, automation can make it easier for the driver to fall out of the loop and become disengaged (Endsley, 2017; Endsley, 1995). Are we just trading off situations in which the technologies can be lifesaving for situations in which the technologies actually create conditions that increase the likelihood that a driver will crash if the technology cannot handle a particular scenario? This is the fundamental paradox of automation. While it can provide unparalleled opportunities, it comes with its own set of challenges.

Perhaps this paradox is best exemplified by a recent study of driver's trust in automation. A field study was run in which the drivers were asked to navigate a network

---

[1] Now Alphabet.

of roads on a closed course using a vehicle with both automatic steering and adaptive cruise control (ACC) (Victor, Tivestan, Gustafsson, Sangberg, & Aust, 2018). The drivers were told that they needed to monitor the driving environment and were warned by the vehicle driver state monitoring system if they did not comply. At the end of the drive, either a car or a large garbage bag was placed in their path. Both were easily crushed (e.g., a balloon car), but not obviously so to the driver before striking them. Driver's trust in automation was measured after the drive on a scale of 0 (no trust) to 7 (high trust). Fully 21 of 76 drivers crashed (28%). All of the drivers who crashed had trust scores of 5 or higher (Victor, 2019). In short, the drivers became so reliant on the technology that they assumed it would avoid obstacles even when the technology encountered situations it was not designed to accommodate.

For readers familiar with the ongoing issues, you will find material in this Handbook which we believe will help set the stage for a human focus on future discussions about ACIVs. To date, the definition that vehicle manufacturers and major standards organizations have put forth concerning automation defines it primarily from a vehicle-centric point of view: as the technology capabilities increase, so too does the level of automation of the vehicle (SAE International, 2018). But this can easily mislead drivers into believing that their role in the driving process decreases as the levels of automation increase, despite warnings to the contrary, to the point where drivers actually feel that they can safely disengage from driving for long periods of time. A more driver-centric viewpoint is essential to extending automation safety benefits, one which defines and supports the new roles that drivers face.

The goal of this Handbook is to identify the real gains that can be achieved by identifying the various human factors, challenges, and opportunities that ACIVs pose and then, whenever possible, to suggest countermeasures (and opportunities for needed research), recognizing that the rapid advances in technology are changing both the challenges and the countermeasures. It is arguably the case that, even if these challenges were not addressed, there would be a net benefit to incorporating advanced technologies into new automobiles (Najm & daSilva, 2000), especially active safety systems like automatic emergency braking (AEB) and electronic stability control. But it is also arguably the case that, if these challenges are addressed, the net benefits will only increase. Moreover, by addressing these challenges, one reduces the real likelihood that the development of the technologies will be hobbled, if not halted, by crashes such as the one that occurred in Phoenix, Arizona (National Transportation Safety Board, 2018).

That a scenario similar to the above might unfold and put temporary brakes on the development and deployment of automated technologies already seems to have occurred, at least in part. At the start of 2018, before the crash in Phoenix, it looked like vehicles with advanced technologies were on the verge of becoming a widespread reality. Uber prepared to launch a robo-taxi service. Waymo indicated that individuals would be able to ride in a driverless car by the end of the year. General Motors touted a demonstration it would undertake in New York City. None of these (and several other similar initiatives) have come to pass, at least yet. In fact, the public has become ever more skeptical about a self-driving vehicle, with some 71% now afraid to ride in such a vehicle compared to 61% before the crash (Edmonds, 2019).

In summary, the paradox of automation is that it can radically reduce human errors while also itself introducing opportunities for new types of human errors. These new types of errors can lead to crashes which slow the introduction of lifesaving technologies. The chapters in this Handbook are just one of many attempts to address this potentially negative feedback cycle, providing both insight into why new types of errors arise and what can be done to overcome them before they do, creating more skepticism about the technology. But before we can speak about the fundamental sources of the automation paradox, some definitions are in order.

1.2   DEFINITIONS

There is an understandable confusion around the terms used in the discussion of automation, human factors, and driving. The terms do change with each passing year, in part because the technologies change and in part because the field becomes more expert and nuanced at understanding how best clearly to differentiate among the various terms. As editors and authors, we have tried with each chapter to make sure that the same terms are used in the same way and, if we deviate, to make it clear how we are refocusing the definition of a term.

1.2.1   Levels of Automation and Active Safety Systems

Several taxonomies have been developed that differentiate among the levels of automation (e.g., SAE International, 2018; NHTSA, 2016). It is arguably the case that the SAE definition of the levels of automation is the one most commonly used. It is worth spending some time on understanding these levels and how they relate to other vehicle systems.

1.2.1.1   Levels of Automation

As conceptualized in vehicle-centric taxonomies (SAE International, 2018), the level of automation at which a vehicle is currently operating depends on three critical factors (Table 1.1): (1) whether only one, both, or neither of the lateral (automatic steering) and longitudinal (ACC) features are activated; (2) the role of the driver (must monitor or need not monitor the forward roadway); and (3) the time that is given to the driver to resume control (immediate to never). The first three levels of automation are easy enough to define: neither lateral nor longitudinal control features are activated (SAE Level 0), just one of these features is activated (SAE Level 1), or both of these features are activated (SAE Level 2). The role of the driver is always to monitor the roadway at each of these levels. The remaining three levels require additional considerations. Specifically, assuming that the vehicles are not geofenced (confined to operating in a particular location), the last three levels are differentiated most clearly on the basis of when it is necessary and how much time the driver has before it is necessary for him or her to resume control: at any moment, somewhere on the order of several seconds warning (SAE Level 3); only when transitioning from inside to outside the operational design domain, somewhere on the order of minutes (SAE Level 4); never needs to take over control (SAE Level 5) (Table 1.1). As described in the sub-sections that follow, there is considerable variability in the

TABLE 1.1
Active Safety Systems and SAE Levels

Active Safety System	Category	SAE Level	Definition	Driver Role		
				Manual Control	Active Monitoring	Take Over
AEB PCW	Driver Support Features (DSF)	0	**No AS No ACC**	Both steering and slowing/accelerating	Yes	N/A
FCW ESC		1	AS or ACC, but **not both**	Steering or slowing/accelerating, but not both	Yes	Yes, limited warning*
BSW		2	**Both** AS and ACC engaged	Hands on wheel or eyes on road.	Yes	Yes, limited warning*
	Automated Driving System (ADS) Features	3	AS and ACC	No inputs while at Level 3	No	Yes, several seconds of warning
		4	AS and ACC	No inputs while at Level 4	No	Yes, minutes of warning (if driver chooses to drive the vehicle outside of the ODD)
A subset of the above		5	AS and ACC	No inputs ever	No	No

Note: AEB: automatic emergency braking; PCW: pedestrian collision warning; FCW: forward collision warning; ESC: electronic stability control; BSW: blind spot warning; AS: automatic steering; ACC: adaptive cruise control.

* Warnings may indicate that the system or systems are no longer functioning, and drivers are responsible for monitoring the traffic environment and taking control when needed.

complexity and range of possible automation types. As just one example, consider a vehicle which has no steering wheel or pedals and is confined to operate only in a circumscribed location. By definition it is a Level 4 vehicle (since it cannot operate in all locations), but the driver is never asked to take over control since the vehicle is confined to an area in which it is assumed to be totally capable.

There are at least four critical, related, details to understand about these definitions in order to keep clear the underlying concepts. First, the level of automation assigned refers to the features which currently are active, not to the vehicle itself. Thus, a car that could operate at Level 3 may also be operating at times at Level 0, being no different in any way in terms of driver inputs than a car with no automated features. Second, the features associated with the first three levels are referred to frequently as *driver support features* (DSF). They are called DSFs because the driver still needs to be continuously monitoring the roadway. Lateral and longitudinal control by themselves do not replace the driver at Levels 1 or 2; there are many other driving functions and tasks that still need to be performed by the driver at these levels. Third, related to this, the features which need to be added in order to achieve the three highest levels are now referred to as Automated Driving System (ADS) features. (SAE International, 2018; Department of Transportation, 2019). Fourth, the term *advanced driver assistance systems* (or *ADAS*) is now no longer used by many researchers because it has become so broad that it is no longer clear to what features an individual is referring when they speak about such systems. In many cases, the systems that are described elsewhere as being ADAS would overlap with many of those portrayed in Table 1.1 under active safety systems, DSF, or ADS.

1.2.1.2 Active Safety Systems

Operating in parallel with or without automatic steering and ACC are what are referred to as *active safety systems*, like AEB. Active safety systems aim to prevent crashes, either warning the driver or taking over the control of the vehicle, often in emergency situations. Thus, active safety systems differ from passive safety systems that aim to mitigate the injury impacts of crashes on vehicle occupants. Moreover, unlike ACC and automatic steering, active safety systems are not continually providing input to the vehicle. Active safety systems can be present at any one of the SAE Levels (left side of Table 1.1). Because, in theory, any one of the SAE Levels 1–4 can revert to any lower level, all active safety systems that interface with the driver must remain operable. However, because SAE Level 5 is imagined to be fully automatic, it will include only a subset of the active safety systems (e.g., it will include AEB, but not blind spot warnings for the driver, because the driver never needs himself or herself to change lanes). Active safety systems are sometimes identified with SAE Level 0, though this can create confusion because, as noted, this technology can operate in parallel with higher levels of automation (e.g., a Level 2 vehicle might also have Level 0 vehicle technology if AEB was labeled as Level 0 rather than as an active safety system).

In this Handbook, we focus not only on the challenges for the driver that are generated at each of the different SAE levels but also on the problems created by the different active safety systems, recognizing that the primary focus is on the entire system with which the driver needs to interact (directly or indirectly as part of the vehicle and larger transportation system). So, for example, the driver is almost never faced with a vehicle which has, for example, just automatic steering and no active safety systems. The driver is instead immersed in a system with many different features, with the human factors challenges becoming correspondingly more complex.

1.2.2 Automated, Connected, and Intelligent Vehicles

When we started work on this Handbook, "automated" usually preceded "connected" and one would refer to automated (or autonomous) connected vehicles and intelligent vehicles. Now one more frequently encounters the reverse, connected and automated vehicles (or CAVs). By the time the Handbook is published, there may be an entirely different acronym. For various reasons, we have chosen to stay with the original order (ACIV) whose motivation is hopefully made clear below.

1.2.2.1 Automated Vehicles

By automated (or autonomous) vehicles we mean those vehicles that have automatic steering, ACC (adaptive speed, braking, acceleration, and deceleration), or both. Depending on the level, they can be activated separately, in concert, or not at all (e.g., if a driver elects to turn them off; Table 1.1). They almost certainly have active safety systems, especially at higher levels, as such systems are an essential ingredient—at least conceptually—to the proper functioning of the higher levels (e.g., AEB is essential, though it may not exist as a separate safety system but as an integrated component in the overall system).

1.2.2.2 Connected Vehicles

Connected vehicles as a term is typically used to refer to vehicles that can communicate with one another via vehicular ad hoc networks (VANET or inter-vehicle connectivity; V2V). But more generally, connected vehicles as entities in and of themselves can have upwards of 200 sensors that need to communicate with each other (intra-vehicular connectivity or vehicle-to-sensor, V2S) (Lu, Cheng, Zhang, Shen, & Mark, 2014). Vehicles can also connect with the roadway infrastructure (V2R), to the internet (V2I), and to the ubiquitous Internet of Things (IOT). Each of these different types of connectivity creates its own human factors challenges.

Note that vehicles can be connected without having any lateral or longitudinal control features. So, for example, intersection collision warning systems require vehicles to communicate with one another and, by definition, warn the driver of a potential collision, but they do not require that the vehicles have automated features. Similarly, a vehicle can be automated without having any connectivity to other vehicles or the changing infrastructure (e.g., signal status).

1.2.2.3 Intelligent Vehicles

We include the term intelligent vehicles to account for the additional features of future vehicles that do not align with the features of automated and connected vehicles. These include driver state monitoring algorithms that can tune the vehicle to the driver's current level of disengagement, the smart human–machine interface (HMI) algorithms that can alter the HMI to provide the driver with a more comfortable and safe journey, and the all-knowing road transportation system networks that can help the driver select the route from one point to another that maximizes the driver's goals, the collective goals of all drivers, or some combination of the two (be it speed, a reduction in carbon emissions, or whatever).

Just as was the case with CAVs, there could be vehicles with any combination of intelligence, automation, and connectivity. For example, an intelligent HMI might appear in a Level 0 vehicle with no connectivity or ADSs.

1.2.3 OPERATIONAL DESIGN DOMAIN

Finally, it is important to describe what is meant by the operational design domain (ODD) of a given feature, whether that feature is a DSF, an ADS feature, or an active safety system feature. First, consider an active driver safety system and, in particular, AEB; or, what is called by Toyota, a pre-collision braking system (Toyota, 2017).[2] The system consists of a pre-collision warning, a pre-collision brake assist, and an actual pre-collision braking when no force has been applied by the driver to the brake pedal and the vehicle detects a possible collision with another vehicle or pedestrian. The components are active at only certain vehicle speeds and at only certain relative speeds between the driver's vehicle and a lead vehicle. Moreover, the speeds differ for pedestrians and vehicles and the speed constraints differ across the components.

The driver not only needs to remember the various conditions in which the system is active but also needs to know the conditions in which the system might suddenly and unexpectedly activate if there is no possibility of a collision. The owner's manual lists a total of 23 different scenarios in which this might occur. Such scenarios include passing under an object (e.g., billboard) at the top of a long, uphill road; approaching a vehicle, pedestrian, or object by the roadside located at the beginning of a sharp curve; and driving on a winding road where there is another vehicle in the opposing lane. These are not uncommon occurrences.

Second, consider a DSF (Table 1.1, SAE Levels 1–2). It should be noted up front that the owner's manuals make clear that the driver has to be monitoring the roadway at all points in time for each of the various DSFs. As an example, consider ACC in a Cadillac CT6.[3] Adaptive cruise allows the driver to select a speed and a following gap. Like AEB, ACC is active only under certain conditions (Cadillac, 2018). In particular, the speed cannot be set using ACC if the vehicle is traveling less than 15 mph, though ACC can be resumed at speeds lower than 15 mph. Just as AEB can unexpectedly activate, so too can ACC if the driver does not understand the ODD. For example, after the ACC is set the driver's vehicle may immediately apply the brakes if the lead vehicle is closer than the time headway that has been selected by the driver. Perhaps, more critically, ACC may not activate when the driver expects it to. For example, if the Traction Control System or the Electronic Stability Control System activates while

---

[2] Note that the systems are in constant flux. New systems may have different operational design domains.
[3] Again, please be aware that the systems are in constant flux. New systems may have different operational design domains.

the ACC is also activated, the ACC could automatically disengage. Additionally, the ACC will not detect or brake for children, pedestrians, animals, or other objects. It may not detect a vehicle ahead on winding and hilly roads. The list goes on.

In summary, it is important for the driver to understand the ODD for all of the reasons listed above. Whether the driver can actually do so is one of the many questions we will address in the chapters in the Handbook, a brief discussion of which follows.

1.3   THE HANDBOOK: A QUICK GUIDE

Up to this point we have focused on the definitions we will need in order to concentrate them in one place (rather than have the reader search throughout the chapter or Handbook). Now that we have a working definition of the levels of automation, active safety features, and ACIVs we can return to a discussion of the automation paradox, to the countermeasures designed to reduce the unintended effects of automation, and to the more general topics that are covered in the Handbook. Perhaps the most central theme running throughout the Handbook centers on whether the driver is or is not engaged in the driving task. Although engagement and disengagement are two sides of the same coin, it is useful to separate the discussion of the primary factors affecting both, realizing that they are necessary complements of one another. User, task, vehicle, and environmental characteristics influence both engagement and disengagement (Figure 1.1). We will refer back to this figure as we make our way through a brief discussion of the chapters in the Handbook.

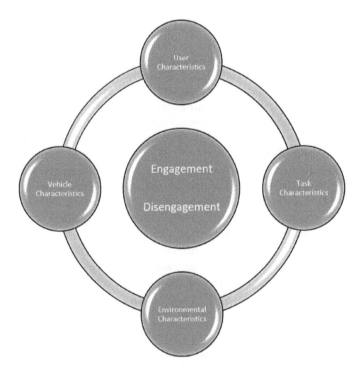

FIGURE 1.1   Engagement and disengagement and the factors which influence them.

We have organized the Handbook into five sections. The first section, including the Introduction, provides the reader with a basic understanding of the human factors issues surrounding ACIVs along with a comprehensive discussion of the many ongoing and future activities targeting the research, development, and deployment of such vehicles. The second section focuses on developing an understanding of the unintended consequences of automation for the human driver. These issues emerge from more fundamental human characteristics that can be used both to identify the source of potential problems and to serve as the foundation for solutions. For example, the mental models that drivers have of automation are fundamental characteristics that can not only lead to problems but also point to how to train people and how to design better HMIs. The third section focuses on the possible solutions to these unintended consequences. The fourth section introduces additional topics that do not neatly fit into the above categories. And, finally, we conclude with a section on the evaluation of the safety benefits of ACIVs.

1.3.1 THE STATE OF THE ART: ACIVS (CHAPTER 2)

Although it will be many years before almost all vehicles will have Level 2 and above capabilities, most people are simply not aware of the extraordinary efforts that are going into making ACIVs a reality. In Chapter 2, the author provides us with a window into the history of, and current efforts centered on, the development and deployment of ACIVs. In addition, detailed and informative discussions are provided that focus on the advent of active safety systems, the different technological advances that have been necessary in order for automated driving to become a reality, the business case for the development and deployment of commercial fleets, and the direction in which the automobile industry is heading in terms of levels of automation both for passenger cars and for commercial vehicles. Both readers familiar with and readers not familiar with advances in ACIVs will find the breadth and depth of the discussion in Chapter 2 truly revealing of the state of the art.

1.3.2 ISSUES IN THE DEPLOYMENT OF ACIVS (PROBLEMS)

Understanding the human factors problems with the deployment of ACIVs is not unique to passenger cars. The problems appear when almost any automation is deployed, whether it be in some other mode of transportation (e.g., aviation) or on the factory floor. Thus, there is a breadth and depth of understanding that would not otherwise be available, and the authors draw on this experience as well as the knowledge generated by the increasing number of experimental, field, and naturalistic studies of automated driving. The first ten chapters in this section focus largely on user characteristics (Figure 1.1), where the user is largely isolated within the vehicle (Chapters 3–12) or is an element in a much larger system (Chapter 13). The last chapter (Chapter 14) in this section focuses on the broader human factors issues in the regulation of deployment.

1.3.2.1 Driver's Mental Model of Vehicle Automation (Chapter 3)

To begin our tour through the forest of human factors problems that may contain landmines for the deployment of ACIVs, we start with a discussion of mental models of driving (user characteristics, Figure 1.1). Mental models inform almost everything the driver does, from deciding at what speed to travel to what route to take and, now, what level of automation to engage, and whether the systems currently activated are operating within the envelope of the ODD. In this chapter, the authors review the various mental models of automation and driving, showing how fundamental psychological biases contribute to their formation, especially when automation is entered into the picture, and then explaining how those mental models govern driver's short (e.g., sudden braking), intermediate (e.g., level of automation), and long-term (e.g., mode choice) behaviors.

1.3.2.2   Driver Trust in ACIVs (Chapter 4)

It is one thing to have technologies that can greatly benefit the driver and society at large and to have an understanding of driver's latent mental models. However, it is an entirely different matter to make sure that these technologies are actually used by the driver. The use of the technologies depends on the trust the driver has in the system. Trust is a complex construct influenced by many different factors. In Chapter 4, the author delves into these different factors. This chapter also addresses the issue of how one goes about building, calibrating, losing, and repairing trust.

1.3.2.3   Public Opinion about ACIVs (Chapter 5)

In the previous chapter, the authors described why the issue of trust among drivers in automation is so important to the adoption and safe operation of automated vehicles. However, it is not only the individual driver's trust in the automated technologies that will influence their widespread adoption but also the public's acceptance of such technologies. The public here includes not only drivers but also vulnerable road users and other stakeholders in the transportation system. The public's acceptance of such technologies includes larger ethical issues as well. In Chapter 5, the authors discuss the factors that influence public acceptance of automated technologies, the measurement issues associated with evaluating acceptance, and the relevant research findings.

1.3.2.4   Workload, Distraction, and Automation (Chapter 6)

Once drivers trust their vehicle's automation enough to actually use it, one must next consider the cognitive, visual, and manual workload required of the driver. The workload requirements will change dramatically as different levels of automation, connectivity, and intelligence along with different types of active safety systems enter the vehicle (task characteristics, Figure 1.1). The basic laws underlying the relation between workload and performance (e.g., modeled after the Yerkes–Dodson law) should presumably remain the same. However, automation does not uniformly decrease workload as might at first be assumed. In fact, automation can both increase and decrease the different types of workload and do so over both very short and very long time horizons. The first question addressed in Chapter 6 by the authors is how the various types of workload are influenced by the different categories of automation and active safety systems. The second question addressed is how to achieve an optimal workload so that drivers are neither over engaged nor under engaged. Distraction figures centrally in an understanding of workload and automation and this concept is woven into the chapter throughout.

1.3.2.5   Situation Awareness in Driving (Chapter 7)

If the driver trusts the system, the driver will presumably use the system. Ideally the workload can be adjusted to the degree needed to keep the driver engaged while not overburdening the driver. But this does not always happen and the driver can lose situation awareness and become disengaged from the driving task. The loss of situation awareness can lead to a spike in automation surprises, either because the driver fails to perceive an event, fails to understand what he or she sees, or fails to predict what actions should be taken based on an understanding of an event. In Chapter 7, the author discusses situation awareness in general, how increases and decreases in cognitive workload can degrade situation awareness and driving performance at each of the six levels (Levels 0–5), and how it can impact the transfer of control from the vehicle to the driver, and conversely, from the driver to the vehicle.

1.3.2.6   Allocation of Function to Humans and Automation and the Transfer of Control (Chapter 8)

Both driving performance (when the driver is in total control or partial control) and the transfer of control can be impacted by the loss of situation awareness. As discussed in Chapter 8, the issues of driver trust, workload, and situation awareness need to be considered (among others) when functions are allocated to the driver and the system at each of the various levels of automation. The correct allocation promotes both safety and efficiency, as well as adding synergistically to trust and situation awareness. Moreover, since in most systems the automation and the human operator will be sharing control and supervisory functions, it is important that the transfer of control be seamless. A discussion is offered of how best to temporally transfer control of different functions between automation and the human operator, both when it is assumed that the operator is fully attentive and when situation awareness is compromised.

1.3.2.7   Driver Fitness in the Resumption of Control (Chapter 9)

The loss of situation awareness can occur for many reasons. When that occurs, driver safety can be greatly impacted, especially when he or she is trying to monitor a vehicle or actually resume control of the vehicle. In Chapter 9, the authors delve into the three primary causes of the loss of situation awareness: distraction, sleepiness, and impairment (user characteristics, Figure 1.1). Information is provided on the fundamental characteristics of distraction, sleepiness, and impairment and how they interact differently with the different levels of automation, especially when control is being transferred between the driver and the vehicle. Research is also summarized on the detection of each of these different states. This research is evolving rapidly and, not surprisingly, has led to an entire cottage industry of driver state monitoring algorithms in order to make sure that the driver is really engaged in the driving task, and may be thought of as a precursor to current automation solutions. Finally, the various countermeasures that are available, or those that will soon be available, are discussed. In some cases, such as sleepiness, the possible countermeasures are counterintuitive, including having drivers engage in what would otherwise be distracting activities.

1.3.2.8   Driver Capabilities in the Resumption of Control (Chapter 10)

Drivers can be capable of resuming control, but temporarily lack fitness to do such as discussed in Chapter 9. In contrast, in Chapter 10 the authors focus on populations of drivers that find themselves in longer term situations that can lead to the loss of the most basic driving skills, which in turn makes it difficult for them to resume control. Such populations include those affected by the normal aging process and various medical conditions. Automated vehicle technology, including in-vehicle information systems, active safety systems, and other automated technologies, hold plausible opportunities for drivers who may be at risk for continued independent and safe driving. The authors frame the discussion in the context of the medical condition, the core clinical characteristics of the condition, the resulting functional performance deficits, their effect on driving, and the hypothetical use of automated vehicle technologies to facilitate fitness to drive abilities in drivers wanting to resume control of the vehicle (i.e., wanting to resume driving).

1.3.2.9   Driver State Monitoring for Decreased Fitness to
Drive (Chapter 11)

While Chapters 9 and 10 discuss the various types of reductions in the fitness and the capability to drive, the question of how one detects the various driver states associated with these reductions is central. The discussion in Chapter 11 is by necessity a more technical discussion, one focusing on the sensors and algorithms used to detect driver state. But it is also a discussion grounded in a deeper understanding of the behaviors which characterize a given state. There have been a number of advances in the detection of driver state, advances due in part to the miniaturization of very powerful sensors and computing capabilities and in part to a better understanding of the physiological and behavioral characteristics that are associated with a given state. In Chapter 11, the authors discuss these advances in driver state monitoring as well as the human factors issues that will arise as these advances work their way into the market.

1.3.2.10   Behavioral Adaptation (Chapter 12)

We have spoken about behavior with regard to the technologies with which the driver must interface at each of the various levels of automation. We have assumed that the behavior remains static across time. In Chapter 12, the author discusses if, and how, the behavior of drivers changes over time after the introduction of a given technology. Research suggests that drivers learn to adapt to safety technologies, sometimes leading to risk homeostasis, a phenomenon where drivers will engage in a more risky behavior when a technology is introduced that is meant to reduce the risk, thereby keeping the overall risk constant. Risk homeostasis is only one of the several theories that are used to explain behavioral adaptation and which are discussed in the chapter.

1.3.2.11   Distributed Situation Awareness (Chapter 13)

As noted, situation awareness is an important consideration in the design of automated vehicles and the road systems in which they will operate. In Chapter 7, it was the situation awareness of the individual driver that was primarily at issue. But it is not only the driver that needs to be situation aware as the driver is embedded in a larger transportation system. In Chapter 13, the authors argue that considering the situation awareness requirements of both human and non-human agents is critical, as well as how they can exchange and connect their awareness with one another in different road environments. To demonstrate their approach, the authors present an overview of their distributed situation awareness model and an analysis of the recent Uber–Volvo fatal collision in Tempe, Arizona.

1.3.2.12 Human Factors Issues in the Regulation of Deployment (Chapter 14)

We have assumed up until this point that automated vehicles will somehow just magically appear on the roads. However, there will arise a number of complex, human factors issues in the pilot testing, lengthy deployment, and regulation of ACIVs. In Chapter 14, the author attempts to answer the question "What human factors issues need to be considered in preparing policy and regulation for the deployment of automated vehicles?".

1.3.3 HUMAN-CENTERED DESIGN OF ACIVs (SOLUTIONS)

A number of automation-induced problems have been identified above along with some solutions (e.g., driver state monitoring). In this part of the Handbook, we turn to solutions based on the design of the HMI.[4] Issues central to the design of the HMI include concerns associated with both the displays that present information to the driver (including warnings) and the controls that facilitate the driver's interactions with the vehicle as a whole and indicate the status of various vehicle components and sub-systems. In the context of vehicle safety systems in particular, the HMI should effectively communicate information while assessing and managing driver workload and minimizing distraction. A broader discussion of how one might actually configure the ODD of the advanced technologies to improve safety or how one might design the control algorithms to reduce the potential of fully automated vehicles causing riders to exhibit symptoms similar to simulator sickness are part of the larger picture of what is meant by human-centric design (Fridman, 2018), but are not covered in this Handbook due largely to issues of space.

1.3.3.1 HMI Design for ACIVs (Chapter 15)

In Chapter 15, the authors discuss the general HMI design requirements for ACIVs. Such design requirements must be determined in the context of many factors, including their influence on safety, public perception, and perceived value; the mix and behaviors of legacy vs. connected vs. automated vehicles over time within the vehicle fleet; and the degree and type of automation associated with the HMI. Within this context, the safe and efficient operation of any motor vehicle requires that the HMI be designed in a manner that is consistent with driver needs, limitations, capabilities, and expectations. As another example of the automation paradox, as automation takes

---

[4] The literature generally uses the terms HMI and Driver–Vehicle Interface (DVI) interchangeably; we will use HMI throughout this paper but view HMI and DVI to be synonymous for our purposes.

on more of the driving task, the HMI might need to become more complex to properly support the driver. A continuing challenge is to identify just what these changes to complexity are amidst the changing and uncertain landscape of advanced vehicle technology. Despite these challenges, the objective of this chapter is to summarize what is known regarding HMI design principles for ACIVs. Many of these principles are aimed at addressing the important issues raised in the preceding chapters.

1.3.3.2 HMI Design for Fitness Impaired Populations (Chapter 16)

The HMI will be used not only by the general population but also by drivers whose fitness is impaired. Assuming one can monitor for driver state, then one could potentially adjust the relationship between automation and the driver dynamically. This concept is referred to as adaptive automation. In Chapter 16, the author considers the application of adaptive automation in the context of driver impairment. First, the author introduces the topic of adaptive automation and important questions that must be addressed by an HMI designer. This discussion is then extended to adaptive vehicle automation in the context of driver impairment, specifically, distraction, drowsiness, and drugs and alcohol. Overall, the chapter provides a framework for the implementation of adaptive interfaces in the automated vehicle, which hinges on understanding the interaction between the nature and degree of driver impairment and the capabilities of vehicle automation.

1.3.3.3 Automated Vehicle Design for People with Disabilities (Chapter 17)

Impairments to a driver's fitness reduce the driver's normal range of safe operations, but generally do not impact the driver's ability to obtain a motor vehicle license. However, drivers with disabilities who cannot obtain a driver's license may be able to obtain one when fully automated vehicles (SAE Levels 4/5) are available, potentially providing them with much greater mobility and independence. In Chapter 17, the author describes what is needed from a human factors standpoint in order for this potential to be realized. In particular, it is argued that accessibility for persons with various disabilities needs to be considered early in the design process. Although accessibility has not traditionally been a design focus for passenger vehicle manufacturers, there are lessons to be learned from other modes of transportation that can be leveraged in the design of fully automated vehicles. The first part of the chapter introduces readers to the social model of disabilities, which is a philosophy that views disabilities as an individual difference similar to height and gender. The second part of the chapter provides an overview of the universal design and its seven principles for developing products and systems that are usable by all. The third and final section of the chapter discusses what aspects of a highly automated vehicle are new and unique and how to make these accessible to persons with disabilities.

1.3.3.4 Importance of Training for ACIVs (Chapter 18)

Even if one could design the best of all possible interfaces for a driver, ACIVs differ enough in their operating requirements from the vehicles currently on the roadway that training is likely to be a recommended, if not a necessary requirement. Despite potentially dramatic changes to the driving task, there has been relatively little work examining how training and learner-centered, HMI design can positively impact the safety of ACIV systems. In Chapter 18, the authors summarize the general concept of learner-centered HMI design for ACIVs and provide information on specific training-related factors. Training is potentially useful for everything from learning how to respond to the warnings, to learning how to monitor the dynamic driving task when in Level 2 and above, and to re-familiarizing oneself with driving skills that might have atrophied. The chapter serves as a foundation for driver training stakeholders, technology developers, consumers, and legislatures to address the growing need to include relevant and effective training for ACIV systems as these technologies are developed and deployed.

1.3.4 SPECIAL TOPICS

The special topics included here are the ones which push the envelope of our understanding of and proposed use for ACIVs, either by looking towards the future or by taking a step backwards and considering what has been learned in other domains.

1.3.4.1 Connected Vehicles in a Connected World: A Sociotechnical Systems Perspective (Chapter 19)

Much of the discussion in the Handbook is focused on understanding the behaviors of the driver isolated from the larger system in which he or she is embedded. Yet, the driver and his or her vehicle will be inevitably integrated within the broader, connected network of smart cars, smart cities, and smart homes, with all of the associated social and technical system components. In Chapter 19, the author discusses the need to regard the connected vehicle within a broader sociotechnical framework. A sociotechnical perspective involves creating a detailed system framework, identifying the various social and technical sub-systems to reveal critical interdependencies, and then defining design requirements and resolving potential conflicts. It is argued that methodologies for system analysis and joint optimization are urgently needed to create the foundations that promote positive emergent properties such as system safety, resilience, and overall effectiveness.

1.3.4.2 Congestion and Carbon Emissions (Chapter 20)

The Handbook has focused on the advances ACIVs can make in the areas of safety and mobility, largely because those are the issues most often considered. However, ACIVs are also proposed as a way to reduce congestion, and hence to reduce carbon emissions. Unless people completely surrendered driving control to ACIVs, researchers must understand how driver behavior is affected by vehicle intelligence, connectedness, and automation. The authors in Chapter 20 review such work. Congestion is the focus of the first part of the chapter. It is shown that knowledge that the vehicle has, such as information about route travel times and parking spot availability, seems to induce driver decisions based on the simple rules of thumb. Vehicle connectedness can be leveraged to decrease congestion if drivers can reason and predict other's behavior—be they human or automated drivers—and, surprisingly, such decision skills should not necessarily follow standard game-theoretic rationality. Carbon emissions are the focus of the second part of the chapter. The authors discuss to what extent such carbon-emissions-reducing ACIVs currently exist, when one takes a life cycle perspective.

1.3.4.3 Automation Lessons from Other Domains (Chapter 21)

There is much that can be learned from accident analysis and research in other domains about human behavior and safety when using automation. This is especially true in aviation, which is arguably the pioneer of the systematic investigation of human automation as it pertains to operators in general. In Chapter 21, the author reviews the lessons that can be learned from human interaction with automated systems other than vehicles, including not only human flight in aviation but also unmanned air vehicles, space, Air Traffic Control, military systems, consumer products, business systems, process control, robotics, and others.

1.3.5 EVALUATION OF ACIVs

The authors in the previous chapters have discussed the details of ACIVs (vehicles, Figure 1.1). They have explored the problems that are fundamental to both a human operator driving in general as well as a human operator driving automated vehicles at each of the different levels of automation (users, Figure 1.1). The discussion has focused both on the general population of users as well as those whose fitness to drive or capability of driving is reduced for whatever reasons. And the discussion has included not only the individual driver but also the driver as part of a larger system. Potential solutions including driver state monitoring and advanced HMI designs (tasks, Figure 1.1) that may solve those problems have also been presented. However, no matter how principled, any solution is only a potential one until it is tested and evaluated.

1.3.5.1 Human Factors Considerations in Testing and Evaluating ACIVs (Chapter 22)

In Chapter 22, the authors provide an overview for testing various aspects of automated and connected driving automation systems (broadly the complete sweep of warning systems, active safety systems, DSFs, and ADSs), including considerations for commercial vehicle testing. The chapter includes reviews of vehicle testing methodologies (e.g., data analytics, experimentation, and naturalistic driving) and advocates for a graduated approach in which systems are tested iteratively throughout the development cycle. In particular, it is necessary to accurately characterize the driving automation system(s) present in a platform, which can inform the training material developed for drivers of the particular vehicles being evaluated and the associated testing scenarios. What is actually tested can also be informed by the analysis of existing naturalistic driving data. Simulator testing provides a method of evaluating early feature designs with a large degree of experimental control, but reduced external validity. Other testing approaches include on-road testing, using a prototype or Wizard of Oz approach to increase the external validity of testing. Finally, late stage testing can include the analysis of large datasets of drivers actively using automation.

1.3.5.2 Techniques for Making Sense of Behavior in Complex Datasets (Chapter 23)

As described in Chapter 22, there are many important considerations when evaluating and testing ACIVs, and many approaches, such as simulation, on-road tests, and naturalistic studies, which can be taken. New and emergent issues concern the management and analysis of data that are derived from these evaluations, especially issues centered on big data. In Chapter 23, the author describes the techniques that can help the analyst make sense of these complex datasets. First, the author describes methods for understanding complex datasets, including how to manipulate, visualize, and examine the data. Second, the author describes the research questions that can be answered about both the operator and the system with large datasets. Finally, the author describes both the techniques and tools needed to make sense of the behavior given the context, user, and the technology itself and the questions that should be asked to check/validate the correctness of the outcomes.

1.4 CONCLUSION

This introduction and overview is not meant to distill the wisdom, experience, and knowledge that are conveyed within the Handbook. Rather, our goal as editors has been to provide the reader with the motivation for the Handbook, which follows from the paradox of automation and the basic terminology. We hope this handbook leaves the reader less rather than more confused about the ACIV landscape as it is relevant to human factors.

We leave where we began, hopefully having provided substance to our initial remarks. Automation has saved, does save, and will continue to save lives. A great majority of the reasons for crashes have been traced back to the driver (National Highway Traffic Safety Administration, 2008). Yet as the chapters make clear, the larger system of which the driver is a part is almost never designed with the characteristics, strengths, and weaknesses of the human front and center. And in those cases where the human driver clearly is at fault, this does not mean that the automation that is being introduced in today's vehicles will necessarily produce the benefits everyone hopes it will. First, the type of automation that has saved so many lives is largely of a different kind (active safety features) than the type of automation which is working its way into Level 1 and 2 cars (automatic steering and ACC). Thus, the automation itself could introduce errors. Second, the type of automation that is being introduced is being used to replace driver operations in areas where the human driver is phenomenally successful, having only 1.25 fatalities per 100 million miles of vehicle travel. Thus, the automation needs to meet an especially high threshold of safety. Third, increasingly autonomous systems do not remove humans and their errors, but simply displace them. Future crashes might be due to programmer, remote operator, or maintenance errors. This is perhaps the most nefarious instance of the automation paradox: automation might eliminate the current driver errors, but might create new opportunities for even more dangerous "human errors." This creates challenges for the driver and others in the sociotechnical system of transportation that do not exist with vehicles in the recent past. Understanding these challenges and providing insight into the potential countermeasures is the purpose of the Handbook. It is a purpose in service of the goal sought by all concerned: achieving the maximum benefits of ACIVs. There is so much promise.

ACKNOWLEDGMENTS

The editors would like to acknowledge the help of the publisher, Taylor & Francis, and everyone with whom we interacted including Cindy Carelli, Erin Harris, and

Assunta Petrone. They were always helpful and quick to respond. We are especially grateful to the fantastic group of authors that contributed both their time and energy in drafting their respective chapters. Without them, such a Handbook would not be possible. We also would like to thank Daniela Barragan and Stanislaw Kolek for their assistance in processing and proofing several chapters. Donald Fisher would like to acknowledge the support of the Volpe National Transportation Systems Center for portions of the preparation of this Handbook. William Horrey is grateful for the support of David Yang and the AAA Foundation for Traffic Safety. The opinions, findings, and conclusions expressed in this publication are those of the authors and not necessarily those of the Department of Transportation, the John A. Volpe National Transportation Systems Center, the AAA Foundation for Traffic Safety, or the University of New South Wales.

REFERENCES

Bonnefon, J., Shariff, A., & Rahwan, I. (2016). The social dilemma of autonomous vehicles. *Science, 24*, 1573–1576.

Cadillac. (2018). *2018-cad-ct6-owners-manual*. Retrieved September 22, 2018, from www.cadillac.com/content/dam/cadillac/na/us/english/index/ownership/technology/super-cruise/pdfs/2018-cad-ct6-owners-manual.pdf

Chang, J. (2015). *Estimated Benefits of Connected Vehicle Applications: Dynamic Mobility Applications AERIS, V2I Safety, and Road Weather Management*. Washington, DC: Department of Transportation.

Department of Transportation. (2019). *Automated Vehicles 3.0. Preparing for the Future of Transportation*. Washington, DC: Department of Transportation.

Edmonds, E. (2019, March 14). *Three in Four Americans Remain Afraid of Fully Self-Driving Vehicles* (AAA). Retrieved April 9, 2019, from https://newsroom.aaa.com/2019/03/americans-fear-self-driving-cars-survey/

Endsley, M. (1995). Toward a theory of situation awareness in dynamic systems. *Human Factors, 37*, 32–64.

Endsley, M. (2017). Autonomous driving systems: A preliminary naturalistic study of the Tesla Model S. *Journal of Cognitive Engineering and Decision Making, 11*, 225–238.

Fridman, L. (2018). *Human-Centered Autonomous Vehicle Systems: Principles of Effective Shared Autonomy*. arXiv, Cornell University. Retrieved October 12, 2019, from *https://arxiv.org/pdf/1810.01835.pdf*

IIHS HLDI. (2018). *GM Front Crash Prevention Systems Cut Police-Reported Crashes*. IIHS, HLDI. IIHS. Retrieved July 14, 2019, from https://www.iihs.org/news/detail/gm-front-crash-prevention-systems-cut-police-reported-crashes

Lu, N., Cheng, N., Zhang, N., Shen, X., & Mark, J. (2014). Connected vehicles: Solutions and challenges. *IEEE Internet of Things Journal, 1*, 289–299.

Najm, W. & daSilva, M. (2000). Benefits estimation methodology for intelligent vehicle safety systems based on encounters with critical driving conflicts. *ITS America 10th Annual Meeting and Exposition*. Boston.

National Highway Traffic Safety Administration. (2008). *National Motor Vehicle Crash Causation Survey (NMVCCS): Report to Congress*. Washington, DC: Department of Transportation.

National Highway Traffic Safety Administration. (2017). *Automated Driving Systems 2.0: A Vision for Safety*. Washington, DC: U.S. Department of Transportation. Retrieved November 23, 2017, from www.nhtsa.gov/sites/nhtsa.dot.gov/files/documents/13069a-ads2.0_090617_v9a_tag.pdf

National Transportation Safety Board. (2018, May 24). *Preliminary Report. Highway. HWY18MH010.* Retrieved from NTSB Investigations. Accident Reports: www.ntsb. gov/investigations/AccidentReports/Reports/HWY18MH010-prelim.pdf

Noy, I., Shinar, D., & Horrey, W. (2017). Automated driving: Safety blind spots. *Safety Science, 102*, 68–78.

SAE International. (2018, December 11). *SAE International Releases Updated Visual Chart for Its "Levels of Driving Automation" Standard for Self-Driving Vehicles.* Retrieved from SAE International: www.sae.org/news/press-room/2018/12/sae-international-releases-updated-visual-chart-for-its-%E2%80%9Clevels-of-driving-automation%E2%80%9D-standard-for-self-driving-vehicles

Toyota. (2017, November 14). *Toyota Owners.* Retrieved from Manuals & Warranty: www.toyota.com/owners/resources/owners-manuals?&srchid=semGOOGLETOO_Resources_Owner_Manuals+-+BMM%2Btoyota+%2Bmanual&gclid=CLvHnsW5v9cCFVCiswo dtzMPmA&gclsrc=ds

Victor, T. (Performer). (2019, January). Transportation Research Board Human Factors Workshop. Guiding the Design of Partially Automated Vehicles. *Annual Meetings of the Transportation Research Board*, Washington, DC.

Victor, T., Tivestan, E., Gustafsson, P. J., Sangberg, F., & Aust, M. (2018). Automation expectation mismatch: Incorrect prediction despite eyes on threat and hands on wheel. *Human Factors, 60*, 1095–1116.

2 Automated Driving
Decades of Research and Development Leading to Today's Commercial Systems

Richard Bishop
Bishop Consulting

CONTENTS

Key Points 24
2.1 Introduction 25
    2.1.1 Automated Driving: From Vision to the Launch of a New
        Industry in 70 Years...25
    2.1.2 Advent of Active Safety Systems...27
    2.1.3 Addressing Safe Testing and Deployment of Automated Driving 27
    2.1.4 Pursuit of Nascent "Holy Grails" 28
2.2 Distinctions within SAE Levels of Automation 29
2.3 Automated Driving: Technology Basis...31
    2.3.1 Understanding the World to Make Proper Driving Decisions 31
    2.3.2 Perception, Mapping, and Localization...31
    2.3.3 Motion Planning and Control 32
    2.3.4 Artificial Intelligence...34
    2.3.5 Off-Board Information Sources..34
    2.3.6 Driver Monitoring as Key to Safety Case...34
    2.3.7 Behavioral Competencies and Remote Support 35
    2.3.8 Design and Test Processes to Ensure Safety 36
        2.3.8.1 Functional Safety 36
        2.3.8.2 Cybersecurity..37
        2.3.8.3 ADS Validation Processes 37
2.4 Automated Driving Commercial Development and Deployment 38
    2.4.1 Automated Fleet Services for Freight and Parcels 39
    2.4.2 Automated Fleet Services for People...40
    2.4.3 Private Ownership: Automation Features in Mass-Market
        Passenger Cars 43
2.5 Regulatory Considerations..47
2.6 Going Forward: ADS Implications for Human Factors 48
References..51

KEY POINTS

- The modern era of automated driving has been underway worldwide since the early 1990s, initially supported by public funding. Private sector funding began to dominate after Google entered the space early in the 2010 decade, greatly increasing the pace of development.
- Active safety systems that intervene via braking or steering to avoid a crash are now available or standard on many passenger vehicles. We can expect the crash rate for human-driven vehicles to begin a distinct downward trend when the number of equipped vehicles reaches an inflection point. This beneficial trend will be largely independent of the deployment pace of automated vehicles.
- The concept of Operational Design Domain (ODD) is essential to describing the capability of driver support features and automated driving systems (ADS). ODD defines the operating conditions under which the ADS is designed to function and can include many factors such as road type, speed, environmental conditions, and traffic conditions.
- Key factors in developing safe roadworthy Level 4 vehicles are functional safety for overall system design, artificial intelligence for perception, simulation to support validation, and robust cybersecurity.
- New approaches and standards are needed to communicate the intention and awareness of driverless trucks and robo-taxis to other road users, i.e., "external human–machine interface (HMI)."
- Off-board information can augment situational awareness for active safety systems, driver support features, and ADSs. This may include GPS, additional data from the cloud (roadworks ahead, weather changes), and low-latency information that may be transmitted directly from the infrastructure or other vehicles. Due to the possibility of wireless communications interruptions or lack of communication from some traffic participants, this data can augment situational awareness but cannot be relied upon. Therefore, while off-board information can enhance performance, it is not necessary for implementing automated driving.
- An exception to the above point is truck platooning systems that rely on vehicle-to-vehicle communications implemented as part of system design, in a manner so that communications integrity is controlled. Level 1 truck platooning is coming to market now, which will evolve to Level 4 driverless follower trucks behind a human-driven lead truck in the coming years.
- Level 4 robo-taxis for street operations and driverless trucks for highway operations are rapidly moving toward commercialization, with on-road testing now underway using safety drivers.
- Regulations governing the operation of driverless vehicles on public roads are evolving and vary significantly around the world; currently the United States is the most open environment.

2.1 INTRODUCTION

Before beginning a detailed discussion of the human factors issues associated with automated, connected, and intelligent vehicles (ACIV), it is important to have some understanding of the current and future technologies that are likely to be introduced into future vehicles, along with the different classification schemes, key concepts, and timelines for deployment. These technologies were described briefly in the previous chapter, and here they are described in more detail. Note: because commercial activities in ACIV are evolving quickly, this chapter should be considered a snapshot of the situation at the time of writing (2019).

The ideal road trip should be safe, smooth, uninterrupted, and expeditious. For the vehicle occupants, the time should be productive, restful, and/or entertaining, and they should be connected with and aware of the information they care about. Accessing this capability needs to be affordable and, for most people, it is important that the trip be "good for society as well as for me," i.e., environmentally friendly. Automated driving promises to bring us closer than ever to this ideal, particularly when joined with vehicle connectivity that enables optimized traffic flow. Tech developers are envisioning a driverless and crashless society and traffic engineers are dreaming about the end of congestion—a tall order indeed: can it really be?

2.1.1 AUTOMATED DRIVING: FROM VISION TO THE LAUNCH OF A NEW INDUSTRY IN 70 YEARS

The concept of automated vehicles was originally introduced by General Motors at the 1939 World's Fair. In the ensuing decades, a series of prototypes was developed in Europe, Japan, and the United States based on available technology, remaining in the realm of research. The pace of development began to increase in the late 1980s when the European PROMETHEUS project work culminated in an automated trip on public roads from Bavaria to Denmark, traveling at highway speeds in automated mode for up to 100 miles at a time (Billington, 2018). At this time the U.S. Defense Department also developed the first automated ground vehicles (AGVs) to support the ground troops, work that continues to this day. During the 1990s, the U.S. Department of Transportation (USDOT), at the direction of Congress, co-funded an ambitious $90M Automated Highway System (AHS) program, in partnership with the National Automated Highway System Consortium (UC Berkeley Institute of Transport Studies, n.d.), led by General Motors. The AHS program culminated in the very successful (in terms of technical performance and media coverage) Demo '97, in which thousands of people experienced automated driving in cars, trucks, and buses on Interstate 15 in San Diego, as depicted in Figure 2.1 (National Automated Highway System Consortium, 1998). This event created a shift in public consciousness: automated driving is not just science fiction; it's actually possible!

Concurrently, the Japanese government and vehicle industry teamed up to create the Advanced Cruise-Assist Highway Research Association, which prototyped and demonstrated automated vehicle capabilities at their Demo 2000 in Tsukuba,

FIGURE 2.1 Demo '97 automated vehicles in platooning formation. (From California Path Program, https://path.berkeley.edu/research/connected-and-automated-vehicles/national-automated-highway-systems-consortium/photo.)

Japan, with an emphasis on infrastructure-supported systems. Late in the 1990s, the European CHAUFFEUR project (European Commission, n.d.) implemented platooning of automated heavy trucks, conducting tests on public roads and examining related business and societal issues. During the 2000s, work continued across all three regions, driven by joint public–private projects with the majority of funding coming from governments (Shladover, 2012).

The traditional vehicle industry was deeply involved in all of these activities, but as noted the primary investments came from the public sector. The tectonic shifts heralding a new order began in the U. S. when the Defense Advanced Research Projects Agency (DARPA) took up the charge with a new approach: the Grand and Urban Challenges (DARPA, n.d.; Voelcker, 2007) offered a significant cash prize to the self-funded tech developers who could best drive the course. While the early off-road Challenges in the desert attracted mainly academics and roboticists, the 2007 Urban Challenge began to attract automotive industry participation and investment. Driverless automated vehicles operated together in a residential street setting during the Urban Challenge, with the top three finishers (Carnegie Mellon University/General Motors, Stanford University, and Virginia Tech, respectively) properly handling traffic circles, four-way stops, parking lots, and road obstacles maneuvering among human-driven vehicles. DARPA's initiatives catalyzed Google's self-driving car program, whose public debut in 2010 fascinated the public and firmly shifted the locus of funding from the public to the private sector. Given the dynamics of venture capital, this brought seemingly unlimited funding and fast pacing based on a highly innovative risk-taking environment.

2.1.2 ADVENT OF ACTIVE SAFETY SYSTEMS

In parallel with this automated driving activity in the research domain, the traditional vehicle industry was introducing a steady stream of active safety systems aimed at assisting humans in avoiding crashes. Beginning in the late 1990s with warning-only systems, systems evolved to active control intervention by the 2000s (Bishop, 2005). By 2010, a well-equipped premium vehicle may have had adaptive cruise control (automatic headway-keeping), automated emergency braking for forward collisions, lane departure prevention, blind spot monitoring, night vision with pedestrian detection, sign recognition, and drowsy driver detection. Over the ensuing years these features rapidly became available across mid-range cars and by now in the United States a $30,000 car can be purchased with all this and more, and even entry-level cars are now equipped with some active safety features that can be purchased. In fact, for several car manufacturers, active safety features are now standard on all of their models. This is good news for the overall road safety picture, regardless of the slower pace at which highly automated driving will become widely available.

Vehicle-makers could have independently built upon active safety and moved their automated driving research into product development at some early point along this timeline. However, the core business of a vehicle-maker is in building and selling vehicles, which is relatively secure as long as they keep up with their competitors. There was no "forcing function" to introduce automated driving into the market and no evidence at the time that their customers were ready for it. In essence, Google and others in the startup space served to "test the waters" for the concept of automated vehicles on our roads. The rapid increase in venture funding plus consumer fascination has opened the thinking of carmakers to consider shifting from an "equipment model" to a "service model"; in fact, many major automotive original equipment manufacturers (OEMs) are now investing in offering automated ride-hailing services (robo-taxis).

Importantly, the activities by the OEMs to develop and field active safety systems during the 2000s resulted in new capabilities in algorithm development focused on road driving, development of automatic steering and braking actuators, and a steady reduction in the cost of sensor hardware as sales volumes increased. While a radar sensor cost about $1,000 USD at the turn of the century, today the cost is in the range of $100 USD for similar capability: an order of magnitude reduction in cost. While it may seem like the startups have done all the innovation, this long process of applied engineering by the traditional vehicle industry has served as a key enabler for bringing automated driving to the public.

2.1.3 ADDRESSING SAFE TESTING AND DEPLOYMENT OF AUTOMATED DRIVING

The period 2017–2018 signaled a turning point due to a tragic crash in which an Uber prototype robo-taxi under test in Arizona collided with and killed a pedestrian, even though a safety driver was at the driver controls (Griggs & Wakabayashi, 2018). Additionally, several Tesla drivers died while using the AutoPilot function

(Lee, 2019). Based on the limited information available, it appears that in each of the cases either the safety driver or the vehicle owner was not adequately fulfilling their "co-pilot" responsibility to monitor the system and intervene when the system capability was exceeded. As will be discussed in the succeeding chapters, this raises significant challenges relating to shared human–machine control, and in part motivates implementation of fully automated vehicles that do not rely on human control. Indeed, it is notable that Google, which started its automated driving program around 2009 and continued it under Waymo, has still not launched a commercial driverless product or service; this is an eternity in Silicon Valley development time! This decade has seen massive investments in getting from a basic working unit to a robust, high availability, fail-safe and fail-operational, cost-effective product that the market would accept. This long incubation period has been due to the need to put in the hard and slow work of "getting it right" in terms of safety, robustness, and service efficiency. While observers may be impatient or disillusioned, developers of highly automated vehicles are focused on detailed engineering and testing within a process permeated by careful functional safety analysis (Japan Automotive Research Institute, n.d.) and implementation of best safety practices to launch commercial products and services. This is a key component of the intellectual property behind safety-critical vehicle technology and is highly proprietary. Although the specifics of the approach are not revealed (spurring much speculation in the news media), completing a comprehensive safety validation process is a key part of achieving regulatory, public, and industry acceptance of new vehicle technology to bring viable solutions to market (see also, this Handbook, Chapter 5).

2.1.4 PURSUIT OF NASCENT "HOLY GRAILS"

From a car industry perspective, the advent of vehicle automation is by now a given. Not only are many of the factors of the "ideal road trip" fulfilled by automation, mobility is expanded for the disabled, elderly, and others who cannot now drive, an especially compelling contention given the aging of the Baby Boom generation (see also, this Handbook, Chapters 10, 17). The current level of investment, testing, and product development across robo-taxis, robo-trucks, robo-buses, and robo-cars is at a remarkably high level. In fact, the automotive industry and investment community have "caught the vision" and fully embraced automated driving, propelling a complete reformation in what it means to be a vehicle manufacturer. Based on the premise that massive new sources of profits will result from individuals extensively using low-cost automated mobility, total investment to date is likely in the tens of billions of dollars, with more to come. The business case for shared and automated mobility was elaborated on by the then President of General Motors Dan Ammann, who asserted that the lifetime revenue generated by one of its automated vehicles could be in the "several hundred thousands of dollars," as compared to their average of $30,000 USD in revenue from one of their current vehicles (Sage & Lienert, 2017). This general viewpoint has been expressed by other automakers as well. Due to the strength of the automated driving business case, OEMs are backing up their aspirational language with substantial investments, joined by the broader tech industry. For example, Toyota has invested hundreds of millions of dollars in Uber (Somerville, 2019). In early 2019, the German Association of the Automotive Industry (VDA) car industry association estimated that Germany's car industry alone will invest 18 billion euros in "digitization and connected and automated driving" by 2021 (Eckert, 2019). Independently, extensive robo-taxi public road testing is underway by numerous startup companies. Though specific introduction dates have not been announced, given the maturity of current systems, it is conceivable that driverless mobility services will be available starting in the 2020 timeframe.

Applying automated driving to truck operations is another very active area, driven by the potential for reduced fuel consumption and labor costs. Fuel use can be reduced via Level 1 "platooning" enabled by vehicle-to-vehicle (V2V) communications. Fully driverless trucking, with the obvious benefits of reduced labor costs, has attracted substantial investment, launching many startups. Truck OEMs have recently become active here as well. For last-mile street-level operations, driverless parcel delivery has also seen an upswing in activity.

The following sections discuss key automated driving concepts and underlying technology, plus provide a snapshot of the current level of testing and deployment activities. The chapter concludes with observations on the implications of these developments for human factors research and development.

2.2 DISTINCTIONS WITHIN SAE LEVELS OF AUTOMATION

The technological underpinnings of automated driving build upon active safety systems that have been developed and deployed widely across the vehicle fleet in the last 20 years. Active safety systems monitor potential obstacles relating to lateral and longitudinal movement, plus lane maintenance, to intervene momentarily if the driver is not responding to a crash imminent situation. For instance, radar-based automatic emergency braking was initially introduced to detect and avoid vehicles ahead and can now respond to pedestrians, bicycles, and animals as well by adding radar–camera sensor fusion. Blind spot monitors use radar or ultrasonic to detect vehicles located in a vehicle's blind spots and provide a stimulus to inform the driver (MyCarDoesWhat.org, n.d.).

These systems have prevented many collisions on our roads (Insurance Institute for Highway Safety/Highway Loss Data Institute, 2019), and we can expect the crash rate for human-driven vehicles to begin a distinct downward trend when the number of equipped vehicles reaches an inflection point. This beneficial trend will be largely independent of the deployment pace of automated vehicles.

In contrast to active safety systems, rather than compensating for driver non-response, automated driving systems (ADS; here referring to vehicles operating at SAE Levels 3 or higher) take over a driving task that the driver or passenger prefers not to do and does so for a sustained period of time. The SAE Levels of Automation, introduced in Chapter 1, focus exclusively on systems providing sustained control. In practice, active safety systems can be expected to be present on all ADS and will be deeply integrated within the technical implementation.

The concept of Operational Design Domain (ODD) is essential to describing the capability of an ADS, which provides the "operating conditions under which a given driving automation system or feature thereof is specifically designed to function; including, but not limited to, environmental, geographical, and time-of-day restrictions, and/or the requisite presence or absence of certain traffic or roadway characteristics" (SAE International, 2018). A particular ADS will only have a single ODD, within which many dimensions and conditions are captured. SAE J3016 provides useful ODD examples as follows (SAE International, 2018):

- "An ADS feature is designed to operate a vehicle only on fully access-controlled freeways in low-speed traffic, under fair weather conditions and optimal road maintenance conditions (e.g., good lane markings and not under construction).
- An ADS-dedicated vehicle is designed to operate only within a geographically defined military base, and only during daylight at speeds not to exceed 25 mph.
- An ADS-dedicated commercial truck is designed to pick up parts from a geofenced sea port and deliver them via a specific route to a distribution center located 30 miles away. The vehicle's ODD is limited to day-time operation within the specified sea port and the specific roads that constitute the prescribed route between the sea port and the distribution center."

At the lower end of the scale (Levels 1–2), Driver Support Features (DSFs) perform a portion of the driving task, controlling lateral or longitudinal control (Level 1) or both (Level 2), with the human having ultimate responsibility to monitor the situation and intervene as needed. Examples are Adaptive Cruise Control and Lane Centering. In Level 3, the ADS is performing the entire driving task within the defined ODD, but a human driver is required to be available to take over control when requested by the system. For Levels 4 and 5, the ADS takes full responsibility for vehicle control; the vehicle, not the driver, is driving. For Level 4 this is conditional, focused on a specified ODD, whereas for Level 5 it is unconditional—i.e., the vehicle can automatically handle all driving situations now handled by human drivers. For the foreseeable future, deployment of highly automated systems will be at Level 4. While Level 5 is useful as a logical endpoint of the scale, there may not be a sufficient business case to actually deploy "anywhere, anytime" Level 5 systems; i.e., society and markets may not see the need for that last 0.0001% of ADS capability.

The levels of automation are intended primarily to distinguish between various degrees of human versus machine control. They do not address implementation aspects such as reliance on connectivity, nor do they address some key aspects that are important to the business case. For instance, whether a vehicle has a driver, or has driver controls, is not addressed for Levels 4 and 5.

It should also be noted that a particular vehicle may operate at different levels of automation depending on the operational environment and task at hand. A vehicle designed for Level 4 operations within a single lane on the highway may revert to Level 2 momentarily if the system requires that the driver approves/monitors a lane change maneuver. The same vehicle may operate at Level 1, providing low-speed Adaptive Cruise Control on suburban and city streets, and automatically execute a Level 3 parking maneuver at the end of the trip.

2.3   AUTOMATED DRIVING: TECHNOLOGY BASIS

2.3.1   UNDERSTANDING THE WORLD TO MAKE PROPER DRIVING DECISIONS

Humans drive vehicles by observing the surrounding traffic and road situation, applying knowledge of driving laws and etiquette, and estimating the likely movement of other entities to make moment-by-moment decisions to operate vehicle controls as well as strategic decisions at the trip level (Brown, 1986; Michon, 1985). ADSs take the same approach and can achieve enhanced situational awareness. On-board physical sensors provide viewpoints beyond that of human eyesight, processing information sensed from every direction if needed. High-definition (HD) maps created specifically to support self-driving systems enable the vehicle to localize and orient itself in the world. Computing prowess assesses the situation and estimates possible trajectories of other traffic participants to determine an optimum course of action. Steering, throttle, and brake actuators are controlled far more precisely than the average driver is capable of. The on-board systems must also be aware of and accurately "read" traffic control devices, such as traffic lights, digital speed limit signs, and railroad crossing gates.

Ideally, an ADS can "see" several seconds ahead in time; in a highway setting this could be in the order of 300 m (Waymo, 2017). Take the example of an automated truck approaching a stopped vehicle in a highway lane: sensing at this range can enable the truck (which has limited maneuverability) to gracefully perform an anticipatory lane change or take other measures to maximize safety.

2.3.2   PERCEPTION, MAPPING, AND LOCALIZATION

The physical sensors enabling automated driving are camera, radar, and Light Detection and Ranging (lidar). Generally speaking, a total 360° field of view is desired. Image processing from camera data supports object identification in the short to medium range, while radar provides range, angle, and velocity of objects for both short and long ranges. Processing the reflected light pulses from lidar sensors provides a quasi-image plus range on each pixel. Most tech developers are fusing data from multiple cameras and radars; many are using lidar as well. While weather and atmospheric conditions can affect the performance of these physical sensors in many ways, their strengths and weaknesses compensate for each other. Sensor performance improvements are occurring at a steady pace; for example, the auto industry is now transitioning radar sensors to the 77 GHz band, which is less susceptible to snow, rain, and dust. Radar developers are also seeking to approach lidar-level performance with new signal modulation techniques (Blanco, 2019). Lidars have recently been introduced in production cars, but for ADS the field is still developing; technical requirements have not converged such that a variety of architectures, sensor fusion strategies, and vehicle integration approaches are in play.

High-definition digital maps include road geometry, direction of travel, lane configurations, crosswalks, traffic lights, stop signs, curbs, and sidewalks, all registered with high accuracy and in three dimensions. Maps.are created in advance by the self-driving vehicle's sensors, as safety operators manually drive test vehicles throughout a city with on-board sensors scanning the entire 3D space including roads, sidewalks, and buildings. The resulting base map is annotated with the relevant traffic regulations and guidance, such as yielding relationships (Ford Motor Company, 2018). Processing of sensor data interacts with the maps; for instance, at a complex signalized intersection with several traffic signal heads hanging above, the known 3D location of the specific traffic light relevant to the ADS vehicle assists the software in detecting signal status of the correct signal head, plus it avoids allocating processing resources to irrelevant information elsewhere in the scene.

Some ADS developers create their own HD maps, while others can access this information from a diverse group of startups or traditional mapping providers. The maps must constantly be refreshed as the world changes. Nevertheless, the ADS must robustly adapt to recent changes in the real world not noted in the map. Any variance between the stored map and the world as detected by on-board sensors is uploaded to the tech developer's cloud. Efforts are underway to establish an industry standard for sharing HD Map data from vehicles back to the cloud, which would enable broader sharing.

ADS developers implement localization techniques to determine the physical location of the vehicle as well as its position relative to nearby objects; this is generally accomplished by correlating digital map data to information coming from perception sensors. While global positioning systems (GPS) are useful for high-level navigation, automated driving must rely on this more direct localization approach. Tech developer Aurora notes that, using these means, they are able to "determine the vehicle's position even in environments that deny or deceive GPS, localizing all six degrees of freedom to within 10 centimeters and 0.1 degree of accuracy" (Aurora, n.d.).

Waymo's Voluntary Safety Self-Assessment Letter, provided to the U.S. National Highway Traffic Safety Administration (NHTSA), offers some excellent examples of how perception, mapping, and location come together to understand a traffic scene. As shown in Figure 2.2, Waymo self-driving vehicle software has detected vehicles (depicted by green and purple boxes), pedestrians (in yellow), cyclists (in red) at the intersection, and a construction zone up ahead (Waymo, 2017).

2.3.3 MOTION PLANNING AND CONTROL

Given an understanding of the world based on the perception system, decisions must then be made as to the desired vehicle trajectory, taking into account not only the current situation but also the probabilities of where other traffic participants are likely to be in the next few moments. Multiple plans and contingencies are created to define and constrain a set of potential actions; these must be predictable to other road users, human-like, and robust relative to other movements and conditions (Ford Motor Company, 2018). Waymo imagery shown in Figure 2.3 demonstrates how their self-driving software assigns predictions to each object in the vicinity of the vehicle, from which an optimum path can be computed and transmitted to the vehicle controls (throttle, brakes, steering; Waymo, 2017).

FIGURE 2.2 (See color insert.) Understanding of a real-world traffic scene in virtual space. (From Waymo 2018.)

FIGURE 2.3 (See color insert.) Assigning predictions to each object in the traffic scene. (From Waymo 2018.)

2.3.4 Artificial Intelligence

Artificial intelligence (AI) techniques incorporating neural networks play a key role in automated driving. Situations that occur on roadways are highly variable, and it is not feasible to write software specific to every possible situation. AI aims to enable ADS software to generalize from a basic understanding of the world and road rules to react safely and appropriately to rare and unusual situations (Waymo, 2017). For instance, based on ingesting data from their 425,000 Hardware 2.0 vehicles on the road, Tesla employs human annotation and predictive algorithms to train their neural network, which enhances understanding of the scene by the ADS (Paukert & Hyatt, 2019).

2.3.5 Off-Board Information Sources

Off-board information can augment situational awareness for the ADS. This may include GPS, additional data from the cloud (roadworks ahead, weather changes), and low-latency information that may be transmitted from the infrastructure or other vehicles (V2X, where X is some other vehicle, element of the infrastructure, or vulnerable road user) (e.g., this Handbook, Chapter 19). This data can augment situational awareness but cannot be relied upon, due to the possibility of wireless communications interruptions or lack of communication from some traffic partici-pants (as it is likely for the foreseeable future that not all vehicles will be equipped). Generally, V2X is seen as important for cooperative safety applications but not nec-essary for implementing automated driving (although see, this Handbook, Chapter 13 for an alternative view). Subsequent generations of ADS will likely incorporate V2X for cooperative automated driving.

For safety applications, the premise of V2X is for vehicles to broadcast basic operating data (position, speed, heading, braking status, etc.) directly to nearby vehicles (on the order of 300 m range). Data received from other vehicles can provide greater situational awareness, particularly in cases such as a large vehicle immedi-ately ahead that restricts the range of on-board sensors. Based on SAE and other standards (Perry, 2017), data are generated and broadcast in less than 100 ms with a broadcast rate on the order of 10 Hz, providing sufficiently rapid updates to support crash avoidance and complex maneuvers such as merging. V2X has been introduced in a small number of production vehicles (Cadillac, 2017), and more are expected in the coming years.

With regard to the physical infrastructure, ADS is designed to operate on roads "as is," such that they do not depend on infrastructure modifications.

2.3.6 Driver Monitoring as Key to Safety Case

For automated vehicles that include a driver role (Levels 1–3), driver monitoring systems may be used to ensure the driver is fulfilling their responsibility to stay attentive to their portion of the driving task (this Handbook, Chapter 11). In a Level 2 system, for instance, the driver is responsible for monitoring the driving scene (while the vehicle system operates the throttle, brakes, and steering), using human judgment to take over driving when needed. Driver-facing cameras assess the driver's head position and/or gaze to "monitor whether the driver is monitoring the technology." If the driver is not attentive, warnings are activated that may lead to the system disabling support or achieving a Minimal Risk Condition (MRC) (see below).

2.3.7 BEHAVIORAL COMPETENCIES AND REMOTE SUPPORT

Behavioral competencies of a road vehicle have been defined to encompass the broad set of basic driving maneuvers and situations. Demonstrating these competencies should be viewed as necessary but not sufficient for the safe operation of an ADS (Nowakowski, Shladover, & Chan, 2016). Examples of specific behavioral competencies are as follows:

- Perform high-speed merge (e.g., onto a freeway)
- Perform lane change or lower-speed merge
- Detect passing and no-passing zones and perform passing maneuvers
- Perform car following (including stop and go)
- Detect and respond to static obstacles in roadway
- Detect and respond as needed to bikes, pedestrians, animals, etc.
- Detect and respond as needed to traffic signals and stop or yield signs
- Navigate intersections and perform turns
- Navigate a parking lot and locate spaces
- Detect work zones or safety officials and navigate appropriately
- Detect and respond as needed to emergency vehicles

There is a key trade-off between having the ADS fulfill all behavioral competencies (incurring additional cost and development time) or taking another approach relying on off-board resources. Remote system support—bringing a human into the loop—can be applied if a driverless vehicle becomes confused or lacks the data to keep driving (due to construction, weather, smoke, etc.) (Nissan Motor Company, n.d.). In this case, the vehicle will stop in-lane and contact a remote human operator who will view the situation and download a new path that the vehicle implements with its on-board ADS. At times this may involve the human authorizing an override of ADS rules, such as "never cross a double yellow line" on the pavement, which could be necessary to proceed beyond a lane closed for maintenance. For robo-taxis in particular, the complexity of endpoints (pickup and drop-off situations) may likely need remote support to provide efficient services to riders.

Remote driving approaches are also being pursued. Here, a remote human driver fully operates a driverless vehicle in a service setting (such as driving tractor trailers from a warehouse to a transfer yard near the highway where a driverless truck will take the load onwards) or to rescue a vehicle designed for driverless operation that had encountered a failure or a situation it cannot handle (such as a sudden snowstorm that is outside the system ODD) (Starsky Robotics, n.d.; Ford Motor Company, 2018).

2.3.8 Design and Test Processes to Ensure Safety

2.3.8.1 Functional Safety

An ADS must be designed to be robust to system failures: a power glitch could disrupt the software, an actuator could fail, and map data could be inaccurate. Functional safety is a discipline foundational to ADS development in which a safety critical system is designed, assessed, and validated in a highly structured process so that the desired level of performance is met for each safety function. ISO standard 26262, titled "Road Vehicles—Functional Safety" (International Organization for Standardization, 2011), provides the basis for functional safety in automotive systems, enabling ADS tech developers to trace all facets of vehicle testing back to high-level safety goals. Other applied hazard analysis techniques, such as Systems-Theoretic Process Analysis (STPA) (Abdulkhaleq et al., 2017) and Safety of the Intended Functionality (SOTIF) (International Organization for Standardization, 2019), are in use as well.

System health monitoring is a fundamental aspect of the overall system approach, as well as redundancies in the power supply, actuators, sensors, and software (Waymo, 2017). These measures greatly reduce but do not eliminate the possibility of a system failure. Therefore, a "fallback" is needed, which in the case of lower level automation (Levels 0–3) is the human driver. For higher levels of automation (Levels 4 and 5), the system itself is the fallback and must detect the problem and achieve a "Safe Stop" MRC (National Highway Traffic Safety Administration [NHTSA], 2017) that the vehicle can accomplish in spite of failures of system elements. Furthermore, safety performance is tightly bound to the system ODD. While an ADS may achieve all performance goals within the ODD, what happens if the vehicle encounters a situation outside the ODD, such as the sudden onset of severe weather for which the system was not designed? First, the ADS needs to be aware it has exited the ODD and then transition to an MRC. For a Level 3 ADS, this may be transition of control to the human driver. If the driver is not responsive, continued mitigation of risk may require bringing the vehicle to a safe stop. This generally entails pulling out of the traffic stream and parking the vehicle in a safe location. If that is not possible, the next step would be stopping the vehicle in the traffic lane and activating flashers (NHTSA, 2017)—not by any means an ideal situation but preferable to the vehicle continuing to travel when the software is confused or a key hardware element is inoperative.

As Ford notes in their Voluntary Safety Self-Assessment letter provided to NHTSA in 2018:

> Our vehicles rely on a comprehensive, robust diagnostics strategy, including a main Virtual Driver System and backup system sharing information. Each system conducts its own onboard diagnostics, as do critical subsystems such as braking and steering, allowing each to not only monitor themselves but each other. Redundant networks, controllers and actuators ensure continued operation in the case of a fault, enabling the vehicle to continuously plan fallback measures and execute them.

Further

> our resulting self-driving vehicle will fail safely or fail functionally depending on the type of the malfunction. Fail-functional systems include braking, steering, electrical

low voltage power and the Virtual Driver System. In the event that driving conditions sufficiently change to violate the ODD's conditions, our vehicles will implement the protocols defined by the Minimal Risk Condition.

(Ford Motor Company, 2018)

2.3.8.2   Cybersecurity

Cybersecurity presents another challenge that must be handled in ADS system design and health monitoring. Most ADS under development is tethered to the manufacturer's or technology developer's cloud for system monitoring, data collection, and, in some cases, active control of the vehicle. This wireless connection creates a "threat surface" that can be attacked by hackers. While news reports continue to emerge about the latest hack of a car (Wired.com, 2017), automobile manufacturers have invested extensive resources in developing new architectures, procedures, and hardware/software to prevent a cyberattack. ADS developers must design cybersecure systems as well. The Alliance of Automobile Makers (AAM) has noted that "vehicles are highly complex with multiple layers of security, and remote access is exceedingly difficult by design. New cars being launched now have a substantial increase in cybersecurity compared to earlier years. Automakers are collaborating in all areas possible, including hardware, software and knowledge sharing with suppliers, government and the research community" (Kunkle, 2019). Collaboration is key to monitoring the evolution of cyberattacks and securing critical vehicle systems. The Automotive Information Sharing and Analysis Center (Auto-ISAC) was established by the auto industry for this purpose, which is open to ADS developers as well. Both the Auto-ISAC (n.d.) and NHTSA (2016) have published reports on automotive cybersecurity best practices.

2.3.8.3   ADS Validation Processes

Testing ADS systems is an extensive process taking various forms (see also, this Handbook, Chapter 22). Systems are initially validated on testing grounds, being put through their paces via specific scenarios designed to emulate everyday driving. An example would be turning across the path of oncoming vehicles when sensor views may be blocked by other vehicles. Track testing is followed by high mileage testing of ADS on public roads, which is necessary to expose an automated vehicle to as many variations in traffic and road conditions as possible; this is an essential part of system development, assessment, and validation. Google/Waymo spent approximately six years (2009–2015) to reach their first million miles of on-road testing. As their fleet has grown, the test miles have increased exponentially: testing miles totaled 4 million by 2017 and another 6 million miles were added in 2018 for a cumulative 10 million miles at that point. Given the current size of Waymo's ADS test fleet, an additional million miles are now being added at a rate of once every six weeks.

Road testing is complemented by simulations, which can take an individual unusual occurrence from the real world (called "edge cases") and expand the key factors across many permutations to stress and evaluate the software beyond the specifics of the original real-world event. Waymo (2017) reports that over 7 billion miles have been traveled in virtual space via simulation. Software modifications that result from this testing and simulation are pushed out to the on-road fleet for further validation, a process that may occur on a daily basis in some cases. This ongoing feedback loop is core to agile and fast development and is a key reason why ADS development has progressed so rapidly.

Another example is BMW's Data-Driven Development process (BMW, 2019), which is being applied to validation of the BMW iNEXT Level 3 system launching in late 2021. To create a broad dataset, the development team is collecting 5 million km (3.1 million miles) of public road driving data from 80 BMW 7 Series cars that are in operation in the United States, Germany, Israel, and China. From this data, 2 million km (1.25 million miles) of highly relevant driving scenarios (including environmental factors) are then extracted. When parameterized and expanded in software, this data provides the simulation system an additional 240 million km (150 million miles) to maximize diversity of the road conditions against which the ADS is tested. The dataset will grow substantially, as the data collection fleet is set to grow to 140 vehicles by the end of 2019.

Because all ADS can be exposed to these edge cases in the real world, ADS developers are discussing ways to share data on edge cases, which would include development of standards to describe on-road scenarios. This process will take time and meanwhile developers are working individually. The recently completed German Pegasus project (Brooke, 2019), involving key automakers, created a database of relevant scenarios that can be implemented in simulation to complement on-road testing.

How well are current systems doing in real-world testing? Californian law requires ADS developers testing on public roads to annually report "disengagements," i.e., situations in which the human safety driver (testing a Level 4 system) took over driving from the ADS to maintain safety. This dataset does not provide a full picture of overall performance, as the complexity of the driving environment can vary dramatically. Nevertheless, this data can at least be indicative. In 2018, GM Cruise reported a takeover rate of approximately once every 5,000 miles, operating in the highly challenging traffic of downtown San Francisco (Ohnsman, 2019). Waymo reported a disengagement rate of approximately once every 11,000 miles, operating in the less complex environment of Silicon Valley suburbs (Waymo, 2019). It should be noted that several ADS Level 4 developers do the majority of their testing in states other than California, where disengagement reports are not required. Keeping in mind the caveat that disengagement data has limited value, from year to year it generally shows a steady improvement in disengagement rates.

2.4   AUTOMATED DRIVING COMMERCIAL DEVELOPMENT AND DEPLOYMENT

The car industry is pursuing an approach that brings low-level automation gradually to mass-market cars (such as Level 2 highway operations with supervised lane changes, or Level 3 Traffic Jam Pilots fully driving the vehicle in low-speed traffic). Because Level 3 operations involve shared responsibility between vehicle and driver yet allow the driver to be "eyes off," the auto industry was initially highly concerned about introducing these types of products. To reduce risk, auto OEMs aimed to "leapfrog" Level 3 to offer Level 4 systems where the driver is out of the control loop. However, given the long tail of Level 4 development (requiring new design approaches and extensive safety validation), Level 3 systems are now coming to market restricted to specific operating conditions, as described below.

As noted in the introduction, though initially highly skeptical about the robo-taxi use case, the a substantial portion of the auto industry has embraced mobility services as key to their future. In particular, offering mobility via robo-taxi services provides the advantage of fleet operation rather than private ownership. This is crucial in bringing such a complex technology to the public; fleet vehicles are not subject to the same cost pressures as a retail car product; return-on-investment is the key metric. Fleets have the advantage of hands-on, skilled staff to conduct software upgrades, system safety certification, maintenance, etc. on a regular basis for every vehicle in the service operation. Their operating area can be limited geographically and mapped in detail, with the complexity of the operating environment matched to the capability of the ADS. For these reasons, higher level automation will be deployed in fleets first; OEMs are viewing fleet operations as incubators for mass-market system development.

Current commercial development activity is focused across the domains of passenger cars, robo-taxi, public transit, parcel delivery, and heavy trucks. The availability of various levels of automation to the public will depend on both product introduction dates and geography. Robo-taxi deployments in particular will begin in highly constrained geographic environments and spread in extent gradually.

2.4.1 AUTOMATED FLEET SERVICES FOR FREIGHT AND PARCELS

Automated driving approaches can be applied across the full range of automation levels to long-haul trucking as well as last-mile delivery. Level 1 truck platooning, enabled by V2V communications, synchronizes longitudinal control across several in-line trucks to operate at much shorter inter-vehicle following distances than human drivers can safely achieve, with the resulting aerodynamic drafting benefits reducing fuel use. Commercial development of Level 1 truck platooning is being pursued by the startup Peloton Technology (n.d.) plus several truck OEMs. The business case for truck platooning is strong, particularly in the United States and Australia, where long-haul trucking on sparsely populated roads is common. Initial deployment in these and similar areas is expected as soon as 2019 (Bishop, 2019). Level 4 "Follow the Leader" operations with a human lead driver and driverless followers are also of high interest in the platooning sphere. This is being pursued by OEMs such as Scania (n.d.), as well as Peloton Technology and the new startup Locomation (n.d.). Deployment is expected within three to five years.

Driverless trucks, which have been operating in closed environments (seaports, mining, etc.) for several years, are now aimed at public road operations, led by a cadre of highly active startups including Einride, Embark, IKE Robotics, Kodiak Robotics, PlusAI, Pronto, Starsky Robotics, TuSimple, and Waymo (Alistair, 2019). In parallel, most truck OEMs have active development programs in Level 4 driverless trucks, particularly Daimler, Volvo Trucks, MAN, and Scania (Alistair, 2019). Many of these companies are currently running a fleet of freight-carrying trucks on public highways collecting performance data, with safety drivers in place to monitor the Level 4-intent system.

By replacing a truck driver with ADS technology for part or all of the journey, the Level 4 business case for trucking is strong and deployment is now at the beginning stages. In early 2019, Einride received permission from the Swedish government to begin operating their driverless vehicles on a 300-m section of non-highway public road (Einride, n.d.). Highway operations in the United States are seen as the strongest market, and truck ADSs are expected to be launched by other startups in the 2019–2020 timeframe in areas of lowest complexity in terms of traffic, topography, and weather. Most tech developers are focused initially on deployment of a "ramp to ramp" system that operates only on limited-access divided highways. Human drivers would bring the load to a transfer yard adjacent to the highway, where the robo-truck would attach to the trailer and then drive the entire highway portion, with the reverse happening at the other end of the journey (Alistair, 2019). However, TuSimple (n.d.) plans to operate from "dock to dock," taking the load driverless from plants, warehouses, or distribution centers via the street network to reach the highway, providing driverless operations for the entire trip. See Table 2.1 for a summary of commercial activity in Automated Freight Movement; note that information in the table is not exhaustive.

Delivery of parcels by automated vehicles operating in street settings is an intriguing use case, because the pressures that come with transporting human passengers (safety, user experience, etc.) are not present. This allows greater design freedom while maintaining the attractive business case of fulfilling deliveries without the need to pay a human driver. Working with retail partners, Ford has tested an automated pizza delivery service (with safety driver) in Miami, with plans to replicate this next in Washington, D.C., and at scale in 2021 in several cities (Ford Motor Company, 2018). Daimler has conducted tests of automated V-Class vans in the Stuttgart area for parcel delivery (Daimler, n.d.). Toyota is aiming to have their e-Palette vehicles (configurable, all-purpose automated platforms) ready in time for the 2020 Olympic and Paralympic Games in Tokyo, with testing in the United States in the early 2020s (Toyota Corporation, n.d.). In 2018, Toyota announced the e-Palette Alliance, which included retail partners such as Pizza Hut and Amazon. Nuro (n.d.) and Udelv (n.d.) are two startups active in this space. See Table 2.2 for a summary of commercial activity in Automated Parcel Delivery; note that information in the table is not exhaustive.

2.4.2 Automated Fleet Services for People

Numerous companies are now testing automated ride hailing in some form. Most are still in the safety driver phase, collecting environmental and performance data as well as HD map information. As one example, Lyft and Aptiv are piloting automated ride hailing in Boston (nuTonomy, n.d.), which includes research addressing aspects of a passenger's experience during the ride. This team also offers robo-taxi rides (with safety driver) in Las Vegas; here, Aptiv mapped the city streets and intersections and is collecting data on the local traffic congestion to enhance performance of the system. Additionally, Ford aims to have 100 self-driving vehicles on the road by the end of 2019 (Fitzgerald, 2019).

TABLE 2.1

A Summary of Commercial Activity in Automated Freight Movement

Fleet Operations Commercial Activity: Automated Freight Movement

Developer	Level	Function	Road Type Highway	Road Type Street	Platform	Public Road Test	Deployment	Comments
Einride	4	Short freight runs		•	Custom vehicle (no driver compartment)		2019	Initial operations in Europe
Peloton	1	Long-haul freight, Platooning	•		Retrofitted production truck	Pre-2019	2020	Initial operations in USA
Multiple startups and OEMs	4	Long-haul freight	•		Retrofitted production truck	Pre-2019		Variety of test environments
Leading startups	4	Long-haul freight	•	•	Retrofitted production truck		2020 (est)	Limited deployment, least complex environments
Startups and OEMs	1, 4	Long-haul freight, platooning w/driverless followers	•		Retrofitted production truck	2020 (est)		Variety of environments
Multiple startups and OEMs	4	Long-haul freight	•	•	Production truck		2025 (est)	Deployment in diverse environments
Startups and OEMs	1, 4	Long-haul freight, platooning w/driverless followers	•		Production truck		2022 (est)	Limited deployment, least complex environments
Startups and OEMs	1, 4	Long-haul freight, platooning w/driverless followers	•		Production truck		2025 (est)	Deployment in diverse environments

TABLE 2.2
A Summary of Commercial Activity in Automated Parcel Delivery

Fleet Operations Commercial Activity: Level 4 Automated Parcel Delivery

Developer	Function	Road Type		Platform	Public Road Test	Deployment	Comments
		Highway	Street				
Nuro	Last-mile delivery: grocery		•	Custom vehicle	2019		Partnership with Kroger
Udelv	Last-mile delivery: grocery		•	Retrofitted production passenger van	2019		Partnership with Walmart
Ford	Last-mile delivery: pizza		•	Retrofitted production passenger car	Pre-2019		Miami, Florida (2018) Washington, DC
Daimler	Last-mile delivery		•	Retrofitted production passenger car	Pre-2019		Stuttgart (2018)
Toyota	Last-mile delivery, multiple functions, reconfigurable		•	Custom vehicle		2020	ePalette multi-purpose platform limited deployment US, Japan
Startups and OEMs	Last-mile delivery		•	Custom vehicle/ production vehicle		2020 (est)	Limited deployment, least complex environments
Startups and OEMs	Last-mile delivery		•	Custom vehicle/ production vehicle		2025 (est)	Deployment in diverse environments

The first launch of robo-taxi services (true driverless, without a safety driver or "minder") is expected in the 2020 timeframe from Waymo through its flagship product, Waymo One (Waymo, n.d.), and from General Motors through their Cruise Automation group (Chokshi, 2018; Colias, 2018). Initial services will be highly limited geographically and expand based on customer demand and increasing technology capability. Toyota and Uber plan to begin pilot-scale deployments on the Uber ride-sharing network at an unspecified time. BMW will be fielding a fleet in late 2021 to test robo-taxi functionality in large-scale trials conducted in defined urban environments (BMW, 2019). Looking longer term, Volvo Cars and Baidu have partnered to develop and mass manufacture "fully electric and autonomous vehicles," estimating 14.5 million units to be sold in China by 2040 for robo-taxi services.

These automated ride hailing services are currently focused only on street operations. While this may serve some markets (such as New York City) very well, many deployment areas will see customer demand for the automated vehicles to also operate on the local highways to optimize the trip time. This has significant implications for the technical approach but is not likely to occur within the first few years of initial deployment.

In addition, high capacity mobility services will be operating in public transit operations. For example, in 2019 five automated buses were slated to begin operation across a 14-mile route including the Forth Road Bridge in Edinburgh, Scotland, on public roads and exclusive busways (BBC, 2018). Initially, true driverless operations will only occur within the bus depot for movements such as parking and moving to the fueling station and bus wash. On road, the presence of a driver will be required "as a back-up for passenger safety and to comply with UK legislation."

See Table 2.3 for a summary of commercial activity in robo-taxi deployment, all of which is focused on Level 4; note that information in the table is not exhaustive.

2.4.3 PRIVATE OWNERSHIP: AUTOMATION FEATURES IN MASS-MARKET PASSENGER CARS

To bring ADSs to the mass market, product developers must make decisions across a multi-dimensional space. One dimension is the driving function, including product definition choices such as driving on highways only or all roads, driving only in certain situations (low-speed traffic jams) or all situations, and operating only in a single lane versus automatically completing lane changes. Across this space is mapped the split between driver and computer-based operation. The level of utility/value to the user must be counterbalanced by the readiness of the ADS to take on additional responsibility for safe driving.

What is on the market now? Many OEMs offer automated longitudinal control plus "lane centering" steering assist that requires hands on the steering wheel. This is true of the Tesla Autopilot system, which despite the public's perception, is not a self-driving system (Tesla, n.d.). Typically, these systems operate only above a speed of 50 mph or more, as they are intended for highway use; Tesla provides this capability at lower speeds as well. Vehicles from several OEMs can also perform an automated lane change: Tesla vehicles can perform lane changes automatically, whereas vehicles from other brands typically suggest a lane change and seek driver

TABLE 2.3
A Summary of Commercial Activity in Robo-Taxi (Level 4)

Fleet Operations Commercial Activity: Robo-Taxi

Developer	Road Type		Platform	Public Road Test	Deployment	Operations Area
	Highway	Street				
May Mobility		•	Custom vehicle		2019	Detroit, now in commercial operation (with on-board "minders")
Transport Scotland	•	•	Modified production bus		2019	12 mile road segment 2019 Edinburgh (with safety driver as required by UK law)
Waymo		•	Retrofitted production passenger car	Pre-2019	2020 (est)	"Waymo Now," Chandler, AZ announced for 2019
Mercedes, Bosch		•	Retrofitted production passenger car	Pre-2019		San Jose, CA
Alliance		•	Retrofitted production passenger car	Pre-2019	2020	2020 Summer Olympics, Japan
Toyota		•	Retrofitted production passenger car		2020	2020 Summer Olympics, Japan
Hyundai		•	Retrofitted production passenger car	Pre-2019	2021	2021 deployment announced
Ford		•	Retrofitted production passenger car	Pre-2019	2021	100 self-driving vehicles on public roads in Miami, Fla., Pittsburgh, PA., and Dearborn, Mich. 2021 deployment announced USA
BMW		•	Retrofitted production passenger car	Pre-2019	2021	China
Volkswagen		•	Retrofitted production passenger car		2021	2021 deployment announced

(Continued)

TABLE 2.3 (*Continued*)
A Summary of Commercial Activity in Robo-Taxi (Level 4)

Fleet Operations Commercial Activity: **Robo-Taxi**

Developer	Road Type Highway	Street	Platform	Public Road Test	Deployment	Operations Area
Multiple		•	Production robo-taxi		2021 (est)	Operations confined to streets
Multiple	•	•	Production robo-taxi		2023 (est)	Natural evolution from street-only services to highways, depending on local road networks
BMW		•	Production robo-taxi		2023	Deployment announced
Mercedes		•	Production robo-taxi		2023	Deployment announced
Aptiv/ nuTonomy		•	Retrofitted production passenger car	Pre-2019		Singapore
GM/Cruise		•	Retrofitted production passenger car	Pre-2019		2019 deployment plan deferred to unspecified later date, USA
Lyft, Aptiv		•	Retrofitted production passenger car	Pre-2019		Las Vegas, Nevada
Pony.ai		•	Retrofitted production passenger car			China, no specific information available
Toyota, Uber		•	Production robo-taxi			No specific information available
Voyage		•	Retrofitted production passenger car	Pre-2019		The Villages (Florida retirement community)
WeRide.ai		•	Retrofitted production passenger car			Remote driving via 5G network, China; no specific information available

permission (examples are BMW, Mercedes, Volvo Cars, and Nissan). For instance, BMW's Automatic Lane Change on the 2019 X5 (BMW, 2018) can be used on highways when the Lane-Keeping Assistant (lane centering) is active. A lane change is activated by the driver holding the turn signal in the desired direction; if the sensors detect that there is space in the adjacent lane and that no other vehicle is approaching at high speed, the lane change is completed.

Driver monitoring is playing an increasingly important role. The 2019 Audi A8 provides lane centering and requires hands on the steering wheel. An escalating series of warnings are triggered if the vehicle detects an inattentive driver while the cruise control is activated; these are audible, visible, and physical interventions (brake taps). Lack of a driver response is interpreted as a driver emergency; the car slows to a complete stop in-lane with the hazard lights on and initiates an emergency call.

True hands-off production systems were pioneered by General Motors (n.d.), which introduced Super Cruise™ on the model year 2018 Cadillac CT6. The ODD of this SAE Level 2 system restricts use to limited-access divided highways based on detailed mapping specifically created for this product. GM speaks of the driver and Super Cruise™ acting as "partners in driving." An infrared camera mounted on the steering column enables head tracking to assess the driver's attentiveness to the road. If driver inattention is detected, visual alerts and haptic seat signals are produced. In the absence of proper driver response, the vehicle will achieve the MRC of slowing down, stopping in the highway, and activating hazard flashers, with the vehicle contacting Onstar Emergency Services (part of the GM telematics system) (Cadillac, n.d.). BMW offers a hands-off Traffic Jam Assistant (Hyatt, 2018) as long as the driver's attention is clearly on the road ahead (using driver monitoring optical cameras placed at the top of the instrument cluster); the system operates on limited-access highways and surface streets at speeds less than 37 mph. The model year 2020 Nissan Skyline, to be available in the Japanese market in late 2019, will offer the ProPilot 2.0 (Nissan Motor Company, 2019) system capable of "navigated" highway driving with hands-off multi-lane driving capabilities. The navigation concept refers to on-ramp to off-ramp highway driving, engaging with the vehicle's navigation system to help maneuver the car according to a predefined route on designated roadways. On the route set by the driver, the system assists with passing, lane diversions, and lane exiting. Here, as in the other Level 2 cases, a monitoring system in the cabin continually confirms that the driver's attention remains on the road.

Tesla stated (somewhat confusingly) that by the end of 2019 its vehicles will be able to operate without any driver control inputs on highways and streets but will still require the driver to pay attention to the road. Sometime during 2020, the company asserts that "the system will be significantly robust to permit users to not pay attention while using it" (Paukert & Hyatt, 2019), which implies a Level 3 or higher system. Similarly, Audi, BMW, Mercedes-Benz, and Volvo have stated they will offer highly automated vehicles in the early 2020s. An interesting Level 3 system example is the 2021 BMW Vision iNEXT (BMW Group, 2019), which will enable drivers to "delegate the task of driving to the car for extended periods of time when driving on the highway at speeds up to 130 km/h (81 mph)."

As product offerings enter Level 3, vehicle manufacturers are adding a new component: pre-approval of roads to enable automated functionality. As noted above, Cadillac only allows use of Super Cruise™ on limited-access divided highways. Level 3 systems now being discussed will likely require specific roads (or road segments) to be approved; in some cases this will be dynamic based on the prevailing weather and traffic conditions. For example, Volvo Cars' model year 2021 XC90 will offer a Highway Pilot system, which would "allow you to eat, sleep, work, watch a movie, relax, do whatever" on "roads that the system was confident it could safely navigate" (Shilling, 2018).

Initial mass-market Level 4 systems, in which the ADS is fully responsible for the driving task within the ODD, will most likely focus on highway operations initially. Audi has announced that by 2021 it will introduce Level 4 "Highway Pilot" for use on limited-access highways. Moreover, startup electric vehicle manufacturers have typically asserted that their offerings will include high levels of automated driving capability. For instance, Byton has announced that Level 4 automated driving is planned as part of their "K-Byte" sport utility vehicle (SUV) launch in 2021 (O'Kane, 2018).

In the past, a product introduction date was synonymous with availability of a feature "everywhere." As can be seen from the above discussion, timing is not everything: it must be paired with the geography and road conditions in which a particular function is made available via the vehicle manufacturer's cloud connection with the vehicle.

2.5 REGULATORY CONSIDERATIONS

Regulations regarding vehicle equipment and usage of vehicles have been largely harmonized with regard to human-driven vehicles (see also, this Handbook, Chapter 14). The advent of automated vehicles presents a massive challenge to regulators, and there are distinct differences in the processes to develop regulatory frameworks for ADS in different parts of the world. The governing structure of countries such as China and Singapore is considered "top down" such that officials can move quickly to set up a regulatory structure. In the near term, Australia has an orderly process underway within their existing vehicle certification framework to allow highly automated cars, trucks, and robo-taxis to operate in the future via a safety assurance system based on mandatory self-certification (Australia National Transport Commission, 2018). Europe is allowing significant testing while enmeshed in a long-term process working with international bodies to enable deployment of ADS. Based on an informal poll conducted by the author, there is a broad agreement among EU experts that Level 4 systems on mass-market vehicles will not be generally allowed for sale or commercial operations in Europe before 2025. However, permits and exemptions processes are expected to allow limited scale and limited time deployment of some fleet services, which could enable small-scale market launches earlier.

In the United States, three legal regimes define the ADS playing field: (1) the 1949 Geneva Convention on Road Traffic, (2) regulations enacted by NHTSA (mainly focused on vehicle "equipment") and the Federal Motor Carrier Safety Administration (FMCSA) (mainly focused on commercial trucks), and (3) the vehicle codes of all 50 U.S. states (mainly focused on how vehicles are used, such as speed limits, mobile phone usage, and seat belt usage). No current NHTSA regulations prohibit or impede sale of automated vehicles. However, the situation in the United States for truckers presents a special case, due to FMCSA regulations covering commercial truck drivers that were written such that "the driver" (assumed to be human) had to perform tasks such as a pre-trip inspection of the truck for any mechanical problems. Truck ADS developers argued that these types of rules should not apply to an ADS truck. In USDOT's Federal Automated Vehicles Policy 3.0 (NHTSA, 2018), the Federal government agreed, explicitly allowing Level 4 driverless trucks to operate on U.S. highways. This pro-industry policy, based on self-certification by tech developers, most likely puts the United States in the lead position worldwide regarding deployment of ADS trucks. Going forward, there may be a need to clarify the authority of state regulators on this topic; some states are very supportive of driverless trucks while others are being more cautious.

First-generation truck platooning represents a special regulatory case. As it relates to a Level 1 system with drivers fully attentive and able to adapt as needed to surrounding traffic, ADS-related regulations are not relevant. However, because trucks follow each other more closely than would be safe without platooning technology, state-level truck-following distance regulations come into play. Motivated by the fuel economy benefits, truck ADS tech developers and early adopter fleets have worked extensively to explain the platooning safety case to state officials. This has resulted in full allowance of commercial truck platooning in 27 U.S. states (Boyd, 2019), with ongoing efforts to amend regulations in many others.

State-level laws in the United States, written before automated driving was envisioned, need to be updated to adapt to ADS operations. For instance, many state regulations consider a vehicle in an active roadway with no human inside to be an abandoned vehicle, which is prohibited—making a robo-taxi without a rider illegal (National Academies of Sciences, Engineering, and Medicine, 2018)! At the local level, the role of law enforcement and emergency response is important in terms of education and developing appropriate practices and norms. For instance, emergency responders need to have confidence that an ADS vehicle involved in a crash will not attempt to start driving when their personnel are nearby (Waymo, 2017). These issues are generally seen as tractable but need to be addressed for full deployment of ADS.

2.6 GOING FORWARD: ADS IMPLICATIONS FOR HUMAN FACTORS

From the preceding discussion of ADS, it is clear that tangible business cases exist, significant resources are being applied, and the pace of technological progress is fairly rapid. This does not imply there will be no bumps in the road. While many challenges lie ahead in terms of technical robustness to achieve an acceptable level of safety in operations, it is vital to examine how these intelligent machines serve the needs of humans as users and beneficiaries. Following are a few thoughts as to human factors implications.

Public trust is fundamental; the average road user has limited or no direct experience with ADS, yet is barraged with endless and confusing media speculation (see this Handbook, Chapters 4, 5). Many ADS developers emphasize the importance of customer trust. As Ford puts it, "We don't believe that the central challenge in the development of self-driving vehicles is the technology. It's trust. Trust in the safety, reliability, and experience that the technology will enable" (Ford Motor Company, 2018).

Traditional human factors investigations apply strongly to Automation Levels 1–3, whereas the emphasis in Level 4–5 systems shifts more toward trust and user experience. Level 1 systems have been available for many years in the form of Adaptive Cruise Control. Truck platooning is a new application that presents interesting issues for follower drivers in particular. In Level 1, the job of such drivers is to cede longitudinal control to the system and retain responsibility for steering, monitoring the road environment, and adjusting to other traffic via lane changes or dissolving the platoon. Will the driver find the follower role to be stressful, monotonous, or both? How will the driver anticipate events ahead? Initial research results based on a field operational test of truck platooning recently completed in Germany found that drivers who were new to the idea of platooning were initially skeptical but very early in the trial came to accept the system, with some preferring it over regular driving due to the "team driving" feeling and high trust in the system. Furthermore, physiological studies showed platoon driving to be no more stressful or monotonous than regular truck driving (Hochschule Fresenius, DB Schenker, & MAN Truck & Bus, n.d.). As deployment of first-generation platooning systems proceeds, additional research would be valuable to assess experiences of platoon drivers as well as those sharing the roads with platoons.

Vehicles equipped for Level 2 and 3 automation are interesting for human factors investigations, because the safety case involves both the ADS and the human driver. Can the driver be trusted to perform their assigned duties, or should they be monitored (see e.g., this Handbook, Chapters 8, 11)? Product decisions about driver monitoring relate to personal responsibility. Initially, the Tesla Autopilot relied on drivers to monitor this system; now, hands-on-wheel detection is installed but there is no attention monitoring. For other automakers, there is a strong trend to implement driver monitoring in some form. Avoiding mode confusion between Level 2 and Level 3 will be of top importance, particularly in a vehicle that may provide Level 3 capability on the highway but Level 2 support after exiting the highway (see also, this Handbook Chapters 7, 15, 21). Or, given that some automakers will enable specific sections of highway for Level 3 support, the user must be aware that the vehicle has proceeded out of the "Level 3 approved section" and is now operating in Level 2 such that the driver's constant attention is required.

ADS delivering "driverless" capability completely changes the human factors paradigm. For robo-taxis, this shifts from understanding driver performance to possibly monitoring the vehicle interior to ensure the service is operating properly and not being abused. Trust will be a key competitive discriminator: service quality will key on providing information from the vehicle (audible, visual, haptic) that reassures riders the ADS knows what it is doing. Additionally, ADS will distinguish themselves competitively by having enough "street smarts" to understand complex streetside dynamics to efficiently pick up and drop off passengers (Figure 2.4). Customer service must encompass situations ranging from informing passengers during a malfunction or crash event to more benign aspects such as allowing passengers to make an unplanned stop.

FIGURE 2.4 Pickup and drop-off points may be challenging for robo-taxis. (Image source: iStock. Photo credit: Anouchka. Lisbon, Portugal - May 2, 2017: People waiting in line to get a taxi in Lisbon Portela Airport, Portugal.).

Those operating in the vicinity of an ADS need to have an understanding of the intentions and likely behavior of that vehicle in the absence of the typical human cues (such as eye contact and waving hands) that are now the norm. Similarly, the ADS needs to understand the intentions and likely behaviors of other road users. The issues here fall under the general term of External HMI, which is now the subject of substantial research and product development. Ford is investigating an External HMI approach that uses a light bar fitted near the robo-taxi windshield to alternately flash, strobe, or pulse in order to indicate vehicle intent—including yielding, accelerating from a stop, and driving (Ford Motor Company, 2018). For instance, a white light slowly passes back and forth when the vehicle is yielding; when the vehicle is stopped and preparing to move, the light blinks rapidly to get people's attention. In these types of communications, it is vital to use a universal language understandable by a variety of cultures. Here, significant additional effort is needed to understand and evaluate various approaches and then converge on an approach that can be standardized across the industry.

For trucking, Level 4 platooning operations present unique human factors challenges. The "Follow the Leader" mode of platooning with a human lead driver and driverless followers presents a driving mode that has never before existed. Issues include assessing stress, how the driver maintains situational awareness of the entire multi-vehicle configuration, and HMI design; these questions are largely unexplored in the literature to date, yet the business case is so strong that investigations are likely to be underway by commercial system developers.

Straddling the domains of inside/outside the vehicle is the concept of remote support to driverless vehicles from humans in vehicle management centers, as well as remote driving of the vehicle; again, related human factors issues are largely unexplored.

REFERENCES

Abdulkhaleq, A., Wagner, S., Lammering, D., Röder, J., Balbierer, N., Ramsauer, L., … Beohmert, H. (2017). A systematic approach based on STPA for developing a dependable architecture for fully automated driving vehicles. *Procedia Engineering, 179,* 51.

Alistair, C. (2019). *These 7 Companies Are Making the Self-Driving Truck a Reality.* Retrieved from www.gearbrain.com/autonomous-truck-startup-companies-2587305809.html

Aurora. (n.d.). *The New Era of Mobility.* Retrieved from https://aurora.tech/vssa/index.html

Australia National Transport Commission. (2018). *Safety Assurance for Automated Driving Systems: Decision Regulation Impact Statement.* Retrieved from www.ntc.gov.au

Automotive Information Sharing and Analysis Center. (n.d.). *Best Practices.* Retrieved from www.automotiveisac.com/best-practices/

BBC. (2018). *First Driverless Edinburgh to Fife Bus Trial Announced.* Retrieved www.bbc. com/news/uk-scotland-edinburgh-east-fife-46309121

Billington, J. (2018). *The Prometheus Project: The Story Behind One of AV's Greatest Developments.* Retrieved from www.autonomousvehicleinternational.com/features/the-prometheus-project.html

Bishop, R. (2005). *Intelligent Vehicle Technology and Trends.* Norwood, MA: Artech House.

Bishop, R. (2019). *The Three Streams of Truck Platooning Deployment.* Retrieved from https://medium.com/@richard_32638/the-three-streams-of-truck-platooning-development-a71edbb8c12a

Blanco, S. (2019). *New Automotive Radars Take Tech from a Blip to a Boom.* Retrieved from www.sae.org/news/2019/03/new-av-radar-sensors

BMW Group. (2018). *The All-New 2019 BMW X5 Sports Activity Vehicle.* Retrieved from www.press.bmwgroup.com/usa/article/detail/T0281821EN_US/the-all-new-2019-bmw-x5-sports-activity-vehicle

BMW Group. (2019). *The New BMW Group High Performance D3 platform. Data-Driven Development for Autonomous Driving.* Retrieved from www.press.bmwgroup.com/global/article/detail/T0293764EN/the-new-bmw-group-high-performance-d3-platform-data-driven-development-for-autonomous-driving

Boyd, S. (2019). *Peloton Technology: Connected Automation for Freight Safety and Efficiency.* Retrieved from www.automatedvehiclessymposium.org/home

Brooke, L. (2019). Autonomy for the masses. *Autonomous Vehicle Engineering,* 2019(March), 8–12.

Brown, I. (1986). Functional requirements of driving. *Paper Presented at the Berzelius Symposium on Cars and Causalities.* Stockholm, Sweden.

Cadillac. (2017). *V2V Safety Technology Now Standard on Cadillac CTS Sedans.* Retrieved from https://media.cadillac.com/media/us/en/cadillac/news.detail.html/content/Pages/news/us/en/2017/mar/0309-v2v.html

Cadillac. (n.d.). *CT6 Super Cruise™ Convenience and Personalization Guide.* Retrieved from www.cadillac.com/content/dam/cadillac/na/us/english/index/ownership/technology/supercruise/pdfs/2018-cad-ct6-supercruise-personalization.pdf

Chokshi, N. (2018). *Mary Barra Says G.M. Is "On Track" to Roll Out Autonomous Vehicles Next Year.* Retrieved from www.nytimes.com/2018/11/01/business/dealbook/barra-gm-autonomous-vehicles.html

Colias, M. (2018). *GM President Dan Ammann to Take New Role as Head of Autonomous-Car Business.* Retrieved from www.wsj.com/articles/gm-president-dan-ammann-to-take-new-role-as-head-of-autonomous-car-business-1543514469

Daimler. (n.d.). *The Vision Van. Intelligent Delivery Vehicle of the Future.* Retrieved from www.daimler.com/innovation/case/electric/mercedes-benz-vision-van-2.html

Defense Advanced Research and Projects Agency (DARPA). (n.d.). *The Grand Challenge.* Retrieved from www.darpa.mil/about-us/timeline/-grand-challenge-for-autonomous-vehicles

Eckert, V. (2019). *German Carmakers to Invest 60 Billion Euros in Electric Cars and Automation: VDA.* Retrieved from www.reuters.com/article/us-autoshow-geneva-germany-vda/german-carmakers-to-invest-60-billion-euros-in-electric-cars-and-automation-vda-idUSKCN1QJ0AU

Einride. (n.d.). *World Premiere: First Cab-Less and Autonomous, Fully Electric Truck in Commercial Operations on Public Road.* Retrieved from www.einride.tech/news/world-premiere-first-cab-less-and-autonomous-fully-electric-truck-in-commercial-operations-on-public-road/

European Commission. (n.d.). *Transport Research and Innovation Monitoring and Information System (TRIMIS), Chauffeur II.* Retrieved from https://trimis.ec.europa.eu/project/promote-chauffeur-ii

Fitzgerald, M. (2019). *Ford Claims It Will Have 100 Self-Driving Cars on the Road by the End of the Year.* Retrieved from www.cnbc.com/2019/04/25/ford-aims-for-100-self-driving-cars-on-the-road-by-the-end-of-2019.html

Ford Motor Company. (2018). *A Matter of Trust: Ford's Approach to Developing Self-Driving Vehicles.* Retrieved from https://media.ford.com/content/dam/fordmedia/pdf/Ford_AV_LLC_FINAL_HR_2.pdf

General Motors. (n.d.). *Giving You the Freedom to Go Hands Free.* Retrieved from www.cadillac.com/world-of-cadillac/innovation/super-cruise

Griggs, T. & Wakabayashi, D. (2018). How a self-driving Uber killed a pedestrian in Arizona. *The New York Times.* Retrieved from www.nytimes.com/interactive/2018/03/20/us/self-driving-uber-pedestrian-killed.html

Hochschule Fresenius, DB Schenker, & MAN Truck & Bus. (n.d.). *EDDI Electronic Drawbar – Digital Innovation: Project Report – Presentation of the Results.* Retrieved from www.deutschebahn.com/resource/blob/4136372/d08df4c3b97b7f8794f91e47e86b71a3/Platooning_EDDI_Project-report_10052019-data.pdf

Hyatt, K. (2018). *BMW's Extended Traffic Jam Assistant System Wants to Stare into Your Eyes.* Retrieved from www.cnet.com/roadshow/news/bmw-driver-monitor-camera-x5/

Insurance Institute for Highway Safety/Highway Loss Data Institute. (2019). *Real-World Benefits of Crash Avoidance Technologies.* Retrieved from www.iihs.org/media/259e5bbd-f859-42a7-bd54-3888f7a2d3ef/e9boUQ/Topics/ADVANCED%20DRIVER%20ASSISTANCE/IIHS-real-world-CA-benefits.pdf

International Organization for Standardization. (2011). *Road Vehicles – Functional Safety* (ISO 26262). Retrieved from www.iso.org/standard/43464.html

International Organization for Standardization. (2019). *Road Vehicles—Safety of the Intended Functionality* (ISO/PAS 21448:2019). Retrieved from www.iso.org/standard/70939.html

Japan Automotive Research Institute. (n.d.). *Functional Safety* (ISO26262). Retrieved from www.jari.or.jp/tabid/223/Default.aspx

Kunkle, F. (2019). *Auto Industry Says Cybersecurity Is a Significant Concern as Cars Become More Automated.* Retrieved from www.washingtonpost.com/transportation/2019/04/30/auto-industry-says-cybersecurity-is-significant-concern-cars-become-more-automated/?wpisrc=nl_sb_smartbrief

Lee, T. B. (2019). *Feds Investigate Why a Tesla Crashed into a Truck Friday, Killing Driver.* Retrieved from https://arstechnica.com/cars/2019/03/feds-investigating-deadly-friday-tesla-crash-in-florida/

Locomation. (n.d.). *Locomation: Driving the Future of Autonomous Trucking.* Retrieved from https://locomation.ai/

Michon, J. A. (1985). A critical view of driver behavior models: What do we know, what should we do? In L. Evans & R. Schwing (Eds.), *Human Behavior and Traffic Safety* (pp. 485–520). New York: Plenum Press.

MyCarDoesWhat.org (n.d.). *All about Today's Car Safety Features*. Retrieved from https://mycardoeswhat.org/

National Academies of Sciences, Engineering, and Medicine. (2018). *Implications of Connected and Automated Driving Systems, Vol. 3: Legal Modification Prioritization and Harmonization Analysis*. Washington, DC: The National Academies Press. doi:10.17226/25293. Retrieved from www.trb.org/NCHRP/Blurbs/178300.aspx

National Automated Highway System Consortium. (1998). *Technical Feasibility Demonstration Final Report*. Retrieved from https://path.berkeley.edu/sites/default/files/part_1_ahs-demo-97.pdf

National Highway Traffic Safety Administration. (2016). *Cybersecurity Best Practices for Modern Vehicles (DOT HS 812 333)*. Washington, DC: National Highway Traffic Safety Administration. Retrieved from www.nhtsa.gov/staticfiles/nvs/pdf/812333_CybersecurityForModernVehicles.pdf

National Highway Traffic Safety Administration (NHTSA). (2017). *Automated Driving Systems 2.0: A Vision for Safety*. Retrieved from www.nhtsa.gov/sites/nhtsa.dot.gov/files/documents/13069a-ads2.0_090617_v9a_tag.pdf

National Highway Traffic Safety Administration. (2018). *Preparing for the Future of Transportation: Automated Vehicles 3.0*. Retrieved from www.nhtsa.gov/vehicle-manufacturers/automated-driving-systems

Nissan Motor Company. (2019). *Nissan to Equip New Skyline with World's First Next Gen Driver Assistance System*. Retrieved from https://newsroom.nissan-global.com/releases/nissan-to-equip-new-skyline-with-worlds-first-next-gen-driver-assistance-system

Nissan Motor Company. (n.d.). *Seamless Autonomous Mobility: The Ultimate Nissan Intelligent Integration*. Retrieved from www.nissan-global.com/EN/TECHNOLOGY/OVERVIEW/sam.html

Nowakowski, C., Shladover, S. E., & Chan, C.Y. (2016). Determining the readiness of automated driving systems for public operation: Development of behavioral competency requirements. *Transportation Research Record: Journal of the Transportation Research Board, 2559*, 65–72.

Nuro. (n.d.). https://nuro.ai/

nuTonomy. (n.d.). *nuTonomy to Test Its Self-Driving Cars on Specific Public Roads in Boston*. Retrieved from www.nutonomy.com/press-release/boston-trial-launch/

O'Kane, S. (2018). *Byton Teases a Fully Autonomous Electric Sedan Due in 2021*. Retrieved from www.theverge.com/2018/6/12/17449996/byton-sedan-concept-k-byte-release-date

Ohnsman, A. (2019). *Waymo Tops Self-Driving Car Disengagement Stats as GM Cruise Gains and Tesla Is AWOL*. Retrieved from www.forbes.com/sites/alanohnsman/2019/02/13/waymo-tops-self-driving-car-disengagement-stats-as-gm-cruise-gains-and-tesla-is-awol/#6a6f623931ec

Paukert, C. & Hyatt, K. (2019). *Tesla Autonomy Investor Day: What We Learned, What We Can Look Forward To*. Retrieved from www.cnet.com/roadshow/news/teslas-autonomy-investor-day-recap/

Peloton Technology. (n.d.). https://peloton-tech.com/

Perry, F. (2017). *Overview of DSRC Messages and Performance Requirements*. Retrieved from www.transportation.institute.ufl.edu/wp-content/uploads/2017/04/HNTB-SAE-Standards.pdf

SAE International. (2018). *Taxonomy and Definitions for Terms Related to Driving Automation Systems for On-Road Motor Vehicles* (J3016_201806). Warrendale, PA: Society for Automotive Engineers. Retrieved from www.sae.org/standards/content/j3016_201806/

Sage, A. & Lienert, P. (2017). *GM Plans Large-Scale Launch of Self-Driving Cars in U.S. Cities in 2019*. Retrieved from www.reuters.com/article/us-gm-autonomous/gm-plans-large-scale-launch-of-self-driving-cars-in-u-s-cities-in-2019-idUSKBN1DU2H0

Scania Group. (n.d.). *Autonomous Transport Solutions*. Retrieved from www.scania.com/group/en/autonomous-transport-solutions/

Shilling, E. (2018). *Volvo Plans Autonomous XC90 You Can "Eat, Sleep, Do Whatever" in By 2021*. Retrieved from https://jalopnik.com/volvo-plans-autonomous-xc90-you-can-eat-sleep-do-what-1826997003

Shladover, S. E. (2012). *Recent International Activity in Cooperative Vehicle–Highway Automation Systems* (FHWA-HRT-12–033). Washington, DC: Federal Highway Administration.

Somerville, H. (2019). *Uber's Self-Driving Unit Valued at $7.25 Billion in New Investment*. Retrieved from www.reuters.com/article/us-uber-softbank-group-selfdriving/ubers-self-driving-unit-valued-at-7-25-billion-in-new-investment-idUSKCN1RV01P

Starsky Robotics. (n.d.). *Voluntary Safety Self-Assessment*. Retrieved from https://uploads-ssl.webflow.com/599d39e79f59ae00017e2107/5c1be1791a7332594e0ff28b_Voluntary%20Safety%20Self-Assessment_Starsky%20Robotics.pdf

Tesla. (n.d.). *Autopilot Section*. Retrieved from www.tesla.com/presskit

Toyota Corporation. (n.d.). *e-Palette Concept*. Retrieved from www.toyota-global.com/pages/contents/innovation/intelligent_transport_systems/world_congress/2018copenhagen/pdf/e-Palette_CONCEPT.pdf

TuSimple. (n.d.). Retrieved www.tusimple.com/

UC Berkeley Institute of Transport Studies. (n.d.). *National Automated Highway Systems Consortium*. Retrieved from https://path.berkeley.edu/research/connected-and-automated-vehicles/national-automated-highway-systems-consortium

Udelv. (n.d.). Retrived from www.udelv.com/

Voelcker, J. (2007). Autonomous vehicles complete DARPA urban challenge. *IEEE Spectrum*. Retrieved from https://spectrum.ieee.org/transportation/advanced-cars/autonomous-vehicles-complete-darpa-urban-challenge

Waymo. (2017). *On the Road to Fully Self-Driving (Waymo Safety Report)*. Retrieved from https://waymo.com/safety/

Waymo. (2019). *An Update on Waymo Disengagements in California*. Retrieved from https://medium.com/waymo/an-update-on-waymo-disengagements-in-california-d671fd31c3e2

Waymo. (n.d.). *We're Building the World's Most Experienced Driver*. Retrieved from https://waymo.com/

Wired.com. (2017). *Car Hacking*. Retrieved from www.wired.com/tag/car-hacking/

3 Driver's Mental Model of Vehicle Automation

Bobbie Seppelt
Massachusetts Institute of Technology

Trent Victor
Volvo Cars Safety Centre, and Chalmers
University of Technology

CONTENTS

Key Points 55
3.1　Importance and Relevance of Mental Models in Driving and Automation 56
3.2　Defining Mental Models 57
3.3　Mental Models under Uncertainty 58
3.4　General and Applied Mental Models 60
3.5　Measurement of General and Applied Mental Models 61
3.6　Supporting Accurate and Complete Mental Models 62
3.7　Conclusion 63
References 63

KEY POINTS

- An operator's expectations of system behavior are guided by the completeness and correctness of his/her mental model.
- The accuracy and completeness of mental models depend on the effectiveness of system interfaces and on the variety of situations drivers encounter.
- Drivers may use biased, uncertain, or incomplete knowledge when operating automation. Good automated system design should strive to account for and mitigate these known human biases and processing limitations.
- Both "general" and "applied" mental models affect reliance action. A mismatch between a driver's general and applied mental models can negatively affect trust and acceptance of vehicle automation.
- Multiple qualitative and quantitative measures can be used to assess the impact of mental models on driver's reliance on automation.
- To support accurate and complete mental model development, drivers need information on the purpose, process, and performance of automated systems.

3.1 IMPORTANCE AND RELEVANCE OF MENTAL MODELS IN DRIVING AND AUTOMATION

As automation is increasingly used to support control, decision-making, and information processing in complex environments, designers must address the question of how to design systems to foster human–automation coordination (Dekker & Woods, 2002). Automation alters the nature of tasks, and forces operators to adapt in novel ways (see this Handbook, Chapters 6, 8). Specifically, operators are required to monitor automated tasks and manage those tasks that remain by using available feedback from the automated system. This feedback is altered in type and extent to that feedback encountered under manual control. A monitoring role creates new attentional and knowledge demands in that operators must understand how different elements of complex systems operate and interact, in addition to the effect of their inputs on system behavior (Sarter, Woods, & Billings, 1997). In order to maintain safety, operators must correctly understand operating conditions to be able to recognize when and how to act in situations in which the automation may be unable to respond appropriately. For example, when lane markings are masked by rain or from the glare of the sun, a lane centering system may deactivate the vehicle's steering support. It is difficult for the driver to maintain correct understanding at all times, particularly when a process is controlled by a complex system with inherent uncertainty—which characterizes use of vehicle automation on public roadways.

The introduction of automation into a complex control task like driving can qualitatively shift an operator from an active processor of information to a passive recipient of information, as well as change the type of feedback s/he receives on the state of the system. As a consequence, operators are at times surprised by the behavior of the automation and uncertain of what the automation is doing, why it behaves in a certain manner, and what it will do next (Wiener, 1989; this Handbook, Chapter 21). Empirical research indicates that operators often have gaps and misconceptions in their mental models of automated systems due to the cognitive demand and complexity of operating such systems in real time (Sarter & Woods, 1994). Other potential consequences of automation's introduction include over- or under-trust of the automation respective to its capabilities, complacent behavior, and behavioral adaptation in safety-degrading ways (Norman, 1990; Sarter et al., 1997; Parasuraman, Mouloua, & Molloy, 1994; Lee & Moray, 1994; this Handbook, Chapter 4, 12).

Until vehicle automation has achieved a level of performance to safely enable drivers to assume the role of a passenger, it is assistance only, and requires drivers to monitor and, more importantly, to understand when and how to act to complement its driving task performance. At present, drivers receive support from forms of vehicle automation, but are still needed to perform object/event detection and response (OEDR), act as a fallback for system malfunctions, and fulfill non-automated operational, tactical, or strategic activities (SAE International, 2018). From a "human-automation coordination" perspective (Lee, 2018), drivers must monitor automated systems as well as the traffic environment because vehicle control and crash avoidance technologies have operational and functional design limitations and may malfunction. In sharing the goal with automation to achieve safe driving in an often unpredictable road environment, monitoring requires more than passive observation; drivers must actively supervise ongoing automation performance and the detection of pre-crash conditions. Supervision of automation in any dynamic, uncertain environment involves information integration and analysis, system expertise, analytical decision-making, sustained attention, and maintenance of manual skills (Bhana, 2010; Casner, Geven, Recker, & Schooler, 2014). Effective development and support of these skills helps operators to troubleshoot and recover if automation fails or if something unexpected happens (Onnasch, Wickens, Li, & Manzey, 2014; Wickens, Li, Santamaria, Sebok, & Sarter, 2010).

For drivers, context-specific system limitations can be difficult to understand (Seppelt, 2009; Larsson, 2012); drivers, consequently, may not recognize their need to act until a conflict is imminent. While events where such detection is needed are rare because current forms of many advanced driver assistance systems (ADAS) technology are relatively robust and react to most threats, to the driver, their occurrence essentially represents an ever-present and unpredictable possibility. Clear, unambiguous understanding of system state and behavior is particularly difficult when detection performance is dependent on sensor quality and environmental conditions (i.e., a dynamic operational design domain; Seppelt, Reimer, Angell, & Seaman, 2017). Common, current limitations may include restrictions in operating speed ranges, operational design domain (e.g. geo-fencing), limits in the amount of steering, braking, and acceleration the system can apply, and limitations in lane and object detection performance (e.g. pedestrians, animals, on-road objects, and oncoming vehicles). Additionally, the driver needs to supervise system failures ranging from sensor blockage to brake pump failure, as these may require immediate intervention. Failures, particularly unpredictable ones or those with an unspecified cause, create uncertainty in the interpretation of automation's behavior and inaccurate mental models. For example, Kazi, Stanton, and Harrison (2004) exposed drivers to different levels of reliability of an adaptive cruise control (ACC) system and found that failures involving random accelerations of ACC led to mental models of its operation that were predominately flawed. Accurate, well-developed mental models are critical to successful interaction between human operators and automated systems; inaccurate mental models lead to operator misunderstandings and inappropriate use.

Of particular interest in this chapter is the influence of mental models on automation use. An operator's expectations of system behavior are guided by the completeness and correctness of his/her mental model (Sarter et al., 1997). Mental models influence the way people allocate their attention to automated systems (Moray, 1986), and in turn influence their response, in which an operator's decision to engage or disengage a system reflects his/her prediction of its behavior (Gentner & Stevens, 1983). Mental models clearly impact how people use automation; but, what is a mental model?

3.2 DEFINING MENTAL MODELS

A mental model is an operator's knowledge of system purpose and form, its functioning, and associated state structure (Johnson-Laird, 1983; Rouse & Morris, 1986). Mental models represent working knowledge of system dynamics and structure of physical appearance and layout, and of causal relationships between and among components and processes (Moray, 1999). Mental models allow people to account for and to predict the behavior of physical systems (Gentner & Stevens, 1983). According to Johnson-Laird (1983), mental models enable people to draw inferences and make predictions, to decide on their actions, and to control system operation. In allowing an operator to predict and explain system behavior, and to recognize and remember relationships between system components and events (Wilson & Rutherford, 1989), mental models provide a source to guide people's expectations (Wickens, 1984).

Further elaborating on this role of mental models for generating predictions, the predictive processing framework (Clark, 2013) has emerged as a powerful new approach to understanding and modeling mental models (Engström et al., 2018). Within the predictive processing framework, a mental model can be seen as a *hierarchical generative model*. The hierarchical generative model is embodied in the brain to generate predictions, after learning over time how states and events in the world, or one's own body, generate sensory input. Predictive processing suggests that frequent exposure to reliable statistical regularities in the driving environment will lead to improvement of generative model predictions and increasingly automatized performance. Further, failures may be understood in terms of limited exposure to functional limitations, and therefore, as an inappropriately tuned generative model in such situations.

Consequently, for the driver's generative model to become better at predicting automation, it needs feedback (sensory input) on statistical regularities. Automation can be designed to provide transparent feedback on the state of the automation and has been shown to improve performance (Bennett & Flach, 2011; Flemisch, Winner, Bruder, & Bengler, 2014; Seppelt & Lee, 2007; 2019; Vicente & Rasmussen, 1992). Dynamic human–machine interface (HMI) feedback can provide statistical regularities of sensory input that the model can tune to. If designed to continuously communicate to drivers the evolving relationship between system performance and operating limits, HMIs can support the development of a generative model to properly predict what the automation can and cannot do (see e.g., this Handbook, Chapter 15). It is in understanding both the proximity to and type of system limitation (e.g., sensor range, braking capability, functional velocity range, etc.) that drivers are able to initiate proactive intervention in anticipation of automation failure (Seppelt & Lee, 2019).

3.3   MENTAL MODELS UNDER UNCERTAINTY

Mental models often include approximations and incompleteness because they serve as working heuristics (Payne, 1991). In functioning as heuristics, meant to enable operators to construct simplified representations of complex systems in order to reduce the amount of mental effort required to use them, mental models are notoriously incomplete, unstable, non-technical, and parsimonious (Norman, 1983). Further, drivers often adopt a satisficing strategy rather than an optimal one in trying to satisfy their needs, and are influenced by motivational and emotional factors (Boer & Hoedemaeker, 1998; Summala, 2007). Thus, there is often no single conceptual form that can be defined as accurate and complete. The merit of a mental model, therefore, is not based on its technical depth and specificity, but in its ability to enable a user to accurately (or satisfactorily) predict the behavior of a system.

The means of organizing knowledge into structured patterns affords rapid and flexible processing of information that translates into rapid responses such as the decision to rely on the automation when operating an automated system (Rumelhart & Ortany, 1977). Mental models are constructed from system-provided information, the environment, and instructions (Norman, 1986). They depend on and are updated from the feedback provided to operators (e.g., as from proprioceptive feedback or in information displays). The accuracy of mental models depends on the effectiveness of system interfaces (Norman, 1988) and the variety of situations encountered. Inaccuracies in mental models are more likely the more unreliable the automation (Kazi et al., 2004). Mental model accuracy may be improved with feedback that informs the operator of the automation's behavior in a variety of situations (Stanton & Young, 2005). For operators who interact with automation that in turn interacts with the environment, such as drivers, it is important that their mental models of the automated system include an understanding of the environment's effect on its behavior. For judgments made under uncertainty, such as the decision to rely on automated systems that control all or part of a complex, dynamic process, two types of cognitive mechanisms—intuition and reasoning—are at work (Kahneman & Frederick, 2002; Stanovich & West, 2002). Intuition (System 1) is characterized by fast, automatic, effortless, and associative operations—those similar to features of perceptual processes. Reasoning (System 2) is characterized by slow, serial, effortful, deliberately controlled, and often relatively flexible and rule-governed operations. System 1 generates impressions that factor into judgments while System 2 is involved directly in all judgments, whether they originate from impressions or from deliberate reasoning.

Consequently, under uncertainty or under time pressure, mental models are subject to cognitive biases—systematic patterns of deviation from norm or rationality in judgment (Haselton, Nettle, & Andrews, 2005). Although there are a large variety of cognitive biases, examples of important biases affecting mental models of automation include

- *Confirmation bias*—the tendency to search for, interpret, focus on, and remember information in a way that confirms one's preconceptions (Oswald & Grosjean, 2004).
- *Anchoring bias*—the tendency to rely too heavily, or "anchor," on one trait or piece of information when making decisions (Zhang, Lewis, Pellon, & Coleman, 2007).
- *Overconfidence bias*—excessive confidence in one's own answers to questions (Hilbert, 2012).
- *Gambler's fallacy*—the mistaken belief that if something happens more frequently than normal during a given period, it will happen less frequently in the future (or vice versa) (Tversky & Kahneman, 1974).

Thus, it can be expected that drivers may use biased, uncertain, or incomplete knowledge when operating automation. Good automated system designs should strive to minimize the effect of these known human biases and limitations, and to measure the accuracy of mental models.

3.4   GENERAL AND APPLIED MENTAL MODELS

Cognitivism theories (e.g., Shiffrin & Schneider, 1977; Rumelhart & Norman, 1981) commonly reference two types of knowledge, which are relevant to how we define mental models: (1) Declarative knowledge: knowledge that is acquired by education; and that can be literally expressed; it is the knowledge of something; and (2) Procedural knowledge: knowledge that is acquired by practice; and that can be applied to something; it is the knowledge of how to do something.

In Figure 3.1, the declarative knowledge type corresponds to the "General Mental Model" and the procedural knowledge type corresponds to the "Applied Mental Model." A general mental model includes general system knowledge (i.e., understanding prior to interaction). An applied mental model is a dynamic, contextually-driven understanding of system behavior and interaction (i.e., understanding during interaction).

Prior to the use of an automated system, a driver's initial mental model is constructed based on a variety of information sources, which may include a vehicle owner's manual, perceptions of other related technologies, word-of-mouth, marketing information, etc. This information is "general" in the sense that it may contain a basic understanding of system purpose and behavior, as well as a set of operating rules and conditions for proper use, but does not prescribe how the driver should behave specific to the interaction of system state with environmental condition. For example, a driver may understand as part of his/her "general" mental model

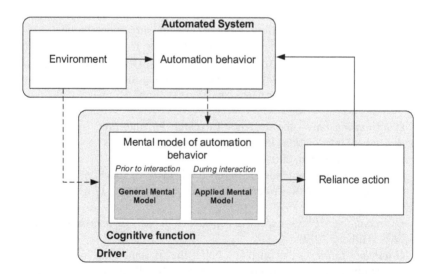

FIGURE 3.1   Conceptual model of the influence of a driver's mental model on automation reliance.

that ACC does not work effectively in rainy conditions, but may not have a well-developed "applied" mental model to understand the combination of road conditions and rain densities that result in unreliable detection of a lead vehicle. As a driver gains experience using an automated system in a variety of situations, s/he develops his/her "applied" mental model, which is conceptually consistent with the idea of "situation models" connecting bottom-up environmental input with top-down knowledge structures (Durso, Rawson, & Girotto, 2007). As depicted in Figure 3.1, both "general" and "applied" mental models affect reliance action (also see in this Handbook Chapter 4). A correct "general" mental model is important in the construction of an adequate "applied" model when experiencing concrete situations on the road and for selecting appropriate actions (Cotter & Mogilka, 2007; Seppelt & Lee, 2007). In turn, experience updates the "general" mental model. A mismatch between a driver's general mental model and experience can negatively affect trust and acceptance (Lee & See, 2004). To help explain drivers' reliance decisions and vehicle automation use behaviors in the short- and long term, it is necessary to measure mental models.

3.5 MEASUREMENT OF GENERAL AND APPLIED MENTAL MODELS

Self-report techniques are commonly used to evaluate and articulate differences in mental models. Operators are asked to explicitly define their understanding of a system via questionnaire or interview (e.g., Kempton, 1986; Payne, 1991). Multiple biases can be introduced in this type of data gathering process relating to social/communication and background/experience from both the participant and the analyst (Revell & Stanton, 2017). However, research on trust in automation has provided a reference for determining which aspects of a general mental model are important for correct system use (Kazi et al., 2007; Makoto, 2012). Healthy, calibrated trust occurs if information on a system's purpose, process, and performance (PPP) are supplied to drivers (Lee & See, 2004). Recently developed questionnaires probe the driver's understanding of these information types. Beggiato and Krems (2013) developed a 32-item questionnaire on ACC functionality in specific PPP situations. Questions covered general ACC functionality (e.g., "ACC maintains a predetermined speed in an empty lane") as well as system limitations described in the owner's manual (e.g., "ACC detects motorcycles"). Seppelt (2009) developed a 16-item questionnaire assessing participants' knowledge of the designer's intended use of ACC (i.e., purpose), of its operational sensing range and event detection capabilities (i.e., process), and of its behavior in specific use cases (i.e., performance). These questionnaires provide a template for assessment of other automated technologies.

As stated, in functioning as heuristics, mental models often have no single conceptual form that can be defined as accurate and complete. It is therefore important to assess a driver's procedural knowledge in addition to his/her declarative mental model, that is, to evaluate the behavioral and performance output from "running" his/her mental model. Such procedural knowledge, or a driver's applied mental model, is accessed through analysis of *driver behavioral measures*, e.g., response time and response quality to events, eye movements, or observation of errors. For an overview of measures of mental models, see Table 3.1.

TABLE 3.1

Example Measures of General and Applied Mental Models

<table>
<tr><td align="center">General Mental Model Measures</td></tr>
</table>

- Questionnaires on purpose, process, and performance (PPP)
- Automation status ratings (i.e., mental model accuracy)
  - Automated system mode(s)
  - Proximity of current automated system state to its capabilities/limits
  - Expected system response and behavior in specific conditions
- Trust ratings

<table>
<tr><td align="center">Applied Mental Model Measures</td></tr>
</table>

- Monitoring of the driving environment
  - Sampling rate and length to the forward roadway
  - Sampling rate and length to designated safety-critical areas (e.g., to intersections & crosswalks)
- Trust-related behaviors during use of the automation (e.g., hands hovering near the steering wheel)
- Presence of safety-related behaviors prior to tactical maneuvers or hazard responses (e.g., glances to rear-view mirror & side mirrors, turn signal use, and over-the-shoulder glances)
- Secondary task use
- Level of skill loss: Manual performance after a period of automated driving compared with manual task performance during the same period of time
- Response time to unexpected events/hazards or system errors/failures
- Use of automation within vs. outside its operational design domain (ODD)

The set of measures listed in Table 3.1 offers a starting point for how to assess general and applied mental models of vehicle automation. Further research is required to assess whether and to what extent the above measures provide practical insight into driver's reliance behavior for the diversity of automated technologies and their real-world use conditions. Further knowledge on how behavior changes in the longer term, after extended use of automated systems, as a function of the type and complexity of a driver's mental model, is also important.

3.6 SUPPORTING ACCURATE AND COMPLETE MENTAL MODELS

In order to safely use automated systems, drivers must fundamentally understand their role responsibility on a moment-by-moment basis. In essence, they must be able to accurately translate their general mental model to the applied mental model question of "*who is responsible and most capable to react to the present vehicle and situation dynamics: me or the automated system?*". Recent research points to both general confusion about the capabilities of deployed forms of vehicle automation (i.e., inaccurate general mental models; e.g., Abraham, Seppelt, Mehler, & Reimer, 2017) as well as an incomplete understanding and expectations regarding limitations of automation (e.g., Victor et al., 2018). Based on current system design practices, there seems to be a fundamental disconnect between driver's general and applied mental models. For example, in Victor et al. (2018), drivers were trained prior to use on the limits of highly reliable (but not perfect) automation. However, in practice, they experienced reliable system operation until the final moment of the study. Regardless of the amount of initial training on system limitations, 30% of drivers crashed into the object they "knew" was outside the detection capabilities of the system. Without reinforcement of the general mental model by the dynamic experience, information decayed or was dominated by dynamic learned trust. Consistent with previous findings on the development of mental models of vehicle automation relative to initial information (Beggiato, Pereira, Petzoldt, & Krems, 2015), this study found that system limitations initially described but not experienced tend to disappear from a driver's mental model. Research to date on mental models of vehicle automation indicate a need to support and/or train driver's understanding of *in situ* vehicle limitations and capabilities through dynamic HMI information (Seppelt & Lee, 2019; this Handbook, Chapters 15, 16, 18), routine driver training, and/or using intelligent tutoring systems.

3.7 CONCLUSION

This chapter described the importance and relevance of mental models in driving and automation, defined how fundamental psychological mechanisms contribute to their formation, provided a new framing of general and applied mental models and how to measure them, and concluded with a review of recent research on how to support accurate and complete mental model development. Future research needs to examine relationships between mental models, trust, and both short- and long-term acceptance across types and combinations of vehicle automation.

REFERENCES

Abraham, H., Seppelt, B., Mehler, B., & Reimer, B. (2017). What's in a name: Vehicle technology branding & consumer expectations for automation. Proceedings of the 9th International ACM Conference on Automation User Interfaces and Interactive Vehicular Applications. New York: ACM.

Beggiato, M. & Krems, J. F. (2013). The evolution of mental model, trust and acceptance of adaptive cruise control in relation to initial information. *Transportation Research Part F: Traffic Psychology and Behaviour, 18*, 47–57.

Beggiato, M., Pereira, M., Petzoldt, T., & Krems, J. (2015). Learning and development of trust, acceptance and the mental model of ACC. A longitudinal on-road study. *Transportation Research Part F: Traffic Psychology and Behaviour, 35*, 75–84.

Bennett, K. B., & Flach, J. M. (2011). Display and interface design: Subtle science, exact art. CRC Press.

Bhana, H. (2010). Trust but verify. *AeroSafetyWorld, 5*(5), 13–14.

Boer, E. R. & Hoedemaeker, M. (1998). Modeling driver behavior with different degrees of automation: A hierarchical decision framework of interacting mental models. In *Proceedings of the XVIIth European Annual Conference on Human Decision making and Manual Control*, 14–16 December, France: Valenciennes.

Casner, S. M., Geven, R. W., Recker, M. P., & Schooler, J. W. (2014). The retention of manual flying skills in the automated cockpit. *Human Factors, 56*(8), 1506–1516.

Clark, A. (2013). Whatever next? Predictive brains, situated agents, and the future of cognitive science. *Behavioral and Brain Sciences, 36*(3), 181–204. doi:10.1017/S0140525X12000477

Cotter, S. & Mogilka, A. (2007). Methodologies for the assessment of ITS in terms of driver appropriation processes over time (HUMANIST Project Deliverable 6 of Task Force E).

Dekker, S. W. A. & Woods, D. D. (2002). MABA-MABA or Abracadabra? Progress on human-automation co-ordination. *Cognition, Technology and Work, 4,* 240–244.

Durso, F. T., Rawson, K. A., & Girotto, S. (2007). Comprehension and situation awareness. In F. T. Durso, R. S. Nickerson, S. T. Dumais, S. Lewandowsky, & T. J. Perfect (Eds.), *Handbook of Applied Cognition* (2nd ed., pp. 163–193). Chichester, UK: John Wiley & Sons.

Engström, J., Bärgman, J., Nilsson, D., Seppelt, B., Markkula, G., Piccinini, G. B., & Victor, T. (2018). Great expectations: A predictive processing account of automobile driving. *Theoretical Issues in Ergonomics Science, 19*(2), 156–194.

Flemisch, F., Winner, H., Bruder, R., & Bengler, K. (2014). Cooperative guidance, control and automation. In H. Winner, S. Hakuli, F. Lotz, & C. Singer (Eds.), *Handbook of Driver Assistance Systems: Basic Information, Components and Systems for Active Safety and Comfort* (pp. 1471–1481). Berlin: Springer.

Gentner, D. & Stevens, A. L. (1983). *Mental Models.* Hillsdale, NJ: Lawrence Erlbaum Associates.

Haselton, M. G., Nettle, D., & Andrews, P. W. (2005). The evolution of cognitive bias. In D. M. Buss (Ed.), *The Handbook of Evolutionary Psychology* (pp. 724–746). Hoboken, NJ: John Wiley & Sons.

Hilbert, M. (2012). Toward a synthesis of cognitive biases: How noisy information processing can bias human decision-making. *Psychological Bulletin, 138*(2), 211–237.

Johnson-Laird, P. (1983). *Mental Models.* Cambridge, MA: Harvard University Press.

Kahneman, D. & Frederick, S. (2002). Representativeness revisited: Attribute substitution in intuitive judgment. In T. Gilovich, D. Griffin, & D. Kahneman (Eds.), *Heuristics and Biases* (pp. 49–81). Cambridge: Cambridge University Press.

Kazi, T. A., Stanton, N. A., & Harrison, D. (2004). The interaction between drivers' conceptual models of automatic-cruise-control and level of trust in the system. *3rd International Conference on Traffic & Transport Psychology (ICTTP).* Nottingham, UK: ICTTP.

Kazi, T., Stanton, N. A., Walker, G. H., & Young, M. S. (2007). Designer driving: drivers' conceptual models and level of trust in adaptive cruise control.

Kempton, W. (1986). Two theories of home heat control. *Cognitive Science, 10*(1), 75–90.

Larsson, A. F. L. (2012). Driver usage and understanding of adaptive cruise control. *Applied Ergonomics, 43,* 501–506.

Lee, J. D. (2018). Perspectives on automotive automation and autonomy. *Journal of Cognitive Engineering and Decision Making, 12*(1), 53–57.

Lee, J. D. & Moray, N. (1994). Trust, self-confidence, and operators' adaptation to automation. *International Journal of Human-Computer Studies, 40,* 153–184.

Lee, J. D. & See, K. A. (2004). Trust in technology: Designing for appropriate reliance. *Human Factors, 46*(1), 50–80.

Makoto, I. (2012). Toward overtrust-free advanced driver assistance systems. *Cognition, Technology & Work, 14*(1), 51–60.

Moray, N. (1986). Monitoring behavior and supervisory control. In K. R. Boff, L. Kaufman, and J. P. Thomas (Eds.), *Handbook of Perception and Human Performance* (Vol. 2, Chapter 40). New York: Wiley.

Moray, N. (1999). Mental models in theory and practice. In D. Gopher & A. Koriat, (Eds.), *Attention and Performance XVII Cognitive Regulation of Performance: Interaction of Theory and Application.* Cambridge: MIT Press.

Norman, D. A. (1983). Some observations on mental models. In D. Gentner & A. L. Stevens (Eds.), *Mental Models* (pp. 7–14). Hillsdale, NJ: Lawrence Erlbaum Associates.

Norman, D. A. (1986). Cognitive engineering. In D. A. Norman & S. W. Draper (Eds.), *User Centered System Design: New Perspectives on Human-Computer Interaction* (pp. 31–61). Hillsdale, NJ: Lawrence Erlbaum Associates.

Norman, D. A. (1988). *The Psychology of Everyday Things.* New York: Basic Books.

Norman, D. A. (1990). The 'problem' with automation: Inappropriate feedback and interaction, not 'over-automation'. *Philosophical Transactions of the Royal Society London, Series B, Biological Sciences, 327*(1241), 585–593.

Onnasch, L., Wickens, C. D., Li, H., & Manzey, D. (2014). Human performance consequences of stages and levels of automation: An integrated meta-analysis. *Human Factors, 56*(3), 476–488.

Oswald, M. E. & Grosjean, S. (2004). Confirmation bias. In R. F. Pohl (Ed.), *Cognitive Illusions: A Handbook on Fallacies and Biases in Thinking, Judgement, and Memory* (pp. 79–96). Hove, UK: Psychology Press.

Parasuraman, R., Mouloua, M., & Molloy, R. (1994). Monitoring automation failures in human-machine systems. In M. Mouloua and R. Parasuraman (Eds.), *Human Performance in Automated Systems: Current Research and Trends* (pp. 45–49). Hillsdale, NJ: Lawrence Erlbaum Associates.

Payne, S. J. (1991). A descriptive study of mental models. *Behavior & Information Technology, 10*(1), 3–21.

Revell, K. M. & Stanton, N. A. (2017). *Mental Models: Design of User Interaction and Interfaces for Domestic Energy Systems.* Boca Raton, FL: CRC Press.

Rouse, W. B. & Morris, N. M. (1986). On looking into the black box: Prospects and limits in the search for mental models. *Psychological Bulletin, 100*, 359–363.

Rumelhart, D. E. & Norman, D. A. (1981). Analogical processes in learning. In J. R. Andersen (Ed.), *Cognitive Skills and Their Acquisition* (pp. 335–359). New York: Psychology Press.

Rumelhart, D. E. & Ortany, A. (1977). The representation of knowledge in memory. In R. C. Anderson & R. J. Spiro (Eds.), *Schooling and the Acquisition of Knowledge* (pp. 99–135). HillsDale, NJ: Lawrence Erlbaum Associates.

SAE International. (2018). *Taxonomy and Definitions for Terms Related to Driving Automation Systems for On-Road Motor Vehicles* (SAE Standard J3016). Warrendale, PA: Society of Automotive Engineers.

Sarter, N. B. & Woods, D. D. (1994). Pilot interaction with cockpit automation II: An experimental study of pilots' model and awareness of the flight management system. *The International Journal of Aviation Psychology, 4*(1), 1–28.

Sarter, N. B., Woods, D. D., & Billings, C. E. (1997). Automation surprises. In G. Salvendy (Ed.), *Handbook of Human Factors & Ergonomics, Second Edition.* Hoboken, NJ: Wiley.

Seppelt, B. D. (2009). *Supporting Operator Reliance on Automation through Continuous Feedback.* Unpublished PhD dissertation, The University of Iowa, Iowa City.

Seppelt, B. D. & Lee, J. D. (2007). Making adaptive cruise control (ACC) limits visible. *International Journal of Human-Computer Studies, 65*(3), 192–205.

Seppelt, B. D. & Lee, J. D. (2019). Keeping the driver in the loop: Enhanced feedback to support appropriate use of imperfect vehicle control automation. *International Journal of Human-Computer Studies, 125*, 66–80.

Seppelt, B. D., Reimer, B., Angell L., & Seaman, S. (2017). Considering the human across levels of automation: Implications for reliance. *Proceedings of the Driving Assessment Conference: 9th International Symposium on Human Factors in Driver Assessment, Training, and Vehicle Design.* Iowa City, IA: University of Iowa, Public Policy Center.

Shiffrin, R. M. & Schneider, W. (1977). Controlled and automatic human information processing: II. Perceptual learning, automatic attending and a general theory. *Psychological Review, 84*(2), 127–190.

Summala, H. (2007). Towards understanding motivational and emotional factors in driver behaviour: Comfort through satisficing. In P. C. Cacciabue (Eds.), *Modelling Driver Behaviour in Automotive Environments.* London: Springer.

Stanovich, K. E. & West, R. F. (2002). Individual differences in reasoning: Implications for the rationality debate. In T. Gilovich, D. Griffin, & D. Kahneman (Eds.), *Heuristics and Biases* (pp. 421–440). Cambridge: Cambridge University Press.

Stanton, N. A. & Young. M. S. (2005). Driver behavior with adaptive cruise control. *Ergonomics, 48*(10), 1294–1313.

Tversky, A. & Kahneman, D. (1974). Judgement under uncertainty: Heuristics and biases. *Science, 185*(4157), 1124–1131.

Vicente, K. J. & Rasmussen, J. (1992). Ecological interface design: Theoretical foundations. *IEEE Transactions on Systems, Man, and Cybernetics, SCM-22*(4), 589–606.

Victor, T. W., Tivesten, E., Gustavsson, P., Johansson, J., Sangberg, F., & Ljung Aust, M. (2018). Automation expectation mismatch: Incorrect prediction despite eyes on threat and hands on wheel. *Human Factors, 60*(8), 1095–1116.

Wickens, C. D. (1984). *Engineering Psychology and Human Performance.* Columbus, OH: Merrill.

Wickens, C. D., Li, H., Santamaria, A., Sebok, A., & Sarter, N. B. (2010). Stages and levels of automation: An integrated meta-analysis. *Proceedings of the Human Factors and Ergonomics Society Annual Meeting* (pp. 389–393). Los Angeles, CA: Sage Publications.

Wiener, E. L. (1989). *Human Factors of Advanced Technology ("Glass Cockpit") Transport Aircraft* (NASA Contractor Report 177528). Mountain View, CA: NASA Ames Research Center.

Wilson, J. R. & Rutherford, A. (1989). Mental models: Theory and application in human factors. *Human Factors, 31*, 617–634.

Zhang Y., Lewis, M., Pellon, M., & Coleman, P. (2007). A preliminary research on modeling cognitive agents for social environments in multi-agent systems. *AAAI Fall Symposium: Emergent Agents and Socialities* (p. 116). Retrieved from:https://www.aaai.org/Papers/Symposia/Fall/2007/FS-07-04/FS07-04-017.pdf

4 Driver Trust in Automated, Connected, and Intelligent Vehicles

John D. Lee
University of Wisconsin-Madison

CONTENTS

Key Points 67
4.1   Introduction 68
4.2   Trust and Types of Automation 68
4.3   Definition and Mechanisms Underlying Trust 72
4.4   Promoting Appropriate Trust in Vehicle Automation 78
        4.4.1   Calibration and Resolution of Trust... 78
        4.4.2   Trustable Automation: Create Simple Use Situations and
                   Automation Structure 80
        4.4.3   Trustable Automation: Display Surface and Depth Indications
                   of Capability 81
        4.4.4   Trustable Automation: Enable Directable Automation and
                   Trust Repair 82
        4.4.5   Trustworthy Automation and Goal Alignment................................ 84
4.5   Trust and Acceptance of Vehicle Technology 84
4.6   Ethical Considerations and the Teleology of Technology 85
4.7   Conclusion 86
Acknowledgements.. 87
References.. 88

KEY POINTS

- Trust complements the concepts of mental models and situation awareness in explaining how and when people rely on automation.
- Trust is a multi-faceted term that operates at timescales of seconds to years and mediates how people rely on, accept, and tolerate vehicle technology.
- Trust mediates micro interactions concerning how people rely on automation to engage in non-driving tasks to macro interactions concerning how the public accepts new forms of transport.

- Trust depends on surface features, such as the look and feel of the human–machine interface (HMI), as well as depth features, such as the reliability of the automation and alignment of the person's and automation's goals.
- The imperfect goal alignment between primary users (e.g., riders in an automated vehicle (AV)) and incidental users (e.g., pedestrians negotiating intersections with AVs) might undermine the trust and tolerance of incidental users.

4.1 INTRODUCTION

In the previous chapters, we described the different types of automated, connected, and intelligent vehicle technologies and the mental models that inform the driver's decisions and behaviors. The discussion of mental models (this Handbook, Chapter 3) shows that it is one thing to develop technologies that can greatly benefit the driver and society, but it is an entirely different matter for these technologies to be used appropriately. Like mental models, trust is one factor that guides how drivers use vehicle automation, as well as the public acceptance and tolerance of driverless vehicles. This chapter addresses how one goes about building, calibrating, and repairing trust for different types of automation.

This chapter begins by describing different types of automation and different roles of people from the perspective of trust. This is followed by a definition of trust, a description of the cognitive mechanisms that underlie it, and the need to promote appropriate trust. Design approaches to promote appropriate trust depend on various relationships between people and vehicle technology, such as drivers monitoring fallible self-driving vehicles, passengers riding in driverless vehicles, people relying on algorithms to choose ride-sharing partners, and even pedestrians who must negotiate intersections with automated vehicles (AV). These human–technology relationships are described in terms of general categories of vehicle technology—shared, traded, and ceded control. These categories are relevant for trust because they reflect increasing degrees to which people make themselves vulnerable to the technology. Considering these categories, we discuss the roles of people, why trust is relevant, the basis of trust, and the design choices that might promote appropriate trust and repair trust after automation mishaps. The chapter concludes by briefly addressing the ethical considerations associated with managing trust and creating trustworthy technology.

4.2 TRUST AND TYPES OF AUTOMATION

Technology is rapidly changing the relationship between people and vehicles. Increasingly, this technology is supporting drivers in new ways and is also taking responsibility for many of the aspects of driving that were once the sole responsibility of drivers. In the extreme, this has produced driverless vehicles that operate with no direct control input from people in the vehicle and the automation takes full responsibility for driving—SAE Level 5 automation. Vehicles with Level 5 automation might not even include a steering wheel or pedals and people *cede control* to the vehicle—drivers become riders. Even more extreme are vehicles that deliver goods and run errands without occupants. Less extreme are intermittently self-driving vehicles where drivers can *trade control* with automation—SAE Level 3 automation. Here, the vehicle might drive itself for part of the journey, but drivers would be able to take control and might be called upon by the vehicle to take control. Many vehicles contain less ambitious automation where drivers *share control* with the driver. Here the driver remains completely responsible for driving, but the automation eases the demands of driving and might allow for short periods of less vigilant attention to the roadway—SAE Levels 1 and 2 automation. Shared control includes adaptive cruise control (ACC) and lane-centering systems that the driver deliberately engages, as well as systems that engage automatically, such as electronic stability control and automatic emergency braking, which activate only in rare and extreme conditions. The terms ceded, traded, and shared are used because they reflect how automation puts people in different positions of vulnerability and uncertainty, which engages trust in guiding behavior (Mayer, Davis, & Schoorman, 1995). Whether in the form of ceded, traded, or shared control, vehicle automation is changing the role and responsibility of drivers, and the theoretical construct of trust will likely play a critical role in mediating how people adapt to this new technology (see also, this Handbook, Chapters 6, 8, 9).

The people for whom vehicle automation is designed—the drivers or riders—are the most obvious users whose trust in it will influence its success, but vehicle automation and people reside in a large interconnected network, see Figure 4.1 (Lee, 2018). In this context, even driverless vehicles, where drivers have ceded control to the vehicle, are not actually autonomous. Driverless vehicles will depend on an array of people to maintain it, repair it, and remotely operate it when the on-board automation encounters situations that exceed its capacity. Here trust might influence whether a remote operator chooses to intervene and "drive" the vehicle, much like trust influences those who manage remotely piloted drones (Guznov, Lyons, Nelson, & Woolley, 2016). Another critical trust relation emerges when people share rides in an AV. Similar to other situations in the sharing economy, trust can inhibit or facilitate people's willingness to accept the suggestion of an algorithm to share a ride with a stranger (Ert, Fleischer, & Magen, 2016; Payyanadan & Lee, 2018). Other nodes on this network that automation may affect are other road users, such as pedestrians and cyclists, as well as drivers of conventional vehicles. These other road users are not the primary users of automation, where the automation has been designed to achieve their goals, but are incidental or passive users, who did not choose to use the automation and whose goals may conflict with those of the AVs and their riders (Inbar & Tractinsky, 2009; Montague, Xu, & Chiou, 2014).

Changes to one node can ripple through the network and affect trust in unanticipated ways. For example, pedestrians who grow distrustful and intolerant of AVs might bully or sabotage them, undermining their efficiency—early demonstrations of self-driving vehicles have encountered hostile people who have vandalized them (Romero, 2018). The diminished efficiency might then undermine the trust of the riders when they fail to arrive at their destinations on time. The many nodes of the network that define the transportation ecosystem mean it is insufficient to consider only the relationship between the person in the vehicle and the vehicle technology (Woods, 2015).

Figure 4.1 shows many trust relationships linking the nodes of a multi-echelon network. The echelons range from the network of sensors and computers that reside

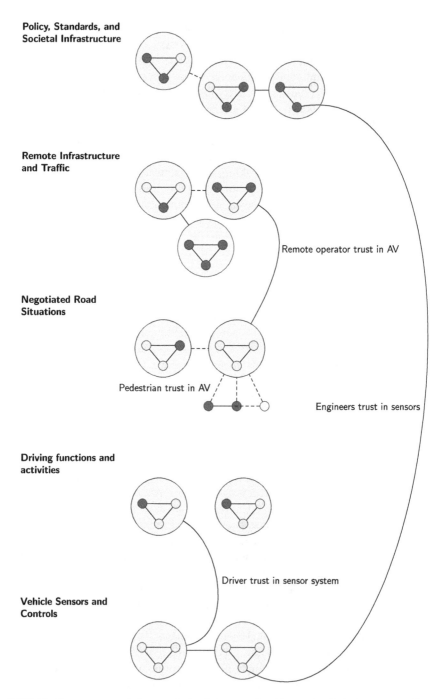

FIGURE 4.1 A multi-echelon network of trust relationships and vehicle technology. Filled circles indicate people and open circles represent automation elements. The lines indicate trust relationships, with the solid lines indicating likely goal alignment and dashed lines indicate relationships where goal alignment is less likely.

in the car to the network of organizations and societal infrastructure that guides technology development, deployment, and regulation. The influence of trust in these relationships depends on the type of vehicle technology and the role of the person. The most commonly studied trust relationships are those between people and the AV system, shown Figure 4.1 as the "Driving Functions and Activities." Other trust relationships span levels, such as between a remote operator of self-driving vehicles at the level of "Remote Infrastructure and Traffic" and the riders involved in the traffic situation at the level of "Negotiated Road Situations." Some trust relationships even span the extremes, as the engineering community at the level of "Policy, Standards, and Societal Infrastructure" and the behavior of sensors at the level of "Vehicle Sensors and Control." Generally, technology serves the goal of the person directly interacting with it, but in this network the goals of some people might be at odds with the automation—shown as dotted lines. For example, the goals of pedestrians negotiating an intersection will often conflict with those of an AV that is serving its rider's goal of minimizing travel time. This figure shows only a few of the many trust relationships that comprise the transportation ecology to highlight the range of situations where trust in technology plays a role.

These trust relationships can suffer from either over-trust or under-trust. For example, with shared control (SAE Levels 1 and 2), technology supports people in their role as drivers (SAE, 2016). Here the danger is that people will over-trust automation and rely on it to perform in ways the designers did not intend. As an example, some drivers over-rely on the Tesla automation, leading to extended periods of disengagement from driving and several fatal crashes (NTSB, 2016). For safety drivers and remote operators who supervise AVs and intervene when the automation encounters unexpected difficulties the situation is similar to shared control: the danger is that they will over-trust the technology and rely on it to perform safety critical driving functions when they should not.

Over-trust with shared control contrasts with under-trust with ceded control (SAE Level 5), where the person's role might be that of a passenger who is unable to intervene and take back the driving task. Here the danger is that people will under-trust the automation, reject it, and fail to enjoy its benefits. For example, people might not trust that self-driving vehicles can drive safely and opt to drive themselves. Trust is needed not just for riding in an AV, but also for requesting an AV, entering and initiating a trip, making trip changes while en route, and safely pulling over and exiting at the destination (Weast, Yurdana, & Jordan, 2016). Figure 4.1 highlights one such relationship, where riders will need to trust the vetting and matching algorithms associated with making shared rides safe. Here people must develop trust in their fellow passengers through the technology used to arrange the ride (Ert et al., 2016).

In contrast with riders and drivers who are primary users of technology, the goals of incidental users, such as pedestrians, might not align with the automation, which could undermine their trust of AVs even if they behave safely. For example, AVs might fail to give way to pedestrians waiting to cross the street, forcing the pedestrians to wait longer than they might with manually driven vehicles. Another class of incidental users is the general public who must share its transportation infrastructure with AVs. Public anger erupted when buses used to shuttle large technology employees in Silicon Valley clogged public bus lanes (De Kosnik, 2014). This anger is an indicator of potential distrust that might emerge if people see AVs as appropriating public resources. The factors affecting trust in technology differ in each of these situations, which can be revealed by considering the psychological mechanisms underlying trust.

4.3 DEFINITION AND MECHANISMS UNDERLYING TRUST

As with most sophisticated technology, acceptance of vehicle automation depends not only on an explicit rational consideration of its features and capabilities but also on an emotional or affective response to its behavior and branding (Lee, 2006; Norman, 2004). Central to this affective response is trust. Trust has emerged as an important concept in explaining how people adapt to and accept new technology in domains as diverse as industrial process automation (Muir & Moray, 1996; Sheridan & Ferrell, 1974; Zuboff, 1988), e-commerce (McKnight, Choudhury, & Kacmar, 2002), weapon systems (Dzindolet, Pierce, Beck, Dawe, & Anderson, 2001), and, at a more macro level, societal acceptance of technology risk (Slovic, 1993; 1999) and participation in the sharing economy (Ert et al., 2016). Many factors influence adaptation and acceptance, but trust is particularly critical in systems that take control in uncertain and risky situations (also see this Handbook, Chapter 5).

A comprehensive review of trust in automation considered the literature on human–human and human–technology trust to synthesize a definition of human–technology trust: "... the attitude that an agent will help achieve an individual's goals in a situation characterized by uncertainty and vulnerability..." (Lee & See, 2004, p. 54). A more recent review of trust in automation reiterated this definition and emphasized that trust formation is a dynamic process (Hoff & Bashir, 2015). These reviews along with several meta-analyses document a substantial empirical base that shows that greater levels of trust lead to greater reliance on and compliance with the automation. These reviews also show that trust is not simply a reflection of the reliability of the automation (Schaefer, Chen, Szalma, & Hancock, 2016; Wu, Zhao, Zhu, Tan, & Zheng, 2011). Trust is an important influence on how people use, accept, and tolerate vehicle technology, but related concepts also play an important role.

Figure 4.2 shows trust in the context of Neisser's perceptual cycle. Trust directs trust-related actions (e.g., reliance, compliance, and tolerance), and these actions enable people to sample the characteristics of the automation (e.g., purpose, process, and performance), and these samples modify trust (through affective, analogical, and analytical cognitive mechanisms). This cycle repeats, with the modified trust, directing trust-related actions, which in turn modifies the level of trust. More specifically, reliability is an aspect of automation performance, which influences trust, and trust influences trust action, such as reliance. Trust depends on other factors than reliability and other factors beyond trust affect reliance (Lee & See, 2004). For example, trust might be influenced by an understanding of how the automation operates (i.e., the automation process) and reliance depends on self-confidence in the ability to perform the task manually. Trust is typically defined in terms of the automation achieving a particular goal of a person, which guides reliance and compliance. With incidental users, whose goals do not align with the automation, tolerance is the willingness to interact with automation that might

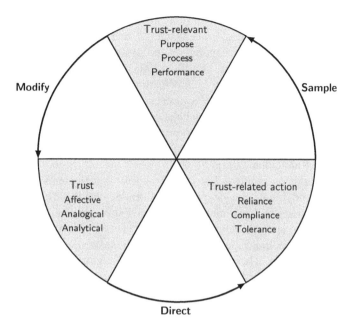

FIGURE 4.2 Trust-related actions, the trust-related information revealed by these actions, and the cognitive mechanisms underlying trust in the context of Neisser's perceptual cycle. (Neisser, 1976.)

interfere with achieving their goals. Tolerance is critical in considering how pedestrians and other drivers might respond to AVs. The balance of this section expands on the cognitive mechanisms of trust and by first relating trust to mental models and situation awareness.

Trust has close connections to other important concepts regarding driver interaction with automation, such as mental models (see this Handbook, Chapter 3) and situation awareness (see this Handbook, Chapter 7). One definition of mental models is an explicit verbalizable mental representation of a system that allows people to anticipate system behavior and to plan actions to achieve desired results (Norman, 1983; Rouse & Morris, 1986). Mental models support simulation of possible future states of the system to guide attention and action. According to this conception of mental models, trust complements them as an implicit, non-verbalizable, representation that guides behavior through an affective response to the system (Lee, 2006). For example, a feeling of diminished trust might lead the driver to suspect that the lane-keeping system is not operating properly and the driver's mental model might lead the driver to recognize that rain is interfering with the sensors. Trust also differs from mental models in that it guides interactions with intentional agents, or with agents whose complexity and autonomy makes them appear intentional, whereas mental models tend to guide interactions with inanimate systems—you might have a mental model of a toaster but trust a voice-based assistant. A toaster requires a mental model to guide specific interactions, but a voice-based assistant requires a belief that it will act and reliably provide information.

Situation awareness can be thought of as the elements of the mental models that are active in the moment, and so guides attention and behavior in a particular situation (Endsley, 2013). It also encompasses the perception, interpretation, and projection of the current system state (Endsley, 1995). All of these reflect the person's mental model and the assimilation of information regarding the situation. Situation awareness, as defined as an explicit awareness of the system, is often critical for guiding decision-making and effective behavior. However, much behavior is guided by implicit knowledge and affective processes that produce advantageous decisions without awareness for the cues that guide them (Bechara, Damasio, Tranel, & Damasio, 1997; Kahneman, 2011).

Figure 4.3 builds on Figure 4.2 to show how trust complements the influence of mental models and situation awareness to influence behavior. The cognitive state of the person, defined by trust, mental model, and situation awareness directs perceptual exploration, specific actions, and, more generally, behavior. This behavior samples the world and, more specifically, the operational domain and device, as well as the specific situation. These samples modify the cognitive state—trust, mental model, and situation awareness.

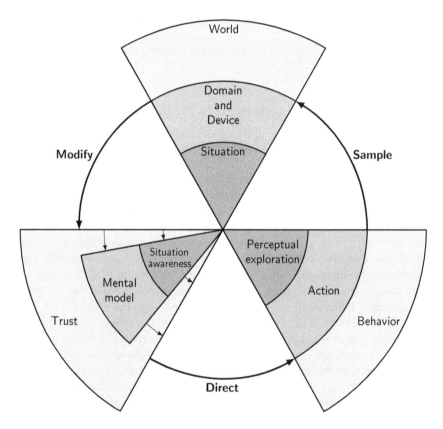

FIGURE 4.3 Links between situation awareness, mental models, and trust based on Neisser's perceptual cycle. (Neisser, 1976.)

Trust influences and is influenced by mental models and situation awareness through an affective process where attitudes and feelings guide response to technology (Gefen, Karahanna, & Straub, 2003; Lee & See, 2004). Trust in a system will shape the mental model and situation awareness of that system, and mental models and situation awareness affect trust. Figure 4.3 shows this joint influence with the small arrows from trust to mental models and situation awareness and from situation awareness and mental models to trust. Trust influences mental model and situation awareness by guiding attention to particular aspects of a system that might require the person's action, such as when the intermittent disengagement of ACC might guide attention of the driver to the display to understand why it disengages (Lee, 2006). This attention to the display contributes to a more complete mental model and better situation awareness. Figure 4.3 shows how mental models, situation awareness, and trust influence each other and jointly guide how people interact with automation.

The theoretical basis of trust in technology is grounded in research addressing trust between people (Muir, 1987; Sheridan & Ferrell, 1974). Trust mediates interactions between individuals and between individuals and organizations (Mayer et al., 1995). In interpersonal relationships, trust acts as a social lubricant that supports cooperation and exchanges in situations where it is not possible to exert complete control over those who are being relied upon to achieve a goal (Rotter, 1980). The substantial research concerning the role of trust in human-to-human relationships indicates that trust is a social emotion that underlies an affective process of responding to social uncertainty and risk (Adolphs, Tranel, & Damasio, 1998; Fehr & Camerer, 2007; Mikolajczak et al., 2010). As an affective process it can act pre-attentively to influence behavior without effort and in a way that can differ from the influence of conscious processes. As an example, a brain imaging study showed that different areas of the brain were active during explicit and implicit judgments of trustworthiness, with the brain activity associated with the implicit judgments indicating an affective process (Winston, Strange, O'Doherty, & Dolan, 2002). This affective process is partially governed by neurochemicals, such as oxytocin, that affect how we feel about others and how we respond to their behavior. Dosing people with oxytocin led to increased social risk-taking, but not risk-taking in general (Kosfeld, Heinrichs, Zak, Fishbacher, & Fehr, 2005). However subsequent studies have failed to replicate these findings, suggesting nasal administration of oxytocin has a small and variable effect on trust (Nave, Camerer, & McCullough, 2015). More generally, trust guides how people relate to other people, teams, organizations, and even governments to achieve their goals in situations characterized by risk and uncertainty (Rousseau, Sitkin, Burt, & Camerer, 1998; Slovic, 1993).

Despite the obvious differences between human–human relationships and human–technology relationships (Madhavan & Wiegmann, 2007), many studies show that people respond to technology in ways that parallel their response to people (Nass & Moon, 2000; Reeves & Nass, 1996). Typically, these responses are not based on people anthropomorphizing technology; but, instead, it seems that simple features of the technology, such as interactive complexity and voice interaction, lead people to respond to the technology as they would a person. Similar to the effect on human–human relationships, exposing people to oxytocin led people to trust and comply with an anthropomorphized aid—an avatar—but had little effect on how people responded to a computer aid without anthropomorphic features (de Visser et al., 2017). As with human–human relationships, trust in automation guides behavior where complexity, effort, and time constraints make it infeasible to completely understand the automation (Lee & See, 2004). The complexity of vehicle automation prevents people from developing a complete understanding of its capabilities which, combined with associated risk and uncertainty of the driving environment, make it very likely that trust will play an important role in how people respond to vehicle automation. Just as trust in people influences the degree to which people are willing to make themselves vulnerable to other people they are unable to monitor or control, trust in automation influences the degree people are willing to make themselves vulnerable to automated systems.

Trust becomes more relevant as technology displays greater degrees of agency. Agency concerns the degree to which technology exhibits behavior that is autonomous, responsive to other agents, and appears to be goal-directed. The perception of agency increases with the degree of autonomy—agents operating without the direct intervention of humans. The perception of agency also increases when technology is reactive, where agents perceive their environment and respond to changes. More than simply being reactive, proactive, and goal-directed, behavior is a particularly powerful indicator of agency (Wooldridge & Jennings, 1995). These characteristics all tend to induce people to consider software agents and similar sophisticated automation as an intentional agent (Lee, 2006; Meyer & Lee, 2013). Surface features of the technology can amplify the perception of automation agency. Even iconic representations of faces and eyes influence emotional response and visual attention (Driver et al., 1999; Langton & Bruce, 1999). Voices can also have a powerful influence. In one study, the emotional quality of the voice in an in-vehicle device interacted with the mood of the driver such that, when the voice matched the driver state (subdued for negative, enthused for positive), drivers drove more safely (Nass et al., 2005). In the context of vehicle technology, shared, traded, and ceded control describes vehicle automation that has increasing agency. Trust is important with highly agentic technology because the complexity of such technology makes a complete understanding difficult.

Agency can refer to a characteristic of technology, or to the sense of control people feel over their influence on the world. The agency of the automation can influence the agency that the person feels in joint activities, and this agency is profoundly important in describing the nature of the trust relationship. People's sense of agency is important because reduced agency can be accompanied by a sense of vulnerability and uncertainty that requires trust to overcome. With human–human relationships, people see the outcomes of interactions with others in terms of *I*, *We*, and *You* agency (Sebanz, Bekkering, & Knoblich, 2006). With *I* agency, people feel in control and take responsibility for outcomes. With *We* agency, people feel they are partnered with another and feel a sense of joint control and joint responsibility, as in a couple dancing. With *You* agency, control over outcomes and the associated responsibility are ascribed to the other. The sense of agency is composed of both a feeling of agency and an explicit judgment of agency, with the feeling of agency being a precondition of the judgment of agency (Haggard, 2017). The feeling of agency derives from the experience of fluent control, whereas the judgment of agency derives from the explicit interpretation of the action and outcomes relative to the goal of the action (Haggard & Tsakiris, 2009). Fluent control requires a direct connection between action and outcome, precise response to movement, and immediate feedback (Sidarus, Vuorre, Metcalfe, & Haggard, 2017). For example, people feel agency in controlling a conventional vehicle because a vehicle responds immediately and smoothly to a driver's movement of the steering wheel. Agency is greatest when people's actions clearly and immediately reflect a goal that they actively choose (Haggard, 2017).

The *I*, *We*, and *You* agencies are relevant to how people trust and respond to vehicle automation because agency influences the capacity to detect errors, accept responsibility, and feel in control (van der Wel, 2015). Through predictive processing of perceptual-motor actions, people are very sensitive to their own errors, in the case of *I* agency; or their partners' errors, in the case of *We* agency (Picard & Friston, 2014). Furthermore, people can predict the goals of others before the action ends, provided that the action conforms to the constraints of biological motion (Elsner, Falck-Ytter, & Gredebäck, 2012). Fluent interactions with automation and a sense of *We* agency depend on these predictive mechanisms that break down when biological motion rules are violated and the automation lacks anthropomorphic characteristics (Sahaï, Pacherie, Grynszpan, & Berberian, 2017). Observable and directable behavior that includes anthropomorphic control patterns will likely engage the mirror neuron system of the person and promote fluent control (Knoblich & Sebanz, 2008; Madhavan & Wiegmann, 2007). For specific examples of how to create such anthropomorphic control patterns see Lasseter (1987). Removing fluent control tends to shift people from *I* to *We* or *You* agency, resulting in diminished error detection, longer latency, and a diminished sense of control and responsibility. Ideally, the *I*, *We*, and *You* agencies should parallel the role of the person in shared, traded, and ceded control.

We agency is sometimes transformed into *I* agency when people develop a sense that they are responsible for control that is actually generated by another. This vicarious control emerges when people are deeply engaged by observation and actively predicting outcomes. Vicarious control tends to occur when instructions precede action but not when the instructions follow the action (Wegner, Sparrow, & Winerman, 2004). In the driving context, this might have implications for the timing of turn-by-turn navigation commands that are being followed by a car with traded control. Instructions that precede the response of the maneuver are likely to induce a greater degree of vicarious control and agency. The opposite of vicarious control—agency collapse—can also occur, leading people who are controlling the vehicle to move from a sense of *I* agency to *You* agency and ascribe agency and responsibility to the vehicle. This seems to occur in some cases of unintended acceleration, where people feel that the car is out of their control despite their continued depression of the accelerator pedal. Such agency collapse might become more prevalent with increasingly autonomous automation (Schmidt, 1989; Schmidt & Young, 2012).

SAE Level 2 automation poses a problem in this regard because removing the driver from the fluent, perceptual-motor, control might lead to a sense of *You* agency where people disengage from driving, fail to detect automation errors, and lose a sense that they are responsible to oversee the automation. A recent test-track study highlighted this possibility by exposing people to obstructions (e.g., a stationary vehicle and garbage bag) that require them to respond because the automation would not. Although people received instructions and had their eyes on the road, 28% failed to respond (Victor et al., 2018). The authors explained the results in terms of an expectation mismatch, which suggests a misunderstanding of system capability due to poor situation awareness and an inaccurate mental model (Engström et al., 2017). The concept of agency suggests a more fundamental mismatch where some people may have developed expectations based on *You* agency rather than the *I* or *We* agency appropriate for SAE Level 2 automation. Shifts of agency are likely driven, not so much by the explicit understanding associated with mental models of automation, but by developing trust based on experiencing the automation.

The research on agency and joint activity suggests that trust is rooted in perceptual-motor experience of acting together (Seemann, 2009). Trust that develops with *We* agency emerges as an attitude of the other that exists and is expressed in the bodily state of the other (embodied) and the intentions of the other can be read from actions of the other (enacted). Trust and *We* agency depends on a sense of joint control that depends on perceptual-motor coupling that some types of automation can enhance by providing additional feedback through the steering wheel and pedals, but that other types of automation can degrade or eliminate leading to *You* agency (Abbink et al., 2017; Mulder, Abbink, & Boer, 2012). More generally, these findings suggest assessing agency might be essential to define the nature of the trust relationship (Caspar et al., 2015; Haggard, 2017). Are people trusting the automation to drive—as in ceded control and *You* agency? Or are they trusting the automation to work with them to drive—as in shared control and *We* agency?

4.4 PROMOTING APPROPRIATE TRUST IN VEHICLE AUTOMATION

As with human-to-human relationships, higher levels of trust in human-to-technology relationships lead people to rely on technology and lower levels lead people to reject it (Lee & Moray, 1992; Muir & Moray, 1996). However, greater trust is not always better. Trusting too much leads to over-reliance and misuse and leaves people vulnerable to automation failures; trusting too little leads to disuse where people reject potentially useful technology (Parasuraman & Riley, 1997). Fostering appropriate trust poses a fundamental design challenge in creating safe and accepted vehicle automation.

4.4.1 CALIBRATION AND RESOLUTION OF TRUST

Appropriate trust is the degree to which the capabilities of automation—trustworthiness—align with the attitude of the person regarding those capabilities—trust. Figure 4.4 shows the relationship between trust and automation capabilities. Well-calibrated trust lies on the diagonal where trust equals trustworthiness (Lee & See, 2004). High-resolution trust is sensitive to changes in the trustworthiness so that changes in trustworthiness lead to similar changes in trust. Poor resolution of trust would be a situation where changes in the capability of the automation do not lead to changes in trust, as shown by the relatively wide range of capability mapping onto a small range of trust.

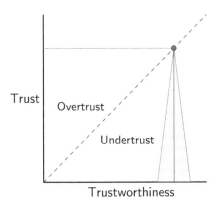

Trust

Overtrust

Undertrust

Trustworthiness

FIGURE 4.4 Calibration and resolution of trust in automation.

Much research examines how automation reliability affects trust and reliance, where reliance is the proportion of time people have the automation engaged. Reliability is often manipulated by inserting random faults in the automation that produce some overall level of reliability, such as 75% of the trials producing error-free performance (Hancock et al., 2011; Schaefer et al., 2016). This simple indication of performance often ignores the important effect of context (Bagheri & Jamieson, 2004). The context or operational domain refers to features of the situation that might influence the automation performance. To the extent that automation reliability depends on the context, the resolution of trust can be improved by highlighting situations where the automation performs well and situations where it does not. Improving the resolution of trust is particularly critical for SAE Level 2 and 3 automation. Such automation might work perfectly within the Operational Design Domain (ODD), which might cover 75% of the driving situations, and not elsewhere. This is a very different situation than the automation operating properly 75% in all driving situations. Designing an ODD that is easy for drivers to understand can greatly enhance the resolution of trust so that drivers trust and rely on the automation when it is within the ODD, and but not when it is not.

Ideally, trust would be highly resolved and well-calibrated so that high levels of trust correspond to high capability to achieve the driver's goal, and low levels of trust correspond to situations where the technology is generally incapable of achieving the driver's goal. Figure 4.4 points to two general approaches in achieving appropriate trust: improve the capability of the automation—improve trustworthiness—to meet people's expectations or make the technology more trustable. Increasing trustworthiness to match people's trust with automation that requires drivers to remain able to take back control—SAE Level 2 or 3—can fail because the better performing automation might engender greater and equally inappropriate trust, leading drivers to be even less likely to take control when needed. This automation conundrum is well documented in other domains (Endsley, 2016; Sebok & Wickens, 2017; see also, this Handbook, Chapter 21). A more promising approach is to enhance the calibration and resolution of trust by making automation more trustable, which is the focus of the balance of this section.

A comprehensive literature review suggested that trustworthiness can be described in terms of the purpose, process, and performance characteristics of the automation and that trust can be calibrated by communicating this information to people (Lee & See, 2004). *Purpose* refers to the goals and intended application of the automation—why the automation was developed. *Process* refers to the mechanisms and algorithms the automation uses to achieve this purpose—how the automation works. *Performance* refers to how consistently and precisely the automation achieves its goals (Lee & See, 2004). These characteristics describe different types of information that guide the development of trust. These dimensions identify information requirements for the HMI of AVs that can promote more appropriate trust. However, these characteristics are not invariant but change with the situation.

4.4.2   TRUSTABLE AUTOMATION: CREATE SIMPLE USE SITUATIONS AND AUTOMATION STRUCTURE

The *use situation* or operational domain can strongly influence the behavior of the automation and defines where and when the automation can be expected to achieve the users' goals. The situation is particularly critical for vehicle automation because a situation may be ill-suited for the automation and likely require a person to intervene—a curving road with poor lane markings—but the moment-to-moment performance of the automation can induce over-trust. For example, the affective processes that guide trust development make it likely that the concrete, immediate feedback, such as the precise lane-keeping of a lane-keeping aid, will outweigh the abstract and general information about how the automation works and the need for the driver to monitor the automation continuously. Furthermore, calibrating trust in the context of precise lane-keeping depends on drivers relating the capabilities of the automation to the situation by extrapolating past experiences, evaluating ongoing experience, and projecting future outcomes (Emirbayer & Mische, 1998); and, in this situation, the salience of the ongoing experience tends to dominate, leading to poorly calibrated trust.

The complexity of the roadway environment and the evolving capacity of vehicle automation may make it particularly difficult for drivers to calibrate their trust. Although drivers might see driving as a routine activity the nearly infinite combination of situations makes it extremely complex for automation. As a consequence, vehicle automation will need to evolve and "learn to drive" over many years. Vehicle automation might update on a weekly or monthly basis, introducing new capabilities and vulnerabilities (Bansal et al., 2019). These changes and the highly variable roadway situations create a *wicked* learning environment that makes it difficult to develop expertise (Hogarth, Lejarraga, & Soyer, 2015). In contrast *kind* environments have stable relationships where people can learn to recognize situations and the appropriate response (Kahneman & Klein, 2009).

Automation can be crafted to be kinder and trustable by considering the *structure* of the automation. One approach is to introduce new features and upgrades in a manner that corresponds to the task chunking that defines driving activities and that logically extends the ODD. For example, an update that extends a lane-keeping system so that it can automatically change lanes should consider that drivers are likely to chunk such a task to include checking the blind spot and so would expect the automation to include that subtask with the update.

4.4.3 TRUSTABLE AUTOMATION: DISPLAY SURFACE AND DEPTH INDICATIONS OF CAPABILITY

An individual can infer the three characteristics of automation—purpose, process, and performance—by observing the automation over time. These characteristics describe the automation in terms of levels of attributional abstraction that help guide trust towards a level appropriate for the situation (Lee & See, 2004). However, people do not experience these characteristics directly, but only through displays and controls that provide an imperfect indication of these underlying characteristics. Specific guidelines for conveying purpose, process, and performance include (Lee & Seppelt, 2012; Seppelt & Lee, 2019; Weast et al., 2016):

Purpose
- Define the operational domain in concrete terms and provide a continuous indicator of whether the vehicle is in the operational domain or not.
- Indicate impending changes in the driving situation relative to the operational domain.
- Monitor the driver's behavior and provide feedback when the driver's behavior, such as attention to the road, deviates from that required by the purpose of the automation.
- Match the goals of the automation to those of the person (e.g., speed or energy efficiency).

Process
- Indicate what the automation "sees" and does not "see."
- Communicate the state of the vehicle and explain its behavior. The vehicle should indicate what events led the vehicle to suddenly brake or change lanes.
- Allow people to request more information to explain behavior and to reduce the flow of information.
- Automation that mimics the process of human driving will produce fewer surprises and be more trustable.

Performance
- Show proximity to control capacity to indicate when performance is likely to suffer, such as in steering through a curve.
- Consider haptic and auditory cues to highlight situations where performance is degraded.

The purpose, process, and performance of automation can be thought of as depth features that define the capability of the automation, which are imperfectly revealed by the surface features of the interface. Sometimes the surface features of the interface themselves can strongly influence trust even though they have no direct connection to the underlying capability of automation. For example, the scent of lavender leads to greater interpersonal trust by inducing a calm, inclusive state (Sellaro, van Dijk, Paccani, Hommel, & Colzato, 2014), and pastel colors enhance trust in cyberbanking interfaces (Kim & Moon, 1998).

Sometimes the surface features of the automation conflict with the depth features that define its true capabilities. One example of this is the label used for various vehicle automation features. These labels can imply that the automation has greater capabilities than it has, but often the labels are ambiguous and confusing (Abraham, Seppelt, Mehler, & Reimer, 2017; Nees, 2018). Although the labeling and description of automation in the owner's manual is an important basis of trust, real-time feedback from the HMI in the vehicle might be more influential (Li, Holthausen, Stuck, & Walker, 2019; this Handbook, Chapter 15). Both visual and auditory displays of performance and process of the automation act as an externalized mental model that helps drivers see its limits (Seppelt & Lee, 2007; 2019).

Moving beyond the HMI, the behavior of automation can make the automation more trustable and promote appropriate trust. The variability of the lane position of the vehicle can be a salient cue that might convey the depth features of automation. Because degraded control behavior of the vehicle engages drivers (Alsaid, Lee, & Price, 2019), less precise lane-keeping can prompt drivers to look to the road (Price, Lee, Dinparastdjadid, Toyoda, & Domeyer, 2019). Precise lane-keeping in situations where the automation is less capable could promote over-trust. Similarly, acceleration cues of ACC helped redirect driver's attention to the road and anticipate conflicts with the vehicle ahead (Morando, Victor, & Dozza, 2016). Similarly, vehicles can announce their intent to change lanes through the roll of the vehicle's body (Cramer & Klohr, 2019). In general, the salient surface features, such as vehicle motion, should be mapped to important depth features of the automation to convey how its capability changes with the evolving situation.

The trust-relevant characteristics of the automation's purpose, process, and performance are often inferred from direct experience with the system, but this might not be possible with the introduction of dramatically new technology, such as self-driving cars. Many people will not have any direct experience on which to base their initial trust of these systems. In the absence of direct experience, people base their trust on relational and societal bases (Lee & Kolodge, 2019). Relational bases include experience with a brand (e.g., GM or Ford) or a type of technology (e.g., computers). Societal bases include the policy and regulatory structures that ensure vehicle safety. Like labels that can lead to inappropriate trust, so can the relational and societal bases of trust (also see this Handbook, Chapter 5).

4.4.4 TRUSTABLE AUTOMATION: ENABLE DIRECTABLE AUTOMATION AND TRUST REPAIR

Revealing the depth features of automation so that they are apparent to the person is sometimes referred to as transparent or observable automation (Endsley, 2016; Klein et al., 2004). Often neglected in the quest to release people of the burden of control is the need to make automation directable by giving people the ability to adjust and guide the automation. As an example, automation that supports lane-keeping might effectively center the vehicle but might make it difficult for the driver to adjust the vehicle's position to accommodate other traffic or potholes. An alternate approach could be for the automation to be more directable and accept steering input from the driver and provide feedback through the steering wheel that indicates when the driver approaches safety boundaries, such as the edge of the road (Abbink et al., 2017; Abbink, Mulder, & Boer, 2012). Such feedback could also indicate to the driver when sensors are providing less precise information. As in the previous discussion of agency, control changes the way we perceive the world and letting the driver direct the automation provides support for appropriate trust that is not easily achieved by simply displaying trust-related information (Wen & Haggard, 2018). More specifically, like active touch and fluent perceptual-motor interaction, perception through direction of automation will likely be more effective in helping drivers detect limitations than perception through observation of automation (Flach, Bennett, Woods, & Jagacinski, 2015; Gibson, 1962). Directable automation may provide a strategy to circumvent one of the ironies of automation: better automation leads to less frequent engagements and the occasional engagements tend to be more challenging (Bainbridge, 1983). Automation that keeps the driver engaged and invites interaction may prepare people for occasional instances where they must take control.

Making the automation directable engages the person in developing trust; trust repair engages the automation in re-developing trust. The process of recovering peoples' trust following an automation failure can be an active process of trust repair (de Visser, Pak, & Shaw, 2018; Kim & Cooper, 2009; Tomlinson & Mayer, 2009). Trust repair describes the process of interacting with the person to recover trust rather than simply relying on the baseline representation of the automation's purpose, process, and performance. Here, additional information is presented to demonstrate that the system is trustworthy. One way to frame this information is in terms of extrapolating past experiences, evaluating the present experience, and projecting future outcomes—explain, show, and promise (Emirbayer & Mische, 1998). This can include an explanation of past failings. It can also include showing information in the present that shows why the automation fails, and it could project to the future with a promise for why the automation will not fail (Kohn, Quinn, Pak, de Visser, & Shaw, 2018). For example, if the automation requests the driver to take back control, it might build trust by describing what about the situation—unusually heavy rain— led to the need for the driver to intervene and why this would be unlikely to occur in the future. More generally, the elements of trust recovery—explain, show, and promise—can convey the locus of causality (e.g., an external event, such as heavy rain versus an internal event, such as a software bug), degree of control over the situation (e.g., the degree to which the rain was unavoidable), and stability of the cause (e.g., the unpredictable nature of intense rain) (Tomlinson & Mayer, 2009).

The aim of trust repair, like the other ways of engineering the relationship with automation, is to promote highly resolved and well-calibrated trust, not to increase trust. Following this logic, thought should be given to the complement of trust repair—trust tempering. Trust tempering would actively monitor situations where the system performed well, but failure was likely, and then explain why such success cannot be counted on in the future can moderate trust and help people avoid relying on automation in such situations in the future.

4.4.5 TRUSTWORTHY AUTOMATION AND GOAL ALIGNMENT

Technically adept vehicles that keep people safe is a necessary, but not a sufficient, condition to ensure people trust and accept automation, particularly when people cede control. Even with ACC, allowing drivers to choose the driving style that aligns with their goals (e.g., fast or economical) can enhance trust (Verberne, Ham, & Midden, 2012). Contrary to most research on trust in automation, the automation in the transportation ecology shown in Figure 4.1 does not always work to achieve the goals of each individual. One example is how connected vehicles might be routed to minimize congestion. The routing might produce shorter trips on average but might delay some drivers. In this case, the goal might be to reduce the trip duration for all vehicles, but this does not necessarily align with the goals of the individual rider. This is particularly true of those who are not directly benefiting from vehicle automation, such as pedestrians and other road users who must negotiate AVs for right-of-way at an intersection (Domeyer, Dinparastdjadid, Lee, & Douglas, 2019). Such incidental users, who are affected by automation, but not served by the automation, might reject automation simply because its goals fail to align with their goals (Inbar & Tractinsky, 2009). In such situations trust and social norms can help avoid the tendency of people to pursue the rational immediate benefit for themselves at the cost of collective benefit—sometimes termed the tragedy of the commons (Hayashi, Ostrom, Walker, & Yamagishi, 1999; Ostrom, 1998) (also see Chapter 19 for a discussion of joint optimization in the transportation system). In these situations, face-to-face interactions can enhance cooperation and mitigate the tragedy of the commons. Unfortunately, increasingly autonomous vehicles might reduce such face-to-face interactions and undermine the tolerance of other road users for autonomous vehicles (Domeyer, Lee, Toyoda, 2020). More generally, the design of the network structure and the associated spatial layout can predict cooperative behavior: small neighborhoods promote more cooperation. Specifically, the benefit/cost ratio for cooperative behavior must exceed 1+ (group size/number of groups) for cooperation to flourish (Miller, 2013).

4.5 TRUST AND ACCEPTANCE OF VEHICLE TECHNOLOGY

The Technology Acceptance Model (TAM) has been used in other domains to determine how features of technology influence acceptance and use of that technology. According to that model, technology acceptance often depends on two perceptions of technology: perceived usefulness and perceived ease of use (Davis, 1989). Perceived usefulness describes how well the functionality meets the person's needs and perceived ease of use describes how accessible that functionality is. Typically, perceived usefulness outweighs perceived ease of use in predicting acceptance, and this might be particularly true in the driving domain. In driving, perceived usefulness might be particularly influential because an AV could turn the 52 minutes of the average daily commute into time the rider could spend working or relaxing (Hu & Young, 1999). People might consider the ability to work while commuting to be very useful and that utility might dominate trust in predicting acceptance of vehicle technology (Ghazizadeh, Lee, & Boyle, 2012). One study that addressed this issue found trust and perceived usefulness influenced the intention to use an autonomous vehicle to a similar extent and that perceived ease of use had a much smaller influence (Choi & Ji, 2015).

Unlike the typical application of TAM to information technology in the workplace, driving involves considerable risk. Although autonomous vehicles promise to greatly reduce the risk of driving, it is not the actual risk and safety that govern trust and technology acceptance but the perceived risk. With perceived risk, people focus on stories and feelings, not statistics and analysis, and so they might neglect the many crashes autonomous vehicles avoided and focus on the few caused by automation (Slovic, Finucane, Peters, & MacGregor, 2004). Consequently, people may perceive autonomous vehicles as much riskier than they actually are. Perceived risk can also deviate from actual risk in situations where the technology is not readily observable, mishaps produce deadly consequences, and people feel they have no control. Such situations produce dread risk (Slovic, 1987; Slovic et al., 2004). An analysis of open-ended items from a large survey found people responding to vehicle automation in terms of dread risk (Lee & Kolodge, 2019). If vehicle automation produces feelings of dread risk, then AVs might need to be 1,000 times safer than conventional vehicles to be perceived as having the same risk (Lee, 2019; Slovic, 1987).

Trust plays a critical role in mediating risk perception (Slovic, 1993). In the context of AVs, trust partially mediated how potential environmental benefits influence risk acceptance, and fully mediated the effect of these benefits on people's intention to use AVs (Liu, Ma, & Zuo, 2019). In other words, people are willing to accept more risk for riding in an environmentally friendly vehicle if they trust automation. Generally, trust is slow to develop and quick to lose (Muir, 1987; Slovic, 1993). This may be even more pronounced with AVs where a failure might lead people to see risk in terms of dread risk. Such a transition might lead to trust collapse and a slow recovery as people monitor the behavior of automation with the expectation of further violations of their trust (Earle, Siegrist, & Gutscher, 2010; Slovic, 1993).

Although TAM suggests that the perceived usefulness of AVs will be a powerful force in their acceptance, the nature of driving makes it likely that risk will likely play an important role. This role might be accentuated by the potential for people to perceive risk associated with vehicle automation as dread risk. Importantly, trust seems to have a strong influence on both perceived usefulness and risk, making it critical to craft trustworthy and trustable automation.

4.6 ETHICAL CONSIDERATIONS AND THE TELEOLOGY OF TECHNOLOGY

The focus of this chapter has been on trust and development of appropriate trust—creating trustable technology. With increasingly autonomous technology, people will ride in vehicles or share the road with the vehicles while having little control over the technology. As such, we need to shift to creating trustworthy automation. Trustworthy automation is automation that should be trusted because it reliably achieves people's goals. This requires a focus not just on ensuring technology can maintain the position of the vehicle in the lane and avoid obstacles, but that it behaves politely with other road users and how it provides service and equitably uses public resources. Vehicle automation can provide mobility and greater agency to those who are poorly served by today's transportation options, such as those that are older, economically disadvantaged, or have vision and mobility limitations (see also, this Handbook, Chapters 10, 17). However, recent history has shown that the algorithms that power social media and public policy tend to exacerbate existing biases and inequities (O'Neil, 2016). To counter this tendency, there is an increasing need for policies to ensure the trustworthiness and ethical implementation of technology (Etzioni & Etzioni, 2016).

Some frame the ethical challenge of AVs in terms of the Trolley Problem, where the algorithm designers face decisions about which people an algorithm causes to die and which it saves in particular crash situations (Jean-Francois, Azim, & Iyad, 2016; Shariff, Bonnefon, & Rahwan, 2017). Such situations and the ethical dilemmas posed are rarely realistic challenges for designers (Bauman, McGraw, Bartels, & Warren, 2014; De Freitas, Anthony, & Alvarez, 2019). A more realistic dilemma is the macro-level "Trolley Problem" that companies and regulatory agencies face: should they pull the lever and enable self-driving cars on the road knowing that the technology is imperfect and their action will be responsible for thousands of people dying or should they wait and let thousands more die in conventional vehicles as is happening today. At a more pragmatic level, developers and policymakers must decide whether it is ethical for fully automated, self-driving vehicles to follow the letter of law. Following the speed limit rather than the speed of the traffic stream might increase the risk of collision; however, over time such behavior might lead others to drive more slowly and follow the speed limit, which might ultimately make driving safer. Chapter 14 describes some of the challenges faced by governmental responses to AVs.

As technology becomes more capable, we confront the threat that increasingly autonomous technology will undermine human agency (Hancock, Nourbakhsh, & Stewart, 2019; Lee, 2019). At the same time, autonomous systems can free us from mundane tasks, such as navigating a congested highway and give us agency to pursue more meaningful activities. The diverse conceptualization of trust presented in this chapter may provide the theoretical perspective needed to identify what level of control people need to achieve their goals and feel enhanced agency rather than diminished agency.

4.7   CONCLUSION

Trust complements the concepts of mental models and situation awareness in explaining how and when people rely on automation. Trust is a multi-faceted term that operates at timescales of seconds to years to describe whether people rely on, accept, and tolerate vehicle technology. Trust mediates micro interactions concerning how people rely on automation to engage in non-driving tasks to macro interactions concerning how the public accepts new forms of transport. Public acceptance may depend on the trust of incidental users, such as pedestrians who must negotiate with AVs at intersections, and drivers who must share the road with AVs. With such incidental users, designers need to consider how to reconcile the goals of the riders of the AVs and the goals of the other road users who must tolerate the AVs if they are to succeed (Domeyer, Lee, & Toyoda, 2020). Unlike traditional users for which the automation was primarily designed, the trust and tolerance of incidental users might play a critical role in the success of increasingly AVs as they use public roadways and resources in ways that conventional vehicles might not. Chapter 5 addresses these issues of public acceptance in more detail.

Harmonizing trust relationships in the transportation network represents a complex and multi-faceted design challenge (this Handbook, Chapter 19). The specific design considerations depend on the particular trust relationships (e.g., driver interaction SAE Level 2 automation or pedestrians interacting with driverless vehicles); however, some specific advice to promote appropriate trust can be offered:

Create Simple Use Situations and Automation Structure
- Create ODDs that are clear, concrete, and simple.
- Minimize the complexity of automation functions and match them to the natural divisions of driving activities.
- Consider human-like driving behavior to make vehicle behavior more predictable.

Display Surface and Depth Indications of Capability
- Create displays that reveal the purpose, process, and performance of automation.
- Show what the automation sees and how it plans to respond.
- Rather than mask control challenges, reveal them through the vehicle control behavior.

Enable Directable Automation and Trust Repair
- Even "fully automated" vehicles should provide people with appropriate control and feedback.
- Provide explanations when mishaps occur.
- Enable people to ask "why" the automation behaved as it did.

While each of these techniques promote appropriate trust, a critical consideration is the context of the transportation ecology; what promotes trust for some might have unintended consequences of undermining trust of others. A network perspective is essential in crafting a trustworthy and cooperative AV system (Miller, 2013).

ACKNOWLEDGEMENTS

This work is partially supported by a grant from Toyota CSRC and the National Science Foundation (NSF 18–548 FW-HTF). The chapter was greatly improved with the comments of the editors, Erin Chiou, Josh Domeyer, and Mengyao Li, and the other members of the Cognitive Systems Laboratory at the University of Wisconsin—Madison.

REFERENCES

Abbink, D. A., Carlson, T., Mulder, M., de Winter, J., Aminravan, F., Gibo, T., & Boer, E. (2017). A topology of shared control systems–Finding common ground in diversity. *IEEE Transactions on Human-Machine Systems*, 48(5), 509–525.

Abbink, D. A., Mulder, M., & Boer, E. R. (2012). Haptic shared control: Smoothly shifting control authority? *Cognition, Technology and Work*, 14(1), 19–28. doi:10.1007/s10111-011-0192-5

Abraham, H., Seppelt, B., Mehler, B., & Reimer, B. (2017). What's in a name: Vehicle technology branding and consumer expectations for automation. *AutomotiveUI 2017–9th International ACM Conference on Automotive User Interfaces and Interactive Vehicular Applications, Proceedings*, 226–234. doi:10.1145/3122986.3123018

Adolphs, R., Tranel, D., & Damasio, A. R. (1998). The human amygdala in social judgment. *Nature*, 393, 470–474.

Alsaid, A., Lee, J. D., & Price, M. A. (2019). Moving into the loop: An investigation of drivers' steering behavior in highly automated vehicles. *Human Factors*. doi:10.1177/0018720819850283

Bagheri, N. & Jamieson, G. A. (2004). The impact of context-related reliability on automation failure detection and scanning behaviour. *2004 IEEE International Conference on Systems, Man and Cybernetics*, 1, 212–217.

Bainbridge, L. (1983). Ironies of automation. *Automatica*, 19(6), 775–779. doi:10.1016/0005–1098(83)90046-8

Bansal, G., Nushi, B., Kamar, E., Weld, D., Lasecki, W., & Horvitz, E. (2019). A case for backward compatibility for Human-AI teams. *arXiv:1906.01148*.

Bauman, C. W., McGraw, P. A., Bartels, D. M., & Warren, C. (2014). Revisiting external validity: Concerns about trolley problems and other sacrificial dilemmas in moral psychology. *Social and Personality Psychology Compass*, 8(9), 536–554. doi:10.1111/spc3.12131

Bechara, A., Damasio, H., Tranel, D., & Damasio, A. R. (1997). Deciding advantageously before knowing the advantageous strategy. *Science*, 275(5304), 1293–1295.

Caspar, E. A., De Beir, A., Magalhaes De Saldanha Da Gama, P. A., Yernaux, F., Cleeremans, A., & Vanderborght, B. (2015). New frontiers in the rubber hand experiment: When a robotic hand becomes one's own. *Behavior Research Methods*, 47(3), 744–755. doi:10.3758/s13428-014-0498-3

Choi, J. K. & Ji, Y. G. (2015). Investigating the importance of trust on adopting an autonomous vehicle. *International Journal of Human-Computer Interaction*, 31(10), 692–702. doi:10.1080/10447318.2015.1070549

Cramer, S. & Klohr, J. (2019). Announcing automated lane changes: Active vehicle roll motions as feedback for the driver. *International Journal of Human-Computer Interaction*, 35(11), 980–995. doi:10.1080/10447318.2018.1561790

Davis, F. D. (1989). Perceived usefulness, perceived ease of use, and user acceptance of information technology. *MIS Quarterly*, 13(3), 319–340.

De Freitas, J., Anthony, S. E., & Alvarez, G. A. (2019). Doubting driverless dilemmas. *PsychArXiv*, 1–5. doi:10.7498/aps.62.064705

De Kosnik, A. (2014). Disrupting technological privilege: The 2013–2014 San Francisco Google bus protests. *Performance Research*, 19(6), 99–107. doi:10.1080/13528165.2014.985117

de Visser, E. J., Monfort, S. S., Goodyear, K., Lu, L., O'Hara, M., Lee, M. R., … Krueger, F. (2017). A little anthropomorphism goes a long way. *Human Factors*, 59(1), 116–133. doi:10.1177/0018720816687205

de Visser, E. J., Pak, R., & Shaw, T. H. (2018). From 'automation' to 'autonomy': The importance of trust repair in human–machine interaction. *Ergonomics*, 0139, 1–19. doi:10.1080/00140139.2018.1457725

Domeyer, J., Dinparastdjadid, A., Lee, J. D., & Douglas, G. (2019). Proxemics and kinesics in automated vehicle-pedestrian communication: Representing ethnographic observations. *Transportation Research Record*. doi:10.1177/0361198119848413

Domeyer, J., Lee, J. D., & Toyoda, H. (2020). Vehicle automation-other road user communication and coordination: Theory and mechanisms. *IEEE Access*, 8, 19860–19872.

Driver, J., Davis, G., Ricciardelli, P., Kidd, P., Maxwell, E., & Baron-Cohen, S. (1999). Gaze perception triggers reflexive visuospatial orienting. *Visual Cognition*, 6(5), 509–540.

Dzindolet, M. T., Pierce, L. G., Beck, H. P., Dawe, L. A., & Anderson, B. W. (2001). Predicting misuse and disuse of combat identification systems. *Military Psychology*, 13(3), 147–164.

Earle, T. C., Siegrist, M., & Gutscher, H. (2010). Trust, risk perception and the TCC model of cooperation. In M. Siegrist, T. C. Earle, & H. Gutscher (Eds.), *Trust in Risk Management: Uncertainty and Scepticism in the Public Mind* (pp. 1–50). London: Earchscan.

Elsner, C., Falck-Ytter, T., & Gredebäck, G. (2012). Humans anticipate the goal of other people's point-light actions. *Frontiers in Psychology, 3*, 120. doi: 10.3389/fpsyg.2012.00120

Emirbayer, M. & Mische, A. (1998). What is agency? *American Journal of Sociology*, 103(4), 962–1023.

Endsley, M. R. (1995). Toward a theory of situation awareness in dynamic systems. *Human Factors*, 37(1), 32–64. doi:10.1518/001872095779049543

Endsley, M. R. (2013). Situation awareness. In J. D. Lee & A. Kirlik (Eds.), The Oxford Handbook of Cognitive Engineering (pp. 88–108). New York: Oxford University Press.

Endsley, M. R. (2016). From here to autonomy. *Human Factors*, 59(1), 5–27. doi:10.1177/0018720816681350

Engström, J., Bärgman, J., Nilsson, D., Seppelt, B., Markkula, G., Piccinini, G. B., & Victor, T. (2017). Great expectations: A predictive processing account of automobile driving. *Theoretical Issues in Ergonomics Science, 19*(2), 154–194. doi:10.1080/14639 22X.2017.1306148

Ert, E., Fleischer, A., & Magen, N. (2016). Trust and reputation in the sharing economy: The role of personal photos in Airbnb. *Tourism Management, 55*, 62–73. doi:10.1016/j. tourman.2016.01.013

Etzioni, A. & Etzioni, O. (2016). AI assisted ethics. *Ethics and Information Technology*, 18(2), 149–156. doi:10.1007/s10676-016-9400-6

Fehr, E. & Camerer, C. F. (2007). Social neuroeconomics: The neural circuitry of social preferences. *Trends in Cognitive Sciences, 11*(10), 419–427. doi:10.1016/j.tics.2007.09.002

Flach, J. M., Bennett, K. B., Woods, D. D., & Jagacinski, R. J. (2015). Interface design: A control theoretic context for a triadic meaning processing approach. *The Cambridge Handbook of Applied Perception Research*, 647–668. doi:10.1017/CBO9780511973017.040

Gefen, D., Karahanna, E., & Straub, D. W. (2003). Trust and TAM in online shopping: An integrated model. *MIS Quarterly, 27*(1), 51–90.

Ghazizadeh, M., Lee, J. D., & Boyle, L. N. (2012). Extending the technology acceptance model to assess automation. *Cognition, Technology & Work, 14*(1), 39–49. doi:10.1007/ s10111-011-0194-3

Gibson, J. J. (1962). Observations on active touch. *Psychological Review*, 69(6), 477–491.

Guznov, S., Lyons, J., Nelson, A., & Woolley, M. (2016). The effects of automation error types on operators' trust and reliance. In S. Lackey & R. Shumaker (Eds.), *Lecture Notes in Computer Science* (Vol. 9740, pp. 116–124). Berlin: Springer. doi:10.1007/978-3-319-39907-2_11

Haggard, P. (2017). Sense of agency in the human brain. *Nature Reviews Neuroscience, 18*(4), 197–208. doi:10.1038/nrn.2017.14

Haggard, P. & Tsakiris, M. (2009). The experience of agency: The experience of agency feelings, judgments, and responsibility. *Current Directions in Psychological Science*, 18(4), 242–246.

Hancock, P. A., Billings, D. R., Schaefer, K. E., Chen, J. Y. C., de Visser, E. J., & Parasuraman, R. (2011). A meta-analysis of factors affecting trust in human-robot interaction. *Human Factors*, *53*(5), 517–527. doi:10.1177/0018720811417254

Hancock, P. A., Nourbakhsh, I., & Stewart, J. (2019). On the future of transportation in an era of automated and autonomous vehicles. *Proceedings of the National Academy of Sciences*, *116*(16), 7684–7691. doi:10.1073/pnas.1805770115

Hayashi, N., Ostrom, E., Walker, J., & Yamagishi, T. (1999). Reciprocity, trust, and the sense of control: A cross-societal study. *Rationality and Society*, *11*(1), 27–46.

Hoff, K. A. & Bashir, M. (2015). Trust in automation: Integrating empirical evidence on factors that influence trust. *Human Factors*, *57*(3), 407–434. doi:10.1177/0018720814547570

Hogarth, R. M., Lejarraga, T., & Soyer, E. (2015). The two settings of kind and wicked learning environments. *Current Directions in Psychological Science*, *24*(5), 379–385. doi:10.1177/0963721415591878

Hu, P. & Young, J. (1999). *Summary of Travel Trends: 1995 Nationwide Personal Transportation Survey*. Washington, DC: Federal Highway Administration.

Inbar, O. & Tractinsky, N. (2009). The incidental user. *Interactions*, *16*(4), 56–59.

Jean-Francois, B., Azim, S., & Iyad, R. (2016). The Social dilemma of autonomous vehicles. *Science*, *352*(6293), 1573.

Kahneman, D. (2011). *Thinking, Fast and Slow*. New York: Macmillan.

Kahneman, D. & Klein, G. A. (2009). Conditions for intuitive expertise: A failure to disagree. *American Psychologist*, *64*(6), 515–526. doi:10.1037/a0016755

Kim, J. & Moon, J. Y. (1998). Designing towards emotional usability in customer interfaces—Trustworthiness of cyber-banking system interfaces. *Interacting with Computers*, *10*(1), 1–29.

Kim, P. H. & Cooper, C. D. (2009). The repair of trust: A dynamic bilateral perspective and multilevel conceptualization. *Academy of Management Review*, *34*(3), 401–422.

Klein, G. A., Woods, D. D., Bradshaw, J. M., Hoffman, R. R., Feltovich, P. J., Hoffman, R. R., … Ford, K. M. (2004). Ten challenges for making automation a "Team Player" in joint human-agent activity. *IEEE Intelligent Systems*, *19*(6), 91–95.

Knoblich, G. & Sebanz, N. (2008). Evolving intentions for social interaction: From entrainment to joint action. *Philosophical Transactions of the Royal Society B-Biological Sciences*, *363*(1499), 2021–2031. doi:10.1098/rstb.2008.0006

Kohn, S. C., Quinn, D., Pak, R., de Visser, E. J., & Shaw, T. H. (2018). Trust repair strategies with self-driving vehicles: An exploratory study. *Proceedings of the Human Factors and Ergonomics Society Annual Meeting*, *62*(1), 1108–1112. doi:10.1177/1541931218621254

Kosfeld, M., Heinrichs, M., Zak, P. J., Fishbacher, U., & Fehr, E. (2005). Oxytocin increases trust in humans. *Nature*, *435*, 673–676.

Langton, S. R. H. & Bruce, V. (1999). Reflexive visual orienting in response to social attention of others. *Visual Cognition*, *6*(5), 541–567.

Lasseter, J. (1987). Principles of traditional animation applied to 3D computer animation. *ACM SIGGRAPH Computer Graphics*, *21*(4), 35–44. doi:10.1145/37402.37407

Lee, J. D. (2006). Affect, attention, and automation. In A. Kramer, D. Wiegmann, & A. Kirlik (Eds.), *Attention: From Theory to Practice* (pp. 73–89). New York: Oxford University Press.

Lee, J. D. (2018). Perspectives on automotive automation and autonomy. *Journal of Cognitive Engineering and Decision Making*, *12*(1), 53–57. doi:10.1177/1555343417726476

Lee, J. D. (2019). Trust and the teleology of technology. *Ergonomics*, *62*(4), 500–501. doi:10.1080/00140139.2019.1563332

Lee, J. D. & Kolodge, K. (2019). Exploring trust in self-driving vehicles with text analysis. *Human Factors*. doi:10.1177/0018720819872672

Lee, J. D. & Moray, N. (1992). Trust, control strategies and allocation of function in human-machine systems. *Ergonomics*, *35*(10), 1243–1270. doi:10.1080/00140139208967392

Lee, J. D. & See, K. A. (2004). Trust in automation: Designing for appropriate reliance. *Human Factors*, *46*(1), 50–80.

Lee, J. D. & Seppelt, B. D. (2012). Human factors and ergonomics in automation design. In G. Salvendy (Ed.), *Handbook of Human Factors and Ergonomics* (pp. 1615–1642). Hoboken, NJ: Wiley. doi:10.1002/9781118131350.ch59

Li, M., Holthausen, B. E., Stuck, R. E., & Walker, B. N. (2019). No risk no trust: Investigating perceived risk in highly automated driving. In *Proceedings of the 11th International Conference on Automotive User Interfaces and Interactive Vehicular Applications* (pp. 177–185). New York: ACM.

Liu, P., Ma, Y., & Zuo, Y. (2019). Self-driving vehicles: Are people willing to trade risks for environmental benefits? *Transportation Research Part A: Policy and Practice*, *125*(March), 139–149. doi:10.1016/j.tra.2019.05.014

Madhavan, P. & Wiegmann, D. A. (2007). Similarities and differences between human–human and human–automation trust: An integrative review. *Theoretical Issues in Ergonomics Science*, *8*(4), 277–301. doi:10.1080/14639220500337708

Mayer, R. C., Davis, J. H., & Schoorman, F. D. (1995). An integrative model of organizational trust. *Academy of Management Review*, *20*(3), 709–734.

McKnight, D. H., Choudhury, V., & Kacmar, C. (2002). Developing and validating trust measures for e-commerce: An integrative typology. *Information Systems Research*, *13*(3), 334–359.

Meyer, J. & Lee, J. D. (2013). Trust, reliance, and compliance. In A. Kirlik & J. D. Lee (Eds.), *The Oxford Handbook of Cognitive Engineering* (pp. 109–124). New York: Oxford University Press.

Mikolajczak, M., Gross, J. J., Lane, A., Corneille, O., de Timary, P., & Luminet, O. (2010). Oxytocin makes people trusting, not gullible. *Psychological Science*, *21*(8), 1072–1074. doi:10.1177/0956797610377343

Miller, H. J. (2013). Beyond sharing: Cultivating cooperative transportation systems through geographic information science. *Journal of Transport Geography*, *31*, 296–308.

Montague, E., Xu, J., & Chiou, E. (2014). Shared experiences of technology and trust: An experimental study of physiological compliance between active and passive users in technology-mediated collaborative encounters. *IEEE Transactions on Human-Machine Systems*, *44*(5), 614–624. doi:10.1109/THMS.2014.2325859

Morando, A., Victor, T., & Dozza, M. (2016). Drivers anticipate lead-vehicle conflicts during automated longitudinal control: Sensory cues capture driver attention and promote appropriate and timely responses. *Accident Analysis & Prevention*, *97*, 206–219. doi:10.1016/j.aap.2016.08.025

Muir, B. M. (1987). Trust between humans and machines, and the design of decision aids. *International Journal of Man-Machine Studies*, *27*, 527–539.

Muir, B. M. & Moray, N. (1996). Trust in automation. Part II. Experimental studies of trust and human intervention in a process control simulation. *Ergonomics*, *39*(3), 429–460.

Mulder, M., Abbink, D. A., & Boer, E. R. (2012). Sharing control with haptics: Seamless driver support from manual to automatic control. *Human Factors*, *54*(2), 786–798.

Nass, C., Jonsson, I. M., Harris, H., Reaves, B., Endo, J., Brave, S., & Takayama, L. (2005). Improving automotive safety by pairing driver emotion and care voice emotion. *Conference on Human Factors in Computing Systems* (pp. 1973–1976). New York: ACM.

Nass, C. & Moon, Y. (2000). Machines and mindlessness: Social responses to computers. *Journal of Social Issues*, *56*(1), 81–103. doi:10.1111/0022–4537.00153

Nave, G., Camerer, C., & McCullough, M. (2015). Does oxytocin increase trust in humans? A critical review of research. *Perspectives on Psychological Science*, *10*(6), 772–789. doi:10.1177/1745691615600138

Nees, M. A. (2018). Drivers' perceptions of functionality implied by terms used to describe automation in vehicles. *Proceedings of the Human Factors and Ergonomics Society Annual Meeting, 62*(1), 1893–1897. doi:10.1177/1541931218621430

Neisser, U. (1976). *Cognition and Reality: Principles and Implications of Cognitive Psychology*. San Francisco, CA: W. H. Freeman and Company.

Norman, D. A. (1983). Some observations on mental models. In D. Gentner & A. L. Stevens (Eds.), *Mental Models*. Hillsdale, NJ: Lawrence Erlbaum.

Norman, D. A. (2004). *Emotional Design*. New York: Basic Books.

NTSB. (2016). *Preliminary Report: HWY16FH018*. Washington, DC: National Transportation Safety Bureau.

O'Neil, K. (2016). *Weapons of Math Destruction: How Big Data Increases Inequality and Threatens Democracy*. New York: Crown.

Ostrom, E. (1998). A behavioral approach to the rational choice theory of collective action: Presidential address, American Political Science Association, 1997. *American Political Science Review, 92*(1), 1–22.

Parasuraman, R. & Riley, V. A. (1997). Humans and automation: Use, misuse, disuse, abuse. *Human Factors, 39*(2), 230–253.

Payyanadan, R. P. & Lee, J. D. (2018). Understanding the ridesharing needs of older adults. *Travel Behaviour and Society, 13*(June), 155–164. doi:10.1016/j.tbs.2018.08.002

Picard, F. & Friston, K. (2014). Predictions, perception, and a sense of self. *Neurology, 83*(12), 1112–1118. doi:10.1212/WNL.0000000000000798

Price, M., Lee, J. D., Dinparastdjadid, A., Toyoda, H., & Domeyer, J. (2019). Effect of automation and vehicle control algorithms on eye behavior in highly automated vehicles. *International Journal of Automotive Engineering, 10*(1), 73–79.

Reeves, B. & Nass, C. (1996). *The Media Equation: How People Treat Computers, Television, and New Media like Real People and Places*. New York: Cambridge University Press.

Romero, S. (2018). *Wielding Rocks and Knives, Arizonans Attack Self-Driving Cars*. Retrieved from www.nytimes.com/2018/12/31/us/waymo-self-driving-cars-arizona-attacks.html

Rotter, J. B. (1980). Interpersonal trust, trustworthiness, and gullibility. *American Psychologist, 35*(1), 1–7.

Rouse, W. B. & Morris, N. M. (1986). On looking into the black box: Prospects and limits in the search for mental models. *Psychological Bulletin and Review, 100*(3), 349–363.

Rousseau, D. M., Sitkin, S. B., Burt, R. S., & Camerer, C. (1998). No so different after all: A cross-discipline view of trust. *Academy of Management Review, 23*(3), 393–404.

SAE. (2016). *Taxonomy and Definitions for Terms Related to Driving Automation Systems for On-Road Motor Vehicles (J3016). Global Ground Vehicle Standards*. Warrendale, PA: Society of Automotive Engineers. doi:10.4271/J3016_201609

Sahaï, A., Pacherie, E., Grynszpan, O., & Berberian, B. (2017). Co-representation of human-generated actions vs. machine-generated actions: Impact on our sense of we-agency? IEEE International Symposium on Robot and Human Interactive Communication, 341–345. Lisbon, Portugal: IEEE.

Schaefer, K. E., Chen, J. Y. C., Szalma, J. L., & Hancock, P. A. (2016). A meta-analysis of factors influencing the development of trust in automation: Implications for understanding autonomy in future systems. *Human Factors, 58*(3), 377–400. doi:10.1177/0018720816634228

Schmidt, R. A. (1989). Unintended acceleration: A review of human factors contributions. *Human Factors, 31*(3), 345–364.

Schmidt, R. A. & Young, D. E. (2012). Cars gone wild: The major contributor to unintended acceleration in automobiles is pedal error. *Frontiers in Psychology, 1*, 1–4.

Sebanz, N., Bekkering, H., & Knoblich, G. (2006). Joint action: Bodies and minds moving together. *Trends in Cognitive Sciences, 10*(2), 70–76.

Sebok, A. & Wickens, C. D. (2017). Implementing lumberjacks and black swans into model-based tools to support human-automation interaction. *Human Factors, 59*(2), 189–203. doi:10.1177/0018720816665201

Seemann, A. (2009). Joint agency: Intersubjectivity, sense of control, and the feeling of trust. *Inquiry, 52*(5), 500–515. doi:10.1080/00201740903302634

Sellaro, R., van Dijk, W. W., Paccani, C. R., Hommel, B., & Colzato, L. S. (2014). A question of scent: Lavender aroma promotes interpersonal trust. *Frontiers in Psychology, 5*(OCT), 1–5. doi:10.3389/fpsyg.2014.01486

Seppelt, B. D. & Lee, J. D. (2007). Making adaptive cruise control (ACC) limits visible. *International Journal of Human-Computer Studies, 65*(3), 192–205.

Seppelt, B. D. & Lee, J. D. (2019). Keeping the driver in the loop: Dynamic feedback to support appropriate use of imperfect vehicle control automation. *International Journal of Human-Computer Studies, 125*, 66–80. doi:10.1016/J.IJHCS.2018.12.009

Shariff, A., Bonnefon, J. F., & Rahwan, I. (2017). Psychological roadblocks to the adoption of self-driving vehicles. *Nature Human Behaviour, 1*(10), 694–696. doi:10.1038/s41562-017-0202–6

Sheridan, T. B. & Ferrell, W. R. (1974). *Man-Machine Systems: Information, Control, and Decision Models of Human Performance.* Cambridge, MA: MIT Press.

Sidarus, N., Vuorre, M., Metcalfe, J., & Haggard, P. (2017). Investigating the prospective sense of agency: Effects of processing fluency, stimulus ambiguity, and response conflict. *Frontiers in Psychology, 8*, 1–15. doi:10.3389/fpsyg.2017.00545

Slovic, P. (1987). Perception of risk. *Science, 236*(4799), 280–285.

Slovic, P. (1993). Perceived risk, trust, and democracy. *Risk Analysis, 13*(6), 675–682.

Slovic, P. (1999). Trust, emotion, sex, politics, and science: Surveying the risk-assessment battlefield. *Risk Analysis, 19*(4), 689–701.

Slovic, P., Finucane, M. L., Peters, E., & MacGregor, D. G. (2004). Risk as analysis and risk as feelings: Some thoughts about affect, reason, risk, and rationality. *Risk Analysis, 24*(2), 311–322.

Tomlinson, E. C. & Mayer, R. C. (2009). The role of causal attribution dimensions in trust repair. *Academy of Management Review, 34*(1), 85–104. doi:10.5465/AMR.2009.35713291

van der Wel, R. P. R. D. (2015). Me and we: Metacognition and performance evaluation of joint actions. *Cognition, 140*, 49–59. doi:10.1016/j.cognition.2015.03.011

Verberne, F. M. F., Ham, J., & Midden, C. J. H. (2012). Trust in smart systems: Sharing driving goals and giving information to increase trustworthiness and acceptability of smart systems in cars. *Human Factors, 54*(5), 799–810. doi:10.1177/0018720812443825

Victor, T. W., Tivesten, E., Gustavsson, P., Johansson, J., Sangberg, F., & Ljung Aust, M. (2018). Automation expectation mismatch: Incorrect prediction despite eyes on threat and hands on wheel. *Human Factors, 60*(8), 1095–1116. doi:10.1177/0018720818788164

Weast, J., Yurdana, M., & Jordan, A. (2016). *A Matter of Trust: How Smart Design Can Accelerate Automated Vehicle Adoption. Intel White Paper.* Santa Clara, CA. doi:10.1016/S1353–4858(19)30060–1

Wegner, D. M., Sparrow, B., & Winerman, L. (2004). Vicarious agency: Experiencing control over the movements of others. *Journal of Personality and Social Psychology, 86*(6), 838–848. doi:10.1037/0022–3514.86.6.838

Wen, W. & Haggard, P. (2018). Control changes the way we look at the world. *Journal of Cognitive Neuroscience, 30*(4), 603–619. doi:10.1162/jocn

Winston, J. S., Strange, B. A., O'Doherty, J., & Dolan, R. J. (2002). Automatic and intentional brain responses during evaluation of trustworthiness of faces. *Nature Neuroscience, 5*(3), 277–283. doi:10.1038/nn816

Woods, D. D. (2015). Four concepts for resilience and the implications for the future of resilience engineering. *Reliability Engineering and System Safety, 141,* 5–9.

Wooldridge, M. & Jennings, N. R. (1995). Intelligent agents: Theory and practice. *Knowledge Engineering Review, 10*(2), 115–152.

Wu, K., Zhao, Y., Zhu, Q., Tan, X., & Zheng, H. (2011). A meta-analysis of the impact of trust on technology acceptance model: Investigation of moderating influence of subject and context type. *International Journal of Information Management, 31*(6), 572–581.

Zuboff, S. (1988). *In the Age of Smart Machines: The Future of Work, Technology and Power.* New York: Basic Books.

5 Public Opinion About Automated and Self-Driving Vehicles
An International Review

Mitchell L. Cunningham
The University of Sydney

Michael A. Regan
University of New South Wales

CONTENTS

Key Points 95
5.1   Introduction 96
5.2   Overall Acceptability..97
    5.2.1   Perceived Benefits..97
    5.2.2   Perceived Concerns..97
    5.2.3   Activities When Riding in an AV 98
5.3   Public Opinion Towards AVs as a Function of Sociodemographic
    Characteristics 99
5.4   Country Differences in Public Opinion Towards AVs 100
5.5   WTP for AVs... 101
5.6   Acceptance of AVs after Experiencing the Technology 103
5.7   Conclusion 104
Acknowledgement 105
References... 105

KEY POINTS

- The predicted benefits of AVs are unlikely to materialize unless there is societal acceptability and acceptance of them.
- To this end, it is important to gauge and understand public opinion about them, before and after use.
- Such knowledge can benefit governments in making future planning and investment decisions; benefit industry in helping technology developers design and refine their products; and benefit research organizations in identifying new directions for research and development.

- Despite some variation across countries, the general public appears largely positive about the potential benefits that may be derived from AVs, although there remain some significant concerns (e.g., in relation to safety) that may hinder their uptake and use.

5.1 INTRODUCTION

Automated vehicles (AVs) are currently being trialed and deployed in many countries around the world. While some predict that AVs capable of driving themselves in all situations will be on our roads in the (relative) near future (e.g., 2030; Underwood, 2014), others predict they will not be ubiquitous until closer to 2050 or afterwards (Litman, 2018). Technologies that automate human driving functions are estimated to yield a variety of societal benefits. These include improved safety, improved mobility (e.g., for the young, elderly, disabled, etc.), reduced congestion, improved productivity, and reduced fuel consumption (see for review Fagnant & Kockelman, 2015). However, these predicted benefits are at present largely speculative and yet to be proven.

In the lead up to the deployment of AVs, it is important to gauge public opinion about them, even if the public has had little or no direct exposure to them (Regan, Horberry, & Stevens, 2014). Gauging public opinion about them can help prepare society for their rollout in a number of ways. For example, an understanding of public opinion about AVs can provide benefits for governments (e.g., to inform future planning and investment decisions), industry (e.g., to help automobile manufacturers design and tailor their products to the perceived needs of end users), and academia (e.g., to flag important directions for further research).

Ultimately, the predicted benefits of AVs will never be realized unless there is societal acceptance of them (Regan et al., 2014). If not, drivers may not use them, may refuse to travel in them, or may use them in ways unintended by designers; and, in doing so, negate some or even all of their predicted benefits (Regan et al., 2014; see also this Handbook, Chapters 6, 9). In the context of this discussion, it is important to distinguish between two terms that are sometimes used interchangeably: "acceptance" and "acceptability". *Acceptance* of a technology is different from a priori *acceptability* of the technology. The latter refers to the judgment, evaluation, and behavioral reactions toward a technology *before* use, whereas the former relates to those reactions towards a technology *after* use (Adell, Várhelyi, & Nilsson, 2014). Since AVs (at Level 2 onwards according to SAE, 2014) are only starting to appear around the world, the examination of public acceptability of these technologies is an important research exercise at this point in time: as the intention to use emerging technology, such as AVs, is likely to be predicted to some degree, by users' a priori acceptability and associated opinions and attitudes towards them (see Payre, Cestac, & Delhomme, 2014; also see Chapter 4 in this Handbook).

In this chapter we review what is known, internationally, about public opinion and attitudes towards AVs.[1] We review, first, empirical research that has examined facets of public opinion towards this technology, including (but not limited to) benefits the

---

[1] From this point on in the Chapter, unless specified otherwise, "AV" will refer to fully automated, fully self-driving vehicles.

public agree/disagree the technology will bring, and concerns and risks the public believe to be associated with the technology which may be curtailing trust and intention to use it. Second, we examine how these AV-related opinions and attitudes vary as a function of key sociodemographic variables such as gender and age, to help us build a profile of the individuals who may be most (or least) receptive to the technology. Third, we review a number of international studies which have investigated cross-cultural differences in AV-related opinions and attitudes, to help us understand which countries/regions may be most ready and prepared for the introduction of AVs from a public acceptance point of view. Fourth, we examine willingness to pay (WTP) for the technology and, specifically, what proportions of the public are likely to pay for the technology; and precisely how much they are willing to pay for different levels of vehicle automation. Fifth, we review the relatively smaller body of literature on AV *acceptance* to gain an understanding of how actually experiencing AV technology influences opinions and attitudes towards the technology. Finally, we conclude the chapter by noting some key findings and limitations that emerge from the literature and suggest some areas for future research.

5.2 OVERALL ACCEPTABILITY

5.2.1 PERCEIVED BENEFITS

The emerging field of research highlights a variety of attitudes and opinions that the public has towards AV technology. Despite the many predicted benefits AVs are expected to bring about, the extant research suggests the public believes that some of these are more likely to come to fruition than others.

In an earlier study by Schoettle and Sivak (2014a), for example, using an online survey of 1,533 participants across the United States, United Kingdom, and Australia, participants were asked *"How likely do you think it is that the following benefits will occur when using completely self-driving vehicles?"* The study found that participants were most likely to believe that AVs would bring about better fuel economy (72% said "likely"), fewer crashes (70.4%), and reduced severity of crashes (71.7%), but were least likely to believe that AVs will bring about shorter travel times (43.3%). Interestingly, in contrast to these findings, Cunningham, Regan, Horberry, Weeratunga, and Dixit (2019a) found that, among 6,133 survey respondents from Australia and New Zealand, the predicted benefits of better fuel economy and enhanced safety were among the least endorsed by participants. Specifically, the study found that less than half (42%) of the respondents believed AVs would be safer than conventional cars, and only 38.9% believed that they would consume less fuel than conventional cars. However, the study found that the benefit participants most strongly believed AVs would bring about is enhanced mobility for people with driving impairments or restrictions (76.7%).

5.2.2 PERCEIVED CONCERNS

The literature suggests that the public has a number of concerns around the use and operation of AVs. In their survey study, Schoettle and Sivak (2014a) found that almost 88% of participants were at least somewhat concerned about driving or riding in an AV, with 29.9% reporting to be "very concerned." Specific issues which attracted most concern included: the safety consequences of equipment or system failure (96.2%), self-driving vehicles getting confused by unexpected situations (94.7%), and legal liability for drivers/owners (92.8%).

These findings, particularly those related to safety, are surprising given the high endorsement of AV safety noted previously (e.g., Schoettle & Sivak, 2014a). Kyriakidis, Happee, and de Winter (2015) found that participants showed the highest levels of concern for issues around software hacking and misuse, as well as issues around legal responsibility and the safety of AVs. Privacy, although the least worried about of issues probed in the survey, was still found to be a considerable concern (Kyriakidis et al., 2015). These findings are consistent with those reported by Cunningham et al. (2019a), which revealed that (a) being legally and financially responsible if the car is involved in an accident or makes mistakes (e.g., speeds) and (b) the ability of their AV to perform safely in all conditions were among the issues which received the most concern.

Interestingly, the issue of highest concern of those probed in the Cunningham et al. (2019a) survey was the theoretical scenario in which the participant was asked if they "would allow your child to ride in the [fully-automated] car by themselves" (85% were at least somewhat concerned). This finding mirrors an earlier survey of 2,000 respondents in which 75% reported that they would not trust an AV to take their children to school (Vallet, 2013). Like Kyriakidis et al. (2015), Cunningham et al. (2019a) found that data privacy (e.g., having your AV's location and destination tracked) was the issue of least concern probed in their survey (but was still of concern to almost 70% of respondents).

5.2.3   ACTIVITIES WHEN RIDING IN AN AV

One of the purported benefits of AV technology is that it will relieve the driver of moment-to-moment driving demands, rendering him or her free to engage in other activities if they wish. Therefore, it is important to understand what activities people may want to undertake when riding in an AV. Together, the research suggests that many of the tasks people may want to engage in when supported by automation are similar to those engaged in by drivers today (in conventional driving conditions). For example, Schoettle and Sivak (2014a) found that 41% of drivers in their survey, if supported by automation, would like to simply "watch the road even though I would not be driving." Interestingly, less than 10% of the participants reported wanting to spend time reading, texting/talking with friends and family, sleeping, watching a movie or TV, or doing work (Schoettle & Sivak, 2014a). Similarly, Kyriakidis et al. (2015) found that tasks such as listening to the radio, eating, interacting with passengers, and just observing what was happening outside were among the most prevalent secondary activities participants would want to engage in when riding in an AV. On the other hand, doing nothing, watching a movie, and reading were among the least prevalent (Kyriakidis et al., 2015).

The above findings mirror those found by Cunningham et al. (2019a) in that the most sought after activities when supported by fully AVs were those that tend, already, to be relatively prevalent in vehicles driven manually (Dingus et al., 2016), such as sitting and observing the scenery, interacting with others in the vehicle, eating and drinking, or doing nothing at all. Conversely, tasks such as doing work, grooming, or sleeping when riding in an AV were among the least likely to be undertaken. These findings are interesting. Increased productivity during commutes is a commonly purported benefit of AVs; however, drivers may not actually be so inclined to engage in work-related activities (for further discussion, see Singleton, 2019).

Understanding what activities people would like to undertake when riding in an AV, or when supported by high levels of vehicle automation, has practical implications for vehicle design. Such findings can be used by designers in designing AVs to optimize acceptance and facilitate uptake. For example, given the research suggesting the high desire to interact with other passengers when riding in an AV, vehicles could be designed to include swivel seats so that front and backseat passengers may more readily interact and/or reclinable seats to support comfort while enjoying the scenery (Large, Burnett, Morris, Muthumani, & Matthias, 2017; Cunningham et al., 2019a). Of course, such design features would also need to consider the cases in which a driver may need to re-engage the vehicle as effectively as possible. On the other hand, reviewed research suggests that using the time riding in an AV to do work may not be a priority for the public at this point in time. Therefore, for designers, designing in-vehicle human–machine interfaces (HMIs) in AVs to support work-related activities may not currently be a priority. Recognizing the different activities that drivers are likely to adopt during periods of automated driving—and incorporating these within the design process as early as possible—may help to ensure the acceptance and successful marketing of AVs (Cunningham et al., 2019b).

5.3 PUBLIC OPINION TOWARDS AVs AS A FUNCTION OF SOCIODEMOGRAPHIC CHARACTERISTICS

While it is instructive to understand public opinion about AVs at a national or population level to help gauge readiness for the technology (as discussed later in the chapter), it is also instructive to have a nuanced understanding of AV acceptance at an individual level. Here, having knowledge of how different sample sociodemographic characteristics (e.g., gender, age) are linked to different AV-related opinions and attitudes helps paint a profile picture of individuals that are most open to, and/or resistant towards, the introduction of AVs—both locally and internationally.

The gender of respondents is a demographic variable that has received considerable attention in the AV acceptability literature. Interestingly, research suggests that males are particularly receptive to AV technology. For example, Schoettle and Sivak (2014a) found that males tend to express lower levels of concern with AV-related issues (the exception being issues around data privacy), and generally felt the expected benefits associated with AVs were more likely to materialize than females. In their online study of 421 French drivers, Payre et al. (2014) found that males were more likely to have a higher intention to both use and buy fully automated driving technology compared with women. More recently, Cunningham et al. (2019a) found that males (compared with females) tended to be less concerned with AV-related issues, exhibited an increased desire to use AVs under most conditions (e.g., in heavy traffic, to drive their children around, when feeling fatigued), and be relatively more comfortable with the AV undertaking different driving functions (e.g., changing lanes, maintaining appropriate speed). However, despite this, men tended not to be any more willing than women to pay for an AV (Cunningham et al., 2019a).

The age of respondents is another demographic variable that has been the focus of investigation in the AV acceptability literature. In their review of this literature, Becker and Axhausen (2017) note that, with the exception of one study, the research concludes that "...younger people are more open to the introduction of automated vehicles" (p. 1299) with multiple studies suggesting younger individuals tend to be less concerned, or worried about, AV-related issues. The study by Cunningham et al. (2019a) found that age showed relatively strong correlations, compared with other sociodemographic variables examined (e.g., gender, driving per week), namely in relation to activities individuals may want to undertake when riding in an AV and WTP for the technology. Specifically, their correlational analysis suggests that younger individuals (compared with older individuals) were more likely to report a desire to engage in other activities when riding in an AV (e.g., rest, sleep, do work), suggesting a higher level of trust in AVs, and were also more willing to pay for an AV (Cunningham et al., 2019a).

To help us understand exactly why age may be a telling indicator of AV acceptability, future research is needed to examine this link further. For example, it may be that, since older individuals are likely to have accrued greater manual driving experience than younger drivers, it may be particularly discomforting for older drivers to hand over control to the AV—as they may believe their manual driving skills are superior, or they simply do not have a nuanced understanding of AV technology and how it works (Abraham et al., 2017). The findings in this AV literature are in line with a large recent meta-analysis suggesting that younger people may be more accepting (due to higher perceived ease of use; although effect sizes were small to moderate) of certain new technologies more generally (e.g., information and learning technologies) (Hauk, Huffmeier, & Krumm, 2018).

5.4 COUNTRY DIFFERENCES IN PUBLIC OPINION TOWARDS AVs

Only a handful of studies have specifically examined how attitudes and opinions towards AVs differ across countries. The evidence reveals that there are some differences that exist in AV acceptability across countries. Schoettle and Sivak (2014a), for example, examined public opinion about AVs—among 1,533 respondents across the United States, United Kingdom, and Australia—and found that the highest proportions of respondents with a positive view of the technology were those from Australia (61.9%), then the United States (56.3%), and then the United Kingdom(52.2%). In line with this, when respondents were asked how concerned they were about riding in an AV, most U.S. respondents reported being "very concerned" (35.9%), and most from the U.K. and Australia reported being "moderately concerned" (31.1%) and "slightly concerned" (31.1%), respectively (Schoettle & Sivak, 2014a). Moreover, the study found that different AV-related issues attracted different levels of concern across the countries, particularly in the United States and United Kingdom. For example, respondents from the United States (compared to U.K. respondents) were more likely to express concern regarding issues such as (1) legal liability/responsibility for drivers/owners, (2) data privacy, and (3) interactions with non-AVs. On the other hand, respondents from the U.K. were less likely to be concerned about issues such as system and vehicle security (from hackers), as well as interactions between the AV and pedestrians/cyclists (Schoettle & Sivak, 2014a).

A follow-up study by Schoettle and Sivak (2014b) investigated differences in public opinion towards AVs among 1,722 respondents across six countries/regions— China, India, Japan, United States, United States Kingdom, and Australia—which exposed some further interesting cross-country differences. For example, some quite large differences were found in the proportions of respondents that had overall positive opinions about AVs, with China showing the largest (87.2%) and Japan showing the lowest (49.8%). Moreover, Indian and Chinese respondents were most likely to believe that AVs will bring about all of the potential AV-related benefits probed in the survey (e.g., fewer crashes, reduced traffic congestion), while Japanese respondents appeared to be the least likely to believe these would come to fruition. However, interestingly, when asked about overall concern associated with riding in an AV, Indian respondents were 3.3 times more likely to report being "very concerned" than Chinese respondents, and Japanese respondents tended to be the least concerned about all AV-related issues probed in the study (e.g., safety consequences of system failure, liability, etc.) (Schoettle & Sivak, 2014b). This study builds upon earlier work (i.e., Schoettle & Sivak, 2014a) in highlighting differences between countries/regions in their attitudes and opinions towards AVs.

In their online survey, Kyriakidis et al. (2015) collected 5,000 responses spanning 109 countries, including 40 with at least 25 respondents, to allow an examination of cross-national differences. Although specific countries were not identified in the publication itself, a number of interesting findings were derived from a correlational analysis of different aspects of AV acceptability and national statistics (e.g., traffic deaths, income, education). The largest correlations were in relation to concern about data being transmitted to different parties (e.g., surrounding vehicles, vehicle manufacturers, roadway organizations). Specifically, respondents from more developed countries (with lower accident death rates, and higher levels of education and income) were more likely to express concern about data transmission to these different parties. Interestingly, these national statistics showed only small to moderate correlations with levels of concern with AV-related issues such as safety, legal liability/responsibly, and misuse.

5.5 WTP FOR AVs

The commercialization of AV technology is imminent (Litman, 2018), and this has triggered empirical investigation into exposing factors which may influence peoples' WTP for the technology (e.g., Cunningham, Regan, Ledger, & Bennett, 2019b). WTP is a particularly important aspect of acceptability as the public may endorse AV technology without actually intending to use or pay for it. If the public is not receptive to the technology to the extent that they are willing to pay for it and related services (e.g., ridesharing), the systems will not be used and their intended benefits will be unlikely to materialize (Regan et al., 2014).

An important line of inquiry in this area of research involves estimating what proportion of the public would actually be willing to pay any more (or even less) for AV technology than for conventional (manually driven) vehicles. In their international survey, Schoettle and Sivak (2014a) asked participants how much more they would be willing to pay for an AV (compared to a car they own or lease in the future). Results showed that more than half (56.6%) of the participants were not willing to pay any extra (i.e., $0) for AV technology, with 43.4% reporting they would pay extra (i.e., >$0). In a recent Australian study (Cunningham et al., 2019a), respondents (5,089 in total) were asked about their WTP for both partially- and fully automated cars. In relation to the former, Cunningham et al. (2019a) found that 34.5% of participants were willing to pay more for a partially automated car than for a car without partial automation, and this proportion increased to 42.4% when probed about a fully automated car. Interestingly, this figure is in contrast to that found in the international survey conducted by Kyriakidis et al. (2015), with 78% of their 5,000-respondent sample reporting they would pay extra for a fully automated vehicle technology to be installed in their vehicle.

This discrepancy between the two studies may be due, in part, to sampling differences; for example, the median age (30 years) of respondents was considerably lower in the Kyriakidis et al. (2015) study than in other studies (e.g., Cunningham et al., 2019a; *Mdn* age = 44 years). As noted earlier in this chapter, younger people appear to be more receptive to AV technology (and therefore may be increasingly willing to pay for it). In addition, a number of recent U.S. survey studies suggest that significant proportions of people were not willing to pay any extra for a fully AV over the price of a conventional vehicle (with no or minimal autonomous capabilities) (e.g., 59%; Bansal & Kockelman, 2017). Together, this research highlights that, while there seemingly is a considerable market for AV technology, there remain significant proportions of the public reluctant to pay for the technology and the capabilities it may afford.

The next empirical question that derives from this line of research relates to precisely how much more people are willing to pay for AV technology. In their survey study, Schoettle and Sivak (2014a) found that 25% of U.S. participants were willing to pay at least $US 2,000 more for fully autonomous capabilities than for a conventional car, with 10% willing to pay at least $US 5,800 more. Their findings also showed that 10% of U.K. and Australian respondents were willing to pay at least $US 5,130 and $US 9,400 extra, respectively (Schoettle & Sivak, 2014a). Bansal and Kockelman (2017) surveyed 2,167 U.S. respondents and estimated the average WTP to add full vehicle automation to their current vehicle was $US 5,857; which increased considerably to $US 14,196 once respondents who were not willing to pay anything for vehicle automation were excluded from their analysis. In an online study of 347 Texans, Bansal, Kockelman, and Singh (2016) compared the WTP for partial- and full vehicle automation, with findings suggesting that people perceive the monetary value of full vehicle automation (WTP = $US 7,253) to be considerably greater than partial automation (WTP = $US 3,300).

These findings are supported by a recent purchase discrete choice experiment of 1,260 U.S. respondents, demonstrating that the average household is willing to pay more for full vehicle automation ($US 4,900) than for partial automation ($US 3,500) (Daziano, Sarrias, & Leard, 2017). In their survey of 5,089 Australians, Cunningham et al. (2019a) found respondents (who were willing to pay more for vehicle automation) were willing to pay a median value of $AU 5,000 (~$US 3,870) more for full vehicle automation than for their current vehicle. This value lies lower than estimates of WTP derived from U.S. studies (e.g., $US 4,900–7,253; Daziano et al., 2017; Bansal et al., 2016). Moreover, the *mean* WTP value found in Cunningham et al. (2019a), $AUD 14,919 (~$US 11,250), is roughly 21% less than the corresponding estimated in a recent large U.S. study ($US 14,196; Bansal & Kockelman, 2017). Although median and mean WTP values for full vehicle automation appear to be lower than those deriving from U.S. studies, there appears to be a small proportion of Australians willing to pay relatively large amounts for the luxury of full automation, with 9% of respondents willing to pay over $US 30,000 more than their current car. Interestingly, corresponding estimates from a large international survey found that only 5% of respondents were willing to pay this amount (Kyriakidis et al., 2015). Together, these findings suggest that there is a market for automated driving technologies across different countries but that there is heterogeneity in WTP estimates for vehicle automation, with large groups of respondents that are not willing to pay any more for vehicle automation.

5.6 ACCEPTANCE OF AVs AFTER EXPERIENCING THE TECHNOLOGY

Level 5 AVs are not yet ubiquitous on our roads, and therefore most studies to date examining public opinion of technologies have had to rely on what people *think* or *believe* the technology will be like without actually experiencing it firsthand (i.e. *acceptability*). However, as AV technology continues to be developed and tested, and more people are able to actually experience and use the technology, research can start to examine peoples' opinions of the technology *after* actually using it (i.e., *acceptance*).

Nordhoff et al. (2018) distributed a questionnaire among 384 individuals in Berlin after riding in a conditionally automated electric shuttle bus. Riders believed the shuttle was useful, was fun, and enjoyable to ride in, and that it would become an important part of the existing public transport system. However, riders still expressed concerns about the technology, particularly in relation to it not being as efficient, fast, or easy to use as current transport options. In a more recent, qualitative, study by Nordhoff, de Winter, Payre, van Arem, and Happee (2019), 30 riders were interviewed after a trip on a conditionally automated shuttle bus in Berlin. Results showed that riders tended to have unrealistically high expectations about the capabilities of the shuttle and expected the shuttle to possess a greater level of autonomy than it did (e.g., reacting to obstacles and route finding without reliance on pre-programmed routes). Other comments conveyed by riders were that the shuttle was perceived to be too slow to be of value during daily commutes, provided reduced personal privacy, and 20% expressed a preference for having a human supervise the operation of the shuttle over unsupervised full automation. However, on a more positive note, riders tended to support the use of automated shuttles as a form of public transport and endorse the possibility of being relieved of manual driving demands (Nordhoff et al., 2019).

A study by Xu et al. (2018), which extends this line of research, examined 300 student riders' attitudes and opinions towards AV technology after riding in a conditionally automated car on a closed test track. In this study, constructs such as trust, perceived usefulness, safety, and intention to use the technology were gauged through a questionnaire before and after the experience in the vehicle. The authors found that levels of trust, perceived usefulness and perceived ease of use, but not the intention to use the technology, increased significantly (although effects were modest) after the ride (Xu et al., 2018). Like Nordhoff et al. (2018), Xu and colleagues (2018) suggest that firsthand experience with the AV did not have a large effect on trust, intention to use the technology, etc., due to riders likely holding unrealistically high expectations of the technology.

5.7 CONCLUSION

The overarching aim of this chapter was to review what is known, internationally, about public opinion and attitudes towards AVs. First, we highlighted some different benefits and concerns the public has associated with the introduction and use of AVs, demonstrating that, although the public tends to be largely accepting AV technology (and believe many AV-related benefits will come to fruition), they still have a number of concerns about the technology (e.g., in relation to safety). Second, we illustrated that AV acceptability may, in part, vary as a function of different sociodemographic variables, such as gender and age. Research suggests that males, and younger individuals, are particularly receptive to the introduction of AVs. Third, in addition to varying AV acceptance at the individual level, public opinion and attitudes towards the technology differ somewhat across countries/regions. This latter information can help us understand which countries/regions may be most ready and prepared for the introduction of AVs from a public acceptance point of view. Fourth, we examined literature on WTP for AV technology, which revealed that, although there appears to be a market for the technology, there are still significant proportions of the public which may be hesitant to pay extra for the technology. Moreover, actual WTP values for the technology seem to vary widely. Finally, we reviewed the relatively smaller body of literature on AV *acceptance* to gain an understanding of how actually experiencing riding in a conditionally or highly automated vehicle may influence opinions and attitudes towards the technology. This literature suggests that, with current automated driving capabilities, experiences riding in such vehicles may have a positive effect on peoples' opinions towards the technology, but in other situations may have a negative effect (e.g., relating to issues of efficiency or usefulness compared to current transport options). The findings from this emerging literature seem to vary from study to study.

The findings reported in this chapter have two main limitations. First, much of the literature reviewed here was survey-based, correlational, and cross-sectional, preventing us from drawing conclusions regarding temporal associations between variables and relationships of causality (e.g., younger age does not cause higher acceptability of AVs). Future research in this area will be greatly benefited from the employment of prospective designs that enable researchers to track how AV-related opinions and attitudes may change over time, and different variables that may influence these changes (e.g., does acceptability of the technology change as people get older or as people become increasingly familiar with the technology?) Second, highly and fully automated passenger vehicles (SAE Levels 3–5) are not yet commercially available in large numbers. Consequently, measurement of public acceptability in most studies to date has been based on people having to imagine how such vehicles might operate in the future. Future research will need to focus on *acceptance* of such vehicles, after people have had direct exposure to them, and to compare the findings with those obtained prior to exposure—to better understand to what extent measures of acceptability are predictive of measures of acceptance.

A proper understanding of public opinion about automated driving systems can benefit different stakeholders in society in a variety of ways: benefit governments, for example, in making future planning and investment decisions; benefit industry, for example, in helping technology developers design and refine their products in response to the perceived needs of end users, locally and overseas; and benefit research organizations in identifying new directions for research.

ACKNOWLEDGEMENT

Sections of this Chapter derive, in part, from Cunningham et al. (2019a; b).

REFERENCES

Abraham, H., Lee, C., Brady, S., Fitzgerald, C., Mehler, B., & Coughlin, J. F. (2017). Autonomous vehicles and alternatives to driving: Trust, preferences, and effects of age. *Proceedings of the Transportation Research Board 96th Annual Meeting*. Washington, D.C.: Transportation Research Board.

Adell, E., Várhelyi, A., & Nilsson, L. (2014). The definition of acceptance and acceptability. In M. A. Regan, T. Horberry, & A. Stevens (Eds.), Driver Acceptance of New Technology Theory, Measurement and Optimisation (pp. 11–21). Surrey, UK: Ashgate.

Bansal, P. & Kockelman, K. M. (2017). Forecasting Americans' long-term adoption of connected and autonomous vehicle technologies. Transportation Research Part A: Policy and Practice, *95*, 49–63.

Bansal, P., Kockelman, K. M., & Singh, A. (2016). Assessing public opinions of and interest in new vehicle technologies: An Austin perspective. Transportation Research Part C: Emerging Technologies, *67*, 1–14.

Becker, F. & Axhausen, K.W. (2017). Literature review on surveys investigating the acceptance of automated vehicles. *Transportation, 44*, 1293–1306.

Cunningham, M. L., Regan, M. A., Horberry, T., Weeratunga, K., & Dixit, V. (2019a). Public opinion about automated vehicles in Australia: Results from a large-scale national survey. *Transportation Research Part A, 129*, 1–18.

Cunningham, M. L., Regan, M. A., Ledger, S. A., & Bennett, J. M. (2019b). To buy or not to buy? Predicting willingness to pay for automated vehicles based on public opinion. *Transportation Research Part F, 65*, 418–438.

Daziano, R. A., Sarrias, M., & Leard, B. (2017). Are consumers willing to pay to let cars drive for them? Analyzing response to autonomous vehicles. Transportation Research Part C: Emerging Technologies, *78*, 150–164.

Dingus, T. A., Guo, F., Lee, S., Antin, J. F., Perez, M., Buchanan-King, M., & Hankey, J. (2016). Driver crash risk factors and prevalence evaluation using naturalistic driving data. *Proceedings of the National Academy of Sciences, 113*(10), 2636–2641.

Fagnant, D. J. & Kockelman, K. M. (2015). Preparing a nation for autonomous vehicles: Opportunities, barriers and policy recommendations for capitalizing on self-driven vehicles. *Transportation Research Part A: Policy and Practice*, 77, 1–20.

Hauk, N., Huffmeier, J., & Krumm, S. (2018). Ready to be a silver surfer? A meta-analysis on the relationship between chronological age and technology acceptance. *Computers in Human Behavior, 84*, 304–319.

Kyriakidis, M., Happee, R., & de Winter, J. C. (2015). Public opinion on automated driving: Results of an international questionnaire among 5000 respondents. *Transportation Research Part F: Traffic Psychology and Behaviour*, 32, 127–140.

Large, D. R., Burnett, G., Morris, A., Muthumani, A., & Matthias, R. (2017). A longitudinal simulator study to explore drivers' behaviour during highly-automated driving. International Conference on Applied Human Factors and Ergonomics (pp. 583–594). Berlin: Springer.

Litman, T. (2018). *Autonomous Vehicle Implementation Predictions: Implications for Transport Planning*. Victoria, BC: Victoria Transport Policy Institute.

Nordhoff, S., de Winter, J., Madigan, R., Merat, N., van Arem, B., & Happee, R. (2018). User acceptance of automated shuttles in Berlin-Schöneberg: A questionnaire study. Transportation Research Part F: Traffic Psychology and Behaviour, *58*, 843–854.

Nordhoff, S., de Winter, J., Payre, W., van Arem, B., & Happeee, R. (2019). What impressions do users have after a ride in an automated shuttle? An interview study. *Transportation Research Part F, 63*, 252–269.

Payre, W., Cestac, J., & Delhomme, P. (2014). Intention to use a fully automated car: Attitudes and a priori acceptability. *Transportation Research Part F: Traffic Psychology and Behaviour, 27*, 252–263.

Regan, M. A., Horberry, T., & Stevens, A. (2014). *Driver Acceptance of New Technology: Theory, Measurement and Optimisation*. Surrey, UK: Ashgate.

Schoettle, B. & Sivak, M. (2014a). A Survey of Public Opinion about Autonomous and Self-Driving Vehicles in the US, *the UK*, and Australia. Ann Arbor, MI: University of Michigan Transportation Research Institute.

Schoettle, B. & Sivak, M. (2014b). *Public Opinion about Self-Driving Vehicles in China, India, Japan, the US, the UK and Australia*. Ann Arbor, MI: University of Michigan Transportation Research Institute.

Singleton, P. A. (2019). Discussing the "positive utilities" of autonomous vehicles: Will travellers really use their time productively? *Transport Reviews, 39*(1), 50–65.

Society of Automotive Engineers (SAE). (2014). *Taxonomy and Definitions for Terms Related to On-Road Motor Vehicle Automated Driving Systems*. Warrendale, PA: SAE International.

Underwood, S. (2014). Automated vehicle forecast: Vehicle Symposium Survey. *Automated Vehicles Symposium 2014*. Burlingame, CA: AUVSI.

Vallet, M. (2013). *Survey: Drivers Ready to Trust Robot Cars?* Retrieved from https://www.foxbusiness.com/features/survey-drivers-ready-to-trust-robot-cars

Xu, Z., Zhang, K., Min, H., Wang, Z., Zhao, X., & Liu, P. (2018). What drives people to accept automated vehicles? Findings from a field experiment. *Transportation Research Part C: Emerging Technologies, 95*, 320–334.

6 Workload, Distraction, and Automation

John D. Lee
University of Wisconsin-Madison

Michael A. Regan
University of New South Wales

William J. Horrey
AAA Foundation for Traffic Safety

CONTENTS

Key Points 108
6.1 Introduction 108
6.2 Workload, Distraction, and Performance 109
    6.2.1 Workload 109
        6.2.1.1 Workload and the Yerkes–Dodson Law 110
        6.2.1.2 Active Workload Management 112
    6.2.2 Distraction 112
        6.2.2.1 Distraction and Types of Inattention 113
        6.2.2.2 The Process of Driver Distraction 113
    6.2.3 Driver Workload and Driver Distraction 114
    6.2.4 Summary 114
6.3 Types of Automation and Workload Implications 114
    6.3.1 Effect of Different Levels of Automation on Workload 114
    6.3.2 The Interaction of Distraction, Workload, and Automation 117
        6.3.2.1 Automation Creating Distraction Directly 118
        6.3.2.2 Automation Creating Distraction Indirectly 118
        6.3.2.3 The Interaction of Other Mechanisms of Inattention,
                Workload, and Automation 120
    6.3.3 Summary 121
6.4 Managing Workload and Distraction in Automated Vehicles 121
6.5 Conclusion 122
References 123

KEY POINTS

- Automation stands to change the fundamental aspects of the driving task as subtasks are added or subtracted, such that the role and responsibilities of the driver will be transformed in the coming years.
- Various types of automation promise to reduce workload by reducing the number of tasks required of the driver; however, automation also changes existing tasks and adds new ones, which can increase workload.
- Automation can be problematic or "clumsy" if it reduces workload in already low workload environments, but increases workload during high workload periods.
- Driver distraction, workload, and automation may interact with one another in ways that can decrease safety. On the one hand, automation may demand drivers' attention at inopportune moments creating a distraction, which can increase the driver's workload that can in turn increase risk. Alternatively, automation may reduce drivers' workload, inducing drivers to engage more often and more deeply in non-driving tasks, thereby creating a distraction and increasing risk.
- System design should consider: (1) how workload is created and re-distributed according to changing driver tasks and roles; (2) avoiding abrupt workload transitions and "clumsy" automation; (3) that monitoring automation is effortful; and (4) support of strategic workload management.

6.1   INTRODUCTION

New technologies have entered the market, and are emerging, which are capable of supporting the driver to perform (or of automating) some, or all, of the functional activities performed traditionally by human drivers (e.g., Galvani, 2019): route finding; route following; velocity and steering control; crash avoidance; compliance with traffic laws; and vehicle monitoring (Brown, 1986). These new technologies are described in Chapters 1 and 2 of this book, and fall into two broad categories: those that support the driver (Driver Support Features (DSF); Levels 1–2) and those that automate—some of the time (Level 3)—or all of the time (Levels 4 and 5) human driving functions (Automated Driving Features (ADF); SAE International, 2018).

In spite of the promise and excitement surrounding these technologies many important human factors considerations have been raised inasmuch as these new technologies stand to impact driver behavior, performance, and safety—as evidenced by the many chapters comprised in this Handbook. Indeed, automation stands to change fundamental aspects of the driving task as subtasks are added or subtracted, such that the role and responsibilities of the driver will be transformed in the coming years (e.g., this Handbook, Chapter 8). This, in turn, will impact driver situation awareness (e.g., this Handbook, Chapters 7, 21) as well as in-vehicle behaviors (e.g., this Handbook, Chapter 9). By extension—or even as a precursor to some of these outcomes—automation will also impact driver workload. The extent of this impact will vary by the type and/or level of automation as well as various situational and environmental factors. For example, various types of automation promise to reduce workload by reducing the number of tasks required of the driver; however, automation also changes existing tasks and adds new ones, which can increase workload (Cook, Woods, McColligan, & Howie, 1990).

This chapter describes the concept of workload and its implications for performance, drawing from existing studies and theories from the human factors literature. The role and impact of driver distraction is also considered in the context of workload. In subsequent sections, the interaction between different levels of automation on the one hand and driver workload and distraction on the other hand is described. Lastly, approaches to manage workload and distraction are discussed in the context of automation.

6.2   WORKLOAD, DISTRACTION, AND PERFORMANCE

6.2.1   WORKLOAD

Workload has long been a hallmark of human factors research across many applied domains, including aviation and aerospace, process control, medicine, education, and training, as well as driving (e.g., O'Donnell & Eggemeier, 1986; Hendy, Liao & Milgram, 1997; Moray, 1979; 1988). Workload has been defined in different ways and, depending on the application, has been linked to different stages of human information processing and action, including cognitive, visual, and manual components (e.g., Lee, Wickens, Liu, & Boyle, 2017). That is, workload can be incurred by the physical aspects of a given task (e.g., the effort required to move a package from a conveyor belt to a pallet) as well as by the mental processing of information related to a task (e.g., the effort required to find a location on a cluttered map and plan out a route to that location).[1] With these distinctions in mind, Proctor and Van Zandt (1994) defined mental workload, relating to the non-physical aspects of workload, as "the amount of mental work or effort necessary to perform a task" (p. 199). Similarly, Gopher and Donchin (1986) defined mental workload as "the difference between the capacities of the information-processing system that are required for task performance to satisfy expectations and the capacity available at any given time" (p. 41; see also Hart & Staveland, 1988).

There are two important points to note regarding workload: (1) most definitions relate to the subjective experience of an individual in light of a given task and (2) workload is not considered performance per se. The first of these (point #1) underscores the importance of individual differences in capacity and performance criteria. The notion of limited attentional resources espoused by Kahneman (1973) elegantly illustrates the importance of capacity. Given the same task and same situation, two individuals might experience markedly different workload, reflecting differences in capacity or differences in how much capacity they choose to invest to achieve a certain level of performance. Similarly, the same individual could experience different levels of workload while performing the same task on separate occasions—especially as other situational factors might vary (e.g., Helton, Funke, & Knott, 2014). That is, individual capacities may vary over time (e.g., Young & Stanton 2002a; b). Moreover, extended exposure to a low demand situation can lead to diminished resources

---

[1] In this chapter, we focus primarily on the mental and cognitive aspects of workload, although the physical components are relevant in some sections (e.g., Table 6.2).

available to respond when demands increase. These diminished resources can under-mine performance when they coincide with a high demand situation. For this and other reasons, workload is a construct that is often associated with performance or, perhaps more specifically, can be mapped onto an individual's capacity or potential to perform in a given context (point #2). This is described further in Section 6.2.1.1.

There are a number of approaches that can be used to assess or quantify the work-load demands associated with a given task (e.g., Wierwille & Eggemeier, 1993). These tools and/or approaches can also distinguish between different types of workload, such as the cognitive, visual, and manual elements of a task (or along similar categoriza-tions). Some tasks might incur high levels of manual workload, but generally require little thought and so do not impose significant cognitive load (e.g., lifting a heavy pack-age). Other tasks might have little or no physical output, but impose significant cogni-tive load (e.g., performing mental rotations of 3-D objects). It is important to consider these nuances, especially for tasks that might appear to be effortless. For example, when automation relieves the burden of performing a task and requires the person do nothing but monitor it, one might think workload would be very low. And, indeed, the physical workload would be low. However, the cognitive workload associated with monitoring and vigilance tasks can be quite high (e.g., Warm, Parasuraman, & Matthews, 2008).

6.2.1.1   Workload and the Yerkes–Dodson Law

Over a century ago, Robert Yerkes and John Dodson described the relationship between stimulus strength and task performance in mice (Yerkes & Dodson, 1908). The Yerkes–Dodson law, as it came to be known, states that performance on a given task increases with arousal up to a certain point, but decreases beyond this point (as depicted in the inverted U functions shown in Figure 6.1). The precise shape of the curve varies as a function of several factors, including task complexity or difficulty; however, the key observation is that there are optimal and sub-optimal levels of arousal when considering performance. Performance can suffer when arousal is too high (i.e., overload), but it can likewise suffer when it is too low (i.e., under-load). Although the original Yerkes–Dodson law related to arousal, the scientific literature suggests that fluctuations in workload appear to have a similar relationship with performance—with performance being optimal when workload is neither too low nor too high, but as a result of different mechanisms (i.e., different from arousal).

In the context of Figure 6.1 and under the assumption that workload follows a similar profile, performance is largely a reflection of the efficacy of an individual's allocation

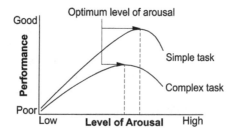

FIGURE 6.1   Depiction of the Yerkes–Dodson law for simple and complex tasks. (From Lee et al., 2017.)

of resources towards a given task (cf. discussion of "capacity" in the definitions of workload, Section 6.2.1). Within the optimal range, the individual is presumed to effectively allocate these cognitive (attentional) and physical resources towards acquiring and processing task-related information, integrating this information and deciding on a course of action, and then acting on the information (e.g., Wickens & Hollands, 2000). At higher levels of workload, the individual may no longer be able to effectively allocate appropriate resources to the task, due to the effect of increased stress and anxiety or due to the competing demands of other activities (see related discussion of distraction in Section 6.2.2). At low levels of workload, appropriate mobilization of resources is also impaired, but this is attributed to a shrinking of the momentary capacity of the individual (i.e., the individual simply does not have the resources necessary to meet the demands of the task, such as in cases of fatigue; e.g., Kahneman, 1973).

With respect to resources, it is important to note that the mobilization of more resources towards a task does not always translate to performance gains. Norman and Bobrow (1975), in describing performance-resource functions (see Figure 6.2), distinguished between tasks that are resource-limited and those that are data-limited. Performance on some tasks (e.g., Task A in Figure 6.2) will improve indefinitely as more resources are allocated to them. These are defined as resource-limited tasks, as the only limit on performance is the availability of resources (i.e., a capacity issue). In contrast, performance on other tasks will improve only to a certain point, beyond which the allocation of more resources will not have any significant impact (e.g., Task B in the figure). These data-limited tasks are constrained by properties of the tasks themselves. The pace at which task-relevant information is delivered can represent a data limit for such tasks.

Often workload does not depend on the demands of a single task but is influenced by multiple tasks competing for attention. Multiple tasks not only impose additional load but also impose the requirement to switch attention and resources between the

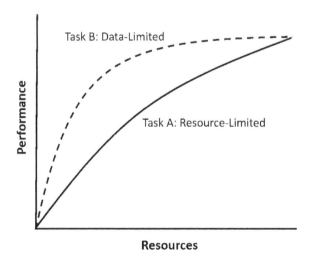

FIGURE 6.2 Performance-resource functions for a resource-limited task (Task A, solid line) and a data-limited task (Task B, dashed line). (Adapted from Wickens & Hollands, 2000.)

tasks. Switching between tasks demands effort that exceeds what might be expected from the demands of either task considered individually (Trafton & Monk, 2007). When multiple tasks are involved, people must adopt strategies to distribute their limited resources to the tasks at hand. Vehicle automation, especially Level 3 automation, will likely magnify workload associated with task switching as automation provides the opportunity to do other tasks, but also requires that drivers monitor automation, and occasionally return to manually controlling the car as well. Active workload management, described in the next section, is a process by which people attempt to prioritize tasks in an attempt to balance attention, workload, and performance. Importantly, these attempts are not always successful as sometimes certain tasks can distract attention away from—in the current context—activities critical for safe driving (discussed in Section 6.2.2).

6.2.1.2   Active Workload Management

Notably, people do not passively respond to workload demands placed on them. Rather, for a given task, or set of tasks, people will attend selectively to tasks and information as a means of strategic workload management and task scheduling (Lee, 2014; Raby & Wickens, 1994). Oftentimes, and depending on their level of workload and arousal, this might come in the form of load or task shedding (Sirevaag et al., 1993; Schulte & Donath, 2011). That is, they abandon their attempt to perform two concurrent tasks, in order to focus their limited resources on a single task—presumably maintaining adequate levels of performance on that task (e.g., a driver might stop trying to adjust the radio when traffic demands increase). Alternatively, individuals might alter their behavior as a means of reducing the demands (and, by extension, their workload) in a given situation (e.g., Fuller, 2005); for example, by slowing down on a highly curved road, a driver will increase the time available to maneuver the curves and reduce their experienced workload. Experts perform well in part because they correctly manage what tasks to do when by prioritizing and scheduling activities and subtasks (e.g., Raby & Wickens, 1994; Tulga & Sheridan, 1980).

Importantly, individuals—drivers in particular—do not always effectively manage their workload, given the opportunity (e.g., Horrey & Lesch, 2009). Other task characteristics or motivational factors can influence the allocation of resources. Distraction from non-driving related activities is one salient example where task management and prioritization may fail (e.g., Regan, Lee, & Young, 2009; this Handbook, Chapter 9). That is, opting to engage in non-driving related tasks can reduce capacity for driving-related tasks and performance can suffer as a result.

6.2.2   Distraction

Driver distraction has been defined as "the diversion of attention away from activities critical for safe driving toward a competing activity" (Lee, Young, & Regan, 2009, p. 34). Chapter 9 discusses the topic of distraction as it relates to automated vehicles, to engagement in non-driving activities, and to the impact of such engagement on the takeover timing and quality. In this chapter, we build on that treatment of the topic and contemplate, more broadly, how driver distraction, and the way we conceptualize it, is likely to change during the journey from partially- to fully automated (self-driving) vehicles.

6.2.2.1 Distraction and Types of Inattention

It is important to note that distraction is only one of several mechanisms of inattention that may result in insufficient or no attention to activities critical for safe driving in automated vehicles. For example, Regan, Hallett, and Gordon (2011) have proposed a taxonomy of inattention that is useful in conceptualizing these different mechanisms of inattention (summarized in Table 6.1). Although the focus in subsequent sections is largely on distraction (driver diverted attention), these other forms of inattention should also be considered in the context of automated vehicles.

6.2.2.2 The Process of Driver Distraction

A typical episode of driver distraction (e.g., driver diverted attention, Table 6.1) can be thought of as involving a number of high-level processes (Regan & Hallett, 2011; Regan et al., 2009): something that triggers distraction; a diversion of attention away from activities critical for safe driving toward a source of distraction; interaction with the source of distraction (the competing activity); and some degree of interference with activities critical for safe driving generated by this interaction. The degree of interference may be moderated to some extent by individual characteristics, the demands of driving, the demands of the secondary activity competing for attention, and the ability of the driver to self-regulate their driving behavior in anticipation of, and in response to, distraction (Young, Regan, & Lee, 2009). In Section 6.3.2, we

---

TABLE 6.1

Mechanisms of Inattention (from Regan et al., 2011)

Mechanism of Inattention	Definition
Driver restricted attention	"Insufficient or no attention to activities critical for safe driving brought about by something that physically prevents (due to biological factors) the driver from detecting (and hence from attending to) information critical for safe driving" (p. 1775); e.g., fatigue or drowsiness.
Driver neglected attention	"Insufficient or no attention to activities critical for safe driving brought about by the driver neglecting to attend to activities critical for safe driving" (p. 1775); e.g., due to faulty expectations.
Driver misprioritized attention	"Insufficient or no attention to activities critical for safe driving brought about by the driver focusing attention on one aspect of driving to the exclusion of another, which is more critical for safe driving" (p. 1775); e.g., a driver who looks over their shoulder while merging and misses a lead vehicle braking.
Driver cursory attention	"Insufficient or no attention to activities critical for safe driving brought about by the driver giving cursory or hurried attention to activities critical for safe driving." (p. 1776); e.g., a driver on the entry ramp to a freeway who is in a hurry, does not complete a full head check when merging, and ends up colliding with a merging car.
Driver diverted attention	"The diversion of attention away from activities critical for safe driving toward a competing activity, which may result in insufficient or no attention to activities critical for safe driving" (synonymous with driver distraction; p. 1776); e.g., a driver reading a text message on a cell phone.

give consideration to the relationship between vehicle automation and these high-level processes, drawing in part on a previous review of the topic (Cunningham & Regan, 2018), and building on some of the findings reported in Chapter 9.

6.2.3 DRIVER WORKLOAD AND DRIVER DISTRACTION

It is important to distinguish briefly between driver workload and driver distraction, as the relationship between the two is not well articulated in the literature (Schaap, van der Horst, van Arem, & Brookhuis, 2013). As noted in the section above, driver workload may be conceptualized as the information processing capacity (or resource) that is used to meet the demands of driving (including activities critical for safe driving). Driver distraction may interact with driver workload in different ways. For example, it has the potential to increase driver workload, as in the case of a persistently flashing low-fuel warning light. As noted by Schaap et al. (2013), however, a driver's motivation to protect the primary task of safe driving may influence the resources available to meet the demands of driving (see Section 6.2.1.2); and, in turn, the amount of resources available for engagement in non-driving activities. Hence, the impact of driver distraction on driver workload might be minimal, as in a driver who chooses not to read an advertising billboard that has involuntarily attracted his or her attention. In a perverse way, driver workload might actually be reduced by distraction; for example, if the distracted driver decides to withdraw processing resources from the driving task in order to focus on a non-driving activity. Such disengagement from driving might reduce workload (Lee, 2014).

6.2.4 SUMMARY

The concepts of workload, task management, distraction, and performance have been studied in many different domains and settings. Various theories in the human factors literature shed some light on how these concepts interact with one another—especially where performance and safety are concerned. In the following section, we discuss how varying degrees of automation impacts these relationships. It is important to note that the implementation of automation will influence how performance is conceived and operationalized (cf. this Handbook, Chapters 8, 9); driver performance will not necessarily be reflected by the same measures assumed in manual conditions and joint-system performance (i.e., driver plus system performance) might become more relevant. Different types of automation can dramatically affect workload and distraction, and associated performance measures.

6.3 TYPES OF AUTOMATION AND WORKLOAD IMPLICATIONS

6.3.1 EFFECT OF DIFFERENT LEVELS OF AUTOMATION ON WORKLOAD

Having defined the general relations between workload, distraction, and performance, the question is how the different types of DSFs and higher-level automation can affect workload. It stands that the cognitive, visual, and manual workload required of the driver will change dramatically as different levels of automation and DSF enter the vehicle. However, the basic laws underlying the relation between workload and performance (i.e., the Yerkes–Dodson law, Figure 6.1 and performance resource functions, Figure 6.2) should presumably remain the same. Understanding how automation moves driver workload away from the optimal level, in either direction, will be critical. Clearly, this is most pressing in situations where drivers maintain some degree of responsibility in the human–automation system (e.g., SAE Levels 1–3).

Much of the variation in driver workload in light of automation will be a result of changes in the underlying tasks as well as the redistribution of responsibility from the driver to the system or vice versa (this Handbook, Chapter 8). For example, for Level 1 and 2 automation, vehicle control (steering and/or speed control) might be offloaded to the system, such that the driver can allocate those resources to other activities, such as maintaining vigilance to monitor the traffic environment as well as the system function (which becomes a more prominent task for drivers using such a system). In some cases, the net impact on workload might vary: a reduction of workload associated with some of the driving subtasks can lead to increased workload on other subtasks. The net impact on workload of changes in the level of automation can also manifest itself by changes in workload across the stages of information processing (e.g., Figure 6.3; Yamani & Horrey, 2018).

Building from the examples provided in Figure 6.3, Table 6.2 illustrates the impact of a variety of automation technologies on specific driving subtasks. As shown, the set of driving subtasks, here based on Brown (1986), incurs a certain amount of workload in normal driving conditions. As new technologies are implemented (e.g., DSFs, Levels 1–2), workload for each of the subtasks might increase,

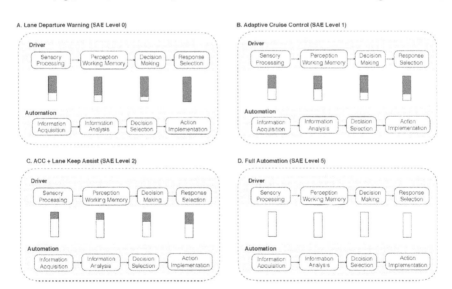

FIGURE 6.3 Variations in driver workload and system responsibility across different levels of automation (from Yamani & Horrey, 2018). Dark portion of the bars represent the driver's contribution to the demands of the task at a particular stage of information processing; the white portion of the bars represents the contribution of the system. ACC = Adaptive cruise control. © Inderscience (used with permission).

TABLE 6.2

Workload Variation across Different Driving Subtasks by Different Levels of Automation. For Illustrative Purposes, Hypothetical Ratings of Workload Range from 0 to 10 Are Provided in Each Cell. Arrows Denote Directional Changes from Manual Driving Conditions

Driving Task[1]	Manual Driving			Level 0: Blind Spot Monitoring System			Level 1: ACC			Level 2: (ACC + LKA)		
	Vis	Cog	Man	Vis	Cog	Man	Vis	Cog	Man	Vis	Cog	Man
Route finding	6	5	2	6	5	2	6	5	2	6	5	2
Route following	5	6	6	5	6	6	5	6	6	5	6	6
Steering and velocity control	5	3	8	5	3	8	↓4	↓2	↓5	↓2	↓1	↓1
Collision avoidance	8	6	7	↓6	↓4	↓5	8	6	↓5	8	6	↓4
Rule compliance	3	3	3	3	3	3	3	3	3	3	3	3
Vehicle/system monitoring	3	3	0	3	3	0	↑5	↑5	0	↑8	↑8	0

[1] After Brown (1986). Vis = visual workload; Cog = cognitive workload, Man = manual workload, ACC = adaptive cruise control, LKA = lane keeping assist.

decrease (in varying degrees), or remain unchanged. For example, the addition of a blind spot monitoring system can reduce the visual, cognitive, and manual demands associated with collision avoidance by providing drivers with additional information regarding the proximity of nearby vehicles. In examining these relationships, it is important to consider where decreases in workload on one subtask are associated with increases on another task. For example, the implementation of adaptive cruise control (ACC) and lane-keeping assist (LKA) are logically associated with reduced demands for steering and velocity control, but with increased demands for vehicle and system monitoring when these systems are in operation.

As noted, the information portrayed in Table 6.2 might be considered to reflect routine operations. However, it is important to consider the impact of systems on workload in other use cases, such as in takeover situations or emergency scenarios. The steady-state distribution of workload might suggest that automation might make routine tasks much easier, but the timeline associated with emergency scenarios might show substantially higher workload that drivers experience without the automation. That is, automation can make easy tasks easier and hard tasks harder (Bainbridge, 1983; Cook et al., 1990). This timeline approach to quantifying task demands according to different mental resources is described elsewhere (e.g., Wickens, 2002).

In considering the impact of different types and levels of automation on driver workload, it is also important to discuss important interaction effects with various individual and situational factors. As automation will not affect every driver in the same manner, the same system might have differential impacts on driver workload depending on the specific traffic context. Automation can be problematic or "clumsy" if it reduces workload in already low workload environments, but increases it in demanding environments (Woods, 1996; Cook et al., 1990). For example, ACC, when used on lightly trafficked highways, can lead to "passive fatigue" (which may lead to reduced attention to activities critical for safe driving) and may encourage drivers to engage in secondary activities. Saxby, Matthews, Warm, Hitchcock, and Neubauer (2013) found that drivers exhibited higher levels of passive fatigue and lower driving task engagement, alertness, and concentration while monitoring an automated driving system. Further, Jamson, Merat, Carsten, and Lai (2013) found that drivers were more likely to engage with in-vehicle entertainment systems when driving with automated systems. It follows that engagement in these activities can exacerbate the workload peaks caused by clumsy automation as well as the effective response to takeover requests or circumstances. That is, Saxby et al. (2013) and others have shown that drivers have slower responses in unexpected takeover situations when monitoring automation compared with similar traffic situations encountered while driving in manual conditions.

Clumsy automation can also increase workload during high workload periods. A common example in aviation is when pilots are required to program flight management systems during landing preparation (see also, this Handbook, Chapter 21). Similarly, automation surprises can lead to spikes in workload which may compromise the performance of activities critical for safe driving. Automation surprises can be manifest in situations where the system reaches its limitations (e.g., the boundaries of its operational design domain) and the driver is either unprepared or unknowing (see this Handbook, Chapters 3, 9, 21). Surprises can also occur when the user misinterprets the system mode or fails to use the system properly, especially under changing traffic conditions or driving goals. For example, a Level 2 system might work perfectly well on a freeway with limited traffic (lower levels of workload) but could make merging for an exit more challenging if the driver forgets to disengage the system. In this case, the resulting increase in workload would be much greater than would be encountered in a routine, non-automated, lane exit. Workload spikes can undermine a driver's effective allocation of resources and workload management strategy.

Automation can both increase and decrease these different types of workload and do so over both very short and long time horizons. The first question addressed in this section was how the various types of workload will be influenced by the different categories of automation and advanced driver assistance systems. In the following section, a more in-depth discussion is provided concerning the interaction of driver distraction, workload, and automation and the effect of that interaction on safety.

6.3.2 THE INTERACTION OF DISTRACTION, WORKLOAD, AND AUTOMATION

As vehicles become increasingly assistive and automated, it is important to understand how distraction will manifest itself with respect to activities critical for safe driving—even as these activities themselves change. As noted earlier, distraction can arise from the automated systems themselves, leading to an increase in workload.

On the other hand, the reduction of workload gleaned from the offloading of some of the driving responsibilities to automation can also lead to increased engagement in non-driving related tasks. Such direct and indirect effects are explored in this section.

6.3.2.1 Automation Creating Distraction Directly

Factors that trigger driver distraction in current-generation vehicles (SAE Levels 0–1) include the state of the driver (e.g., whether they are fatigued or bored), their need to communicate with others and internal stimuli within the mind itself (that, for example, trigger mind wandering). It is likely that these same factors will continue to induce distraction in vehicles equipped with DSFs (Levels 0–2). However, with the emergence of these technologies comes the possibility that the technologies themselves may become a trigger for driver distraction. There is already evidence that automation actions and alerts that are unexpected, through lack of training, lack of situational awareness, or some other mechanism, may create "automation surprises" (e.g., Hollnagel & Woods, 2005), and, in doing so, trigger a diversion of attention away from activities critical for safe driving being performed by the driver. Even routine alerts and indicators that highly automated vehicles produce might draw attention away from the road at inopportune moments.

In vehicles equipped with SAE Level 3 automation, which is capable of driving the vehicle in limited conditions, the automation is considered to be driving the vehicle (SAE International, 2018). The driver is, however, expected to resume control of the vehicle if requested by the vehicle, for example, if the automation fails or if the driver considers that it would be safer to take back control. Here, the frame of reference for distraction is different: the requirement to supervise vehicle automation could itself become a source of distraction (Hancock, 2009). The requirement to take back control of the vehicle, when this arises, will require the performance of a new activity critical for safe driving not neatly encapsulated by Brown's (1986; Table 6.2) taxonomy of functional driving activities—resumption of vehicle control. This is equally relevant for Level 4 automation; however, the timeframe for such transitions is likely to be prolonged compared to Level 3 (see this Handbook, Chapter 1).

6.3.2.2 Automation Creating Distraction Indirectly

The propensity for drivers to engage in secondary activities when supported by automation is well documented, in both driving simulators (see, for example, Carsten, Lai, Barnard, Jamson, & Merat, 2012; Jamson et al., 2013; this Handbook, Chapter 9) and instrumented vehicles on test tracks (Llaneras, Salinger, & Green, 2013; Dingus et al., 2016), with the propensity to do so being greater for technologies that provide high levels of automation (e.g. SAE Level 2 versus Level 1). In this sense, automation has the potential to induce drivers to exploit attentional resources freed by the automation (e.g., Yamani & Horrey, 2018).

6.4.2.2.1 Effects of Diversion of Attention from Driving Toward a Source of Distraction

The factors described above may trigger a diversion of attention away from activities critical for safe driving toward a competing activity. In vehicles equipped with SAE Level 0–2 automation, the driver is considered to be driving the vehicle and is assisted in performing activities critical for safe driving by a variety of driver support systems (e.g., LKA). By reducing the momentary demands on the driver these systems might encourage drivers to engage in potentially distracting activities. Distraction, when it occurs, may undermine driver performance and, in particular, monitoring, although other technologies within the vehicle may help to mitigate any detrimental impacts this distraction may have (e.g., Automatic Emergency Braking; Tingvall, Eckstein, & Hammer, 2009).

Additionally, a comprehensive analysis of the resumption of control is necessary because drivers will often be called upon to resume control in situations that the automation cannot handle. Consequently, these situations might also be situations that challenge drivers. In addition, the behavior of the automation leading up to the takeover could put the vehicle in a state that is unfamiliar to the driver, further increasing the workload the driver might face. Understanding the specific workload demands of the situation is an important complement to understanding how driver state influences takeover quality.

There is evidence, reviewed in this book (see this Handbook, Chapter 9), that takeover quality is impaired when drivers are distracted; although, interestingly, the execution speed of motor actions to commence the takeover (e.g., to grab the steering wheel or activate the brakes) appears to be somewhat immune. There is also some evidence (e.g., Merat, Jamson, Lai, Daly, & Carsten, 2014) that manual driving performance may be compromised for a substantial period of time after the handover of control has been completed. Both a lack of situational awareness (Zeeb, Buchner, & Schrauf, 2015; de Winter, Happee, Martens, & Stanton, 2014; Samuel, Borowsky, Zilberstein, & Fisher, 2016; this Handbook, Chapter 7) and the out-of-the loop (OOTL) problem (Louw, Kountouriotis, Carsten, & Merat, 2015; Louw, Madigan, Carsten, & Merat, 2017) have been posited as mechanisms by which the quality of handover is degraded by driver distraction (see Cunningham & Regan, 2018 for a review; and this Handbook, Chapter 9).

6.4.2.2.2   Factors Moderating the Impact of Distraction in Automated Vehicles

Just as drivers of manually operated vehicles often self-regulate their behavior to manage distraction (e.g., Tivesten & Dozza, 2014), there is also some evidence they self-regulate in automated vehicles. Jamson et al. (2013), for example, found that drivers with automation support self-regulated their behavior to reduce the likelihood of them diverting attention away from the forward roadway in conditions of higher traffic density.

Large individual differences in the nature and frequency of secondary task engagement in automated vehicles have been found (e.g., Llaneras et al., 2013; Clark & Feng, 2017; this Handbook, Chapter 9). Clark and Feng (2017), for example, investigated the impact of driver age on secondary task engagement during periods of automated driving. Both younger and older drivers were found to engage in secondary activities when supported by automation. However, while younger drivers mostly used an electronic device, older drivers mostly conversed. Körber and Bengler (2014) have reviewed a number of individual differences that may moderate the involvement and impact of driver inattention and distraction in automated vehicles, including complacency and trust in automation, driver experience, and the propensity to daydream and become bored.

6.3.2.3   The Interaction of Other Mechanisms of Inattention, Workload, and Automation

As noted in Section 6.2.2.1, distraction and workload are only two of the mechanisms that may result in insufficient attention to activities critical for safe driving (Regan et al., 2011). And as described in Section 6.3.2, the interaction of distraction and workload with automation is complex. There are other forms of inattention that may interact with workload and automation. Table 6.3 shows the potential effect of Level 3 automation on other mechanisms of inattention. Further study needs to be made of their interaction with workload.

For example, as noted above, as vehicles become increasingly automated, the role of the driver is expected to shift from being that of an active controller of the vehicle to that of a more passive supervisor of the automated driving system (Desmond & Hancock, 2001; see also this Handbook, Chapter 8). There is evidence that this reduction in task engagement can induce passive fatigue (reduced attentional capacity arising from driving task demands which are too low (Desmond & Hancock, 2001;

TABLE 6.3

Impact of Level 3 Automation on Different Mechanisms of Inattention (following Regan et al., 2011)

Mechanism of Inattention	Impact of Level 3 Automation
Driver restricted attention	For vehicles equipped with Level 3 automation operating autonomously, the activity most critical for safe driving will be the requirement to take back of control of the vehicle—when requested by automation or when deemed necessary by the driver. Whilst there is to our knowledge no research on the impact of this mechanism of inattention per se on takeover ability, there is some research, reviewed in Chapter 9, on the impact of sleepiness on takeover ability.
Driver neglected attention	In the present context, one might imagine that, if the requirement to take back control of a Level 3 vehicle operating autonomously is rarely or never encountered, and hence not expected, drivers may over time become inattentive to this activity critical for safe driving.
Driver misprioritized attention	In a vehicle equipped with Level 3 automation operating autonomously there is only one activity critical for safe driving—the requirement for the human to take back control of the vehicle if necessary (and all the sub-activities associated with it). Hence, here, there would seem to be no scope for misprioritized attention.
Driver cursory attention	In a vehicle equipped with Level 3 automation operating autonomously, this form of inattention is likely to manifest itself in the driver giving only cursory attention to elements of the traffic environment that is critical in facilitating the timely and successful resumption of vehicle control.
Driver diverted attention	As noted, there is for this mechanism of inattention, unlike other mechanisms, an accumulating body of research on its impact on the resumption of control in vehicles equipped with Level 3 automation (see also, this Handbook, Chapter 9).

Saxby et al., 2013) and, in turn, driver inattention (Saxby et al., 2013; Körber, Cingel, Zimmermann, & Bengler, 2005). In this case, inattention is brought about not by distraction per se, but by other mechanisms.

6.3.3 SUMMARY

Section 6.3 describes many important considerations related to the intersection of workload, distraction, and vehicle automation. The introduction of different levels of automation can reduce workload for some driving-related tasks, but could increase workload on other tasks or could introduce new tasks that were not previously required of drivers (e.g., Table 6.2)—automation doesn't simply replace the human role in performing tasks, but changes the nature of tasks performed. Even when automation succeeds in its aim of reducing driving demands, performance and safety might not improve because low workload can undermine performance just as high workload can. Distraction and inattention are other noteworthy concerns when considering workload and the advent of automation, especially where drivers fail to attend to activities critical for safe driving either because automation demands driver's attention at inopportune moments (thereby creating a distraction which increases drivers' workload because they need to ignore the request in order to pay attention to the safety-critical elements) or automation induces drivers to engage more often and more deeply in non-driving tasks (the low workload encourages involvement in non-driving related tasks which, in turn, increases distraction). In both cases distraction, automation and workload interact with each other and influence a driver's performance. The following section is aimed at trying to understand how to address the workload and distraction issues so that drivers are neither over-engaged nor under-engaged when using these automation technologies.

6.4 MANAGING WORKLOAD AND DISTRACTION IN AUTOMATED VEHICLES

Chapters 15–17 of this Handbook provide a detailed discussion of some of the human–machine interface (HMI) design considerations for automated, connected, and intelligent vehicles. In this section, we provide a brief summary of design considerations or remediation strategies for workload and distraction, in particular.

From a workload management perspective, the major question is how to produce the optimal level of workload. As illustrated by the Yerkes–Dodson curve (Figure 6.1), performance can suffer when workload is too high or too low. Determining the effects of different types of automated systems on overall workload as well as on different types of mental resources or at different stages of processing (e.g., Wickens, 2002) is an area in need of further study. More fundamentally, what degree of agency and engagement should the automation support so that the joint vehicle-driver system is able to comfortably accommodate the various safety threats posed by the roadway? Furthermore, in doing so, we also need to consider what monitoring systems might be included in automated vehicles in order to measure and quantify workload (see this Handbook, Chapters 11, 9). In general, system designers should:

- Design to avoid creating new automation management tasks that compete for resources with manual driving tasks, that is, to avoid inadvertently increasing workload.
- Design considering that monitoring automation is effortful.
- Design to avoid diminished resources and abrupt workload transitions.
- Design to consider other tasks that reduced effort with driving enables (distraction).
- Design to avoid clumsy automation.
- Design to support strategic workload management.

Efforts to mitigate the effects of distraction and other mechanisms of inattention on the performance of activities critical for safe driving in vehicles equipped with automation are in their infancy. Some have been reviewed in this book (see Chapter 9) and elsewhere (e.g., Cunningham & Regan, 2018). Currently, these efforts focus on education and training (e.g., Casner & Hutchins, 2019; see Chapter 18), HMI design (e.g., Carsten & Martens, 2019; see Chapter 15), legislation and enforcement (see Chapter 14), and use of driver state monitoring technologies and driver feedback to detect distraction and re-orient attention (Lee, 2009; this Handbook, Chapters 9, 11).

There are, of course, other countermeasures that have been implemented to miti-gate the effects of driver distraction in vehicles not equipped with ADFs (see Regan et al., 2009; European Commission, 2015). These include vehicle fleet management policies and programs, driver licensing (e.g., testing knowledge around distraction and the ability to manage it), and road design (e.g., forgiving roadways that cushion drivers in the event that distraction-related crashes occur and cannot be avoided). Countermeasure development in these areas, with automated vehicles as the frame of reference, however, is yet to materialize.

6.5 CONCLUSION

Automation promises to reduce workload and make driving easier. As in other domains, automation does not simply replace the person's role of driving; it changes the nature of driving and can make driving harder. Automation design should:

- Avoid inadvertently increasing rather than reducing workload to some appropriate threshold
- Avoid inadvertently decreasing workload below some threshold and thereby enabling drivers to engage in potentially distracting activities
- Support monitoring automation, which can be surprisingly effortful
- Avoid creating clumsy automation that can make the hard tasks harder and the easy tasks easier
- Avoid high workload periods or peaks associated with transitions of control

Even with these design considerations at hand, it is important to underscore that, with increasing technological support and automation, the driving functions and tasks performed by drivers will change, and this will change the repertoire of knowledge, skills, and behaviors required by drivers to maintain safe driving performance (see Chapter 18). Even now, a modern driver has a unique skill set compared with drivers two or three decades ago; many drivers today might never have had to pump their brakes on slippery roads, but they might need to understand the distinction between a traction control system and a stability control system (Spulber, 2016). Likewise, drivers of increasingly autonomous vehicles might need to develop workload management skills that are not necessary in less sophisticated vehicles. Design of vehicle automation that considers the implications for workload and distraction might reduce the need for such workload management skills.

REFERENCES

Bainbridge, L. (1983). Ironies of automation. *Automatica, 19*(6), 775–779.

Brown, I. (1986). Functional requirements of driving. *Berzelius Symposium on Cars and Causalitie*. Stockholm, Sweden.

Carsten, O. & Martens, M. H. (2019). How can humans understand their automated cars? HMI principles, problems and solutions. *Cognition, Technology & Work, 21*(1), 3–20.

Carsten, O., Lai, F. C., Barnard, Y., Jamson, A. H., & Merat, N. (2012). Control task substitution in semiautomated driving: Does it matter what aspects are automated? *Human Factors, 54*(5), 747–761.

Casner, S. M. & Hutchins, E. L. (2019). What do we tell the drivers? Toward minimum driver training standards for partially automated cars. *Journal of Cognitive Engineering and Decision Making, 13,* 55–66.

Clark, H. & Feng, J. (2017). Age differences in the takeover of vehicle control and engagement in non-driving-related activities in simulated driving with conditional automation. *Accident Analysis & Prevention, 106,* 468–479.

Cook, R. I., Woods, D. D., McColligan, E., & Howie, M. B. (1990). Cognitive consequences of "clumsy" automation on high workload, high consequence human performance. SOAR 90, Space Operations, Applications and Research Symposium. Houston, TX: NASA Johnson Space Center.

Cunningham, M. L. & Regan, M. A. (2018). Driver distraction and inattention in the realm of automated driving. *IET Intelligent Transport Systems, 12,* 407–413.

de Winter, J. C., Happee, R., Martens, M. H., & Stanton, N. A. (2014). Effects of adaptive cruise control and highly automated driving on workload and situation awareness: A review of the empirical evidence. *Transportation Research Part F: Traffic Psychology and Behaviour, 27,* 196–217.

Desmond, P. A. & Hancock, P. A. (2001). Active and passive fatigue states. In P. A. Hancock & P. A. Desmond (Eds.), *Stress, Workload, and Fatigue* (pp. 455–465). Hillsdale, NJ: Lawrence Erlbaum.

Dingus, T. A., Guo, F., Lee, S., Antin, J. F., Perez, M., Buchanan-King, M., & Hankey, J. (2016). Driver crash risk factors and prevalence evaluation using naturalistic driving data. *Proceedings of the National Academy of Sciences, 113*(10), 2636–2641.

European Commission. (2015). *Study on Good Practices for Reducing Road Safety Risks Caused by Road User Distractions*. Brussels, Belgium: European Commission.

Fuller, R. (2005). Towards a general theory of driver behaviour. *Accident Analysis & Prevention, 37*(3), 461–472.

Galvani, M. (2019). History and future of driver assistance. *IEEE Instrumentation & Measurement Magazine, 22*(1), 11–16.

Gopher, D. & Donchin, E. (1986). Workload: An examination of the concept. In K. R. Boff, L. Kaufman, & J. P. Thomas (Eds.), *Handbook of Perception and Human Performance, Vol. 2- Cognitive Processes and Performance* (pp. 1–49). Oxford: John Wiley & Sons.

Hancock, P. A. (2009). On the philosophical foundations of the distracted driver and driving distraction. In M. A. Regan, J. D. Lee, & K. Young (Eds.), *Driver Distraction: Theory, Effects and Mitigation* (pp. 11–30). Boca Raton, FL: CRC Press.

Hart, S. G., & Staveland, L. E. (1988). Development of NASA-TLX (Task Load Index): Results of empirical and theoretical research. *Advances in Psychology, 52,* 139–183.

Helton, W. S., Funke, G. J., & Knott, B. A. (2014). Measuring workload in collaborative contexts: Trait versus state perspectives. *Human Factors, 56*(2), 322–332.

Hendy, K. C., Liao, J., & Milgram, P. (1997). Combining time and intensity effects in assessing operator information-processing load. *Human Factors, 39*(1), 30–47.

Hollnagel, E. & Woods, D. D. (2005). *Joint Cognitive Systems: Foundations of Cognitive Systems Engineering.* Boca Raton, FL: CRC Press.

Horrey, W. J. & Lesch, M. F. (2009). Driver-initiated distractions: Examining strategic adaptation for in-vehicle task initiation. *Accident Analysis & Prevention, 41*(1), 115–122.

Jamson, A. H., Merat, N., Carsten, O. M., & Lai, F. C. (2013). Behavioural changes in drivers experiencing highly-automated vehicle control in varying traffic conditions. *Transportation Research Part C: Emerging Technologies, 30,* 116–125.

Kahneman, D. (1973). *Attention and Effort.* Englewood Cliffs, NJ: Prentice-Hall.

Körber, M. & Bengler, K. (2014). Potential individual differences regarding automation effects in automated driving. *Proceedings of the XV International Conference on Human Computer Interaction.* New York: ACM.

Körber, M., Cingel, A., Zimmermann, M., & Bengler, K. (2015). Vigilance decrement and passive fatigue caused by monotony in automated driving. *Procedia Manufacturing, 3,* 2403–2409.

Lee, J. D. (2009). Can technology get your eyes back on the road? *Science, 234*(April), 344–346.

Lee, J. D. (2014). Dynamics of driver distraction: The process of engaging and disengaging. *Annals of Advances in Automotive Medicine, 58,* 24–32.

Lee, J. D., Wickens, C. D., Liu, Y., & Boyle, L. N. (2017). *Designing for People: An Introduction to Human Factors Engineering.* Charleston, SC: CreateSpace.

Lee, J. D., Young, K. L., & Regan, M. A. (2009). Defining driver distraction. In M.A. Regan, J. D. Lee, & K. Young (Eds.), *Driver Distraction: Theory, Effects and Mitigation* (pp. 31–40). Boca Raton, FL: CRC Press.

Llaneras, R., Salinger, J., & Green, C. (2013). Human factors issues associated with limited ability autonomous driving systems: Drivers' allocation of visual attention to the forward roadway. *Proceedings of the Seventh International Driving Symposium on Human Factors in Driver Assessment, Training and Vehicle Design* (pp. 92–98). Iowa City, IA: University of Iowa.

Louw, T., Kountouriotis, G., Carsten, O., & Merat, N. (2015). Driver inattention during vehicle automation: How does driver engagement affect resumption of control? *4th International Conference on Driver Distraction and Inattention (DDI2015).* Sydney: ARRB Group.

Louw, T., Madigan, R., Carsten, O., & Merat, N. (2017). Were they in the loop during automated driving? Links between visual attention and crash potential. *Injury Prevention, 23*(4), 281–286.

Merat, N., Jamson, A. H., Lai, F. C., Daly, M., & Carsten, O. M. (2014). Transition to manual: Driver behaviour when resuming control from a highly automated vehicle. *Transportation Research Part F: Traffic Psychology and Behaviour, 27,* 274–282.

Moray, N. (1979). *Mental Workload: Its Theory and Measurement.* New York: Plenum.

Moray, N. (1988). Mental workload since 1979. *International Reviews of Ergonomics, 2,* 123–150.

Norman, D. A. & Bobrow, D. G. (1975). On data-limited and resource-limited processes. *Cognitive Psychology, 7*(1), 44–64.

O'Donnell, R. D. & Eggemeier, F. T. (1986). Workload assessment methodology. In K. R. Boff, L. Kaufman, & J. P. Thomas (Eds.), *Handbook of Perception and Human Performance, Vol. 2- Cognitive Processes and Performance* (pp. 1–49). New York: Wiley.

Proctor, R. W. & Van Zandt, T. (1994). *Human Factors in Simple and Complex Systems.* Needham Heights, MA: Allyn & Bacon.

Raby, M. & Wickens, C. D. (1994). Strategic workload management and decision biases in aviation. *The International Journal of Aviation Psychology, 4*(3), 211–240.

Regan, M. A. & Hallett, C. (2011). Driver distraction and driver inattention: Definitions, mechanisms, effects and mitigation. In B. Porter (Ed.), *Handbook of Traffic Psychology* (pp. 275–286). Amsterdam, The Netherlands: Elsevier.

Regan, M. A., Hallett, C., & Gordon, C. P. (2011). Driver distraction and driver inattention: Definition, relationship and taxonomy. *Accident Analysis & Prevention, 43*, 1771–1781.

Regan, M. A., Lee, J. D., & Young, K. (2009). *Driver Distraction: Theory, Effects and Mitigation.* Boca Raton, FL: CRC Press.

SAE International. (2018). *Taxonomy and Definitions for Terms Related to Driving Automation Systems for On-Road Motor Vehicles* (J3016). Warrendale, PA: Society for Automotive Engineers.

Samuel, S., Borowsky, A., Zilberstein, S., & Fisher, D. L. (2016). Minimum time to situation awareness in scenarios involving transfer of control from an automated driving suite. *Transportation Research Record, 2602*(1), 115–120.

Saxby, D. J., Matthews, G., Warm, J. S., Hitchcock, E. M., & Neubauer, C. (2013). Active and passive fatigue in simulated driving: Discriminating styles of workload regulation and their safety impacts. *Journal of Experimental Psychology: Applied, 19*(4), 287–300.

Schaap, T. W., Van der Horst, A. R. A., Van Arem, B., & Brookhuis, K. A. (2013). *The Relationship Between Driver Distraction and Mental Workload* (Vol. 1, pp. 63–80). Farnham: Ashgate.

Schulte, A. & Donath, D. (2011). Measuring self-adaptive UAV operators' 'load-shedding strategies under high workload. *International Conference on Engineering Psychology and Cognitive Ergonomics.* Berlin: Springer.

Sirevaag, E. J., Kramer, A. F., Wickens, C. D., Reisweber, M., Strayer, D. L., & Grenell, J. F. (1993). Assessment of pilot performance and mental workload in rotary wing aircraft. *Ergonomics, 36*(9), 1121–1140.

Spulber, A. (2016). *Impact of Automated Vehicle Technologies on Driver Skills.* Ann Arbor, MI: Centre for Automotive Research.

Tingvall, C., Eckstein, L., & Hammer, M. (2009). Government and industry perspectives on driver distraction. In M. A. Regan, J. D. Lee, & K. Young (Eds.), *Driver Distraction: Theory, Effects and Mitigation* (pp. 603–620). Boca Raton, FL: CRC Press.

Tivesten, E. & Dozza, M. (2014). Driving context and visual-manual phone tasks influence glance behavior in naturalistic driving. *Transportation Research Part F: Traffic Psychology and Behaviour, 26*, 258–272.

Trafton, J. G. & Monk, C. A. (2007). Task interruptions. *Reviews of Human Factors and Ergonomics, 3*(1), 111–126. doi:10.1518/155723408X299852

Tulga, M. K. & Sheridan, T. B. (1980). Dynamic decisions and work load in multitask supervisory control. *IEEE Transactions on Systems, Man, and Cybernetics, 10*(5), 217–232.

Warm, J. S., Parasuraman, R., & Matthews, G. (2008). Vigilance requires hard mental work and is stressful. *Human Factors, 50*(3), 433–441.

Wickens, C. D. (2002). Multiple resources and performance prediction. *Theoretical Issues in Ergonomics Science, 3*(2), 159–177.

Wickens, C. D. & Hollands, J. G. (2000). *Engineering Psychology and Human Performance.* New York: Psychology Press.

Wierwille, W. W. & Eggemeier, F. T. (1993). Recommendations for mental workload measurement in a test and evaluation environment. *Human Factors, 35*(2), 263–281.

Woods, D. D. (1996). Decomposing automation: Apparent simplicity, real complexity. In R. Parasuraman & M. Mouloua (Eds.), *Automation and Human Performance: Theory and Applications* (pp. 3–17). Hillsdale, NJ: Lawrence Erlbaum.

Yamani, Y. & Horrey, W. J. (2018). A theoretical model of human-automation interaction grounded in resource allocation policy during automated driving. *International Journal of Human Factors and Ergonomics, 5*(3), 225–239. doi: 10.1504/IJHFE.2018.095912

Yerkes, R. M. & Dodson, J. D. (1908). The relation of strength of stimulus to rapidity of habit-formation. *Journal of Comparative Neurology and Psychology, 18*(5), 459–482.

Young, K. L., Regan, M. A., & Lee, J. D. (2009). Factors moderating the impact of distraction on driving performance and safety. In M. A. Regan, J. D. Lee, & K. Young (Eds.), *Driver Distraction: Theory, Effects and Mitigation.* Boca Raton, FL: CRC Press.

Young, M. S. & Stanton, N. A. (2002a). Malleable attentional resources theory: A new explanation for the effects of mental underload on performance. *Human Factors, 44*(3), 365–375.

Young, M. S. & Stanton, N. A. (2002b). Attention and automation: New perspectives on mental underload and performance. *Theoretical Issues in Ergonomics Science, 3*(2), 178–194.

Zeeb, K., Buchner, A., & Schrauf, M. (2015). What determines the take-over time? An integrated model approach of driver take-over after automated driving. *Accident Analysis & Prevention, 78*, 212–221.

7 Situation Awareness in Driving

Mica Endsley
SA Technologies

CONTENTS

Key Points 127
7.1   Introduction 128
7.2   SA Requirements for Driving 128
7.3   SA Model 133
    7.3.1   Individual Factors 134
        7.3.1.1   Limited Attention 134
        7.3.1.2   Limited Working Memory 134
        7.3.1.3   Goal-Driven Processing Alternating with Data-Driven
                  Processing 135
        7.3.1.4   Long-Term Memory Stores 136
        7.3.1.5   Expertise 137
        7.3.1.6   Cognitive Automaticity 138
    7.3.2   Vehicle and Driving Environment 139
        7.3.2.1   Information Salience 139
        7.3.2.2   Complexity 140
        7.3.2.3   Workload, Fatigue, and Other Stressors 140
        7.3.2.4   Distraction and Technology 141
    7.3.3   Automation and Vehicle Design 142
7.4   Conclusions 146
References 147

KEY POINTS

- Situation awareness (SA), as a driver's integrated understanding of what is happening in the driving environment, is critical for successful performance, with poor SA being implicated as a significant cause of vehicle crashes.
- A goal-directed task analysis (GDTA) was used to systematically determine the requirements for driver SA, including perception, comprehension, and projection elements, needed for key decisions.
- The many factors that affect driver SA in the dynamic road transportation environment are presented through the lens of a model of SA showing the cognitive processes involved, including attention, memory, goal-driven

and data-driven processing, mental models, automaticity, and the role of expertise.

- Features of the road and vehicle system and environmental factors that impact SA are also discussed, including information salience, complexity, work-load, fatigue and other stressors, distractions and in-vehicle technologies.
- The effects of autonomous and semi-autonomous vehicle technology on driver SA are presented based on this foundation, showing opportunities for enhancing SA, along with many hazards that are endemic to some forms of vehicle autonomy.
- Guidelines for supporting SA via improved automation design are discussed, and enhanced driver training, including hazard awareness, is highly recommended.

7.1 INTRODUCTION

Situation awareness (SA) forms the central organizing mechanism for a driver's understanding of the state of the vehicle and environment that forms the basis for ongoing decision-making in the rapidly changing world of road transportation. While much driving research examines aspects of driver performance in isolation (e.g., distraction, expertise), SA research considers all of these factors and processes in a more holistic fashion, shedding light on human performance strengths and limitations within the context of the driving environment.

SA has been found to be central for successful performance in driving (Horswill & McKenna, 2004; Ma & Kaber, 2005). Poor SA has been implicated as a significant cause of vehicle crashes (Gugerty, 1997), with improper lookout and inattention as two salient examples (Treat et al., 1979). Distractions and recognition errors (in which the driver "looks but does not see") are also significant causes of vehicle crashes that point to problems with SA (Sabey & Staughton, 1975). An analysis of critical reasons for driver-related pre-crash events shows that over 40% of crashes are related to poor Level 1 SA (perception), including inadequate surveillance, internal or external distractions, inattention, and other failures (National Highway Traffic Safety Administration, 2008). Other problems cited, more indicative of failures of Level 2 SA (comprehension) or Level 3 SA (projection), include false assumption of other's actions (4.5%) and misjudgment of gap or others' speed (3.2%).

This chapter will examine what SA means from the standpoint of road transportation, based on a model of SA that describes the cognitive mechanisms involved in driving, as well as external environmental and system factors that impact on SA, including the advent of new technologies and vehicle autonomy. Based on this overview, directions for the design of vehicles and training to improve driver SA are recommended.

7.2 SA REQUIREMENTS FOR DRIVING

SA is defined as "the perceptions of the elements in the environment within a volume of space and time, the comprehension of their meaning, and the projection of their status in the near future" (Endsley, 1988). This definition of SA has been used across many different domain areas including piloting, air traffic control, power system operations, military command and control, space flight, and driving.

An important factor for its application to driving is a clear delineation of the specific perception (Level 1 SA), comprehension (Level 2 SA), and projection (Level 3 SA) requirements. This is typically accomplished via a goal-directed task analysis (GDTA) that establishes the key goals for a given operational role in the domain (e.g., driver, pedestrian, mechanic), the key decisions that need to be made to reach each goal, and the SA requirements that are needed to accurately make each decision (Endsley, 1993; Endsley & Jones, 2012). The GDTA provides a systematic method for understanding the cognitive requirements associated with any job or task, including performance under both routine and unusual conditions.

As an example, a hierarchical goal tree for an automobile driver is presented in Figure 7.1. This shows the different goals and subgoals that are involved in the task of driving. The overall goal is to maneuver a vehicle from the point of origin to the point of destination in a safe, legal, and timely manner. There are a number of associated main goals, including (1) ensuring that the vehicle is safe for driving, (2) selecting the optimum path to the destination, (3) executing the chosen route in a safe, legal, and timely manner, and (4) minimizing the impact of abnormal situations. Further, each of these goals breaks down into relevant subgoals as needed.

It should be noted that not all goals and subgoals are active at all times. For example, the driver may *ensure that the vehicle is safe for driving* and *determine the best route to the destination* at the beginning of a trip, and only reactivate those goals later on during the trip if needed. The goal of *minimizing the impact of abnormal situations* may only become active infrequently, when those abnormal situations occur. The majority of the time the driver will likely be focused on subgoals associated with *executing the chosen route*, but may need to dynamically switch to other goals in response to abnormal situations, or the need for a reroute, for example. Further, many of these subgoals may interact with each other.

The GDTA then breaks out the detailed decisions and SA requirements for each subgoal, an example of which is shown in Figure 7.2, including perception, comprehension, and projection elements. This provides a systematic analysis that determines not just the information that drivers need to be aware of but also how they need to combine that data to form relevant situation comprehension and projections of changes in the situation that are relevant to making decisions. Table 7.1 provides a partial listing of the SA requirements for drivers, summarized across the various goals, subgoals, and decisions contained in the full GDTA.

GDTAs in general strive to be technology free. That is they describe what decisions need to be made, and the information needed to make them, but not how that information will be obtained, which could change dramatically with technology. For example, vehicle navigation could be accomplished via a compass, map, or Global Positioning System (GPS)-enabled navigation system. Nonetheless, the same questions will be relevant and the same SA considerations present. As vehicles become more automated, it may be easier to determine some of these SA requirements. For example, computers can easily determine many distance and time calculations such as distance to other vehicles, time until refueling, and projected time to destinations. Even with semi-autonomous vehicles, as long as the

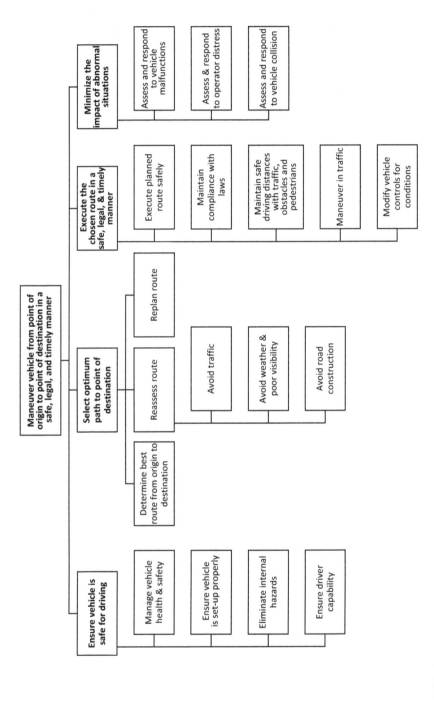

FIGURE 7.1 Goal tree for driving in road transportation.

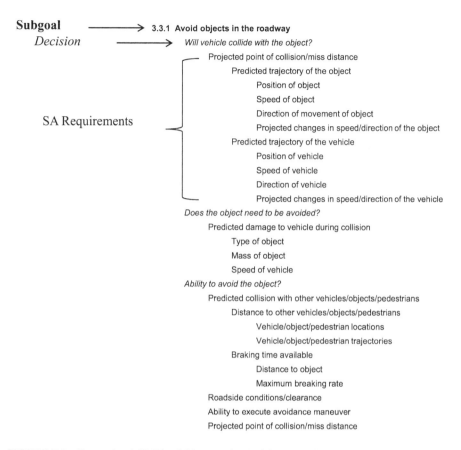

Subgoal ⟶ **3.3.1 Avoid objects in the roadway**

Decision ⟶ *Will vehicle collide with the object?*

Projected point of collision/miss distance
Predicted trajectory of the object
Position of object
Speed of object
Direction of movement of object
Projected changes in speed/direction of the object
SA Requirements
Predicted trajectory of the vehicle
Position of vehicle
Speed of vehicle
Direction of vehicle
Projected changes in speed/direction of the vehicle

Does the object need to be avoided?
Predicted damage to vehicle during collision
Type of object
Mass of object
Speed of vehicle

Ability to avoid the object?
Predicted collision with other vehicles/objects/pedestrians
Distance to other vehicles/objects/pedestrians
Vehicle/object/pedestrian locations
Vehicle/object/pedestrian trajectories
Braking time available
Distance to object
Maximum breaking rate
Roadside conditions/clearance
Ability to execute avoidance maneuver
Projected point of collision/miss distance

FIGURE 7.2 Example of GDTA of driver goals, decisions, and SA requirements.

driver remains responsible for the safe operation of the vehicle, he or she will need to remain aware of this key information to ensure that the automation is operating properly, in accordance with the driver's desires and taking into account the many unforeseen circumstances that can occur. (This will be discussed in more detail later.) Only if functions become fully automated will the driver's SA needs change.

Michon (1985) describes three types of driving behaviors: (1) strategic, which focuses on high-level goal-related decisions such as navigation, (2) tactical, which focuses on maneuvering, and (3) operational, which focuses on low-level actions such as steering and braking. In comparing this taxonomy to the goal tree in Figure 7.1, strategic behaviors would map to three of the major goals: *ensure vehicle is safe for driving, select optimum path to point of destination,* and *minimize the impact of abnormal situations.* Tactical behaviors would map primarily to *executing the chosen route.* While operational behaviors like steering and braking require perception of situational information (such as brake lights, stoplights, and lane markings), these low-level behaviors are subsumed under subgoals such as *maintain safe driving*

TABLE 7.1
SA Requirements for Driving

Level 1 SA: Perception	Level 2 SA: Comprehension	Level 3 SA: Projection
Location of nearby objects (vehicles, pedestrians, cyclists, other objects)	Distance to other objects, vehicles, pedestrians, cyclists	Projected trajectory of own vehicle, other vehicles, objects, pedestrians, cyclists
Relative speed of traffic in adjacent lanes	Preferred lane for traffic avoidance/speed	Projected collision/miss distances
Open areas in adjacent lanes	Vehicle in blind spot	Projected effect of evasive maneuver/braking
Planned route	Compliance with planned route	Projected distance to turns/exits
Planned destination(s) location	Traffic lane needed for route execution	Projected distance & time remaining to destination
Traffic density along route(s) (crashes, construction, major events, time of day, road/exit closures)	Areas of congestion	Projected trip delay time
	Alternate routes	Projected time to destination on alternate routes
Emergency vehicles	Avoidance of emergency vehicles	Projected traffic stops/slowdowns ahead
Emergency personnel	Compliance with safety personnel	
Hazardous weather along route (rain, snow, icing, fog, high winds, areas of flooding)	Impact of weather on vehicle safety, systems, & route time	Projected changes in weather
		Projected safety of route(s)
Daylight/dusk/night	Visibility of road and vehicle	
Road conditions along route (road size/paving, construction, frequency of stop signs/lights, security)	Impact of road conditions on route time	Projected time and distance to destination on route(s)
	Impact of road conditions on route safety	Projected cost/benefit of change in route
Speed limit	Vehicle compliance with laws	Projected locations of police
Stoplight status		
Traffic control measures		
Lane markings		
Direction of traffic		
Vehicle parameters (speed, gear, fuel level, position in lane, headlights, wipers)	Fuel sufficiency & usage	Projected time until refueling is needed
	Fuel to reach destination	
	Road worthiness	
Vehicle malfunctions	Vehicle safety	Projected ability of vehicle to make trip
Location of fuel stations	Distance to refueling stations	Projected refueling points
Location of restaurants	Distance to restaurants	Projected stop points
Parking place(s) (location, size)	Distance to vehicles, curbs	
Driver status (fatigue, injury, hunger, thirst, intoxication)	Need for rest break, assistance, alternate driver	Projected time until stop is needed

distances with traffic, obstacles, and pedestrians, and *maneuvering in traffic*. These low-level operational behaviors (often categorized as skill based) may be conducted in a conscious, deliberate way, or may become cognitively automatized (to be discussed in more detail later). Even the execution of heavily practiced routes (e.g., the route home from work) can become highly automatized.

The driving GDTA includes the need for all three levels of SA for making decisions under each goal. Kaber, Sangeun, Zahabi, and Pankok (2016) found evidence for the importance of all three levels of SA for the performance of tactical driving tasks (e.g., passing) and that Level 3 SA, as well as overall SA, is important for strategic performance (e.g., arrival time). Other research also supports the importance of Level 3 SA for strategic driving performance (Ma & Kaber, 2007; Matthews, Bryant, Webb, & Harbluk, 2001) and for tactical driving performance (Ma & Kaber, 2005). Kaber et al. (2016) found that while operational performance was generally highly cognitively automatized and not affected by the addition of a secondary task during normal operations, after exposure to a hazard it became more conscious and correlated with Level 1 SA.

7.3 SA MODEL

A cognitive model of SA is shown in Figure 7.3 (Endsley, 1995) that can be used to understand the factors that affect driver SA in the dynamic road transportation environment, each of which will be discussed. Based on the factors in this model, the role and impact of automation and vehicle design on SA are then considered.

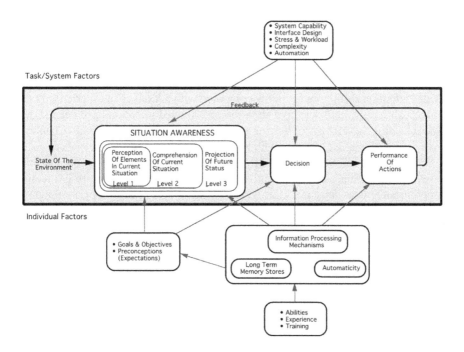

FIGURE 7.3   Model of SA in dynamic decision-making. (Endsley, 1995.)

7.3.1 INDIVIDUAL FACTORS

Key cognitive processes and constructs that are functions of the individual driver are shown across the bottom section of Figure 7.3. Relevant information processing mechanisms include attention, working memory, and goal-directed and data-driven processing approaches.

7.3.1.1 Limited Attention

In many cases, there may be multiple things to pay attention to at once, requiring divided attention from the driver. For example, the driver may need to attend to multiple surrounding vehicles, the vehicle ahead, and upcoming signage when attempting a lane change. Likewise the driver making a left turn in an intersection may need to attend to oncoming traffic as well as pedestrians crossing at a cross walk. Since people have a limited amount of attention, this creates a major constraint on the driver's ability to perceive multiple pieces of information accurately in parallel. While the ability to share attention across different types of information can moderate this difficultly to some degree (Damos & Wickens, 1980; Wickens, 1992), attention is a major limiting factor on the driver's ability to maintain SA (Fracker, 1989).

In other cases, attention may not be fully engaged by the driving task (e.g., on a highway with light traffic), and the driver may be prone to mind-wandering (thinking of other tasks, events, or just daydreaming) with the excess capacity. Yanko and Spalek (2014) found that drivers reported mind-wandering 39% of the time, and it was significantly related to increases in speed and slower breaking reaction times.

Kircher and Ahlstrom (2017) argue that only some minimum amount of attention may be required at any given time to gather the SA needed to satisfice task requirements. While this may be partially true, particularly in low event environments, it also neglects the important role of effective deployment of limited attention in more information dense driving environments. Endsley and Bolstad (1994) showed divided attention abilities to be highly correlated with SA scores in experienced individuals. Moreover, both selective attention and divided attention have been shown to be related to Level 1 SA in driving tasks and to driving performance (Chaparro, Groff, Tabor, Sifrit, & Gugerty, 1999; Kaber et al., 2016).

Shortcomings in attention are a frequent target of many automation attempts; however, automation can also negatively affect attention by redirecting attention inappropriately (e.g., nuisance alarms distracting drivers at inappropriate times) or by encouraging the driver to redirect attention to other more compelling things (such as when steering control becomes automated).

7.3.1.2 Limited Working Memory

Similarly, working memory is highly constrained and creates a limit on novice drivers' ability to gather and integrate data to formulate the higher levels of SA, as well as for drivers when dealing with novel situations (e.g., when lost in a new city or driving on unfamiliar roads) (Endsley, 1995). Gugerty and Tirre (1997) showed working memory to be significantly related to SA scores in a driving task. Kaber et al. (2016) found that Level 1 SA was related to working memory span in a simulated driving task, with a higher correlation following a hazard event, but not Level 2 or 3 SA.

As people gain experience, they are able to draw on long-term memory stores (such as mental models) to significantly overcome working memory limitations, which has been demonstrated in a number of studies (Endsley, 1990; 2015; Sohn & Doane, 2004; Sulistayawati, Wickens, & Chui, 2011). Bolstad (2001), for example, found no relationship between working memory and driver performance in her study with highly experienced drivers. So both the experience level of the driver and the presence of abnormal or usual situations that require more conscious and focused information processing affect the exact role of working memory in driving.

7.3.1.3 Goal-Driven Processing Alternating with Data-Driven Processing

For the most part, in complex driving environments drivers are forced to dynamically switch their attention between information that is relevant for current goals (i.e., goal-driven processing) and information that catches their attention, which can be important for triggering a dynamic change in current goals (i.e., data-driven processing) (Casson, 1983). Drivers who are purely data driven are very inefficient at processing the complex driving environment. As there is too much information to process in parallel, they primarily react to the cues that are most salient based on their pre-attentive sensory characteristics (Neisser, 1967; Treisman & Paterson, 1984). Bright lights, colors (e.g., flashing lights), and sounds (e.g., car horns) will grab attention in a busy environment for example.

With experience, drivers move towards more goal-driven processing. This allows them to search for information that is relevant to their goals (based on associated mental models; see this Handbook, Chapter 3), providing for a more efficient information search and a mechanism for determining the relevance of the information that is perceived. So drivers who are goal driven will seek out signs or landmarks that indicate they are on the correct route, for example. A driver who is considering a lane change will look for information about traffic in the adjacent lane based on this new goal, whereas, otherwise, she may not pay attention to traffic behind her in the next lane. Goal-driven processing is far more efficient overall in gathering relevant information in a prioritized manner as needed for goal attainment.

If drivers are only goal-driven, however, they may miss key information that indicates they need to change goals. For example, if a driver is intent on changing lanes, he/she still needs to be alert to the car in front hitting the brakes. This should trigger a rapid switch from the "lane change" goal to the "avoid other vehicles goal." Good SA is characterized by alternating between these modes: using goal-driven processing to find and efficiently process the information needed for achieving goals, and using data-driven processing to regulate the selection of which goals should be most important at any given time (Endsley, 1995).

Significant problems with SA occur when this alternating process breaks down. People who remain too goal driven will miss important information, such as the emergency vehicle in their rearview mirror, the low gas gage on their dashboard, or the stalled vehicle in the lane ahead while they are focused on interacting with their on-board navigation system. People who are only data driven may fail to direct their attention to relevant information when needed. For example, they may get distracted by a cell phone's ring and neglect to maintain their attention on relevant information for avoiding other vehicles and maintaining lane following. While sometimes these highly salient cues are non-driving related (i.e., distracted driving related to cell phones, text messages, and passenger conversations), it should be pointed out that in many cases they are relevant to the driving task. For example, drivers can be distracted by looking at pedestrians near a busy intersection and miss a sudden stop in traffic ahead. In busy driving environments, it can be quite challenging to maintain awareness of all the relevant information due to limited attention.

7.3.1.4 Long-Term Memory Stores

Novice drivers tend to operate in a largely data-driven manner and will be highly overloaded in seeking to gather information, understand what it means, and formulate correct responses (Endsley, 2006). With experience, drivers gain long-term memory stores in the form of mental models that are very important for overcoming the challenges of limited attention and working memory. Mental models include an understanding of the road system and rules of the road, driving norms, and patterns of events and behaviors that are important for understanding and projecting effectively. So for example, drivers will learn not only the formal laws and traffic signs that they should adhere to, but also driving norms, such as who yields the right-of-way in traffic circles or the likelihood of traffic stopping for pedestrians in different cities. Mental models may also include relevant information about how pedestrians, cyclists, or roadside animals will behave.

Mental models are important for reducing the cognitive workload associated with processing and integrating information in real time. They allow for (Endsley, 1995):

- A means of rapidly interpreting and integrating various pieces to form an understanding of their meaning (Level 2 SA). This allows a driver to understand the criticality of a warning light on the dash, for example, or the impact of road conditions and weather on vehicle performance and safety.
- A mechanism for projecting future states of the environment based on its current state and an understanding of its dynamics (Level 3 SA). This allows a driver to project roads and times that are likely to have heavy traffic, for example, or the likely behavior of vehicles in a parking lot.
- Knowledge regarding which aspects of the environment are relevant and the critical cues that have significance. By including information about critical cues, mental models can be very effective for rapidly understanding situations and for directing attention. For example, a ball on residential street would trigger an experienced driver to slow down and search for children who might run into the street.

In addition to long-term memory in the form of mental models, there is also evidence that people develop schema of prototypical states of the mental model. These schema are patterns consisting of the state of each of the relevant elements of the mental model (Endsley, 1995). By pattern matching between critical cues in the current situation and known schema, people can instantly recognize many classes of situations. Schema can be learned through direct experience or vicariously through training or storytelling. For example, drivers may have typical schema for "traffic hour in Atlanta," "drunk drivers," or "skidding on icy roads." Pattern matching to learned schema provides a considerable shortcut for SA and decision-making. Rather than processing data to determine Level 2 or 3 SA (requiring working memory or exercising the mental model), that information is already a part of the schema and must merely be recalled.

Thus mental models, when correct, create highly effective mechanisms for assisting rapid information processing in complex environments. When they are incorrect, however, they can lead to misunderstandings of situations, or incorrect expectations about how to direct attention. For example, Pammer, Sabadas, and Lentern (2018) showed that drivers were more likely to miss seeing motorcycles than other vehicles, and this was directly related to internalized assessments of their threat level.

The development of mental models for correctly understanding and projecting the actions of automated devices is extremely important for successful joint driver-automation operations, and will be a major challenge with new forms of in-vehicle automation. Without effective mental models of automation, drivers are likely to have significant challenges in directing their attention, understanding the behaviors and capabilities of the automation (including its limitations and states), and in taking proactive actions based on projected automation behaviors (see also this Handbook, Chapter 3).

Further, these mental models, developed through experience may lead drivers to have a very different understanding of the importance of events and objects as compared to automated systems that do not have these rich models, and may lead them to have different priorities and projections as compared to automated systems. This sets the stage for conflicting views of appropriate actions when drivers are overseeing new automated technologies, which can create confusion and incorrect expectations, as well as a new workload burden.

7.3.1.5 Expertise

Data-driven processing by novice drivers is highly inefficient (Endsley, 2006). Since novice drivers lack knowledge on which information is most important, their scan patterns tend to be sporadic and non-optimal (Chapman & Underwood, 1998; Underwood, 2007). They may neglect key information or over-sample information unnecessarily. The novice driver will not know where to look for important information. Novices also fail to direct their attention to as wide a range of information in the environment as experts, possibly due to more impoverished mental models directing their search (Underwood, Chapman, Bowden, & Crundall, 2002).

Properly understanding the significance of what is perceived may also pose a problem, as novice drivers do not have the experience base that is needed for interpreting and prioritizing cues. Novice drivers often fail to recognize potentially hazardous cues and to anticipate how hazards may develop (Borowsky, Shinar, & Oron-Gilad, 2010; Borowsky & Oron-Gilad, 2013; Finn & Bragg, 1986; Parmet, Borowsky, Yona, & Oron-Gilad, 2015). Lack of understanding of hidden hazards has also been found to underlie the inferior speed management of novice drivers, both of which are major factors associated with crashes in this group (Parmet et al., 2015).

Luckily, as drivers gain expertise through experience and training, they significantly reduce these problems through a number of mechanisms, including mental models, schema, and goal-driven processing. Mourant and Rockwell (1972) showed that experts were better able to adjust their scan patterns to the road type. Expert drivers have better search models regarding where to look for hazards (Underwood et al., 2002) that can guide their scanning behaviors.

A number of researchers have examined the ability of drivers to project future hazards in the driving environment (McKenna & Crick, 1991; 1994), a part of Level 3 SA, showing that experienced drivers are significantly faster at detecting and reacting to hazards. This superior level of hazard projection improves with both driving time and advanced driver training (McKenna & Crick, 1991). Horswill and McKenna (2004) showed improved hazard awareness to be significantly related to a reduced likelihood of being involved in a crash.

Horswill and McKenna (2004) found that while experienced drivers are better at anticipating hazards than less experienced drivers, they were even more negatively affected by secondary tasks that tap into the central executive function of working memory (such as discussions on a cell phone or mental calculations). This implies that the projection of hazards in the driving environment (a part of Level 3 SA) is cognitively demanding, requiring working memory, and is not automatic. It involves effortful processes as experienced drivers consciously search for hazards using a dynamic mental model of the driving environment.

Of significant concern is the role that vehicle automation will have on the development of expertise in novice drivers. If many of the tasks of driving are handed off to automation, will new drivers ever develop the deep knowledge (mental models, schema and goal-directed processing) that is critical for SA? For example, as GPS navigation devices have become common, the ability to read maps to navigate has become a lost skill. Further, evidence suggests that even experienced drivers can lose skills if they primarily rely on automation (Wiener & Curry, 1980). When vehicles provide automated blind spot warnings, drivers may no longer look behind them, becoming dependent on automation that may not always be reliable. Significant attention will need to be paid to how vehicle automation affects the development of expertise in future drivers, as well as its effect on skill atrophy in experienced drivers.

7.3.1.6   Cognitive Automaticity

With experience people can also acquire a form of cognitive automaticity. Automatic cognitive processing tends to be fast, autonomous, effortless, and unavailable to conscious awareness in that it can occur without attention (Logan, 1988) or with a minimum amount of attention as needed for bringing appropriate schema into play, executing scripts and processes (Reason, 1984).

Automaticity can significantly benefit SA by providing a mechanism for overcoming limited attention capacity, leaving attention available for other tasks; however, it can also result in lowered SA on well-learned features in routine situations (Endsley, 1995). Charlton and Starkey (2011) showed that with increased experience in driving, ratings of task difficulty lowered as performance became more automatized, and detection of objects is more proceduralized. Further, drivers become more "attention blind" to many objects, indicative of driving without awareness (Charlton & Starkey, 2013).

So for example, a driver who is operating automatically may see stoplights and brake lights, and press the accelerator or brake pedal appropriately, but with little conscious awareness as his attention is directed towards daydreaming or other mental tasks. While this extra mental capacity may be advantageous and allows for generally good performance within the bounds of well-learned situations, automaticity also leaves the driver prone to missing novel cues that are outside of the automatized behavior pattern. So they may miss a new stop sign or the turn to stop at the store on the route home, for example.

Thus, automaticity is generally characterized by low SA for the relevant automatized processes, which could include horizontal and lateral control as well as navigation on well-known routes (Endsley, 1995). Kaber et al. (2016) found that lower-level driving skills associated with operational driving may frequently be automatized and thus not reflect much Level 1 SA under normal driving conditions; however, they also found increased Level 1 SA following a hazardous event. McKenna and Farrand (1999) found that hazard awareness is not generally automatized, however, and requires conscious effort and processing.

Of significance with the introduction of vehicle automation is its effect on highly automatized cognitive processes. For example, if semi-autonomous vehicles take over steering control (horizontal and vertical), drivers will no longer be steering the vehicle via these low-workload automatized processes. This means that conscious attention will instead be required to respond to stoplights and road debris, which previously would have happened automatically from a human performance standpoint. And that conscious attention may be diverted or slow to determine that human action will be needed, creating new challenges.

7.3.2 VEHICLE AND DRIVING ENVIRONMENT

Achieving high levels of SA is not just a function of the capabilities of driver, but also of several key features of the driving environment, as shown in the upper section of Figure 7.3, that affect the processes involved in developing and maintaining SA.

7.3.2.1 Information Salience

The salience of relevant information in the driving environment significantly affects drivers' ability to accurately perceive important information, particularly when multi-tasking across multiple competing goals. Examples include readability of signs and road markers, with faded or obscured signage often missed; rumble strips on the edge of roads that provide highly salient cues for even visually saturated drivers; brightness of vehicle headlights and taillights; and darkness or adverse weather reducing visibility.

Known incidences of crashes directly attributable to poor information salience are low—2.1% of crashes involve obstructions, 0.3% involve missing traffic signs/signals, and 0.2% involve problems with lane delineations (National Highway Traffic Safety Administration, 2008). In environments overloaded with salient features (e.g., city streets with many flashing lights and signs, or in vehicles with brightly colored, moving displays), it can also be challenging for people to attend to the most critical information.

In-vehicle technologies and automation create a significant source of competition for attention, in that they often contain highly salient features, such as bright lights, moving displays, and audible tones or communications. Thus, the high salience of these technologies can compete directly for attention with external information that may be critical, but of low salience.

7.3.2.2 Complexity

As driving environments become more complex, there are more things to keep track of, making Level 1 SA more difficult, and it is more difficult to obtain an accurate mental model of that environment, making Level 2 and 3 SA more challenging. For example, National Highway Traffic Safety Administration (NHTSA) found that 36% of crashes involve turning or crossing at intersections (National Highway Traffic Safety Administration, 2008). Factors that increase environmental complexity include intersections with the potential for crossing traffic; an increased number of lanes of traffic; the presence of pedestrians and cyclists; heavy traffic; complex roadways and interchanges (e.g., five-way intersections, exits in unexpected directions); and adverse weather creating more diverse driver behaviors. Vehicle automation presents a new form of complexity for drivers, compounded by the numbers of features, modes, and logic patterns inherent in its design.

7.3.2.3 Workload, Fatigue, and Other Stressors

Driving may often be accompanied by a number of stressors that may be put on drivers by themselves or their environment, such as anxiety, time pressure, mental workload, uncertainty, noise or vibration, excessive heat or cold, poor lighting, physical fatigue, and driving at times that are not aligned with one's circadian rhythms.

These stressors all act similarly to reduce SA by further reducing limited working memory (Hockey, 1986), by reducing the efficiency of information gathering, and by increasing the incidence of premature closure (arriving at a decision without taking into account all available information; Janis, 1982; Keinan & Friedland, 1987). People tend to pay less attention to peripheral information, become more disorganized in scanning information, and are more likely to succumb to attentional tunneling when affected by these stressors (Bacon, 1974; Baddeley, 1972; Eysenck, 1982; Hockey, 1986). Drivers with attentional tunneling will lock in on certain information, and either intentionally or inadvertently drop normal scanning behaviors. For example, the driver who is fatigued may fail to check in his blind spot before changing lanes, remaining locked onto the view ahead. Drivers with attentional tunneling may have good SA on the part of the environment they are focused on, but not other neglected parts. Similarly anxiety producing conversations have been shown to create higher workload and attentional narrowing in driving (Briggs, Hole, & Land, 2011).

Fatigue is associated with 7% of vehicle crashes, and roughly doubles the likelihood that a person will be involved in a crash (National Highway Traffic Safety Administration, 2008). Similar effects are found due to other types of stressors. For example, anger has been shown to reduce SA in driving (Jeon, Walker, & Gable, 2014). Both low workload (e.g., long drives producing low levels of vigilance) and high workload can be challenges for driving (see also, this Handbook, Chapter 6).

Lee, Lee, and Boyle (2007) showed that increases in cognitive workload affected both data-driven (exogenous) and goal-driven (endogenous) attentional control processes. Engström, Markkula, Victor, and Merat (2017) reviewed a number of studies to show that high cognitive workload tends to impair the performance of novel, unpracticed, or highly variable driving tasks, but that well-automatized tasks are often unaffected. Under high workload conditions, experienced drivers are far less negatively affected than inexperienced drivers (Patten, Kircher, Ostlund, Nilsson, & Svenson, 2006), therefore, due to their increased ability to automatize certain tasks and draw on long-term memory structures such as mental models and schema. This apparently does not extend to tasks such as hazard awareness, however, which is not automatized, but requires central executive control (Horswill & McKenna, 2004).

On the low workload side, drivers may suffer from performance deficits when placed in low attention, demanding conditions that decrease vigilance (Davies & Parasuraman, 1980). Schmidt et al. (2009), for example, showed progressive increases in reaction times and perceived subjective monotony, which were correlated with physiological indicators of lowered arousal (as measured by changes in electroencephalography (EEG) and heart rate) in drivers over each segment of a 2-hour trip during daytime driving. Vigilance decrements are more likely to occur when the need for driver action is low. For example, LaRue, Rakotonirainy, and Pettitt (2011) found both road design (degree of curvature and changes in elevation) and lack of variability in roadside scenery increased the probability of vigilance-related performance decrements.

The 23% of vehicle crashes that are attributed to driver inattention and inadequate surveillance (National Highway Traffic Safety Administration, 2008) may in part be due to these challenges with vigilance, workload, and stress, leading to attentional narrowing and inefficiencies associated with information gathering.

New vehicle automation may affect driver workload in different ways. Similar to automation in aviation, it can significantly reduce workload during low workload periods leading to vigilance problems (e.g., driving on highways with little traffic), but also increase it during high workload periods, such as maneuvering in heavy traffic (Bainbridge, 1983; Endsley, 2017a; Ma & Kaber, 2005; Schmidt et al., 2009; Wiener & Curry, 1980; this Handbook, Chapter 6).

7.3.2.4 Distraction and Technology

The problem of distracted driving has received significant attention (see e.g., this Handbook, Chapters 6, 9, 16). Distractions internal to the vehicle were noted in 10.7% of driver-related crashes and external distractions in 3.8% (National Highway Traffic Safety Administration, 2008). External distractions include natural factors in the driving environment, such as the actions of other vehicles or pedestrians, crashes, and other events. Internal distractions include a variety of activities inside the car, with conversation being the most prevalent problem. People engage in a number of additional activities in addition to driving, including conversations with passengers and on cell phones, as well as texting.

Engaging in these secondary tasks is the natural result of excess mental capacity or boredom, and may actually be an attempt by the driver to overcome the negative effects of low-vigilance conditions. The distractibility of mobile devices is a chronic problem due their highly salient and attention demanding nature. Bright lights and ringtones from handheld devices are naturally salient features that are, by design, attention demanding.

Caird, Simmons, Wiley, Johnston, and Horrey (2018) conducted a meta-analysis of 93 studies which showed that conversations with passengers and conversations on cell phones (both handsfree and handheld) all equally negatively affect driving performance. Similarly, texting has a negative effect on driver performance, with decrements far exceeding those of cell phone conversations (Drews, Yazdani, Godfrey, Cooper, & Strayer, 2009).

Parkes and Hooimeijer (2000) found low levels of SA for drivers talking on cell phones. These effects are more severe for less experienced drivers, who are often also younger drivers more likely to use mobile technologies. Kass, Cole, and Stanny (2007) showed that inexperienced drivers had lower SA than experienced drivers when talking on cell phones; however, both groups showed performance decrements.

Conversations and talking on cell phones may negatively affect SA by taxing the central executive function of working memory, thus impairing driver SA (Heenan, Herdman, Brown, & Robert, 2014) by leaving insufficient capacity for projection, an important aspect of driving, as well as lowered engagement. When Lee et al. (2007) simulated the effects of looking away from the road (e.g., distraction), this resulted in greater effects on data-driven processes, such as noticing new information outside of the current goal. Ma and Kaber (2005) found lowered Level 3 SA when drivers talked on cell phones. Kaber et al. (2016) showed that SA was 10–20% worse when drivers were exposed to a secondary task, with this reduction related to driver's abilities in both selective and divided attention.

Johannsdottir and Herdman (2010) and Heenan et al. (2014) demonstrated that the visual–spatial scratchpad component of working memory was related to drivers' ability to maintain SA of vehicles in front, and the phonological loop or executive function of working memory was more related to the SA of vehicles to the rear of the driver, as they are only viewed sporadically. This explains why they found conversations on the cell phone more detrimental to SA for vehicles to the rear of the driver. It is also possible that the additional workload of talking on the cell phone interfered with normal scan patterns due to attentional narrowing.

In a review of the literature on driver distraction, Strayer and Fisher (2016) showed that as drivers' involvement in secondary tasks increases, SA lowers due to drivers' reduced scanning of periphery and side- or rearview mirrors, and a reduction in anticipatory glances towards areas of potential hazards. Strayer, Drews, and Johnston (2003) also found a 50% increase in inattention blindness, where people looked at objects but failed to register them cognitively when talking on a cell phone, showing the negative effects of dual tasking on central executive processing. New forms of vehicle automation will introduce new sources of distraction, as people interact with and oversee them, that will need to be considered in future system designs.

7.3.3 AUTOMATION AND VEHICLE DESIGN

The effect of vehicle automation on SA is largely dependent on the type of automation involved and the functions that it automates (Endsley, 2017b; Onnasch, Wickens, Li, & Manzey, 2014). As shown in Figure 7.4, vehicle automation can be directed towards functions that assist driver SA, driver decision-making, or the performance of driver actions, all examples of SAE Level 0 automation (SAE, 2018). Automation

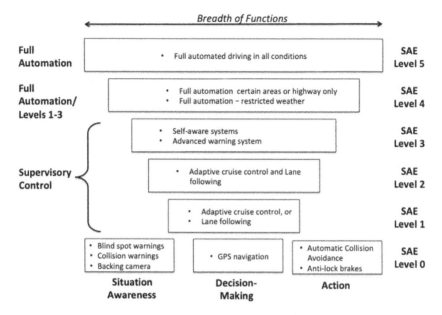

FIGURE 7.4   SAE 2018 levels of vehicle autonomy with examples.

can also be directed at a combination of all three, with the driver expected to act as a supervisory controller (SAE Levels 1–3). The higher levels of automation (SAE Levels 4 and 5) lean towards full automation in which human supervision is not necessary. While SAE Level 4 is considered full automation when operating within some prescribed set of circumstances (e.g., a highway, a certain section of a city, or in a certain weather), it reverts to a lower level of automation if operated outside of those conditions.

Based on an analysis of some 30 years' worth of research, it is expected that each of these SAE levels of automation vary significantly in terms of driver SA and performance (Endsley, in press). The majority of forms of SAE Level 0 automation, which provide assistance on some tasks but leave the driver with dynamic driving control, are expected to be beneficial. Automation directed at driver SA should be particularly helpful for SA, as long as it is reliable and has a low false alarm rate. Forward collision warning systems, for example, reduce crashes and do not lead to increases in taking on secondary tasks (Muhrer, Reinprecht, & Vollrath, 2012). Automation involving decision aids will be beneficial if it provides correct advice, but will lead to decision biasing, and loss of SA when wrong. Automation involving low-level manual tasks is also expected to be primarily positive.

However, SAE Level 1–3 automation that requires human oversight and intervention is expected to significantly lower driver SA, and will form the remainder of the focus in this chapter. Supervisory control automation can significantly lower SA, increasing the likelihood of being out-of-the-loop when driver action is required

(Endsley, 2017b; Endsley & Kiris, 1995; Onnasch et al., 2014). This can be attributed to three main factors (Endsley & Kiris, 1995):

1. *Information Display*. Many automated systems provide insufficient information on the state of the automation, poor automation transparency, and limited feedback on state of the system it is controlling to assist people in properly overseeing it (see also, this Handbook, Chapters 15, 21). For example, the acoustical characteristics of in-vehicle collision avoidance warning systems are critical for their effectiveness, and poorly designed auditory warnings actually decrease driver performance in comparison to no warnings at all (Lewis, Eisert, & Baldwin, 2018). Automation transparency is also important for allowing people to understand what the automation is doing and properly project what it will do in any given situation (Endsley & Jones, 2012; Selkowitz, Lakhmani, & Chen, 2017).

2. *Monitoring, Vigilance, and Trust*. The increased requirement for automation monitoring associated with supervisory control creates a significant problem due to vigilance problems. Hancock provides a detailed discussion of the negative effects of automation on monitoring performance under vigilance conditions (Hancock, 2013; Hancock & Verwey, 1997). Greenlee, DeLucia, and Newton (2018) showed vigilance decrements of about 30% over a 40-min drive when monitoring automation, with slower response times and an increase in missed detection of hazards. Further, their drivers rated subjective mental workload high under such conditions. Attempting to remain vigilant during long periods of monotony has been found to be hard work (Grubb, Miller, Nelson, Warm, & Dember, 1994).

   This problem is exacerbated as trust in automation increases. As trust in automation grows, people show a decrease in attention to critical information (Hergeth, Lorenz, Vilimek, & Krems, 2016; Lee & See, 2004; Muir & Moray, 1996). Additionally, people are more likely to trust automation when they have competing tasks to attend to (Wickens & Dixon, 2007), further decreasing SA of the automated functions (Carsten, Lai, Barnard, Jamson, & Merat, 2012; Kaber & Endsley, 2004; Ma, Sheik-Nainar, & Kaber, 2005; Sethumadhavan, 2009). Additionally, as trust in automation increases, so too does driver's willingness to engage in other tasks, including daydreaming, operating in-vehicle technologies, talking or texting on cell phones, eating, and grooming (Carsten et al., 2012; Hergeth et al., 2016; Ma & Kaber, 2005; Sethumadhavan, 2009), further lowering SA.

3. *Engagement*. Loss of driver engagement also occurs when passively processing information (as when watching automation or another person doing a task) as compared to when they are actively engaged in doing a task themselves (Endsley & Kiris, 1995; Manzey, Reichenbach, & Onnasch, 2012; Metzger & Parasuraman, 2001; Roediger & Karpicke, 2006; Slamecka & Graf, 1978). Endsley and Kiris (1995) were able to rule out complacency and differences in information presentation between automated and non-automated conditions in an automobile navigation task, and still found lower SA to occur due to passive processing.

A major review of automation research studies was summarized by the *automation conundrum*: "The more automation is added to a system, and the more reliable and robust that automation is, the less likely that human operators overseeing the automation will be aware of critical information and able to take over manual control when needed" (Endsley, 2017b). As vehicle automation becomes more capable, increasing its levels of reliability and robustness for performing a wider range of the driving tasks, driver's SA will become significantly lower, and they will fail in monitoring and providing intervention when needed. Even when people attempt to be vigilant, the degrading effects of reduced engagement are very difficult to overcome.

Petermeijer, Abbink, and de Winter (2015) investigated automated steering systems and found that while they improve driver satisfaction and performance, they also increase the time to recover from a system shutdown, demonstrating an out-of-the-loop problem. The reduced SA associated with automation can be significant. Young and Stanton (2007) found that approximately 2 seconds was required to respond to an event when drivers were using adaptive cruise control, an increase of between 1 and 1.5 seconds compared with manual driving conditions. Eriksson and Stanton (2017) also found that drivers performing other tasks took significantly longer to take control from automation, and that there was considerable variability in response times, from 3 to 21 seconds.

Biondi et al. (2018) showed that drivers of SAE Level 2 automation suffered reduced cognitive engagement as measured by physiological arousal measures, which correlated with poorer performance on peripheral event detection tasks, both of which got worse over time. Conversely, Endsley (2017a) found both periods of increased SA (due to freed up resources to look around) and decreased SA (due to reduced levels of engagement) when operating with SAE Level 2 automation over a six-month period, finding that it increased the variability of SA. Low SA occurred due to an increased tendency to engage in secondary tasks, passive processing which led to a delay in responding to events (i.e., becoming more like a passenger than a driver), and challenges in understanding and projecting what the automation would do in any given situation (Level 2 and 3 SA).

A key problem with vehicle automation is that it significantly increases the likelihood that drivers will engage in other non-driving-related tasks. De Winter, Happee, Martens, and Stanton (2014) found that highly automated driving resulted in a 261% increase in drivers performing other secondary tasks. Carsten et al. (2012) likewise showed that drivers in highly automated vehicles increasingly engage in other tasks, leading to significant workload spikes when intervention is required. These secondary tasks can range from mind-wandering to conversations with passengers or use of mobile technologies. In that automation frees up the driver to do other things, the fact that they do so should not be surprising.

The increase in willingness to direct attention to other tasks, particularly during periods of under-load, leads to slower response times to critical incidents (Merat, Jamson, Lai, & Carsten, 2012). Further, the willingness to engage in secondary tasks with automation increases significantly over time (Endsley, 2017a; Large, Burnett, Morris, Muthumani, & Matthias, 2017). The added negative impact on SA is significant. Ma and Kaber (2005) found improved SA with adaptive cruise control, due to an increase in mental capacity for attending to traffic; however, SA was significantly lowered when the drivers also used cell phones, with the driver's ability to project future events (Level 3 SA) most affected.

Solving this challenge is not easy. While some vehicle manufacturers have attempted to maintain driver engagement with alerts for drivers to keep their hands on the wheel or their eyes on the road, these are likely to be inadequate. Drivers using SAE Level 2 automation have been found to generally remove their hands from the steering wheel during most driving and only return them due to system warnings (Banks, Eriksson, Donoghue, & Stanton, 2018). However, lowered driver engagement due to automation is primarily cognitive. Interventions such as making drivers keep their hands on the wheel or eyes on the road have been found to be ineffective, with 28% of drivers still crashing into unexpected objects (Victor et al., 2018). The challenge is more of keeping one's mind on the road.

Overall, a wide body of research shows that vehicle automation placing drivers in the role of a supervisory controller (SAE Levels 1–3), with the expectation that they should be able to monitor and intervene, will significantly lower driver SA due to reduced cognitive engagement and vigilance, accompanied by an increased likelihood of conducting secondary tasks, both of which will lead to a resultant increase in automation-induced crashes. While some problems have been seen with SAE Level 1 automation, these problems will be greatly exacerbated with SAE Levels 2 and 3 automation that take on greater amounts of the driving task.

If and when vehicles achieve full automation (SAE Level 5), concerns about driver SA may become moot. However, it will take significant gains in automation software reliability and robustness to achieve acceptable levels of safety equal to or exceeding that of average human drivers. Software capable of full automation will likely take many years, and a number of constraining factors will act to reduce the likelihood of its adoption, including issues of who will bear the costs, willingness of manufacturers or consumers to take on the responsibility and liabilities of automated system performance, lack of trust, and the realization of actual safety advantages and disadvantages (as compared to current promises) (Endsley, 2019).

7.4   CONCLUSIONS

SA is fundamental to successful driving. People are extremely good at learning and adapting to highly variable driving conditions, resulting in over 495,000 miles between crashes and over 95 million miles between fatal crashes, a record that vehicle automation is far from matching (Endsley, 2018). Thus, it will be many years before driving becomes fully automated. In the meantime, finding ways to keep driver SA high will be vital. A number of recommendations can be made towards this goal.

1. Efforts to minimize the in-vehicle use of mobile technologies are required, due to their compelling nature, ubiquitous presence, and deleterious effect on driver SA. Technical interventions such as Apple's "Do not disturb while driving" should be very helpful. However, as it is easy to over-ride this feature, even more techniques may be needed.
2. Vehicle automation that acts to assist drivers, leaving them active in steering the vehicle, but providing protective features (i.e., automation limiters)

will provide a more effective approach than SAE Level 2/3 supervisory control approaches (Endsley, in press).

3. Implementations of vehicle automation need to pay close attention to the need to support driver SA. Guidelines for supporting SA on automation functioning and transparency, allowing drivers to better understand and project automation actions, can be used to improve driver-automation interactions (Endsley, 2017a; Endsley & Jones, 2012).

4. Enhanced driver training programs will be needed to help drivers better understand the behaviors and limitations of new automated features. Because automation can learn and change over time, developers will need to constantly update this training and deliver it to drivers at home as well as when a vehicle is purchased. Given that even very knowledgeable automation researchers can fall prey to automation induced loss of SA, however, it is unlikely that automation reliability training will be sufficient to compensate for its inherent vigilance challenges (Endsley, 2017a).

5. Enhanced driver training focused on improving SA, including hazard awareness, is highly recommended. This training would be very useful for beginner drivers, helping them to more rapidly develop the mental models that characterize expert driving (Endsley & Jones, 2012).

REFERENCES

Bacon, S. J. (1974). Arousal and the range of cue utilization. *Journal of Experimental Psychology, 102*, 81–87.

Baddeley, A. D. (1972). Selective attention and performance in dangerous environments. *British Journal of Psychology, 63*, 537–546.

Bainbridge, L. (1983). Ironies of automation. *Automatica, 19*, 775–779.

Banks, V. A., Eriksson, A., Donoghue, J. A., & Stanton, N. A. (2018). Is partially automated driving a bad idea? Observations from an on-road study. *Applied Ergonomics, 68*, 138–145.

Biondi, F., Lohani, M., Hopman, R., Mills, S., Cooper, J. L., & Strayer, D. L. (2018). 80 MPH and out-of-the-loop: Effects of real-world semi-automated driving on driver workload and arousal. *Proceedings of the Human Factors and Ergonomics Society Annual Meeting* (pp. 1878–1882). Santa Monica, CA: Human Factors and Ergonomics Society.

Bolstad, C. A. (2001). Situation awareness: Does it change with age? *Proceedings of the Human Factors and Ergonomics Society 45th Annual Meeting* (pp. 272–276). Santa Monica, CA: Human Factors and Ergonomics Society.

Borowsky, A. & Oron-Gilad, T. (2013). Exploring the effects of driving experience on hazard awareness and risk perception via real-time hazard identification, hazard classification, and rating tasks. *Accident Analysis & Prevention, 59*, 548–565.

Borowsky, A., Shinar, D., & Oron-Gilad, T. (2010). Age and skill differences in driving related hazard perception. *Accident Analysis & Prevention, 42*, 1240–1249.

Briggs, G. F., Hole, G. J., & Land, M. F. (2011). Emotionally involving telephone conversations lead to driver error and visual tunnelling. *Transportation Research Part F: Traffic Psychology and Behaviour, 14*, 313–323.

Caird, J. K., Simmons, S. M., Wiley, K., Johnston, K. A., & Horrey, W. J. (2018). Does talking on a cell phone, with a passenger, or dialing affect driving performance? An updated systematic review and meta-analysis of experimental studies. *Human Factors, 60*(1), 101–133.

Carsten, O., Lai, F. C. H., Barnard, Y., Jamson, A. H., & Merat, N. (2012). Control task substitution in semiautomated driving: Does it matter what aspects are automated? *Human Factors, 54*(5), 747–761.

Casson, R. W. (1983). Schema in cognitive anthropology. *Annual Review of Anthropology, 12*, 429–462.

Chaparro, A., Groff, L., Tabor, K., Sifrit, K., & Gugerty, L. J. (1999). Maintaining situational awareness: The role of visual attention. *Proceedings of the Human Factors and Ergonomics Society 43rd Annual Meeting* (pp. 1343–1347). Santa Monica, CA: Human Factors and Ergonomics Society.

Chapman, P. R. & Underwood, G. (1998). Visual search of driving situations: Danger and experience. *Perception, 27*, 951–9964.

Charlton, S. G. & Starkey, N. J. (2011). Driving without awareness: The effects of practice and automaticity on attention and driving. *Transportation Research Part F: Traffic Psychology and Behaviour, 14*(6), 456–471.

Charlton, S. G. & Starkey, N. J. (2013). Driving on familiar roads: Automaticity and inattention blindness. *Transportation Research Part F: Traffic Psychology and Behaviour, 19*, 121–133.

Damos, D. & Wickens, C. D. (1980). The acquisition and transfer of time-sharing skills. *Acta Psychologica, 6*, 569–577.

Davies, D. R. & Parasuraman, R. (1980). *The Psychology of Vigilance*. London: Academic Press.

de Winter, J. C., Happee, R., Martens, M. H., & Stanton, N. A. (2014). Effects of adaptive cruise control and highly automated driving on workload and situation awareness: A review of empirical evidence. *Transportation Research Part F: Traffic Psychology and Behaviour, 27*, 196–217.

Drews, F. A., Yazdani, H., Godfrey, C. N., Cooper, J. M., & Strayer, D. L. (2009). Text messaging during simulated driving. *Human Factors, 51*(5), 762–770.

Endsley, M. R. (1988). Design and evaluation for situation awareness enhancement. *Proceedings of the Human Factors Society 32nd Annual Meeting* (pp. 97–101). Santa Monica, CA: Human Factors Society.

Endsley, M. R. (1990). A methodology for the objective measurement of situation awareness. *Situational Awareness in Aerospace Operations* (AGARD-CP-478) (pp. 1/1–1/9). Neuilly Sur Seine, France: NATO - AGARD.

Endsley, M. R. (1993). A survey of situation awareness requirements in air-to-air combat fighters. *International Journal of Aviation Psychology, 3*(2), 157–168.

Endsley, M. R. (1995). Toward a theory of situation awareness in dynamic systems. *Human Factors, 37*(1), 32–64.

Endsley, M. R. (2006). Expertise and situation awareness. In K. A. Ericsson, N. Charness, P. J. Feltovich, & R. R. Hoffman (Eds.), *The Cambridge Handbook of Expertise and Expert Performance* (pp. 633–651). New York: Cambridge University Press.

Endsley, M. R. (2015). Situation awareness misconceptions and misunderstandings. *Journal of Cognitive Engineering and Decision Making, 9*(1), 4–32.

Endsley, M. R. (2017a). Autonomous driving systems: A preliminary naturalistic study of the Tesla Model S. *Journal of Cognitive Engineering and Decision Making, 11*(3), 225–238.

Endsley, M. R. (2017b). From here to autonomy: Lessons learned from human-automation research. *Human Factors, 59*(1), 5–27.

Endsley, M. R. (2018). Situation awareness in future autonomous vehicles: Beware the unexpected. *Proceedings of the 20th Congress of the International Ergonomics Association* (pp. 303–309). Florence, Italy: Springer.

Endsley, M. R. (2019). The limits of highly autonomous vehicles: An uncertain future. *Ergonomics, 62*(4), 496–499.

Endsley, M. R. (in press). Human-automation interaction and the challenge of maintaining situation awareness in future autonomous vehicles. In M. Mouloua & P. Hancock (Eds.), *Automation and Human Performance: Theory and Applications* (2nd ed.). Boca Raton, FL: CRC Press.

Endsley, M. R. & Bolstad, C. A. (1994). Individual differences in pilot situation awareness. *International Journal of Aviation Psychology, 4*(3), 241–264.

Endsley, M. R. & Jones, D. G. (2012). *Designing for Situation Awareness: An Approach to Human-Centered Design* (2nd ed.). London: Taylor & Francis.

Endsley, M. R. & Kiris, E. O. (1995). The out-of-the-loop performance problem and level of control in automation. *Human Factors, 37*(2), 381–394.

Engström, J., Markkula, G., Victor, T., & Merat, N. (2017). Effects of cognitive load on driving performance: The cognitive control hypothesis. *Human Factors, 59*(5), 734–764.

Eriksson, A. & Stanton, N. (2017). Takeover time in highly automated vehicles: Noncritical transitions to and from manual control. *Human Factors, 59*(4), 689–705.

Eysenck, M. W. (1982). *Attention and Arousal: Cognition and Performance.* Berlin: Springer-Verlag.

Finn, P. & Bragg, B. W. E. (1986). Perception of the risk of an accident by young and older drivers. *Accident Analysis & Prevention, 18*, 289–298.

Fracker, M. L. (1989). Attention allocation in situation awareness. *Proceedings of the Human Factors Society 33rd Annual Meeting* (pp. 1396–1400). Santa Monica, CA: Human Factors Society.

Greenlee, E. T., DeLucia, P. R., & Newton, D. C. (2018). Driver vigilance in automated vehicles: Hazard detection failures are a matter of time. *Human Factors, 60*(4), 465–476.

Grubb, P. L., Miller, L. C., Nelson, W. T., Warm, J. S., & Dember, W. N. (1994). Cognitive failure and perceived workload in vigilance performance. In M. Mouloua & R. Parasuraman (Eds.), *Human Performance in Automated Systems: Current Research and Trends* (pp. 115–121). Hillsdale, NJ: LEA.

Gugerty, L. J. (1997). Situation awareness during driving: Explicit and implicit knowledge in dynamic spatial memory. *Journal of Experimental Psychology: Applied, 3*, 42–66.

Gugerty, L. J. & Tirre, W. (1997). Situation awareness: A validation study and investigation of individual differences. *Proceedings of the Human Factors and Ergonomics Society 40th Annual Meeting* (pp. 564–568). Santa Monica, CA: Human Factors and Ergonomics Society.

Hancock, P. (2013). In search of vigilance: The problem of latrogenically created psychological phenomena. *American Psychologist, 68*(2), 97–109.

Hancock, P. & Verwey, W. B. (1997). Fatigue, workload and adaptive driver systems. *Accident Analysis & Prevention, 29*(4), 495–506.

Heenan, A., Herdman, C. M., Brown, M. S., & Robert, N. (2014). Effects of conversation on situation awareness and working memory in simulated driving. *Human Factors, 56*(6), 1077–1092.

Hergeth, S., Lorenz, L., Vilimek, R., & Krems, J. F. (2016). Keep your scanners peeled: Gaze behavior as a measure of automation trust during highly automated driving. *Human Factors, 58*(3), 508–519.

Hockey, G. R. J. (1986). Changes in operator efficiency as a function of environmental stress, fatigue and circadian rhythms. In K. Boff, L. Kaufman & J. Thomas (Eds.), *Handbook of Perception and Performance* (Vol. 2, pp. 44/41–44/49). New York: John Wiley.

Horswill, M. S. & McKenna, F. P. (2004). Drivers hazard perception ability: Situation awareness on the road. In S. Banbury & S. Tremblay (Eds.), *A Cognitive Approach to Situation Awareness: Theory, Measurement and Application* (pp. 155–174). Aldershot, UK: Ashgate Publishing.

Janis, I. L. (1982). Decision making under stress. In L. Goldberger & S. Breznitz (Eds.), *Handbook of Stress: Theoretical and Clinical Aspects* (pp. 69–87). New York: The Free Press.

Jeon, M., Walker, G. N., & Gable, T. M. (2014). Anger effects on driver situation awareness and driving performance. *Presence, 23*(1), 71–89.

Johannsdottir, K. R. & Herdman, C. M. (2010). The role of working memory in supporting drivers' situation awareness for surrounding traffic. *Human Factors, 52*(6), 663–673.

Kaber, D., Sangeun, J., Zahabi, M., & Pankok, C. (2016). The effect of driver cognitive abilities and distractions on situation awareness and performance under hazard conditions. *Transportation Research Part F: Traffic Psychology and Behaviour, 42*(1), 177–194.

Kaber, D. B. & Endsley, M. R. (2004). The effects of level of automation and adaptive automation on human performance, situation awareness and workload in a dynamic control task. *Theoretical Issues in Ergonomic Science, 5*(2), 113–153.

Kass, S. J., Cole, K. S., & Stanny, C. J. (2007). Effects of distraction and experience on situation awareness and simulated driving. *Transportation Research Part F: Traffic Psychology and Behaviour, 10*, 321–329.

Keinan, G. & Friedland, N. (1987). Decision making under stress: Scanning of alternatives under physical threat. *Acta Psychologica, 64*, 219–228.

Kircher, K. & Ahlstrom, C. (2017). Minimum required attention: A human-centered approach to driver inattention. *Human Factors, 59*(3), 471–484.

Large, D. R., Burnett, G. E., Morris, A., Muthumani, A., & Matthias, R. (2017). Design implications of drivers' engagement with secondary activities during highly-automated driving – A longitudinal simulator study. *Proceedings of the Road Safety and Simulation International Conference (RSS2017)*, The Hague, Netherlands: RSS.

LaRue, G. S., Rakotonirainy, A., & Pettitt, A. N. (2011). Driving performance impairments due to hypovigilance on monotonous roads. *Accident Analysis & Prevention, 43*(6), 2037–2046.

Lee, J. D. & See, K. A. (2004). Trust in automation: Designing for appropriate reliance. *Human Factors, 46*(1), 50–80.

Lee, Y. C., Lee, J. D., & Boyle, L. N. (2007). Visual attention in driving: The effects of cognitive load and visual disruption. *Human Factors, 49*(4), 721–733.

Lewis, B. A., Eisert, J. L., & Baldwin, C. L. (2018). Validation of essential acoustic parameters for highly urgent in-vehicle collision warnings. *Human Factors, 60*(2), 248–261.

Logan, G. D. (1988). Automaticity, resources and memory: Theoretical controversies and practical implications. *Human Factors, 30*(5), 583–598.

Ma, R. & Kaber, D. (2005). Situation awareness and workload in driving while using adaptive cruise control and a cell phone. *International Journal of Industrial Ergonomics, 35*, 939–953.

Ma, R. & Kaber, D. (2007). Situation awareness and driving performance in a simulated navigation task. *Ergonomics, 50*(8), 1351–1364.

Ma, R., Sheik-Nainar, M. A., & Kaber, D. B. (2005). Situation awareness in driving while using adaptive cruise control and a cell phone. *Proceedings of the Human Factors and Ergonomics Society 49th Annual Meeting* (pp. 381–385). Santa Monica, CA Human Factors and Ergonomics Society.

Manzey, D., Reichenbach, J., & Onnasch, L. (2012). Human performance consequences of automated decision aids: The impact of degree of automation and system experience. *Journal of Cognitive Engineering and Decision Making, 6*, 57–87.

Matthews, M. L., Bryant, D. J., Webb, R. D., & Harbluk, J. L. (2001). Model of situation awareness and driving. *Transportation Research Record, 1779*, 26–32.

McKenna, F. & Crick, J. L. (1991). *Hazard Perception in Drivers: A Methodology for Testing and Training*. Reading, UK: University of Reading, Transport and Road Research Laboratory.

McKenna, F. & Crick, J. L. (1994). *Developments in Hazard Perception*. London, UK: Department of Transport.

McKenna, F. & Farrand, P. (1999). The role of automaticity in driving. In G. B. Grayson (Ed.), *Behavioural Research in Road Safety IX*. Crowthorne, UK: Transport Research Laboratory.

Merat, N., Jamson, A. H., Lai, F. C. H., & Carsten, O. (2012). Highly automated drving, secondary task performance, and driver state. *Human Factors, 54*, 762–771.

Metzger, U. & Parasuraman, R. (2001). The role of the air traffic controller in future air traffic management: An empirical study of active control vs. passive monitoring. *Human Factors, 43*(4), 519–528.

Michon, J. A. (1985). A critical view of driver behavior models: What do we know, what should we do? In L. Evans (Ed.), *Human Behavior and Traffic Safety* (pp. 485–524). Boston, MA: Springer.

Mourant, R. R. & Rockwell, T. H. (1972). Strategies of visual search by novice and experienced drivers. *Human Factors, 14*(4), 325–335.

Muhrer, E., Reinprecht, K., & Vollrath, M. (2012). Driving with partially autonomous forward collision warning system: How do drivers react? *Human Factors, 54*(5), 698–708.

Muir, B. M. & Moray, N. (1996). Trust in automation: Part 2. Experimental studies of trust and human intervention in a process control simulation. *Ergonomics, 39*, 429–460.

NHTSA (2008). *National Motor Vehicle Crash Causation Survey* (HS 811 059). Washington, DC: National Highway Traffic Safety Administration.

Neisser, U. (1967). *Cognitive Psychology*. New York: Appleton-Century, Crofts.

Onnasch, L., Wickens, C. D., Li, H., & Manzey, D. (2014). Human performance consequences of stages and levels of automation: An integrated meta-analysis. *Human Factors, 56*(3), 476–488.

Pammer, K., Sabadas, S., & Lentern, S. (2018). Allocating attention to detect motorcycles: The role of inattentional blindness. *Human Factors, 60*(1), 5–19.

Parkes, A. & Hooijmeijer, V. (2000). The influence of the use of mobile phones on driver situation awareness. *Driver Distraction Internet Forum*. www-nrd.nhtsa.dot.gov/departments/nrd-13/driver-distraction/PDF/2.PDF.

Parmet, Y., Borowsky, A., Yona, O., & Oron-Gilad, T. (2015). Driving speed of young novice and experienced drivers in simulated hazard anticipation scenes. *Human Factors, 57*(2), 311–328.

Patten, C. J. D., Kircher, A., Ostlund, J., Nilsson, L., & Svenson, O. (2006). Driver experience and cognitive workload in different traffic environments. *Accident Analysis & Prevention, 38*(5), 887–894.

Petermeijer, S. M., Abbink, D. A., & de Winter J. C. F. (2015). Should drivers by operating with an automation-free bandwidth? Evaluating haptic steering support systems with different levels of authority. *Human Factors, 57*(1), 5–20.

Reason, J. (1984). Absent-mindedness and cognitive control. In J. E. Harris & P. E. Morris (Eds.), *Everyday Memory, Action and Absent-Mindedness* (pp. 111–132). London: Academic Press.

Roediger, H. L. & Karpicke, J. D. (2006). The power of testing memory: Basic research and implications for educational practice. *Perspectives on Psychological Science, 1*(3), 181–210.

Sabey, B. E. & Staughton, G. C. (1975). Interacting roles of road environment, vehicle and road user in accidents. *Proceedings of the 5th International Conference of the International Association of Accident and Traffic Medicine* (pp. 1–5). London: IAATM.

Schmidt, E. A., Schrauf, M., Simon, M., Fritzsche, M., Buchner, A., & Kincses, W. E. (2009). Drivers' misjudgement of vigilance state during prolonged monotonous daytime driving. *Accident Analysis & Prevention, 41*, 1087–1093.

Selkowitz, A. R., Lakhmani, S. G., & Chen, J. Y. C. (2017). Using agent transparency to support situation awareness of the autonomous squad member. *Cognitive Systems Research, 46,* 13–25.

Sethumadhavan, A. (2009). Effects of automation types on air traffic controller situation awareness and performance. *Proceedings of the Human Factors and Ergonomics Society 53rd Annual Meeting* (pp. 1–5). Santa Monica, CA: Human Factors and Ergonomics Society.

Slamecka, N. J. & Graf, P. (1978). The generation effect: Delineation of a phenomenon. *Journal of Experimental Psychology: Human Learning and Memory, 4*(6), 592–604.

Sohn, Y. W. & Doane, S. M. (2004). Memory processes of flight situation awareness: Interactive roles of working memory capacity, long-term working memory and expertise. *Human Factors, 46*(3), 461–475.

Strayer, D. L. & Fisher, D. L. (2016). SPIDER: A framework for understanding driver distraction. *Human Factors, 58*(1), 5–12.

Strayer, D. L., Drews, F. A., & Johnston, W. A. (2003). Cell phone induced failures of visual attention during simulated driving. *Journal of Experimental Psychology: Applied, 9,* 23–52.

Sulistayawati, K., Wickens, C. D., & Chui, Y. P. (2011). Prediction in situation awareness: Confidence bias and underlying cognitive abilities. *International Journal of Aviation Psychology, 21*(2), 153–174.

Treat, J. R., Tumbas, N. S., McDonald, S. T., Shinar, D., Hume, R. D., Mayer, R. E., … Catellan, N. J. (1979). *Tri-level Study of the Causes of Traffic Accidents: Final Report Volume I: Causal Factor Tabulations and Assessments. Institute for Research in Public Safety* (DOT HS-805). Bloomington, IN: Indiana University.

Treisman, A. & Paterson, R. (1984). Emergent features, attention and object perception. *Journal of Experimental Psychology: Human Perception and Performance, 10*(1), 12–31.

Underwood, G. (2007). Visual attention and the transition from novice to advanced driver. *Ergonomics, 50*(8), 1235–1249.

Underwood, G., Chapman, P., Bowden, K., & Crundall, D. (2002). Visual search while driving: Skill and awareness during inspection of the scene. *Transportation Research, Part F, 5*(2), 87–97.

Victor, T. W., Tivesten, E., Gustavsson, P., Johansson, J., Sangberg, F., & Ljung Aust, M. (2018). Automation expectation mismatch: Incorrect prediction despite eyes on threat and hands on wheel. *Human Factors, 60*(8), 1095–1116.

Wickens, C. D. (1992). *Engineering Psychology and Human Performance* (2nd ed.). New York: Harper Collins.

Wickens, C. D. & Dixon, S. R. (2007). The benefits of imperfect diagnostic automation: A synthesis of the literature. *Theoretical Issues in Ergonomics Science, 8,* 201–212.

Wiener, E. L. & Curry, R. E. (1980). Flight deck automation: Promises and problems. *Ergonomics, 23*(10), 995–1011.

Yanko, M. R. & Spalek, T. M. (2014). Driving with the wandering mind: The effect that mind-wandering has on driving performance. *Human Factors, 56*(2), 260–269.

Young, M. S. & Stanton, N. (2007). Back to the future: Brake reaction times for manual and automated vehicle. *Ergonomics, 50,* 46–58.

8 Allocation of Function to Humans and Automation and the Transfer of Control

Natasha Merat and Tyron Louw
University of Leeds

CONTENTS

Key Points 153
8.1　Introduction 154
8.2　Defining FA 154
　　8.2.1　Allocating Responsibility 155
　　8.2.2　Allocating Authority to Take Responsibility for a Function 157
8.3　Defining the Driving Task: How Automation Changes FA 158
8.4　The *Can* and *Why* of Allocating Functions 160
8.5　The Consequences of Inappropriate FA 163
8.6　Transfer of FA in AVs 165
8.7　Summary and Conclusions 166
References 167

KEY POINTS

- Traditional allocation of functions to machines has historically been used in static environments, which is challenging for the dynamic driving domain.
- As more functions are allocated to automated vehicles, it is important to understand the dynamic and fluid relationship that exists between humans and machines, especially in SAE Level 2–4 vehicles
- For highly automated vehicles to deliver the promise of reducing human error in road-related crashes, it is important that system designers are aware of the limitations and expectations of human users, to minimize the unexpected consequences of inappropriate function allocation.
- Safe and successful transfer of control from automated vehicles to humans requires better knowledge of how human attention and vigilance can be maintained during prolonged periods of automation, and how factors such as fatigue, distraction and complacency can be mitigated and managed.

8.1   INTRODUCTION

Over its entire history, the motor vehicle has probably had its most fundamental structural change during the past 20 years, with the addition of many primary (active) and secondary (passive) safety systems, which have mostly been implemented due to the need to improve road safety, and reduce vehicle-related injuries and deaths (Peden et al., 2004). Examples of primary (active) safety systems include electronic stability control, automatic emergency braking, and antilock brakes, which help reduce the likelihood of crashes (Scanlon, Sherony, & Gabler, 2017; Page, Hermitte, & Cuny, 2011). Secondary (passive) safety systems include airbags, seatbelts, and more advanced vehicle body engineering solutions, which mitigate the impact of crashes (Richter, Pape, Otte, & Krettek, 2005; Frampton & Lenard, 2009).

In addition to these safety systems, today's vehicles incorporate features that provide drivers with higher levels of assistance in performing the driving task, guiding, warning and informing the driver, as well as taking over particular driving functions, which can replace drivers' actual physical control of the vehicle, for certain time periods. As more tasks are taken away from the driver, and controlled by the vehicle's various systems, a number of human factors implications need to be considered regarding this change of role, to fully understand the implication of these additional systems on driver's behavior and performance, and the effect of any consequent changes on driver and road safety (see e.g., this Handbook, Chapters 1, 2).

To understand whether, and how, such allocation of function(s) to the vehicle's various systems is likely to influence the driving task, this chapter begins by providing a short overview of Function Allocation (FA) between humans and machines, initially considered during human–machine interaction studies in other domains. We then set the scene by outlining the functions required by humans in a conventional driving task, using several well-established models in this context, before discussing how, when, and why, different functions are allocated in an automated vehicle (AV). Specifically, we examine the rationale used by designers and engineers to allocate functions to each actor in an AV, followed by an overview of how, and when, drivers are informed about this allocation. Furthermore, we discuss what the consequences of such allocation of function might be on driver behavior and performance, how these may be managed, as well as the broader effect such consequences might have on road safety.

8.2   DEFINING FA

When considering the interaction and cooperation between humans and machines, function (or task) allocation simply considers whether, why, and how a function/task, or a series of related functions/tasks, are allocated to, and must be managed by the human, the machine, or a combination of the two, in order for a particular goal to be achieved (Bouzekri, Canny, Martinie, Palanque, & Gris, 2018). According to Pankok and Bass (2017), FA is *"a process which examines a list of functions that the human–machine system needs to execute in order to achieve operational requirements, and determines whether the human, machine (i.e., automation), or some combination should implement each function"* (p. A-7). Historically, this allocation of function

TABLE 8.1
The original Fitts List

Humans Appear to Surpass Present-Day Machines with Respect to the Following:	Present-Day Machines Appear to Surpass Humans with Respect to the Following:
1. Ability to detect a small amount of visual or acoustic energy	1. Ability to respond quickly to control signals and to apply great force smoothly and precisely
2. Ability to perceive patterns of light or sound	
3. Ability to improvise and use flexible procedures	2. Ability to perform repetitive, routine tasks
4. Ability to store very large amounts of information for long periods and to recall relevant facts at the appropriate time	3. Ability to store information briefly and then to erase it completely
	4. Ability to reason deductively, including computational ability
5. Ability to reason inductively	5. Ability to handle highly complex operations, i.e., to do many different things at once.
6. Ability to exercise judgment	

Source: Fitts et al. (1951), p. 10.

has been fixed in nature, with the MABA-MABA ("Men Are Better At"—"Machines Are Better At") lists (Price, 1985) revealing the assumption that either the human or the machine would be superior for a particular function. As initially outlined by Fitts et al. (1951; see Table 8.1), this allocation is partly determined by the ability of each actor to successfully achieve the required task, with humans being generally better at tasks requiring judgment, reasoning, and improvization, while machines are generally better at repetitive tasks, or those that require force, precision, and/or a quick response (de Winter & Hancock, 2015). However, although this type of FA continues to be considered (de Winter & Dodou, 2014), it has been widely criticized (Jordan, 1963; Fuld, 1993; Hancock & Scallen, 1996; Sheridan, 2000; Dekker & Woods, 2002), because it assumes that FA is static, and is considered acontextual, because it is insensitive to the influence of environmental variables (Scallen & Hancock, 2001). This view solidified as we began to understand the nature of work better, including its context and environment, while also appreciating that users of a system have different needs and information processing capabilities. Today, human factors investigations demonstrate that, for tasks requiring cooperation between machines and humans, although assigning the right function to the right actor is important, other factors must also be considered, to ensure safe and efficient task completion. These include the number of functions assigned, as well as the frequency and sequence of allocation of these functions. Also, it is essential that each actor assumes the appropriate *responsibility*, and *authority*, for taking control of, or assigning, a function.

8.2.1 ALLOCATING RESPONSIBILITY

Flemisch et al. (2012) suggest that by assuming responsibility for a task an actor becomes *"accountable for his actions."* With regards to humans, an awareness of this responsibility may be determined before they start a task, for example by reference to appropriate information and training about the task, or by the relay of suitable messages from the system during task completion, for instance, via relevant Human–Machine Interfaces (HMI). Assigning and assuming the right degree, and type, of responsibility is likely to reduce errors and confusion. The transfer of this responsibility between actors must also be achieved in a timely manner, and under the correct circumstances, in order to avoid or reduce task error. For example, safety may be affected if the human resumes responsibility for a task unnecessarily, when it is being well-controlled and managed by the system (Noy, Shinar, & Horrey, 2018). Equally, passing responsibility back to the human by the system in a timely manner is important, to ensure that adequate mental and physical resources are available to assume such responsibility (Louw, Kountouriotis, Carsten, & Merat, 2015).

Therefore, it is important that system engineers consider this allocation of responsibility to the system and human carefully, communicating this information clearly, to ensure the user is aware of their role, versus that of the system. Any allocation of responsibility to the human is also done under the assumption that the human honors that responsibility, and that there is a minimal likelihood of misuse or abuse of the system's functionality, which would result in reductions of their own and others' safety. However, the automation must also be designed with the human's limitations in mind, to ensure that any likely failures can be appropriately managed by the user.

Part of the designers' challenge is ensuring that users have the correct *mental model* of system functionality (Sarter & Woods, 1995; see also, this Handbook, Chapter 3) so that their responsibility for every stage of task completion is clear. Research shows that the ability to assume responsibility for a particular task, or a series of related tasks, also relies on users' expectation, training, and experience (Flemisch et al., 2012). This responsibly may also shift, be interrupted, or neglected, if the user is engaged in other competing tasks, which may or may not be related to the user's primary goal outcome. For example, in a recent driving simulator study, we investigated driver response to "silent" failures in SAE Level 2 and 3 automated driving, where automation failure during a simulator drive was not preceded by a takeover warning (Louw et al., 2019). This is an example of when automation hands responsibility of the function back to the driver, due to an unexpected/unknown failure, perhaps because the system encounters a scenario not anticipated by its designer, such as absence/obstruction of lane markings used for keeping the vehicle in its lane. Drivers completed two drives in a counterbalanced order. In one drive, they were required to monitor the road and driving environment during automation, where attention to the road was maintained by asking them to read the words on a series of road-based Variable Message Signs (VMS, Level 2). For the other condition (Level 3), drivers performed an additional non-driving related task (NDRT) during automation, the visual search-based Arrows task (see Jamson & Merat, 2005). This task was presented on a screen near the gear shift, obliging drivers to look down and away from the road. The VMS task was also required in this drive, which meant that the drivers divided their attention between the road, the VMS, and the NDRT.

When considering performance, results showed that, after the silent failure, a significantly higher number of lane excursions were observed during the NDRT drive (Level 3), and participants took longer to take over control. Participants also had a more erratic pattern of eye movements after silent automation failure in the NDRT drive, presumably because they were attempting to gather the most useful visual information from the road ahead and the dash-based HMI, which contained information about automation status (on/off). This example of a fluid, and un-signalized, shift of task control between machine and human, illustrates the detrimental effects of unclear responsibilities, especially if the driver's attention is directed away from the forward road, and if the accompanying HMI is confusing, rather than assisting, the driver.

One main reason for striving towards clearer allocation of responsibility for each actor is to ensure that the source of any possible errors during task engagement can be rapidly identified. Therefore, the main human factors challenge here is not only for system designers to assign the correct responsibility to each actor, but also for humans to be aware, and capable, of honoring this responsibility, through to completion of the task. Of course, problems arise when this allocation is unreasonable, or when sustained commitment by the human is not possible, for example, when fatigue and distraction creep in.

8.2.2 Allocating Authority to Take Responsibility for a Function

In addition to the allocation of the correct level and type of responsibility to each actor, when humans and machines are cooperating, successful and safe accomplishment of tasks must also consider the *rights* of each actor in assuming a function, or allocating authority for the function to the other actor. Flemisch et al. (2012) describe this "authority" as what each actor is, or is not, *"allowed"* to do. Here, it is assumed that the authority to take over a function, or allocate it to the other actor, must be closely linked to the capabilities of either actor, and will be partly related to the MABA-MABA list (Fitts et al., 1951; Price, 1985). For systems, this authority to manage the task, or take over from the human, must be accompanied by a guarantee that the system can function in all foreseeable scenarios. From a human factors perspective, the consideration for functions that can be shared between humans and machines is that, following prolonged assignment of function responsibility to the system, humans will likely suffer from certain limitations, which can affect goal completion, especially if there is a system failure. The likelihood of these limitations arising must, therefore, be taken into consideration, when deciding whether to give humans the authority to take control of a function. Examples of such limitations include (1) reduced attention to, and monitoring of, the function (Carsten, Lai, Barnard, Jamson, & Merat, 2012; Körber, Cingel, Zimmermann, & Bengler, 2015); which leads to (2) loss of situation awareness towards the system and surroundings, diminishing ability to resume responsibility and control (if required) (Sarter & Woods, 1995; this Handbook, Chapter 7), as well as (3) onset of drowsiness/fatigue (Goode, 2003; this Handbook, Chapter 9), and (4) loss of skill to complete the function, due to prolonged non-use (Hampton, 2016). Therefore, system designers must decide whether or not an agent should be given the authority to assume and/or assign control of a particular function, when a task is being performed adequately by the other agent. Here, it seems sensible that the human should be given less authority following longer periods of system functionality.

A key challenge in this context is understanding how humans can remain vigilant and engaged during prolonged periods of automation use, sustaining the ability to successfully resume responsibility of the function (for instance, due to failures; see also, this Handbook, Chapter 6). A consideration of how this vigilance and capability of humans can be determined by the system is also important, as is the ethical consequences of incorrect FA, or authority to assume control. For example, should an appropriately functioning system cede control to an impaired driver, if asked to do so?

Having provided a generic overview of FA and summarized the implications of allocating responsibility and authority to each agent in complex scenarios involving human–machine interactions, it is now important to understand how this general overview relates specifically to the interaction of humans with higher levels of vehicle automation. This is especially important when considering functions that are currently *shared* between the human and the vehicle, where responsibility and authority are assigned to/assumed by either agent. Before considering this FA in AVs, the next section provides a brief overview of models developed for describing the driver's role during manual control of driving and describes how these roles are likely to change as a result of allocating functions to the vehicle.

8.3 DEFINING THE DRIVING TASK: HOW AUTOMATION CHANGES FA

Models of driver behavior, originally developed for manual control of the vehicle, generally consider three main categories of driver responsibility and control, for achieving the driving task (Hollnagel & Woods, 2005; Michon, 1985). The first category involves a physical (perceptual-motor) engagement with the vehicle controls, to manage the appropriate lateral and longitudinal position of the vehicle in the intended driving lane (maintained via moment-to-moment steering, brake/accelerator control). The next category of control by the human involves tactical functions, such as negotiating intersections, or detecting and avoiding obstacles, and changing lane accordingly, which also relies on perceptual-motor engagement. Finally, strategic control involves higher levels of (cognitive) engagement, for tasks such as navigation and route planning. As outlined in Merat et al. (2019), in manual driving, some level of monitoring is required by the driver, for each of these levels of control (see Figure 8.1; SAE, 2016a; b). However, as these functions are (either progressively, intermittently, or completely) assumed by the vehicle's automated driving system (Level 1–5), the role of the driver will change from that of an *active controller* of individual functions that operate the vehicle during a designated journey, to one that *monitors* and supervises some of these functions, while perhaps continuing to engage in, and be responsible for, others.

However, it has long been shown that humans are not especially effective at supervising or monitoring a system for long periods (Mackworth, 1948; this Handbook, Chapters 6, 21) and that as the length of such monitoring increases, human factors problems arise, which can lead to reduced safety. Examples from aviation show that this is due to issues such as user distraction (Endsley & Garland, 2000; Loukopoulos, Dismukes, & Barshi, 2001; this Handbook, Chapter 9), fatigue (Goode, 2003; this Handbook, Chapter 9), over-trust of system capabilities—followed by

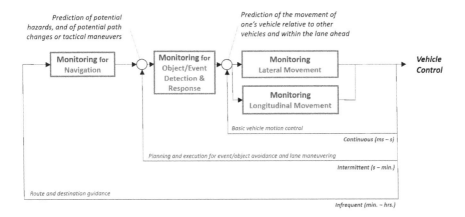

FIGURE 8.1   Driver's monitoring role in manual control of the driving task. (From Merat et al., 2019; based on Michon's model, 1985; © 2019 Springer. Reprinted with Permission of Springer Publications.)

user complacency (Parasuraman & Riley, 1997; this Handbook, Chapter 4), loss of skill (Hampton, 2016; this Handbook, Chapter 10), and degraded situation awareness (Salmon, Walker, & Stanton, 2016; this Handbook, Chapters 7, 13). Some of these errors are thought to be exacerbated by lack of suitable feedback from the system via its HMI (Lee & Seppelt, 2009). Here, we distinguish between monitoring of systems controlled by the human, where the perceptual-motor (physical) link is still preserved in manual driving (as shown in Figure 8.1), compared to where an automated system's performance is monitored without this physical link. Indeed, there is also a need to describe precisely what monitoring refers to in this context. While "monitoring" is considered synonymous with "checking" and "observing" a system's performance, it is not simply a case of verifying that some level of physical/perceptual-motor/cognitive engagement is maintained with the system (such as establishing whether eyes are on the road and hands are on the steering wheel). Instead, as Victor et al. (2018) highlight, an additional cognitive element (an element beyond paying attention) may also be required as part of such monitoring, to ensure that the user is capable of *"understanding in the mind the need for action control."*

Therefore, as more functionality is taken over by the automated system, and the role of the driver changes to that of a supervisor of these functions, there is a need for additional *aides* in the vehicle, to help the human controller with their altered role. These include interfaces that offer intuitive and accurate information about the automated system's functionality, informing the driver of likely changes in this functionality. This information should also be timely, provided with adequate notice, and should not surprise, distract, or overload the user (see also, this Handbook, Chapter 15). Drivers may also need assistance in managing the function that is being transferred back from the automated system, since skill degradation and reduced situation awareness are known to accompany such FA to the vehicle (Endsley, 2017), especially after longer periods of system use (Trösterer et al., 2017). Here, driver monitoring systems will be a useful addition to the vehicle

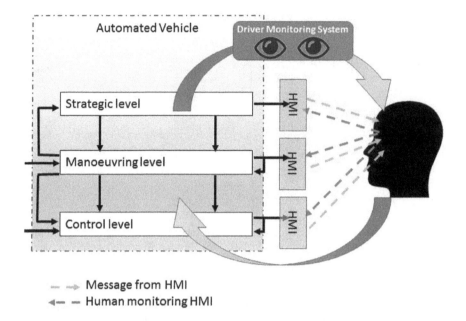

FIGURE 8.2 **(See color insert.)** A proposed model showing the changing position and role of the human due to the introduction of automated functions in the driving task.

(this Handbook, Chapter 11), to ensure that drivers are vigilant, and capable of honoring their responsibilities (this Handbook, Chapters, 9, 10). Figure 8.2 illustrates the effect of these changes on the driver's role, altering the original models of driver behavior, developed for conventional driving. As more and more functions are allocated to systems, the driver's physical control of the vehicle decreases. Depending on the level of automation engaged, drivers' reliance on good warning and communication from the different HMIs will increase. To ensure there is a suitable degree of monitoring of this HMI, and that important information and warnings are not missed by the driver, an informative and accurate driver monitoring system (DMS), which manages the human–HMI-vehicle link is required. In addition to acquiring more accurate data about driver state for such DMS, future research must consider opportunities for informative and intuitive HMI for highly AVs (Carsten & Martens, 2018). This knowledge can also be used to inform extensive, and regular, training of the human driver, to ensure they have a good understanding and mental model of system capabilities and limitations.

8.4   THE *CAN* AND *WHY* OF ALLOCATING FUNCTIONS

Whether a function *can* be performed by an agent, and therefore allocated, has traditionally been considered simply by assessing strengths and weaknesses, or capabilities, of each agent (Hollnagel & Bye, 2000). However, even if there is a good argument that an individual function *can* be allocated, it does not necessarily provide compelling justification or guidance for *why* a function should be allocated, *when* it is suitable for this allocation to take place, and *who* should be responsible for this allocation. Some regard of how different functions interact with each other, and their effect on human performance, is also important, where it can be argued that the sum effect on performance is not equal to all of its parts.

As discussed above, when considering the allocation of functions between humans and machines, it is important to establish *what* task is being allocated (i.e., the nature of the task/function), as well as the *degree* of involvement in task management for each agent (i.e., how is responsibility assumed or shared between agents). However, as part of this discussion, it is also essential to establish whether or not a function *can* be allocated (i.e., can the machine perform the function/task, at least as well as, or better than, the human?). There are also safety, ethical, and moral issues when deciding whether a function *should* be allocated, that go beyond the system's technical capabilities. This is also linked to how much authority a designer gives the system for taking responsibility of the task. Here, it seems essential for engineers and system designers to have a good appreciation of the unintended consequences of this FA on humans, which, as outlined above (and further below), may lead to user confusion, distraction, fatigue, loss of skill, or complacency. These unintended consequences are also closely tied to system failures, which may occur due to unforeseen technological limitations (not yet known by designers), or an unintentionally lax testing protocol, as well as user (mis)understanding of system capabilities.

In the case of vehicle automation, the manufacturer's motivations, and rationale, for FA is partly motivated by the challenges facing our congested and polluted cities, and a desire to reduce transport-related emissions and increase throughput, while also enhancing driver comfort and improving road safety. However, based on our understanding of how automation affects human performance, the aspiration to increase road safety by removing "the human element" is currently a growing irony. For example, the "out-of-the-loop" problems associated with a lack of engagement in the driving task (Merat et al., 2019; this Handbook, Chapter 21), especially when sparse and uncomplicated road conditions provide drivers with a false sense of security about system capabilities, are leading to real-world crashes (Vlasic & Boudette, 2016; Silverstein, 2019; Stewart, 2019).

Another important motivation for allocation of more functions to the system, which in turn increases the level of vehicle automation, is the desire to release humans from the monotonous aspects of the driving task, providing freedom for engagement in other (more productive) tasks. This freedom to attend to other tasks is linked to economic benefits, estimated to be worth tens of billions of dollars (Leech, Whelan, Bhaiji, Hawes, & Scharring, 2015). Again, if not planned and implemented well, this task substitution can lead to the same human factors problems outlined above, in addition to an eventual loss of skill, with prolonged system engagement (Carsten et al., 2012).

Different approaches have been used for categorizing the capabilities of automated driving systems. For example, original categorization of levels of automation, proposed by control engineers, typically accounts for the locus of control (human or automation) and how information is presented to the human (cf. Sheridan & Verplank, 1978). There are also more driving-specific levels of automation, which describe, at each level of automation, what aspects of the primary driving task are performed by the human or the system (SAE, 2016a). Finally, when considered within a human factors context, systems can also be classified based on their correspondence to models of human information processing, such as sensory perception, working memory, decision-making, and response selection (Parasuraman, Sheridan, & Wickens, 2000; this Handbook, Chapter 21).

While each of these approaches has a particular application, and implies some pre-determined FA, the main failure of such categorization is that, for the most part, they focus on the capabilities of systems, rather than that of their users. They also fail to identify the influence of system and human on one another, when the two have to work together, as is currently the case for vehicle automation. Therefore, while simply considering a vehicle's ability based on its stated levels of automation would be a desirable solution, it does not represent the ideal approach for determining an appropriate FA.

Here, it can be argued that the MABA-MABA type lists (Price, 1985; see Table 1) represent the skills and abilities of machines and humans in *the best-case* scenarios. However, in reality, these abilities cannot be maintained in perpetuity. For example, systems will have specified Operational Design Domains (ODD), which will determine whether, where, and when they reach their limitations. Due to shortcomings in technological advances, this limitation is not an issue that will be resolved in the near future, even though substantial developments are being achieved in this context, on a daily basis, with both automotive companies, Tier 1 suppliers, and big and small newcomers in the market investing heavily in this area. However, widespread penetration of vehicles with Level 5 (SAE, 2016a) automated driving capability, for all road types and environments, is not likely for some decades, before which humans will still be involved in, and responsible for, different aspects of the driving task.

The continuous nature of driving, and the constantly changing environment in which it is performed, means that the moment-to-moment driving tasks and responsibilities will also change. Consequently, the allocation of responsibility for some functions/tasks will need to transfer between the AV and the human driver, depending on the capability of the system, and the particular driving environment. To illustrate, using an example from limited ODD, an AV may be able to operate at Level 2 in most areas, at Level 3 in some areas, and at Level 4 in only a few areas. If this vehicle moves from an area where it can function at Level 4, to one where it can function at Level 2, the human is required to be aware of this change, and start monitoring system and road environment for Level 2 functionality. However, if the vehicle moves from a Level 4 to a Level 0 area, this change in ODD would require a fundamental shift in the human driver's responsibilities, which will not only require monitoring the road and vehicle, but also resuming lateral and longitudinal control of the vehicle. This transition may also involve the responsibility for obstacle detection and avoidance. Therefore, functional capability alone is not adequate, with different (and changing) environmental settings also playing a role in this relationship. The need for dynamic allocation of responsibility and authority between human and machines is therefore necessary for some functions, until the system can satisfactorily perform in all possible driving conditions. Here, an ideal solution for system functionality is its ability to *recognize* its own limitations, as the environment changes, informing the human, in sufficient time (Figure 8.2). System functionality should therefore only be available for the correct environmental setting, and otherwise, not operational.

Another consideration in this context is that the hardware and software utilized in automated driving systems can change rapidly and frequently. For example, some OEMs allow "over-the-air" download updates of automated driving software (Barry, 2018). This type of, instant, change may alter the vehicle's behavior in certain scenarios, also changing system capability, for example, by activating latent hardware, which may enable new features. The nature of this update creates problems with "type approval" of vehicles, as well as presenting significant human factors challenges. For example, this approach requires users to update their mental model of the functionality of the system, which presents a higher risk of mode confusion. These issues can be of concern for driver training, especially novices, since research on training in other domains, such as aviation, has shown that some novice pilots are biased towards trusting automation over their own judgment (Parasuraman & Riley, 1997). Casner and Hutchins (2019) argue that, for successful use of new automated systems in the driving domain, a comparable level of consideration, to that used in aviation, should be given to the training protocols developed for human drivers, to ensure they are familiar with the capabilities of the system, appreciating their own capabilities and limitations, as well as having a good understanding of the "human-automation team."

In sum, the key point to consider here is that, while a function may be allocated in good faith for a particular system, or in a particular context, as soon as that context changes, the reallocation of a function may actually cause more harm than good, if it is not properly understood or implemented by its user. It is therefore important to fully appreciate these aspects of the technology's fallibility and propensity to change rapidly, when deciding who does what, and when, and also how the authority for resuming this responsibility is determined. At the moment, the rapid implementation of automation in driving means that humans are left to do the tasks that machines are either not yet, or perhaps ever, able to achieve, a concept known as the *leftover approach* (Bailey, 1989). As Chapanis (1970) aptly argues, it is our job as human factors researchers/engineering psychologists to ensure that these tasks are manageable within human capabilities.

8.5   THE CONSEQUENCES OF INAPPROPRIATE FA

As system designers assign functions to humans or machines, in addition to considering the capabilities of each, they must also reflect on how the number of functions, and the timing of FA to the human, affects performance, with factors such as user impairment (distraction and fatigue; e.g., Horberry, Anderson, Regan, Triggs, & Brown, 2006; Brown, 1994), under/overload (Hancock & Warm, 1989), and skill degradation (FAA, 2013; Louw et al., 2017a), a few of the unintended consequences of inappropriate FA (Lee, Wickens, Liu, & Boyle, 2017).

A long history of driver behavior research has demonstrated that (apart from the very young/inexperienced and the very old) humans are generally competent drivers, with around two fatal crashes, per 100 million miles driven (Tefft, 2017). However, up to 90% of road vehicle accidents are thought to be caused by human error (Treat et al., 1979). Indeed, as highlighted above, this accountability of humans in crashes is one of the many rationales for introducing automated functions in vehicles. However, it has also been aptly argued that, if automation is not designed with the human in mind, replacing humans with automation does not necessarily remove the error, but simply *changes* it, "To the extent that a system is made less vulnerable to operator error through the application of automation, it is made more vulnerable to designer error" (Parasuraman & Riley, 1997, p. 249).

In today's 24/7 hyper-connected society, drivers are still as, if not more, susceptible to fatigue, distraction (Gary, Lakhiani, Defazio, Masden, & Song, 2018; this Handbook, Chapter 9) and loss of awareness, than they were when so many driving functions were not automated even ten years ago (Endsley, 2017; this Handbook, Chapter 7). As long as drivers have some safety-critical role, the effects of these limitations on performance and safety will remain, and are likely to be, magnified in some cases by the addition of automation. Moreover, the introduction of automation exposes "new" human factors limitations, based on the requirement by humans to interact, arbitrate, communicate, and cooperate with a system that has been developed without consideration of human's need for continued training in this context. For example, if drivers have to constantly update their mental model of the automated driving system, *errant mental models* (Saffarian, de Winter, & Happee, 2012) or inappropriate communication could develop, giving rise to *mode confusion* (Sarter & Woods, 1995), especially if a vehicle has multiple modes of operation or levels of automation that have narrowly defined ODDs.

The extent to which users *trust* a system is also an important factor when considering FA (e.g., this Handbook, Chapter 4). When assigning a task or function to a machine, it is important that systems can help users develop an appropriate (well-calibrated) level of trust in, or reliance on, the machine, where it performs the task they believe it will perform. However, in their interaction with the machine, users may either not trust the system enough (*distrust*) or trust it too much (*over-trust*). *Distrust* can lead to users perceiving the automation as imperfect, resulting in *disuse* of the system (Parasuraman & Riley, 1997). Disuse can occur if the system is too susceptible to being incorrectly configured, if the procedure for operating it is too convoluted (Carsten & Martens, 2018), or if there is a high rate of false alarms (Parasuraman & Riley, 1997). Distrust can also occur if a system does not adequately communicate its behavior or intention to its users, resulting in *automation induced surprises* (De Boer & Dekker, 2017) or misclassification of system errors (Wickens, Clegg, Vieane, & Sebok, 2015), and thus unsafe, or inappropriate, attempts to intervene (Louw et al., 2019).

On the other hand, *over-trust,* sometimes referred to as *complacency* or *automation bias*, occurs when users trust a system more than is warranted, and is exacerbated when users rarely encounter failures (Bainbridge, 1983; Moray, 2003; Parasuraman & Manzey, 2010). One of the consequences of complacency is *out-of-the-loop* behavior, where users are no longer actively attending to/monitoring or controlling the process, or significantly reduce their monitoring of the system (Merat et al., 2019). This leads to users with reduced *situation awareness* (Ma & Kaber, 2005) during automated driving, and an increased likelihood of *skill loss* in manual control, as a consequence of long-term automation use (Trösterer et al., 2017;

Ward, 2000). Of course, these issues may also arise in compliant drivers (i.e., drivers who are appropriately monitoring their Level 2 vehicle). Therefore, the primary concern of inappropriate allocation of tasks is that errors and risks are not detected and dealt with, either by the human or the system, compromising safety. Safety is also affected if there is inappropriate intervention by the human. For instance, if the efficient operation of the system is unnecessarily interrupted by the human resuming control of steering, disturbing the safe trajectory of the vehicle, resulting in a collision. This would also be of concern if the human is incapacitated (e.g., due to fatigue or intoxication). Research needs to establish the risk of the above occurrence in the context of when, and how, drivers will have to interact with the system.

8.6   TRANSFER OF FA IN AVs

One of the unique characteristics of allocating functions for automated driving systems, and in particular Levels 1–3 (SAE, 2016a), is that driving is a continuous task, which means that if the allocation changes, the human driver must resume manual vehicle control (with little or no warning), in order to maintain an appropriate level of task performance. This type of failure is perhaps dissimilar to some control room tasks, which are serialized and, therefore, can be stopped and restarted when the allocation changes, avoiding a deleterious impact on overall task performance. The driving environment can also be more complex than, for example, aviation, with the chance of an error being more catastrophic on a congested road. With fluid or dynamic allocation of functions (Byrne & Parasuraman, 1996; Hancock & Scallen, 1996; this Handbook, Chapter 16), there are inevitably situations in which drivers will have to either resume manual control or cede control to the automated driving system. For example, if a driver, who is using a Level 2 automated driving system, is not monitoring the road environment appropriately, the automated driving system may request the driver to resume manual control, although recent research in this context has illustrated that safe resumption of control by drivers is not always possible (Louw, Madigan, Carsten, & Merat, 2017b; Zhang, de Winter, Varotto, Happee, & Martens, 2019) or, if possible, is not always initiated (Victor et al., 2018).

For safe resumption of control, drivers must have enough information about the vehicle's operation and the surrounding traffic situation, and there must be a clear means of communication between the two actors. However, despite intense work on trying to understand and solve this example of a "hand-over" in recent years (c.f. Gold, Damböck, Lorenz, & Bengler, 2013; Merat, Jamson, Lai, Daly, & Carsten, 2014; Louw, Merat, & Jamson, 2015; Zeeb, Buchner, & Schrauf, 2015; Madigan, Louw, & Merat, 2018; Louw et al., 2017a; Zhang et al., 2019; Mole et al., 2019; McDonald et al., 2019; Gonçalves, Louw, Madigan, & Merat, 2019), a widely applicable solution is not yet available. This is due to the complex set of scenarios present in the driving environment which can lead to hand-overs, making it difficult to specify the correct timing, location, modality, legibility, and sequence of the information that must be presented by the vehicle's HMI, and managed by the driver. The value of this guidance also interacts with the driver's own expectations, mental models, experience, and capabilities. Therefore, when considering whether to create a machine to perform a function, it is important to establish whether a driver can reasonably be expected to intervene in time, appropriately, and safely. In this consideration, it should also be clear to the designer what the expectations are of the driver, what a successful response is, how this is measured, and whether these expectations and assessment methods apply uniformly across different situations. A fluid allocation of functions, therefore, requires engineers and designers to balance the benefits of introducing automation carefully against the risk of errors accompanying allocation switching.

Aside from the particular hand-over problems discussed above, there are a number of other factors that should be considered when allocating functions that will be controlled simultaneously by the human and automation, for example, in a Level 2 automated driving system. These include the likelihood that drivers will misuse or abuse the automated driving system, whether they are flexible enough to adapt to dynamically changing roles and responsibilities during different journey types, and whether they actually want to have joint responsibility for the task or function. Further, there may be training requirements for users to learn to operate the automated driving system, depending on its design characteristics, which could bring additional costs, research needs, and barriers to automated driving system rollout (see e.g., this Handbook, Chapter 18). For example, it may be possible for a car dealership to train the driver of a newly purchased vehicle, but this is not realistic for users of a rental car. As the introduction of such systems is relatively rapid, new and more agile methods of driver training, examination, and driver licensing need to be developed, which is especially relevant given situations where drivers may need to operate new, borrowed, or rental cars equipped with systems they may be unfamiliar with. Finally, a move to more standardized HMI must be considered (Carsten & Martens, 2018).

8.7 SUMMARY AND CONCLUSIONS

AVs with some SAE Level 2 and Level 3 functionality are already available to consumers in the developed world, and manufacturers and researchers are currently testing human interaction with higher levels of automation (see for example: www. l3pilot.eu, http://agelab.mit.edu/avt). As the frequency and duration of drivers' use of automation systems increase, designers must be aware of the concomitant human factors challenges, especially if occasional system limitations require human intervention and control.

This chapter highlights the challenges that still exist with respect to FA for a dynamic task such as driving and provides some possible mid-term solutions for system engineers. In an era where such systems are not yet 100% capable of replacing the human in all road environments, and at all times, it is important for system engineers to be aware of the possible unintended consequences of their developed systems, ensuring that the responsibility of each agent in this human–automation relationship is clear throughout a drive, and that suitable mitigation processes are in place for managing system and human limitations. The journey towards achieving fully self-sufficient AVs, safe to be used in all infrastructures, requires human factors specialists and engineers to work together to provide the knowledge needed for suitable training, communication, and interaction protocols, better HMI, and more advanced driver monitoring systems.

REFERENCES

Bailey, R. W. (1989). *Human Performance Engineering: Using Human Factors/Ergonomics to Achieve Computer System Usability.* Upper Saddle River, NJ: Prentice-Hall.

Bainbridge, L. (1983). Ironies of automation. In G. Johannsen & J. E. Rijnsdorp (Eds.), *Analysis, Design and Evaluation of Man–Machine Systems* (pp. 129–135). Oxford: Pergamon.

Barry, K. (2018). *Automakers Embrace Over-the-Air Updates, But Can We Trust Digital Car Repair?* Retrieved from www.consumerreports.org/automotive-technology/automakers-embrace-over-the-air-updates-can-we-trust-digital-car-repair/

Bouzekri, E., Canny, A., Martinie, C., Palanque, P., & Gris, C. (2018). Using task descriptions with explicit representation of allocation of functions, authority and responsibility to design and assess automation. *IFIP Working Conference on Human Work Interaction Design* (pp. 36–56). Berlin: Springer.

Brown, I. D. (1994). Driver fatigue. *Human Factors, 36*(2), 298–314.

Byrne, E. A. & Parasuraman, R. (1996). Psychophysiology and adaptive automation. *Biological Psychology, 42*(3), 249–268.

Carsten, O., Lai, F. C., Barnard, Y., Jamson, A. H., & Merat, N. (2012). Control task substitution in semiautomated driving: Does it matter what aspects are automated? *Human Factors, 54*(5), 747–761.

Carsten, O. & Martens, M. H. (2018). How can humans understand their automated cars? HMI principles, problems and solutions. *Cognition, Technology & Work, 21*(1), 3–20.

Casner, S. M. & Hutchins, E. L. (2019). What do we tell the drivers? Toward minimum driver training standards for partially automated cars. Journal of Cognitive Engineering and Decision Making, 13(2), 55–66. doi:10.1177/1555343419830901

Chapanis, A. (1970). Relevance of physiological and psychological criteria to man-machine systems: The present state of the art. *Ergonomics, 13*(3), 337–346.

De Boer, R. & Dekker, S. (2017). Models of automation surprise: Results of a field survey in aviation. *Safety, 3*(3), 20.

de Winter, J. C. F. & Dodou, D. (2014). Why the Fitts list has persisted throughout the history of function allocation. *Cognition, Technology & Work, 16*(1), 1–11.

de Winter, J. C. F. & Hancock, P. A. (2015). Reflections on the 1951 Fitts list: Do humans believe now that machines surpass them? *Procedia Manufacturing, 3*, 5334–5341.

Dekker, S. W. & Woods, D. D. (2002). MABA-MABA or abracadabra? Progress on human–automation co-ordination. *Cognition, Technology & Work, 4*(4), 240–244.

Endsley, M. R. (2017). Autonomous driving systems: A preliminary naturalistic study of the Tesla Model S. *Journal of Cognitive Engineering and Decision Making*, 11(3), 225–238.

Endsley, M. R. & Garland, D. J. (2000). Pilot situation awareness training in general aviation. *Proceedings of the Human Factors and Ergonomics Society Annual Meeting, 44*, 357–360.

Federal Aviation Administration (2013). *Safety Alert for Operators* (No. 13002). Retrieved from www.faa.gov/other_visit/aviation_industry/airline_operators/airline_safety/safo/all_safos/media/2013/SAFO13002.pdf

Fitts, P. M., Viteles, M. S., Barr, N. L., Brimhall, D. R., Finch, G., Gardner, E. ,... Stevens, S. S. (1951). *Human Engineering for an Effective Air-Navigation and Traffic-Control System.* Columbus, OH: Ohio State University Research Foundation.

Flemisch, F., Heesen, M., Hesse, T., Kelsch, J., Schieben, A., & Beller, J. (2012). Towards a dynamic balance between humans and automation: Authority, ability, responsibility and control in shared and cooperative control situations. *Cognition, Technology & Work, 14*(1), 3–18.

Frampton, R. & Lenard, J. (2009). The potential for further development of passive safety. *Annals of Advances in Automotive Medicine/Annual Scientific Conference* (Vol. 53, p. 51). Chicago, IL: Association for the Advancement of Automotive Medicine.

Fuld, R. B. (1993). The fiction of function allocation. *Ergonomics in Design, 1*(1), 20–24.

Gary, C. S., Lakhiani, C., Defazio, M. V., Masden, D. L., & Song, D. H. (2018). Caution with use: Smartphone-related distracted behaviors and implications for pedestrian trauma. *Plastic and Reconstructive Surgery, 142*(3), 428e.

Gold, C., Damböck, D., Lorenz, L., & Bengler, K. (2013). "Take over!" How long does it take to get the driver back into the loop? *Proceedings of the Human Factors and Ergonomics Society Annual Meeting, 57*, 1938–1942.

Gonçalves, R., Louw, T., Madigan, R., & Merat, N. (2019). Using Markov chains to understand the sequence of drivers' gaze transitions during lane-changes in automated driving. *Proceedings of the 10th International Driving Symposium on Human Factors in Driver Assessment, Training, and Vehicle Design* (pp. 217–223). Santa Fe, NM: Iowa Public Policy Center.

Goode, J. H. (2003). Are pilots at risk of accidents due to fatigue? *Journal of Safety Research, 34*(3), 309–313.

Hampton, M. E. (2016). *Memorandum: Enhanced FAA Oversight Could Reduce Hazards Associated with Increased Use of Flight Deck Automation*. Washington, DC: U.S. Dept. of Transportation.

Hancock, P. A. & Scallen, S. F. (1996). The future of function allocation. *Ergonomics in Design, 4*(4), 24–29.

Hancock, P. A. & Warm, J. S. (1989). A dynamic model of stress and sustained attention. *Human Factors, 31*, 519–537.

Hollnagel, E. & Bye, A. (2000). Principles for modelling function allocation. *International Journal of Human-Computer Studies, 52*(2), 253–265.

Hollnagel, E. & Woods, D. D. (2005). *Joint Cognitive Systems: Foundations of Cognitive Systems Engineering*. Boca Raton, FL: CRC Press.

Horberry, T., Anderson, J., Regan, M. A., Triggs, T. J., & Brown, J. (2006). Driver distraction: The effects of concurrent in-vehicle tasks, road environment complexity and age on driving performance. *Accident Analysis & Prevention, 38*(1), 185–191.

Jamson, A. H. & Merat, N. (2005). Surrogate in-vehicle information systems and driver behaviour: Effects of visual and cognitive load in simulated rural driving. *Transportation Research Part F: Traffic Psychology and Behaviour, 8*(2), 79–96.

Jordan, N. (1963). Allocation of functions between man and machines in automated systems. *Journal of Applied Psychology, 47*(3), 161.

Körber, M., Cingel, A., Zimmermann, M., & Bengler, K. (2015). Vigilance decrement and passive fatigue caused by monotony in automated driving. *Procedia Manufacturing, 3*, 2403–2409.

Lee, J. D. & Seppelt, B. D. (2009). Human factors in automation design. In S. Nof (Ed.), Springer *Handbook of Automation* (pp. 417–436). Berlin: Springer.

Lee, J. D., Wickens, C. D., Liu, Y., & Boyle, L. N. (2017). *Designing for People: An Introduction to Human Factors Engineering*. Charleston, SC: CreateSpace.

Leech, J., Whelan, G., Bhaiji, M., Hawes, M., & Scharring, K. (2015). *Connected and Autonomous Vehicles - The UK Economic Opportunity*. Amstelveen, The Netherkands: KPGM.

Loukopoulos, L. D., Dismukes, R. K., & Barshi, I. (2001). Cockpit interruptions and distractions: A line observation study. *Proceedings of the 11th International Symposium on Aviation Psychology* (pp. 1–6). Columbus, OH: Ohio State University Press.

Louw, T., Kountouriotis, G., Carsten, O., & Merat, N. (2015). Driver inattention during vehicle automation: How does driver engagement affect resumption of control? *4th International Conference on Driver Distraction and Inattention*. Sydney: ARRB Group.

Louw, T., Kuo, J., Romano, R., Radhakrishnan, V., Lenné, M., & Merat, N. (2019). Engaging in NDRTs affects drivers' responses and glance patterns after silent automation failures. *Transportation Research Part F: Traffic Psychology and Behaviour, 62*, 870–882.

Louw, T., Madigan, R., Carsten, O., & Merat, N. (2017b). Were they in the loop during automated driving? Links between visual attention and crash potential. *Injury Prevention, 23*(4), 281–286.

Louw, T., Markkula, G., Boer, E., Madigan, R., Carsten, O., & Merat, N. (2017a). Coming back into the loop: Drivers' perceptual-motor performance in critical events after automated driving. *Accident Analysis & Prevention, 108*, 9–18.

Louw, T., Merat, N., & Jamson, H. (2015). Engaging with highly automated driving: To be or not to be in the loop? *Proceedings of 8th International Driving Symposium on Human Factors in Driver Assessment, Training and Vehicle Design* (pp. 190–196). Snowbird, UT: Iowa Public Policy Center.

Ma, R. & Kaber, D. B. (2005). Situation awareness and workload in driving while using adaptive cruise control and a cell phone. *International Journal of Industrial Ergonomics, 35*(10), 939–953.

Mackworth, N. H. (1948). The breakdown of vigilance during prolonged visual search. *Quarterly Journal of Experimental Psychology, 1*(1), 6–21.

Madigan, R., Louw, T., & Merat, N. (2018). The effect of varying levels of vehicle automation on drivers' lane changing behaviour. *PloS ONE, 13*(2), e0192190.

McDonald, A. D., Alambeigi, H., Engström, J., Markkula, G., Vogelpohl, T., Dunne, J., & Yuma, N. (2019). Toward computational simulations of behavior during automated driving takeovers: A review of the empirical and modeling literatures. *Human Factors, 61*(4), 642–688.

Merat, N., Jamson, A. H., Lai, F. C., Daly, M., & Carsten, O. M. (2014). Transition to manual: Driver behaviour when resuming control from a highly automated vehicle. *Transportation Research Part F: Traffic Psychology and Behaviour, 27*, 274–282.

Merat, N., Seppelt, B., Louw, T., Engström, J., Lee, J. D., Johansson, E., ... McGehee, D. (2019). The "out-of-the-loop" concept in automated driving: Proposed definition, measures and implications. *Cognition, Technology & Work, 21*(1), 87–98.

Michon, J. A. (1985). A critical view of driver behavior models: What do we know, what should we do? In L. Evans & R. C. Schwing (Eds.), *Human Behavior and Traffic Safety* (pp. 485–524). Boston, MA: Springer.

Mole, C., Lappi, O., Giles, O., Markkula, G., Mars, F., & Wilkie, R. (2019). Getting back into the loop: The perceptual-motor determinants of successful transitions out of automated driving. *Human Factors, 61*(7), 1037–1065.

Moray, N. (2003). Monitoring, complacency, scepticism and eutactic behaviour. *International Journal of Industrial Ergonomics, 31*(3), 175–178.

Noy, I. Y., Shinar, D., & Horrey, W. J. (2018). Automated driving: Safety blind spots. *Safety Science, 102*, 68–78.

Page, Y., Hermitte, T., & Cuny, S. (2011). How safe is vehicle safety? The contribution of vehicle technologies to the reduction in road casualties in France from 2000 to 2010. *Annals of Advances in Automotive Medicine/Annual Scientific Conference* (Vol. 55, p. 101). Chicago, IL: Association for the Advancement of Automotive Medicine.

Pankok, Jr. C. & Bass, E. J. (2017). Appendix A – Function allocation literature review. In Jr. C. Pankok, E. J. Bass, P. J. Smith, J. Bridewell, I. Dolgov, J. Walker, ... Spencer, A. (Authors), *A7—UAS Human Factors Control Station Design Standards (Plus Function Allocation, Training, and Visual Observer)*. Washington, DC: Federal Aviation Administration. Retrieved from https://rosap.ntl.bts.gov/view/dot/36213/dot_36213_DS1.pdf

Parasuraman, R. & Manzey, D. H. (2010). Complacency and bias in human use of automation: An attentional integration. *Human Factors, 52*(3), 381–410.

Parasuraman, R. & Riley, V. (1997). Humans and automation: Use, misuse, disuse, abuse. *Human Factors, 39*(2), 230–253.

Parasuraman, R., Sheridan, T. B., & Wickens, C. D. (2000). A model for types and levels of human interaction with automation. *IEEE Transactions on Systems, Man, and Cybernetics-Part A: Systems and Humans, 30*(3), 286–297.

Peden, M., Scurfield, R., Sleet, D., Mohan, D., Hyder, A. A., Jarawan, E., & Mathers, C. (2004). *World Report on Road Traffic Injury Prevention*. Geneva, Switzerland: World Health Organization.

Price, H. E. (1985). The allocation of functions in systems. *Human Factors*, *27*(1), 33–45.

Richter, M., Pape, H. C., Otte, D., & Krettek, C. (2005). Improvements in passive car safety led to decreased injury severity–A comparison between the 1970s and 1990s. *Injury*, *36*(4), 484–488.

SAE. (2016a). *Taxonomy and Definitions for Terms Related to Driving Automation Systems for On-Road Motor Vehicles* (J3016 201609). Warrendale, PA: Society of Automotive Engineers.

SAE. (2016b). *Human Factors Definitions for Automated Driving and Related Research Topics* (J3114 201612). Warrendale, PA: Society of Automotive Engineers.

Saffarian, M., de Winter, J. C., & Happee, R. (2012). Automated driving: Human-factors issues and design solutions. *Proceedings of the Human Factors and Ergonomics Society Annual Meeting*, *56*, 2296–2300.

Salmon, P. M., Walker, G. H., & Stanton, N. A. (2016). Pilot error versus sociotechnical systems failure: A distributed situation awareness analysis of Air France 447. *Theoretical Issues in Ergonomics Science*, *17*(1), 64–79.

Sarter, N. B. & Woods, D. D. (1995). How in the world did we ever get into that mode? Mode error and awareness in supervisory control. *Human Factors*, *37*(1), 5–19.

Scallen, S. F. & Hancock, P. A. (2001). Implementing adaptive function allocation. *The International Journal of Aviation Psychology*, *11*(2), 197–221.

Scanlon, J. M., Sherony, R., & Gabler, H. C. (2017). Injury mitigation estimates for an intersection driver assistance system in straight crossing path crashes in the United States. *Traffic Injury Prevention*, 18 (sup1), S9–S17.

Sheridan, T. B. (2000). Function allocation: Algorithm, alchemy or apostasy? *International Journal of Human-Computer Studies*, *52*(2), 203–216.

Sheridan, T. B. & Verplank, W. L. (1978). *Human and Computer Control of Undersea Teleoperators*. Cambridge, MA: Massachusetts Institute of Technology, Man-Machine Systems Lab.

Silverstein, J. (2019). *Driver Says Tesla Car Gets "Confused" and Crashes on Highway*. Retrieved from www.cbsnews.com/news/tesla-autopilot-car-gets-confused-and-crashes-on-highway/

Stewart, J. (2019). *Tesla's Self-Driving Autopilot Involved in Another Deadly Crash*. Retrieved from www.wired.com/story/tesla-autopilot-self-driving-crash-california/

Tefft, B. C. (2017). *Rates of Motor Vehicle Crashes, Injuries and Deaths in Relation to Driver Age, United States, 2014–2015*. Washington, DC: AAA Foundation for Traffic Safety.

Treat, J. R., Tumbas, N. S., McDonald, S. T., Shinar, D., Hume, R. D., Mayer, R. E., ... Castellan, N. J. (1979). *Tri-Level Study of the Causes of Traffic Accidents. Final Report, Vol. I. Causal Factor Tabulations and Assessments*. Bloomington, IN: Indiana University Institute for Research in Public Safety.

Trösterer, S., Meschtscherjakov, A., Mirnig, A. G., Lupp, A., Gärtner, M., McGee, F., ... Engel, T. (2017). What we can learn from pilots for handovers and (de) skilling in semi-autonomous driving: An interview study. *Proceedings of the 9th International Conference on Automotive User Interfaces and Interactive Vehicular Applications* (pp. 173–182). New York: ACM.

Victor, T. W., Tivesten, E., Gustavsson, P., Johansson, J., Sangberg, F., & Ljung Aust, M. (2018). Automation expectation mismatch: Incorrect prediction despite eyes on threat and hands on wheel. *Human Factors*, *60*(8), 1095–1116.

Vlasic, B. & Boudette, N. (2016). *Self-Driving Tesla Was Involved in Fatal Crash, U.S. says*. Retrieved from www.nytimes.com/2016/07/01/business/self-driving-tesla-fatal-crash-investigation.html

Ward, N. J. (2000). Task automation and skill development in a simplified driving task. *Proceedings of the Human Factors and Ergonomics Society Annual Meeting, 44*(3), 302–305.

Wickens, C. D., Clegg, B. A., Vieane, A. Z., & Sebok, A. L. (2015). Complacency and automation bias in the use of imperfect automation. *Human Factors, 57*(5), 728–739.

Zeeb, K., Buchner, A., & Schrauf, M. (2015). What determines the take-over time? An integrated model approach of driver take-over after automated driving. *Accident Analysis & Prevention, 78,* 212–221.

Zhang, B., de Winter, J., Varotto, S., Happee, R., & Martens, M. (2019). Determinants of take-over time from automated driving: A meta-analysis of 129 studies. *Transportation Research Part F: Traffic Psychology and Behaviour, 64,* 285–307.

9 Driver Fitness in the Resumption of Control

Dina Kanaan and Birsen Donmez
University of Toronto

Tara Kelley-Baker
AAA Foundation for Traffic Safety

Stephen Popkin
Volpe National Transportation Systems Center

Andy Lehrer
Changeis, Inc.

Donald L. Fisher
Volpe National Transportation Systems Center

CONTENTS

Key Points 174
9.1 Introduction 175
9.2 Distraction 176
    9.2.1 Definitions and Effects 176
        9.2.1.1 What Is Driver Distraction? 177
        9.2.1.2 Potential Sources of Distraction and Their Effects on
                Non-Automated Driving 178
        9.2.1.3 Effects of Automation on Distraction 179
        9.2.1.4 Effects of Distraction on Driver-Automation
                Coordination 179
    9.2.2 Detection.. 182
    9.2.3 Remediation... 183
9.3 Sleepiness 186
    9.3.1 Definitions and Effects 186
        9.3.1.1 What Is Sleepiness? 186
        9.3.1.2 Potential Sources of Sleepiness and Their Effects on
                Non-Automated Driving 187
        9.3.1.3 Effects of Automation on Sleepiness 188
        9.3.1.4 Effects of Sleepiness on Driver-Automation Coordination... 188

      9.3.2   Detection... 189
            9.3.2.1   Detecting and Predicting Sleepiness 189
            9.3.2.2   Detecting and Predicting Microsleeps.............................. 190
            9.3.2.3   Detecting and Predicting Sleep 190
      9.3.3   Remediation... 190
            9.3.3.1   Technologies and Practices: Opportunities to Develop,
                        Test, and Implement Solutions.. 191
            9.3.3.2   Gaps in Knowledge and Related Opportunities 192
9.4   Alcohol and Other Drugs (AOD).. 194
      9.4.1   Definitions and Effects 194
            9.4.1.1   Alcohol Prevalence and Crash Risk.................................. 194
            9.4.1.2   Other Drugs Prevalence and Crash Risk 194
            9.4.1.3   Prevalence vs. Impairment.. 195
            9.4.1.4   Varying Drug Impairing Behaviors.................................. 195
             9.4.1.5   Effects of Automation on AOD-Impaired Driving........... 196
            9.4.1.6   Effects of AOD-Impaired Driving on
                        Driver-Automation Coordination...................................... 197
      9.4.2   Detection... 197
            9.4.2.1   Alcohol Breath Testers and Sensors, Alcohol-Ignition
                        Interlocks, and DADSS... 197
            9.4.2.2   Breathalyzers for Drugs Other than Alcohol.................... 198
            9.4.2.3   Transdermal Sensors and Other Biosensors for Alcohol
                        and Drug Detection... 198
            9.4.2.4   Behavioral Indicators for AOD-Impaired Driving 199
      9.4.3   Remediation 200
9.5   Motion Sickness..201
      9.5.1   Definitions and Effects 201
            9.5.1.1   What Is Motion Sickness and What Are Its Symptoms?.... 201
            9.5.1.2   Contributors to Motion Sickness Experienced by
                        Drivers of Automated Vehicles...202
            9.5.1.3   Effects of Motion Sickness on Drivers of Automated
                        Vehicles 202
      9.5.2   Detection...202
            9.5.2.1   Self-Report Measures...202
            9.5.2.2   Other Measures...203
      9.5.3   Remediation...204
9.6   Conclusion 204
Acknowledgments..205
References...205

KEY POINTS

- For non-automated driving, research is conclusive that distracting activities that claim visual/manual resources are particularly detrimental to safety as these resources are also required for vehicle control.

- For automated driving, distraction has been found to be detrimental to both takeover time and quality, with visual–manual distraction being most detrimental. Further research is needed to more consistently quantify measures of takeover quality and to investigate the moderating effects of individual and contextual differences.
- A multi-dimensional problem, sleepiness requires data-driven research and solutions that focus not only on drivers falling asleep or succumbing to brief microsleep periods, but also periods when eyes are open yet sleepiness still threatens cognitive and psychomotor driver competencies.
- Sleepiness interacts with other impairment factors in complex ways. Broadly, sleepiness can amplify the negative impact of other impairment(s), while conversely, distraction may actually reduce sleepiness risk and serve to counter impairment.
- Automated driving creates a redistribution of safety risk with greater potential for sleepiness under conditions of low arousal, such as could occur during SAE Levels 2 and 3 prior to driver resumption of control. Drivers are especially vulnerable to impairment effects more broadly during these periods of disproportionately higher risk.
- Until SAE Level 5 driving automation, locking alcohol and other drug impaired drivers from driving may be the only appropriate solution. At lower levels, the vehicle can detect and provide warnings, which require driver compliance. Higher levels of automation may present complications as drivers may over-rely on automation and be more likely to choose to drive when impaired with alcohol and other drugs.
- Motion sickness is an emerging concern for automated vehicle drivers as the driver's role becomes closer to that of a passenger; however, the variety of symptoms that individuals may experience makes this state hard to define and detect.

9.1 INTRODUCTION

Although automation can relieve drivers from performing lateral and longitudinal vehicle control tasks, as currently implemented, even the most advanced driving automation technologies—namely, SAE Level 2, 3, or 4 automation (SAE International, 2018)—require drivers to take over, or resume, control when the automation is unable to safely drive the vehicle. Given the limitations of these technologies, drivers are expected to either constantly monitor both environment and automation, and to identify and step in when there are situations that the automation is unable to handle (SAE Level 2), or take over control when the automation identifies a situation that it cannot handle and notifies the driver (SAE Level 3), or perform scheduled takeovers (SAE Level 4). Therefore, even with state-of-the-art systems, the drivers need to be fit to perform these activities when needed. That is, the drivers need to be in a state where they have (SAE Level 2) or can achieve (SAE Levels 3 and 4) adequate levels of situation awareness (SA, Endsley, 1995; this Handbook, Chapter 7) that would translate to proper takeover performance, as indicated by response time and quality.

The fitness of the driver to take over vehicle control may be degraded due to various states that the driver may experience, including distraction, sleepiness, impairment from alcohol and other drugs (AOD), and motion sickness. Drivers may recover from some of these states more rapidly (e.g., distraction) and some states may be more prolonged throughout a drive (e.g., motion sickness, and impairment from AODs). All of these states can degrade the driver's information processing abilities, including perception, cognition, and action, but manifest in different ways, at different rates, and for different durations. The sources that lead to these impaired states are also varied; for example, carried-in devices such as mobile phones may be a source of distraction (e.g., Caird, Simmons, Wiley, Johnston, & Horrey, 2018), while extended work hours and night and irregular shifts may be a source of sleepiness (e.g., McCartt, Rohrbaugh, Hammer, & Fuller, 2000). The remedies for these impairments may depend on the source, the nature of any interactions among impairment types, as well as the individual driver. One source of impairment may even be used to prevent another type of impairment. For example, a momentary impairment such as distraction can be mitigated with an alert, whereas a more prolonged impairment such as from alcohol would require prevention; sleepiness due to under-arousal may be prevented by keeping the driver engaged in arousal-enhancing activities that themselves could be distracting. Although detection is important for remediating all these states, prediction of a degraded state would be an ideal strategy to more proactively remediate certain states, such as sleep and motion sickness.

This chapter provides an overview of four major driver impairments to takeover performance, namely, distraction, sleepiness, impairment due to AOD, and motion sickness. Each impairment type is discussed in detail with regard to its sources, effects, detection, and remediation. The focus is on the former two as Chapter 11 of this handbook provides a more detailed review of driver state detection and remediation of impairment. Figure 9.1 provides an overview of the chapter content by presenting an information-processing-centric driver model as well as the potential interactions between the driver and automation.

9.2 DISTRACTION

9.2.1 Definitions and Effects

Driver distraction has been recognized as a major traffic safety concern particularly over the past two decades, starting with the proliferation of mobile phone technologies. The National Highway Traffic Safety Administration (NHTSA) reported that in 2015, 391,000 people in the United States were injured in motor vehicle crashes that involved distracted drivers, and 3,477 traffic fatalities (10% of all traffic fatalities) were attributed to distraction (National Highway Traffic Safety Administration, 2017). The Traffic Injury Research Foundation (2018) reported that 25% of motor vehicle fatalities in Canada in 2015 involved at least one distracted driver.

There is a large body of literature for non-automated vehicles on the effects of driver distraction on driving performance and crash risk, as well as ways to mitigate the negative effects of distraction (Donmez, Boyle, & Lee, 2006; 2007; 2008b; Regan, Lee, & Young, 2008). In the past few years, efforts have started to shift

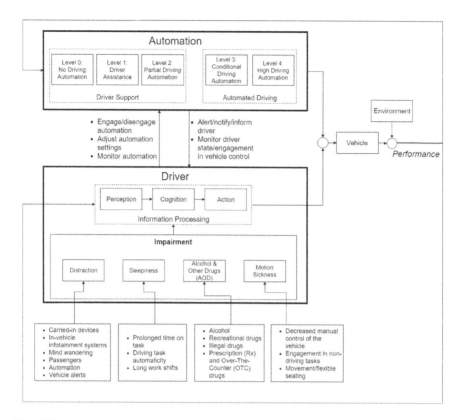

FIGURE 9.1 Driver-automation coordination affected by different types and sources of impairment.

toward investigating the prevalence and effects of distraction in automated driving, with a more recent, yet relatively limited focus on mitigation strategies. An overview of these topics is provided in this section. Overall, there is a need for further research on distraction in automated vehicles, in particular, on mitigating its effects on resumption of control (or takeover performance).

9.2.1.1 What Is Driver Distraction?

Driver distraction has been defined in many diverse ways (Lee, Young, & Regan, 2008; Regan, Hallett, & Gordon, 2011). Building on different definitions, Lee et al. (2008, p. 34) defined driver distraction as "a diversion of attention away from activities critical for safe driving toward a competing activity," which is the definition adopted in this chapter. It should be noted that other impaired states discussed in this chapter can also lead to a diversion of attention away from driving-related activities; however, they are not included within our view of distraction, as distraction is a specific case of inattention that involves an activity competing with the driving task for attentional resources.

Although the definition of distraction proposed by Lee et al. (2008) was developed in an era of lower driving automation levels (SAE Levels 0 and 1), it applies to all levels of automation where the driver is expected to take part in the driving task

(SAE Levels 0–4). However, for different SAE Levels, "activities critical for safe driving" mean different things. For example, for lower SAE Levels (0 and 1), these activities relate to the manual control of the vehicle; for SAE Level 2, they include monitoring the automation, identifying the need for takeover, and performing the takeover; and for higher SAE Levels (3 and 4), these activities include noticing and acting upon takeover requests. Because of this changing notion of "activities critical for safe driving" based on the level of driving automation, distraction may affect the driving task differently at different automation levels, and thus may need to be mitigated using different strategies. Further, as constant monitoring of the environment is not required in SAE Level 3 and above, it can be argued that non-driving activities, usually referred to as "secondary tasks" in the literature, may no longer be considered secondary to driving (see also this Handbook, Chapters 6, 8). In this chapter, we adopt the term "non-driving tasks" rather than "secondary tasks" to refer to non-driving activities, with the understanding that non-driving activities are still secondary to driving in SAE Level 2 and below.

9.2.1.2 Potential Sources of Distraction and Their Effects on Non-Automated Driving

There is a wealth of literature on the different sources of distraction and their effects on driving (e.g., Regan et al., 2008). These sources may interfere with the driver's information processing abilities at all stages, including perception, cognition, and action (Figure 9.1). However, different sources may interfere with these stages in different ways and to varying extents. Ranney, Garrott, and Goodman (2000) categorized driver distraction into visual (e.g., looking away from the roadway), auditory (e.g., listening to music), manual (e.g., eating), and cognitive (e.g., mind-wandering). In general, drivers experience a combination of these categories when they are distracted, and many forms of distraction may include a cognitive component. Given that driving is mainly a visual perception—manual control (or action) task, distracting activities that require visual perception and manual control cause the highest detriment to safety (Dingus et al., 2016; Wickens, 2008).

Though several different sources of driver distraction have been reported in the literature (some of which are included in Figure 9.1), many recent driver distraction studies have focused on driver interactions with in-vehicle infotainment systems and carried-in devices such as cell phones and tablets given the proliferation of these technologies. NHTSA (2019) estimated that in 2017, 416,000 U.S. drivers were using handheld cellphones during daylight hours, with the highest prevalence of mobile device use while driving being among younger drivers (ages 16–24). Meta-analyses of different on-road and simulator studies reveal the driving performance decrements resulting from the use of mobile phones (e.g., Caird, Johnston, Willness, Asbridge, & Steel, 2014; Caird et al., 2018; Horrey & Wickens, 2006). Whether these results translate to higher crash risks for different sources of distraction is an active research area.

Despite the fact that various distracting activities have been shown to degrade driving performance and some to increase crash risk, drivers engage in some of these activities very often without getting into a crash (e.g., Dingus et al., 2016). One reason why drivers can engage in some distracting activities with no ramifications is that driving is a skill-based task that is mostly automatic and thus does not require a driver's full attention, allowing for spare capacity to engage in non-driving activities. Another reason is that drivers are known to demonstrate adaptive and compensatory behaviors while distracted (e.g., Metz, Schömig, & Krüger, 2011; Oviedo-Trespalacios, Haque, King, & Washington, 2017; Platten, Milicic, Schwalm, & Krems, 2013; Reimer et al., 2013).

Although drivers can have spare capacity to engage in distracting activities and can to some extent adapt their engagement and driving behaviors, they may not always be successful in doing so, as they may fall into attention traps (Lee, 2014). Further, drivers may not always have control over their non-driving task engagement behaviors: distraction can also be involuntary, with drivers' attention automatically captured by an external stimulus or internal thoughts (i.e., mind-wandering) (Chen, Hoekstra-Atwood, & Donmez, 2018). Individual differences between drivers (e.g., age, driving experience) play a significant role in their susceptibility to both voluntary and involuntary distractions (e.g., Chen & Donmez, 2016) and how their driving is affected by distraction (e.g., Haque & Washington, 2014; He & Donmez, 2018). These individual differences can inform personalized distraction mitigation strategies.

In summary, extensive research investigated the sources and effects of distraction on non-automated driving in the past two decades. In comparison, there is relatively little research conducted to date on distraction in the context of automated driving, though this research is expanding.

9.2.1.3   Effects of Automation on Distraction

With increasing levels of automation, as the driver's role transitions to mostly supervision, drivers get even more spare capacity to engage in non-driving tasks. In fact, results from simulator studies have suggested that drivers are more likely to engage in non-driving tasks with increasing levels of automation (Carsten, Lai, Barnard, Jamson, & Merat, 2012). Drivers may decide to engage in non-driving activities in order to fight boredom and increase their arousal levels (de Winter, Happee, Martens, & Stanton, 2014; this Handbook, Chapter 6). The decision to engage in non-driving activities also depends on other factors, such as driver objectives and trust in automation. Mis-calibrated trust in automation may lead to inappropriate distraction engagement behaviors, with drivers allocating their attention to non-driving tasks at the expense of monitoring performance. With higher levels of driving automation, drivers may also engage in an even larger variety of non-driving tasks than they do in vehicles with lower or no driving automation (e.g., the non-driving tasks reported by Dingus et al., 2016) and may create new non-driving task use cases.

9.2.1.4   Effects of Distraction on Driver-Automation Coordination

To achieve successful driver-automation coordination, drivers must maintain a sufficient level of SA, which involves knowledge of their surrounding environment, current and possible events, the capabilities of automation, and the changes in its reliability. Increased engagement in non-driving tasks may impair not just the driver's awareness of the environment but also their ability to physically control the vehicle as they transition back from the supervisory task to the visual–manual driving task. This phenomenon, named the out-of-the-loop performance problem (Cunningham & Regan, 2018; Endsley & Kiris, 1995; Merat et al., 2019; this Handbook, Chapters 7, 21), describes driver's reduction in SA due to breakdowns in perception and cognition, as well as the reduction in physical engagement with the vehicle.

Research has reported a variety of effects of distraction on driver's ability to take back vehicle control from the automation, focusing mostly on SAE Level 2 and 3 driving automation. For example, Shen and Neyens (2017) found that while driver's reaction times to lane-keeping failures were slower with combined adaptive cruise control and lane-keeping assistance relative to non-automated driving, performance decrements were further exacerbated when participants were engaged in a video watching task. Vogelpohl, Kühn, Hummel, Gehlert, and Vollrath (2018) observed that while reaction times of distracted drivers (including brake reaction times and the time to deactivate the automation and take back control) were relatively unaffected, there were delays in the first gaze to the side mirror and speedometer after a takeover request, suggesting a longer time taken to regain SA after takeover. Similar effects have been suggested by Zeeb, Buchner, and Schrauf (2015; 2016) in terms of takeover time and quality. Whereas interaction with a non-driving task was shown by Zeeb et al. (2015) to result in slower brake reaction times and higher collision rates while responding to takeover requests, the time needed to make first contact with the steering wheel was not affected by distraction. Drivers who were distracted showed greater lane deviations upon taking back control of the vehicle after a takeover request, while the time needed to take back control was unaffected (Zeeb et al., 2016). Louw et al. (2017) also suggest that a slower takeover time is not necessarily associated with decrements in takeover quality. Gold, Körber, Lechner, and Bengler (2016), in fact, argued that a longer takeover time could indicate better takeover quality because it implies that drivers are taking more time to regain SA before taking over, a notion similar to "speed-accuracy trade-off" (Wickelgren, 1977).

Because the process of resuming control from automation involves visual, manual, and cognitive components (Zeeb et al., 2015), measures of takeover quality assess a variety of these aspects. It should be noted, however, that different studies have used different measures to define takeover quality—for instance, some have defined it in terms of gaze behavior (e.g., Vogelpohl et al., 2018) while others have defined it in terms of measures of driving performance after takeover (e.g., Zeeb et al., 2016) or even suggested that takeover time is an indicator of takeover quality (e.g., Gold et al., 2016). In general, measures of driving performance and visual attention that are used in a non-automated driving context may not be suitable or adequate in defining quality of takeovers in automated driving. Further, the interpretation of these measures may differ based on the level of driving automation (e.g., reaction time assessed as response to a takeover request in SAE Level 3 automation versus as response to a failure in SAE Level 2 automation).

When considering the effects of non-driving task engagement on driver-automation coordination, in addition to the level of driving automation, individual and contextual factors need to be considered. In a simulator study, He and Donmez (2019) observed that both novice and experienced drivers demonstrated a greater level of interaction with a non-driving task in automated driving in general than non-automated driving (indicated by the number of taps on the non-driving task display and percent of time looking at the non-driving task). However, experienced drivers' non-driving task engagement was less affected by automation as they had shorter and fewer long (>2 seconds) glances and a lower rate of manual interaction with the non-driving task than novice drivers. Further, Jamson, Merat, Carsten, and Lai (2013) suggest that when given the option of interacting with non-driving tasks or taking over vehicle control at any time, drivers of a highly automated vehicle preferred to keep the automation in control and continue to engage in non-driving tasks in light traffic, but demonstrated higher attention to the road and the driving task in heavier traffic. In contrast, Gold et al. (2016) observed a negative effect of traffic density on reaction time as well as measures of takeover quality such as acceleration, time to collision, and risk of crashes, when drivers were engaged in a verbal non-driving task. It appears that although drivers may adapt their attention allocation based on roadway demands in automated vehicles as they do in non-automated vehicles, they still may not be able to fully compensate for the degradation in their monitoring performance when engaged in a non-driving task.

Patterns and types of non-driving task engagement are other contextual factors that can affect driver behavior and, specifically, takeover performance. Wandtner, Schömig, and Schmidt (2018b) found that when drivers were aware of an impending takeover and were not already engaged in a non-driving task, they were less likely to voluntarily engage in one and were able to take over safely; however, when they were already engaged in a non-driving task, drivers tended to continue their interaction despite the need for takeover. Yoon and Ji (2019) observed that the type of non-driving task (searching for a new radio channel using the entertainment console, watching a video on a smartphone, or playing a game on a smartphone) had varying effects on different takeover performance measures: the video task resulted in the highest takeover time and longest first road glance after a takeover request. Vogelpohl et al. (2018) and Zeeb et al. (2016) also found differences in takeover performance based on the type of non-driving task, e.g., an increased time to first road glance while performing a gaming task compared with a reading task or not performing any task (Vogelpohl et al., 2018) and longer time to deactivate the automation after a takeover request while watching a video compared to while responding to an email, reading the news, or not performing any non-driving task (Zeeb et al., 2016).

In line with our knowledge about the effects of different distraction types on non-automated driving, Wandtner, Schömig, and Schmidt (2018a) found that an auditory–vocal version of a non-driving task was least detrimental to takeover performance compared with visual–vocal or visual–manual versions, with the slowest response times being associated with the visual–manual version of the task. Roche, Somieski, and Brandenburg (2019) also found similar results regarding the effects of visual versus auditory non-driving tasks in terms of takeover performance and attention to the roadway. Dogan, Honnêt, Masfrand, and Guillaume (2019) found that the type of the automation failure (obstacle avoidance vs. lane-keeping failure) had more of an effect on takeover performance than the type of non-driving task (watching videos vs. writing emails). These results demonstrate that there is a need to further study the differences in driver behaviors in automated driving based on individual characteristics and driving and non-driving task demands. A recent meta-analysis (Zhang, de Winter, Varotto, Happee, & Martens, 2019) suggests that urgency of the takeover situation, the modality of the non-driving task and its medium of presentation (e.g., handheld device), and the modality of the takeover request are factors that can influence mean takeover time. McDonald, Alambeigi, et al. (2019) further identified time budget, non-driving tasks performed using a handheld device, repeated exposure to takeovers, and silent failures (i.e., failures without a warning) as factors that affect takeover time. In addition to these factors, the authors identified the driving environment, takeover request modality, level of automation, trust, fatigue, and alcohol impairment to influence takeover quality, and stated a research need for examining the interacting effects of these factors as well as the link between delayed takeover time and degraded takeover quality.

Another direction for further research is investigating the effects of non-driving tasks in mitigating the effects of under-arousal during automated driving, how these effects may be different in different levels of driving automation, and how non-driving task interactions can be designed to keep drivers engaged in the driving task without diverting their attention to an extent that compromises safety. Miller et al. (2015) found that drivers who were engaged in a reading task or a video watching task were less likely to experience drowsiness in SAE Level 3 automated driving, and Naujoks, Höfling, Purucker, and Zeeb (2018) found that engaging in a variety of non-driving tasks during a long drive (1–2 hours) in SAE Level 2 automated driving kept drowsiness at relatively low levels. However, Saxby, Matthews, and Neubauer (2017) found that a verbal conversation did not fully counteract the effects of under-arousal resulting from SAE Level 3 driving automation as drivers demonstrated slower responses to a critical event after they were handed back vehicle control.

9.2.2 DETECTION

Reviews of different driver state monitoring techniques, including distraction, can be found in Young, Regan, and Lee (2008), Victor, Engström, and Harbluk (2008), Dong, Hu, Uchimura, and Murayama (2011), Aghaei et al. (2016), Kircher and Ahlstrom (2018), He, Risteska, Donmez, and Chen (in press), and McDonald, Ferris, and Wiener (2019) (see also, this Handbook, Chapter 11). These techniques can utilize measures that are vehicle-based (e.g., speed), physiological (e.g., heart rate), and facial and body expression based (e.g., eyes-off-road time as implemented in Cadillac Super Cruise automated driving feature in the 2019 Cadillac CT6 (General Motors, 2018) or hands-off wheel time as implemented in Tesla Autopilot (e.g., Tesla, 2019)), or more ideally a combination of these categories given that different measures tend to have different limitations (Aghaei et al., 2016). Further, to assess readiness for resumption of control, they can also leverage information about road demands. Based on these measures, algorithms need to be developed, setting criteria for classifying the driver to be distracted, or even more ideally, not fit to resume control in the context of automated driving.

For non-automated driving, various types of algorithms and measures have been explored for detecting distraction. Earlier algorithms were based on analytical methods that mainly focused on eye-tracking measures (e.g., eyes off forward roadway (Klauer, Dingus, Neale, Sudweeks, & Ramsey, 2006), AttenD (Kircher, Kircher, & Ahlström, 2009), risky visual scanning patterns (Donmez et al., 2007, 2008b), and multi-distraction detection (Victor, 2010)). Recent efforts have shifted towards machine learning methods, utilizing and comparing different machine learning techniques on different data sources, such as vehicle-based measures, head position, eye movements, and electroencephalography (EEG), collected from both simulator and on-road studies (e.g., He, Risteska, et al., in press; Kanaan, Ayas, Donmez, Risteska, & Chakraborty, 2019; Li, Bao, Kolmanovsky, & Yin, 2018; Liang & Lee, 2014; Schwarz, Brown, Lee, Gaspar, & Kang, 2016).

Because of the changing nature of the driving task and the changing interpretations of distraction under different automated driving conditions, the methods of distraction detection developed for non-automated driving may not directly apply to automated driving. In an automated driving context, the driver is either sharing or completely delegating the driving task to automation, rendering driving performance measures at times unavailable and less useful, unless the automation hands back control to the driver in a controlled manner to assess fitness to resume control based on takeover performance. Physiological and face and body expression measures can provide a more continuous assessment of distraction; however, they also have their disadvantages as some physiological sensors may be intrusive to the driver, signal quality may be affected by environmental factors, and some measures may present privacy concerns due to the use of video recording. Further, criteria for distraction detection developed for non-automated driving may be unsuitable for automated driving. For example, as part of its guidelines for evaluating visual–manual distraction caused by in-vehicle (NHTSA, 2013) and portable (NHTSA, 2016) devices in non-automated vehicles, NHTSA recommends that the tasks performed on these devices should not require off-road glances longer than 2 seconds at a time. This threshold is based on Klauer et al.'s (2006) finding that glances longer than 2 seconds away from the road are associated with doubled crash risk. While this threshold could apply in vehicles with lower automation, where drivers are still expected to monitor the road throughout their drive, it may not apply for higher levels of automation (i.e., SAE Level 3 and above), where longer glances away from the road and towards non-driving tasks would be expected and acceptable.

Recent findings on the sensitivity of different physiological and eye-tracking measures to distraction in automated driving suggest promise for the use of these measures in distraction detection in the context of automation (e.g., Louw & Merat, 2017; Sibi, Ayaz, Kuhns, Sirkin, & Ju, 2016). Further efforts have begun to explore distraction detection algorithms in automated vehicles (e.g., Braunagel, Geisler, Rosenstiel, & Kasneci, 2017; Gaspar, Schwarz, Kashef, Schmitt, & Shull, 2018; Pech, Enhuber, Wandtner, Schmidt, & Wanielik, 2018); however, research on this topic is still in its infancy.

9.2.3 REMEDIATION

For non-automated driving, a number of distraction mitigation frameworks have been proposed over the past two decades. For example, Donmez, Boyle and Lee (2003; 2006) proposed a taxonomy consisting of three dimensions: type of task (driving- or non-driving-related), source of initiation (driver or system), and level of automation of the mitigation strategy (low, medium, or high). This taxonomy focused on strategies that are implemented pre-drive or during a drive. However, feedback can also be presented post-drive and cumulatively over time to help change behavior. Therefore, Donmez, Boyle, and Lee (2008a) later considered feedback timing as another dimension that can inform the design of distraction mitigation strategies, and classified feedback timescales into concurrent (real-time), delayed (by a few seconds), retrospective (post-drive), and cumulative.

Driving-related strategies are remediation strategies that target the driving task and support the driver in the control of the vehicle, while non-driving-related strategies target driver interactions with non-driving tasks. These mitigation strategies can be either initiated by the system or the driver and may differ according to the level of automation of the strategy. Under driving-related, system initiated strategies, intervening (high automation) is the process whereby the system takes control of the driving task when the driver is too distracted to react to critical events. Warning (moderate automation) refers to alerts that are provided by the system to the driver to take a needed action, while informing (low automation) refers to the display of relevant information to the driver. Driving aids ranging from notifications to recommendation to automated driving capabilities (e.g., emergency braking) fall under this category of driving-related, system-initiated strategies. Strategies similar to these may also be initiated by the driver. Interaction with non-driving tasks may be modulated by the system through locking the driver out of or interrupting their non-driving tasks, automatically prioritizing and filtering the most important, urgent, or relevant ones, or giving feedback to the driver (i.e., advising) about their degree of interaction with non-driving tasks. Driver-initiated, non-driving related strategies include the driver pre-setting task features, place keeping during engagement with the non-driving task (e.g., using a bookmark), and choosing methods of interaction with the non-driving task that impose lower cognitive or visual–manual demand.

As driving becomes a more automated task, traditional views of distraction mitigation need to be re-evaluated and modified. As automation takes over more of the driving task, driving may come to be regarded at the same level of importance with non-driving tasks and may even be regarded as secondary to them, which may necessitate a shift away from lockout strategies and towards strategies that support time-sharing between non-driving and driving tasks. He, Kanaan, and Donmez (2019) proposed a revised taxonomy of strategies (Table 9.1) for automated vehicles that focuses on supporting time-sharing between the driving and non-driving tasks. The updated taxonomy classifies strategies based on their timing into pre-drive/drive strategies (by re-interpreting entries in the original taxonomy), post-drive (or retrospective) strategies, and cumulative strategies. The taxonomy classifies retrospective strategies into driving-related (risk evaluation strategies) and non-driving-related (engagement assessment): these strategies provide post-drive feedback to drivers about their takeover performance and non-driving task engagement in a completed drive. Cumulative strategies are also classified into driving and non-driving related strategies, respectively named as education and informing social norms. While education strategies target driver's awareness of automation capabilities and limitations so that they could learn when to pay more attention to the driving environment, strategies that inform social norms attempt to influence driver's non-driving task engagement behaviors through social norms interventions.

TABLE 9.1

Taxonomy of Time-Sharing Strategies

		Driving-Related		Non-Driving-Related	
		Automation-Initiated	*Driver-Initiated*	*Automation-Initiated*	*Driver-Initiated*
Pre-drive/ drive	*High intervention*	Intervening	Delegating	Locking & interrupting	Controls pre-setting
	Moderate intervention	Warning	Warning tailoring	Prioritizing & filtering	Place keeping
	Low intervention	Informing	Perception augmenting	Advising	Demand minimizing
Post-drive (retrospective)		Risk evaluation		Engagement assessment	
Cumulative		Education		Informing social norms	

Source: Recreated from He et al. (2019).

The pre-drive/drive strategies are re-interpreted in the revised taxonomy (He et al., 2019) as mechanisms that support time-sharing between driving and non-driving tasks, rather than adopting the view that distraction should ideally be prevented, a view appropriate for non-automated vehicles or vehicles with lower levels of automation (SAE Levels 0–2). For example, takeover requests (under the category of warning) can warn the driver about an impending event or a possible need for taking over vehicle control, while allowing the driver to engage in non-driving tasks in non-takeover situations. In addition to discrete and relatively infrequent warnings, more continual information about automation can also be provided to the drivers, which may guide their non-driving task engagement in a more informed manner. For example, the driver may be informed continually about the reliability of the automation (Stockert, Richardson, & Lienkamp, 2015; Wulf, Rimini-Doring, Arnon, & Gauterin, 2015), which can help drivers allocate their attention between non-driving tasks and monitoring of the automation. At a higher intervention level, the level of control authority between the driver and the automation (Benloucif, Sentouh, Floris, Simon, & Popieul; 2017) or the state of the automation (Cabrall, Janssen, & de Winter, 2018) can be dynamically changed in a system-initiated manner based on driver distraction state: a re-interpretation of the intervening strategy from the original taxonomy (see also, this Handbook, Chapter 16). For example, if the driver is detected to improperly monitor automation, the driver may be prevented from engaging automation. However, such methods may cause drivers to be susceptible to mode confusion errors (Sarter & Woods, 1995).

The taxonomy presented in He et al. (2019) highlights areas of future research. Most research on supporting time-sharing in automated vehicles has focused mainly on driving-related, driver-initiated strategies, with relatively more focus on the design of warning and informing displays (e.g., Gold, Damböck, Lorenz, & Bengler, 2013; Seppelt & Lee, 2019; Stockert et al., 2015). Research is still ongoing on the effects of different design parameters (such as timing and modality) related to warning and informing strategies, with results suggesting the potential usefulness of combining warnings and information displays, and utilizing multi-modal displays and alerts (e.g., Politis, Brewster, & Pollick, 2017; Seppelt & Lee, 2019; see also in this Handbook, Chapters 15, 16).

Relatively speaking, non-driving related strategies have attracted less attention from the automated driving research community. Köhn, Gottlieb, Schermann, and Krcmar (2019) found that frequently interrupting a video watching task to provide an image of the driving scene (i.e., locking and interrupting strategy combined with informing) improved takeover performance and awareness of the driving environment. Wandtner et al. (2018a) explored locking out the non-driving task during take-over requests and found that lockouts resulted in faster hands-on-wheel time and were rated as highly accepted. In general, such strong intervention strategies should be designed in a way that facilitates the acceptance of the strategy, while support-ing the resumption of the non-driving task after the interruption (e.g., Altmann & Trafton, 2002; McFarlane & Latorella, 2002; Naujoks, Wiedemann, & Schömig, 2017). Further, as mentioned earlier, Wandtner et al. (2018a) found that an auditory–vocal version of a non-driving task was least detrimental to takeover performance compared with visual–vocal or visual–manual versions, lending support to the effec-tiveness of demand minimizing strategies.

An alternative method of mitigating the effects of distraction is introducing gami-fication to guide monitoring and non-driving task engagement behaviors, which has been investigated in non-automated driving (e.g., Schroeter, Oxtoby, & Johnson, 2014; Steinberger, Schroeter, Foth, & Johnson, 2017; Xie, Chen, & Donmez, 2016). In general, further research is needed to help inform the design of interface mecha-nisms to support time-sharing between driving and non-driving tasks that are tai-lored to the individual driver characteristics, as individual differences can influence the effectiveness and acceptance of such mechanisms. There is also a need for devel-oping standards for safety in automated driving (e.g., what an acceptable duration is for long off-road glances, or what types of non-driving tasks present higher risk levels) as remediation strategies would depend on relevant risk thresholds.

9.3   SLEEPINESS

Despite consensus among scientists on the basic human biology underlying sleepiness and its safety impact (Dinges, 1995; Wesensten, Belenky, & Balkin, 2005), it remains a national safety and health issue. Crash statistics would prompt national focus where similar data are reported due to illness or disease. Sleepiness is estimated to be a factor in 100,000 police-reported crashes each year, including over 71,000 injuries and 1,550 fatalities (National Safety Council, 2019). These are likely underestimates given the difficulty in determining driver sleepiness at the time of crash.

9.3.1   DEFINITIONS AND EFFECTS

9.3.1.1   What Is Sleepiness?

Physiological sleepiness is the body's susceptibility to sleep (Dinges, 1995), which can differ from individual subjective perception or awareness of sleepiness while engaged in tasks such as driving. Physiological sleepiness depends on two

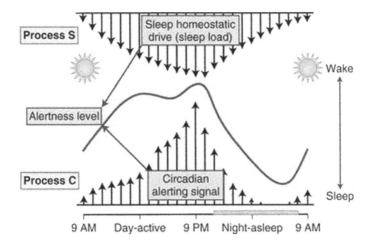

FIGURE 9.2   (See color insert.) The two-process model of sleep–wake regulation (Borbély, 1982; Borbély, Daan, Wirz-Justice, & Deboer, 2016) depicting the interaction between homeostatic sleep drive (Process S) and sleep-independent circadian arousal drive (Process C) to produce S+C alertness level (solid line).

independent processes. The onset, duration, and end of a sleep period are guided by both sleep and wake pressures as described in Borbély's (1982) two-process model of sleep regulation (see Figure 9.2).

The model continuously pits homeostatic sleep pressure (Process S; time awake) against sleep-independent circadian arousal drive (Process C; time of day) to produce a combined alertness level (solid line). Ideal opportunities for restorative sleep are available once sleep pressure is sufficiently greater than arousal drive. Sleep pressure builds as one remains awake and can be compounded through intentional sleep deprivation. The circadian curve, however, operates outside of conscious awareness and imparts a cyclical drive. Both circadian and homeostatic pressures directly and continuously mediate sleepiness. Åkerstedt and Folkard's (1995) "alertness nomogram" further reviews and validates the S and C model components.

Sleepiness can occur over extended periods of time without the driver falling asleep. However, on occasion it precedes brief sleep periods, called microsleeps, where the driver is asleep for up to several seconds before spontaneously reawakening. Finally, the driver can fully fall asleep at the wheel. Sleepiness is also conceptually distinct from moderating factors such as boredom—a state of low arousal due to low or no task demand or interest, as can occur with vigilance-only driving under automated control, as the driver may over-rely on technology and lose interest in sustaining vigilance (see e.g., this Handbook, Chapter 6).

9.3.1.2 Potential Sources of Sleepiness and Their Effects on Non-Automated Driving

Some potential sources of sleepiness are listed in Figure 9.1. Degree of sleepiness impacts how well a person senses and processes information to make informed and appropriate judgments, decisions, and actions. Sleepiness influences all three stages of Michon's (1985) model of driving: strategic, tactical, and operational. At the strategic level, sleepiness contributes to cognitive impairment risk including lapses in judgment and decision-making (e.g., see Dinges, 1995; Moore-Ede, Sulzman, & Fuller, 1982; Wesensten et al., 2005). At the tactical level, as with distraction, sleepiness can impair SA (Cuddy, Sol, Hailes, & Ruby, 2015; Dinges et al., 1997; Lim & Dinges, 2008; also see this Handbook, Chapter 7), impeding capacity to scan, predict, identify, make decisions, and execute safe responses (Fisher & Strayer, 2014). Finally, at the operational level, sleepiness relates to performance declines in simple motor functions (Owens et al., 2013). Not surprisingly, driving performance deteriorates during microsleeps. Driving simulator studies indicate that, during a microsleep, drivers show a significant deterioration in their vehicle control performance, correlated with the duration of the microsleep, particularly on curved roads (Boyle, Tippin, Paul, & Rizzo, 2008).

9.3.1.3 Effects of Automation on Sleepiness

Within highly automated environments, a redistribution of safety risk results in greater relative risk during specific periods of time. Pockets of increased risk emerge with the potential for driver sleepiness and related inattention that could result in equal or greater risk than in less automated environments, contrary to the promise of continuously reduced risk in automated systems. Specifically, driver engagement in the driving task shifts from active to supervisory to disengaged passenger as automation level increases. Over-trusting the system could encourage drivers to give in to sleep pressure at higher automation levels. Drivers could then unexpectedly be required to resume vehicle control under potentially increased and dangerous levels of impairment/sleepiness.

9.3.1.4 Effects of Sleepiness on Driver-Automation Coordination

Sleepiness also affects automated driving, as drivers have a finite capacity to selectively attend to, process, and act on information (Fawcett, Risko, & Kingstone, 2015). An already sleep-deprived driver struggling to monitor or execute within an automated environment could reach attentional capacity limits and thus render incoming automation cues ineffective. There are also chronic and acute instances of sleepiness (Costa, Gaffuri, Ghirlanda, Minors, & Waterhouse, 1995; Zijlstra, Cropley, & Rydstedt, 2014). Chronic (permanent) night work is associated with both reduced performance given ongoing reduced day sleep and impaired arousal without sufficient recovery over extended periods of time (Rosa et al., 1990; Tepas & Carvalhais, 1990), as well as increased crashes (Mitler et al., 1988), raising monitoring and takeover performance concerns.

Acute sleepiness is also problematic given near-term impairment including intentional acute sleep deprivation such as an *all-nighter* (i.e., staying awake the entire night), which is a significant risk for drivers in manual, mixed, and automated environments. As well, environmental and organizational factors including light levels and sleep timing and duration interact with physiology to impact sleepiness (e.g.; Dijk & Czeisler, 1994; Thomas, Raslear, & Kuehn, 1997). Individual differences also interact with sleepiness, impacting safety and performance (Van Dongen, 2006).

Yet, prophylactic napping is not recommended even in SAE Level 4 due to significant risk of loss of situational awareness and *sleep inertia* effects in case the driver was suddenly awoken to resume control. Abrupt awakening has been shown to cause subjective grogginess and diminished motor skills (Tassi & Muzet, 2000) as well as significantly impaired cognitive performance, which often recedes only over a period of tens of minutes (Wertz, Ronda, Czeisler, & Wright, 2006). This imposes a significant risk in dynamic driving ecosystems that could require the driver to understand the situation and what is required for immediate or near-term driver resumption of control. Thus driver state monitoring systems and the various countermeasures these monitoring systems are intended to activate need to be tuned to environmental and individual variables (this Handbook, Chapter 11 and below), as preemptive driver napping enroute is not recommended.

9.3.2 DETECTION

Detection technologies of all three states of driver sleepiness have been available for some time. What is needed are prediction technologies that, ideally, can alert the driver ahead of time when the driver is sleepy, when the driver will have a microsleep, and when the driver will fall asleep (Jacobé De Naurois, Bourdin, Stratulat, Diaz, & Vercher, 2019). It should be acknowledged up front that it is not always possible to determine whether a reduced state of arousal is due to sleepiness or rather boredom.

Detection and prediction of any of the above three states of sleepiness require both sensors and algorithms to classify the real-time data being provided by the sensors (see this Handbook, Chapter 11). The complexity of the algorithms can vary greatly, as can the complexity of the sensors, focusing on operator physiology, driving behavior, or both. As for the sensors, they can be off-the-shelf technologies which record steering wheel and yaw angles (Li, Chen, Peng, & Wu, 2017), more complex sensors that record measures of eye behavior such as eyelid closure, or still more sophisticated sensors such as electrocardiogram (EKG) capable heart rate monitors (Watson & Zhou, 2016). As for algorithms, given that a driver's operational state involves a complex set of psychological, physiological, and physical parameters, it is not surprising that some of the better performing algorithms integrate the data across all three parameters (Jacobé De Naurois et al., 2019).

9.3.2.1 Detecting and Predicting Sleepiness

Perhaps one of the most popular methods of detecting sleepiness is PERCLOS (Wierwille, Wreggit, Kirn, Ellsworth, & Fairbanks, 1994). The measure used in PERCLOS is a function of the percentage of the time during a given interval that the driver's eyes are closed beyond some threshold (80% is typically used; blinks do not count). Given the success of PERCLOS in detecting sleepiness, early studies focused on the use of vehicle variables to predict PERCLOS (taken as ground truth for sleepiness). However, the evidence that vehicle variables could be used to detect sleepiness was inconclusive (Tijerina et al., 1998). As the cost of eye behavior measures have come down, more recent studies have used machine vision to measure PERCLOS, with generally good success (Junaedi & Akbar, 2018).

Prediction is the ultimate objective. A more recent study used artificial neural networks both to detect and predict driver sleepiness (Jacobé De Naurois et al., 2019). The best models (those whose rates of successful detection or prediction are the highest) used information about eyelid closure, gaze and head movements, and driving time. The performance of the model relative to prediction was promising, since the model could predict to within 5 minutes when the driver's performance would become impaired (moderately sleepy). Interestingly, knowledge about the individual participant (e.g., age, quality of sleep, caffeine consumption, driving exposure) did not significantly improve the predictions. Thus, the algorithm can be used across individuals (at least the individuals in the study), an important issue when scaling is considered.

9.3.2.2 Detecting and Predicting Microsleeps

It has been notoriously difficult to predict microsleeps, and it still remains a real challenge as there seems to be a number of individual difference factors that come into play. Still, progress has been made. In a recent paper, Watson and Zhou (2016), using an EKG capable heart rate monitor, were able to detect the first microsleep with 96% accuracy and were able to predict, between 15 s and 5 min in advance, the time when the next microsleep will occur. Although the time when the first microsleep occurs cannot be predicted by such methods, the fact that it is possible to predict a second and subsequent microsleep is an important and critical advance, both from the standpoint of practice and engineering.

9.3.2.3 Detecting and Predicting Sleep

Ultimately the driver will fall asleep, and when they do the results are often catastrophic. A number of studies have been conducted that show that it is possible to predict whether an individual driver is likely to fall asleep at the wheel from information based on measures that are collected at or before an individual begins driving (Heaton, Browning, & Anderson, 2008). However, it is important to predict in real time the likelihood that a driver will fully fall asleep. Heart rate variability (and functions of heart rate variability) are critical, as are a number of other measures, including eyelid closure, gaze and head movements, and driving time. Overall, the current state of the art would suggest that artificial neural networks which are a function of a multiplicity of inputs perform best when the objective is to predict sleep (Jacobé De Naurois et al., 2019). Moreover, to further improve accuracy, external information such as driving time or a driver profile can be added to the model.

9.3.3 Remediation

Sleepiness research reveals that a driver need not be fully asleep—nor even in microsleep—for sleepiness to negatively impact safety risk and outcomes (Jacobé De Naurois et al., 2019); impairment to judgment and decision-making happens before the eyes close. The challenge then is how to apply the current understanding of sleep to innovative technologies and practices while addressing knowledge gaps to further reduce risk.

9.3.3.1   Technologies and Practices: Opportunities to Develop, Test, and Implement Solutions

9.3.3.1.1   Alerts

With rapidly advancing connected data capabilities, detection and prediction tools can be combined and applied through data fusion/algorithm development to integrate approximate (and ultimately exact, via monitoring) driver sleep–wake history and relevant inter/intra-individual differences to deliver an alert at an appropriate time and in an appropriate modality for a given driver state and level of automation. Combined, calibrated inputs could further improve validity in triggering and warning on current and prospective sleepiness-induced thresholds after which risk of impairment/bad decisions significantly increase in onset, severity, or duration.

Warning algorithms could be tuned to anticipated cognitive demands as well as projected traffic, road, and weather conditions. Importantly, context-specific intra-individual differences could be integrated, including task levels that shift among passenger, supervisory, and active driver roles. Over-reliance/trust in the automated system though could lead to driver transitions between roles in states of reduced alertness, whereas mistrust of system measures, including driver alertness, could prompt drivers to dismiss or turn off the system. Feedback from increasingly accurate, context-specific algorithms will likely foster readiness and use.

Several sleepiness-related fatigue and alertness algorithms exist and could be further tuned to predict sleepiness-related cognitive and physical impairment, providing predicted driver states to the vehicle. This could prompt meaningful feedback to the driver (e.g., that the driver will become increasingly sleepy, or have a microsleep, or will fall asleep, for example, in the next 5 minutes), or change how the vehicle operates (e.g., switching automation level). These models however do not integrate more specific individual operator performance and cognitive or psychomotor functions (Dijk & Larkin, 2004) required in a driver-automation ecosystem. For a nuanced review of existing alertness and fatigue models (Lehrer & Popkin, 2014), as well as a systems-based approach to understand and reduce fatigue and related sleepiness and cognitive/physical impairment, see Lehrer's (2015) eight-state model. Integrating internal factors such as individual arousal level, and external factors such as task demands, such an approach could assess driver capacity, predict risk, and trigger warnings or control actions. Warnings could vary in timing, frequency, duration, and intensity, tuned and communicated based on patterns of sleepiness and risk to provide targeted alarms delivered in time, but without being too far ahead of a potential state to reduce effectiveness (see Handbook, Chapter 15).

9.3.3.1.2   Training

Education and training could leverage realistic, immersive simulation exercises as part of a licensing and renewal curriculum to integrate sleepiness-specific challenges within and across SAE levels. Such training has already demonstrated success with nurses (Hamid, Samuel, Borowsky, Horrey, & Fisher, 2016). As well, both group and individual differences are important to recognize and support through periodic training as new technologies enter the driver-automation ecosystem.

9.3.3.1.3   Human–Machine Interface Design

To improve the driver-automated ecosystem, visual displays could show both driver state and vehicle status as well as immersive views of the outside driving environment. To further protect driver vigilance, stimulus types and required discriminations matter (Proctor & Zandt, 2018). As more information is served to the driver, the ability to selectively monitor and act on essential information in real time—while not engaging inappropriate actions or controls—becomes a challenge. As one becomes sleepy, tunneling occurs and information focus narrows.

It is important then that information presented to the driver be kept to a reasonable level and salient to safety given a finite capacity to selectively attend. Presenting actionable visual, aural, or tactile information triggers decision-making processes that require significant neural resources regardless of the decision's importance (Levitin, 2014). Real-time driver workload assessments could identify optimal cognitive loads required for vigilance during SAE Level 2, for example, and safely adjust complexity to support ideal arousal levels for sleepy drivers (see this Handbook, Chapter 6).

9.3.3.1.4   Decision Support Tools

Bugarski, Bačkalić, and Kuzmanov (2013) noted that both normal and emergency procedures in supervisory control could be aided by *"decision-support* tools *based on expert knowledge"* to better consider the complete system model and rethink how systems fail, such as driver actions to take and *not to take* when suddenly jarred out of sleepiness-induced cognitive impairment. Since a driver typically has had experience with a relatively limited range of system states, the driver's mental model of the system will be a reduced version of the complete system model (see this Handbook, Chapter 3). Though useful in normal driving scenarios drawing on less mental capacity, in an emergency situation the driver could benefit from comprehensive system model considerations and refined decision-support tools delivered through automation.

9.3.3.2   Gaps in Knowledge and Related Opportunities

In addition to building on existing areas of knowledge, opportunities exist to further explore gaps and target risk-reduction. Some ideas are common sense next steps while others suggest a larger leap. Both approaches innovate and advance safety, as lessons learned could improve driver-automation environments and offer cross-modal application.

9.3.3.2.1   Understanding/Changing Failure Modes

As more is learned about automated warning and action thresholds customized to individual driver scenarios in real time, changing failure modes could enhance safety through special automation, adjusting based on understood driver states. It is possible, for example, to further parse SAE events, including driver takeovers, into partial events based on driver preparedness as a function of net impairment due to sleepiness and other impairment factors and interactions. Instead of an all or none takeover, a driver could be given partial, initial control if automated assessment determined that this would minimize an overall risk.

For example, a driver could be allowed to resume braking control but not steering given that suddenly awoken drivers could overcompensate on steering and even flip the vehicle if recovering from drifting over a small berm or curb, for example. Automation can calculate the precise combination of safest driver/automation control and adjust that control along a continuum as the driver becomes less sleepy and thus more cognitively and physically fit to resume full control. The automation would capture real-time sleepiness-related risk and continuously solve for optimal proportions of driver/automation takeover in SAE Levels 2 and 3, for example, as long as proportional control imparts less current risk than full driver control.

9.3.3.2.2   Big Data

To further understand gaps and reduce risk, a private sector initiative could collect and analyze voluntary driver sleepiness data via dash-mounted sensors, for example, from millions of drivers over time in a collaborative, transparent effort to better understand sleeping driving factors, risks, and outcomes. A key challenge remains to understand and prevent sleepiness, microsleep, and falling asleep while driving, as each has demonstrated a significant safety risk. The ability to capture events in real time before, during, and after each aspect of sleepiness will substantially advance the understanding of risk and mitigation. The DRIVE - TRUST (Driver Real-time Integrated Vehicle Ecosystem) could be a name for such an initiative. Driver status could be continuously tracked and mapped to risk algorithms, countermeasures, and safety data. Data analysis could determine, for example, if driver sleepiness results in more accidents, injuries, and fatalities than actually falling asleep. An expert panel could further apply the data to create a tiered "sleepiness" rating scale with relative risk and countermeasure assigned to each state.

9.3.3.2.3   Alternative Routing

Driver sleepiness status could activate several prevention strategies. Automated systems could sense sleepiness and either not start, operate at reduced speed in a designated slow lane, pull over, temporarily reroute to centrally located driver *"safe-nap"* facilities, or alert the driver of being "intoxicated" by sleepiness. The latter is an apt comparison—Dawson and Reid (1997) showed that after 24 hours of sustained wakefulness, cognitive psychomotor performance deficits were similar to those observed at a blood alcohol concentration (BAC) of roughly 0.10%, more than the least stringent intoxication level of 0.08% in the United States.

9.3.3.2.4   Additional Possibilities

Considering interaction effects with other impairment risks, there remains much to be learned, assessed, and applied. Interestingly, ongoing studies of hyper-aroused states will likely further inform sleepiness risk as an aspect of the same challenge to allocate cognitive and physical resources to the demands of automated driving. As well, attention shifting declines in older drivers with various types of age-related dementia (Parasuraman & Nestor, 1991), suggesting the need to integrate aging with other interacting effects, such as distraction, which can further moderate the impact of sleepiness and vice versa. Sleepiness is also a common possible side effect of medication and drugs (National Sleep Foundation, 2019). Excessive sleepiness can be caused by the sedating effects of prescribed and non-prescribed medications and illegal substances. Improved individual sleepiness predictions will better tune targeted alerts.

Impairment factor interactions are not always negative. For example, Schömig, Hargutt, Neukum, Petermann-Stock, and Othersen (2015) showed that sleep pressure was countered somewhat by adding a secondary task. Interaction effects thus suggest the value of a *net impairment algorithm* integrating the multiple contributing factors discussed in this chapter, accessible to the automation as a governing guide to reduce risk.

9.4 ALCOHOL AND OTHER DRUGS (AOD)

9.4.1 DEFINITIONS AND EFFECTS

9.4.1.1 Alcohol Prevalence and Crash Risk

In 2017, 10,874 people died in a motor vehicle fatality involving a driver with BACs of 0.08 g/dL or higher (the legal limit in the Unites States where drivers are considered to be alcohol-impaired). Among these people, 6,618 (61%) were the drivers themselves (National Center for Statistics and Analysis, 2018).

Although a BAC of 0.08 g/dL is the legal limit, research has repeatedly established that crash risk from alcohol starts below the legal set limit. In 1964, the Grand Rapids Study (Borkenstein, Crowther, & Shumate, 1974; Borkenstein, Crowther, Shumate, Ziel, & Zylman, 1964) provided evidence that a BAC level of 0.04 g/dL was associated with increased crash risk, and the risk grows exponentially at higher BACs. Most recently, a case–control study in Virginia found drivers with a BAC of 0.05 g/dL were more than twice as likely to crash compared to drivers with no alcohol. Drivers with a BAC of 0.08 g/dL were almost four times at risk than drivers with no alcohol (Lacey et al., 2016).

9.4.1.2 Other Drugs Prevalence and Crash Risk

In 2007, NHTSA's National Roadside Survey included for the first time procedures for collecting samples of both breath (for alcohol), and oral fluid and blood (for other potentially impairing drugs) from drivers on the roadway. They repeated this in 2013–2014. These surveys were designed to provide estimates of AOD use by weekday daytime and weekend nighttime drivers. In 2007 and 2013–2014, 12.4% and 15.1% of drivers tested positive for any illegal drug. Moreover, 3.9% and 4.9% tested positive for a potentially impairing medication in the two time periods (Berning, Compton, & Wochinger, 2015).

Today, there are an increasing number of states in the U.S. legalizing cannabis (aka marijuana) recreationally and/or medicinally. Thus, there is a considerable interest in its prevalence among drivers and its contribution to crash risk. Delta-9-tetrahydrocannabinal (THC), the psychoactive ingredient in cannabis, was found in 8.6% of drivers in 2007. This rate increased significantly to 12.6% in 2013–2014 (Berning et al., 2015).

Crash risk related to cannabis and other illegal or legal (prescription or over-the-counter) drugs remains relatively unconfirmed. In laboratory studies, many drugs have been shown to impair driving. Simulator and closed-course studies assessing reaction time, divided attention, lane-keeping, and speed control, among other driving-related functions, have found that individual drugs and drugs in combinations produce different impairing effects and are likely to increase driving risk (Couper & Logan, 2014; Ogden & Moskowitz, 2004).

The most recent crash-risk study conducted in the United States found crash-involved drivers to be significantly more likely to test positive for THC and sedatives, to have used more than one class of drug, and to have used any type of drug compared with control drivers (Compton & Berning, 2015). THC was associated with a 25% elevated risk of crashing. However, once adjusted for age, gender, and race/ethnicity, the increases associated with risk were no longer significant (Compton & Berning, 2015). It should be noted that the results of this study should not imply drugs do not increase crash risk, but rather highlight the difficulties in obtaining drug measures that can accurately and consistently assess prevalence, potential impairment, and subsequent crash risk.

9.4.1.3 Prevalence vs. Impairment

Alcohol impairs many brain processes affecting driving-related skills. When one consumes alcohol, it is readily absorbed into the blood system, and the effects of alcohol on behavior (balance, coordination, reaction time), attention (divided attention, vigilance), cognition (decision-making), and other tendencies (risk taking, judgment, etc.) correlate well with BAC. Impairment increases with rising alcohol concentration and declines with dropping alcohol concentration.

Technology to detect alcohol in a driver and to determine whether its presence is at or above legal limits is widely available and used by law enforcement. This is not the case for drugs other than alcohol. Although technologies are under development to aid in the detection of other drugs (oral fluid devices, breathalyzers, etc.), significant challenges lay in assessing driving impairment. It cannot be overemphasized that drug prevalence does not necessarily imply impairment.

For many drugs, presence can be detected long after the passing of measurable signs of impairment. For example, unlike alcohol, which is water soluble and eliminates immediately as it is being ingested, cannabis is fat soluble (i.e., stored in the fatty tissues of the body). It can be released back into the blood, sometimes long after ingestion. Further, how cannabis metabolizes is different from alcohol, and levels of THC do not necessarily correspond with impairment (Compton, 2017). Given these differences, we cannot rely on a "per se" level for drugs other than alcohol. Behavioral signs of impairment are extremely important to consider, as detection of a drug (presence alone) is not enough to deem a driver impaired.

9.4.1.4 Varying Drug Impairing Behaviors

Research establishing that alcohol affects driving performance is well known and accepted. Information processing abilities affected by alcohol (albeit to different extents at certain BAC levels) include psychomotor (coordination and balance), vision, and cognition (including perception, tracking, and attention). Drugs other than alcohol have also been shown to impact driving skills, but differently by route of administration, dose, duration of therapy, tolerance, and user experience.

Cannabis has been shown to impair performance on driving simulator tasks and on open and closed driving courses. Specifically detected in a number of studies include decreased car handling performance, attention deficits, inappropriate reaction times, poor distance estimation, inability to maintain headway, subjective sleepiness, motor incoordination, poor hazard perception, and impaired sustained vigilance (Couper & Logan, 2014; Ogden & Moskowitz, 2004). Interestingly, unlike alcohol—where drivers tend to driver faster, follow at closer distances, and take risks—some drivers using cannabis drive slower, follow other cars at greater distances, and take fewer risks than when sober. However, research has suggested that these drivers may be attempting to compensate for the subjective effects of the drug (Compton, 2017; Couper & Logan, 2014).

There are a number of illegal drugs, such as cocaine, gamma-hydroxybutyrate (GHB), ketamine, lysergic acid diethylamide (LSD), methamphetamine, 3,4-methyl enedioxy-methamphetamine (MDMA), phencyclidine (PCP), and toluene, that have been shown to impair driving by drivers speeding, actively being aggressive (turning in front of other vehicles and other high-risk behaviors), being inattentive, and showing poor impulse control (Couper & Logan, 2014). Medicinal drugs, either prescribed or over-the-counter (OTC), that can impair driving include antidepressants (can cause sedation); stimulants (can influence attention, aggressiveness, and risk taking); sedatives like barbiturates, benzodiazepines, muscle relaxants, and sleep aids (can cause drowsiness and impair cognitive and motor function), and opioids/narcotics (can cause drowsiness and cognitive and motor impairment) (Kelley-Baker, Waehrer, & Pollini, 2017). Of course, not all medications are used as prescribed and are sometimes used when not prescribed.

Given the multitude of driving abilities potentially affected by various drugs, developing behavioral measures of impairment for drugs other than alcohol is critical. This may be a challenge as the level of automation changes, both in terms of detection and remediation.

9.4.1.5 Effects of Automation on AOD-Impaired Driving

As previously noted, increasing levels of automation gives rise to issues of drivers' reduced engagement with driving tasks and over-reliance on automation. Like distracted and sleepy drivers, AOD-impaired drivers may over-rely on automation and increasingly disengage with the handling of the vehicle. At lower levels of automation, this could be particularly problematic, as already evidenced in cases with Level 2 technology. In 2018, an alcohol-impaired driver operating his car's Level 2 system fell asleep at the wheel. While the vehicle's system operated appropriately, it took time for the police to safely slow down, and eventually stop the vehicle (presenting new challenges for law enforcement). Although to date there have been no documented fatalities related to AOD-impaired driving and automated vehicles, there have been non-fatal crashes with impaired drivers relying on automated systems.

Unique to AOD use is the opportunity to depend upon the vehicle to drive autonomously to a location, leading to increased drinking and/or other drug use. The vehicle essentially becomes the "designated driver." Prior research has found that passengers of designated drivers consumed more alcohol than when they drove themselves (Barr & Mackinnon, 1998) and that their alcohol levels were higher than those of the designated driver (Ditter et al., 2005). Though touted as an opportunity to reduce AOD-impaired crashes, full automation may present other public health concerns.

9.4.1.6   Effects of AOD-Impaired Driving on
Driver-Automation Coordination

In addition to the inherent dangers of over-reliance on lower-level automated driving systems, there are also dangers of AOD impairment presented to drivers who find themselves in a situation where the car requires them to take over. Many drugs, and certainly alcohol, adversely affect reaction times. Thus, a system's warning for the driver to engage with the system may not yield sufficient time for an impaired driver to "safely" resume control. (This assumes that the AOD-impaired driver is even capable of directing attention, processing information, and has the visual functions and other skills to re-engage with the vehicle.) Deterioration in reaction times after using alcohol vary considerably depending on the levels of alcohol, but have been identified in drivers with alcohol levels as low as 0.02 BAC, and repeatedly identified in drivers with BACs lower than 0.05 (Blomberg, Peck, Moskowitz, Burns, & Fiorentino, 2005; Zador, Krawchuk, & Voas, 2000), significantly below the current U.S. legal level of 0.08. The complexity of the driving task and how it will affect a driver's ability to safely resume control also needs further consideration. These are issues standard to any driver that certainly present additional challenges with AOD-impaired drivers.

Relying on these systems may also result in a driver not acknowledging internal signs of impairment related to drugs taken. Drugs that impair driving also impair a person's ability to judge the extent of their impairment and driving skills. Thus, driver self-perceptions are likely insufficient to establish when to rely on the automation and when to take over from the automation. In addition, AODs often distort judgment and result in poor decision-making. One area of concern for automated vehicles is the disabling of systems by drivers. Lack of familiarity and frustration with a system may result in a typical driver disabling it. However, the consequences of disabling technology may be significantly pronounced for AOD-impaired drivers.

9.4.2   Detection

9.4.2.1   Alcohol Breath Testers and Sensors,
Alcohol-Ignition Interlocks, and DADSS

The first alcohol breath-testing device was invented in 1954. This technology is used to detect alcohol presence with a quantitative result for indicating level of impairment (i.e., BAC). Akin to the alcohol breath-testing device that requires a mouthpiece for collecting the air specimen is the passive alcohol sensor (PAS). This device activates a small electrical pump that pulls the expired air from the front of the driver's face and provides an indication of the presence of alcohol and its detected level (Cammisa, Ferguson, & Wells, 1996; Lund & Jones, 1987).

Utilizing breath testing is the alcohol ignition interlock device. The device is connected to the vehicle's engine ignition system. It requires the driver to blow into the mouthpiece of the device (like a breathalyzer) to start and operate the vehicle. If the sample contains alcohol greater than the programmed BAC, the device prevents the engine from starting. Further, the device is also programmed to require "rolling re-tests" (i.e., random breath samples throughout the trip) to ensure drinking does not occur after the car is started. Many interlocks are equipped with cameras to ensure that the person taking the test is the driver. In the United States, these devices are typically used as a sanction for Driving While Intoxicated (DWI). However, they are being used in other countries as a preventative mechanism in professional and commercial transportation (Magnusson, Jakobsson, & Hultman, 2011).

Today, the Driver Alcohol Detection System for Safety (DADSS) program is working to automatically detect when a driver is intoxicated and prevent a vehicle from moving (Ferguson, Traube, Zaouk, & Strassburger, 2009). The DADDS program is working with two different technologies. The first is a breath-based system and the second is a touch-based system. Similar to the PAS, the breath-based system measures alcohol from the exhaled air of the driver (Ljungblad, Hök, Allalou, & Pettersson, 2017). The touch-based system measures alcohol levels under the skin's surface by shining an infrared light through the fingertip of the driver (using discrete semiconductor laser diodes; Ver Steeg et al., 2017). Integrated into the vehicle controls through the start button or steering wheel, both technologies prevent the engine from starting should the driver have a set alcohol level. It will take time for this technology to become available commercially, but it is thought that in the near future vehicles will be deployed with these systems. As with advanced driver assistance systems (ADAS), DADSS could be integrated into future fleets of vehicles (automated at various levels) to aid with the detection, and eventually, remediation of alcohol-impaired driving. One concern for programs like DADSS is the disabling of the system by the driver, especially if it is regarded as a preventative mechanism.

An alcohol detection system such as DADSS is likely most effective in a vehicle with Level 2 and 3 driving automation, provided it is not disabled. Here the vehicle simply would not start and thus prevent the driver from driving alcohol impaired. A similar detection system in a vehicle with Level 4 automation, where the driver still has the option to control the vehicle, would need to be equipped with remediation strategies should alcohol be detected and the driver demonstrates impaired behavior.

9.4.2.2 Breathalyzers for Drugs Other than Alcohol

In addition to alcohol, cannabis has become more prevalent among drivers on the roads (Kelley-Baker, Berning et al., 2017), and an increasing number of states have now legalized cannabis for medical and recreational use. Further, a number of other potentially impairing drugs have been detected at the roadside and in fatal crashes (Li, Brady, & Chen, 2013; Romano & Voas, 2011). Similar to breathalyzers for alcohol, devices using breath detection for other drugs are currently under development (see for example SensAbuses, Cannabix Technologies, Hound Labs). These noninvasive devices are intended to be used at the roadside to quickly identify drivers using cannabis and other drugs. Recent studies have indicated support for breath as a matrix for drug testing although further testing is needed (Beck, Sandqvist, Dubbelboer, & Franck, 2011).

9.4.2.3 Transdermal Sensors and Other Biosensors
for Alcohol and Drug Detection

Wearable, electronic, transdermal alcohol sensors collect data through expelled perspiration (e.g., Systems Continuous Alcohol Monitor/SCRAM CAM®, BACtrack, etc.). Typically used to measure alcohol consumption over long periods, they have been employed for monitoring patient treatment outcomes (Swift, Martin, Swette, Laconti, & Kackley, 1992), and results correlate well with breath alcohol measurements (Leffingwell et al., 2013). More recent studies have examined their utility for alcohol-use self-monitoring (Kim et al., 2016; Simons, Wills, Emery, & Marks, 2015).

Wearable biosensor devices also can test for drugs other than alcohol. Some use electrodermal activity, skin temperature, and locomotion (Wang, Fang, Carreiro, Wang, & Boyer, 2017). Several studies have reported sweat as a reliable matrix for detecting recent drug use in controlled studies (de la Torre & Pichini, 2004; Huestis et al., 2008). Roadside testing for drugs using sweat have also demonstrated success-ful detection for amphetamine-type stimulant drugs, although improvements with cannabis and benzodiazepines may be needed. Finally, a system currently under development that could be employed in automated vehicles uses an "Electronic Nose System" that obtains body odor from the skin surface (Voss et al., 2014). Although still in the early stages, these and similar detection devices could be equipped on future automated vehicles to aid in drug use detection. Even the DADSS touch-based system may be open to adaptation for eventually detecting drugs.

9.4.2.4 Behavioral Indicators for AOD-Impaired Driving

As noted in the previous sections, AOD use may lead to a number of behavioral indi-cators or performance degradations. Identifying these is important because just rely-ing on the detection of a drug could compromise any safety benefits an automated vehicle would have to offer.

For many people, certain drugs are life sustaining (e.g., blood pressure medi-cations, diabetes management drugs, etc.) and other drugs aid in retaining one's driving privilege (e.g., attention-deficit/hyperactivity disorder (ADHD), epilepsy, etc.). The quantity or concentration of the drug is an important consideration, as is tolerance and drug use duration. Further, similar to alcohol, the effects of drugs can vary significantly by individual. As the presence of a drug does not equate to impairment, behavioral indicators of degraded driving performance are a necessity. Unfortunately, specific behavior indicators that reliably identify impairment inde-pendent of the drug levels in the driver's system are currently inconclusive, but the need has been acknowledged and is under investigation.

With the case of alcohol, although the legal BAC limit in the United States is currently set at 0.08 g/dL, we know impairment begins at lower levels. Further, impairment by alcohol is transitional (at each level, some skill becomes impaired), and there is no set level that distinguishes non-impaired from impaired. Further, although drivers are impaired at any level, individual drivers vary in when and how they demonstrate performance degradations. Thus, safety features relying only on a BAC of 0.08 g/dL would be misguided.

The behavioral indicators related to AOD impairment that could be detected via in-vehicle technology are likely to be similar to sleepiness and distraction and can be guided by research discussed earlier. For example, with AOD, drowsiness is a typical behavioral indicator. Thus, one detection measure would include eye tracking (eye-lid closure—open/closed/sleepy, pupil detections) as might body imaging/position (e.g., slumped over) and face imaging/dense face tracking (i.e., head collapsed down or to the side). Other indicators include high-risk behaviors such as speeding, tailgating, swerving, and issues in maintaining lane control. Driving automation at lower levels can attempt to alert and even attempt to prevent these behaviors (high-speed alert, lane warning and assist, crash avoidance, automatic emergency braking, etc.), but it will be the higher levels of driving automation that will likely have the greatest impact on safety when the driver is impaired.

Recently, Volvo announced that their newest vehicles will be able to slow and even stop cars being operated with alcohol-impaired and distracted drivers. The vehicles are said to be able to identify long periods without steering input, weaving across lanes, slow reaction times, and identify closed eyes via in-car driver-monitoring cameras.

9.4.3 Remediation

Detecting a driver who is impaired by AOD poses distinct challenges as impairment thresholds and behaviors are not necessarily standard. Mitigating risk and intervening with an AOD-impaired driver present an entirely different set of challenges both programmatically and socially. AOD use can create a prolonged state of impairment similar to motion sickness (e.g., dizziness and nausea) and some forms of sleepiness (Roehrs & Roth, 2001). However, unlike other impaired states, even when alerted and warned, the driver is not likely to be in a suitable condition to take control of the vehicle (and/or may not relinquish the control voluntarily as many drugs, including alcohol, alter judgment).

As noted in earlier sections, the prevalence of AODs by itself does not indicate impairment. Therefore, in addition to identifying minimum thresholds for detecting alcohol and drugs (i.e., screening levels), behavioral signs must accompany these detection levels (for example, positive for alcohol at 0.04 and exhibiting lack of lane control). Once an agreed standard has been established, a number of remediation strategies have to be determined and evaluated. These will likely vary by driving automation level.

At the highest levels, driving automation may provide the greatest opportunity to prevent crashes. This type of automation may be the "designated driver" of the future. However, should an AOD-impaired driver refuse to relinquish control or disable the vehicle automation and take control, a mitigation plan becomes necessary. Depending on the situation, several remediation strategies might be available at various points in the driving experience/trip.

At a recent National Academies of Sciences Transportation Research Board meeting (January, 2019), a committee of researchers and experts, in both impaired driving and automation, arrived at an example of what a staged AOD-impaired mitigation plan might look like. Though we are a long way from saying definitively, it could entail the following. At the onset of the drive (pre-drive), the system (at virtually any automation level) can be programmed to warn the driver that AOD has been detected. This not only provides information and advises the driver about their potential impaired state but also alerts the system to activate other detection methods (i.e., behavioral signs) for impairment. Depending upon programmed AOD screening levels, the vehicle could also "not start." Much like today's alcohol ignition interlock, immediate detection of the drug at some set level prevents the driver from operating the vehicle. Another remediation strategy might include notifying an emergency contact that the driver is assuming control and may be impaired. When the driver begins the trip (post-start), the system can not only detect the presence of AOD but also activate other monitoring systems to detect driver behavioral cues of impairment. The system can again provide information or warning to the driver about impairment, and even attempt to assist with the vehicle control.

Should the driver continue the trip and ignore the warnings and refuse the assistance (disabling or attempting to disable the system), a number of potential strategies might activate. For example, the system might slow down the vehicle until it reaches a safe point to stop. Additionally, the system could activate the vehicle horn and lights should the driver persist. This strategy is similar to the alcohol ignition interlock, when a driver repeatedly does not provide a re-test sample during the ride. A more extreme remediation strategy might be rerouting the driver either back "home" or safer still, a hospital, fire station, etc. Should the system not be able to stop the vehicle or reroute the driver, notifying law enforcement might be an option.

The final mitigation stage would occur in the event of a crisis, such as an impending crash. Here, the system would need to be prepared to minimize collision and appropriately brace for impact.

Similar staged mitigation plans can be supposed. However, there needs to be significant programming and testing, in both the lab and field. Further, even if a suitable plan can be developed and programmed into the system, getting the public on-board may require considerable marketing and legislation.

9.5   MOTION SICKNESS

9.5.1   DEFINITIONS AND EFFECTS

With increasing levels of driving automation, motion sickness is expected to become a significant issue for drivers. Motion sickness can induce a variety of symptoms and can impair resumption of control.

9.5.1.1   What Is Motion Sickness and What Are Its Symptoms?

Motion sickness is generally a complex phenomenon with multiple possible symptoms and causes that vary based on the type of environment and the type of motion to which the individual is subjected, among other factors like individual propensity. Motion sickness is caused by "exposure to real or apparent, unfamiliar motion to which the individual is not adapted" (Benson, 2002, p. 1049), and has most commonly been attributed to a "cue conflict theory" (Reason & Brand, 1975), which describes an incongruency between how the body's motion is perceived by the vestibular (inner ear) system and how it is perceived visually. Some of the most common symptoms of motion sickness include nausea, vomiting, and headaches, along with other, more visible symptoms like pallor, sweating, and sleepiness, with varying levels of susceptibility and manifestations among different individuals (Lackner, 2014). The design of a vehicle (e.g., the suspension system) and the driving style of the driver (e.g., an aggressive or accelerated driving style) can contribute to or exacerbate motion sickness in passengers, as they can prompt the type of motions that can induce motion sickness (Diels, 2014). For more detailed reviews about the etiology, symptoms, and treatment of motion sickness in general, see Reason and Brand (1975), Reason (1978), Benson (2002), and Lackner (2014).

9.5.1.2 Contributors to Motion Sickness Experienced by Drivers of Automated Vehicles

In non-automated vehicles, vehicle dynamics is the main contributor to carsickness. However, the symptoms of carsickness are mostly prevalent among passengers rather than drivers, who can better anticipate the vehicle's trajectory, and thus would experience less of a mismatch between their vestibular and visual systems. Passengers who have less of a view of the surrounding environment and the visual cues of motion (e.g., passengers in rearward facing seats on public transport or passengers who engage in non-driving tasks) are more susceptible to motion sickness (Turner & Grifffin, 1999).

With increasing levels of driving automation, the driver's role involves less manual control of the vehicle and, as mentioned previously, provides more spare capacity for drivers to engage with non-driving tasks, allowing the driver to essentially become a passenger for many portions of the trip. Vehicles with higher levels of automation (e.g., SAE Levels 3 and 4) may even allow drivers added flexibility in terms of seats that swivel or rotate. These changes in driver tasks and seat position (captured in Figure 9.1) would result in reduced visual attention to the road and increase motion sickness (Diels, 2014; Diels & Bos, 2016; Diels, Bos, Hottelart, & Reilhac, 2016).

9.5.1.3 Effects of Motion Sickness on Drivers of Automated Vehicles

Motion sickness can considerably hinder reaping the promised benefits of driving automation for increasing trip productivity and enjoyment, and hence may decrease their acceptance. Some drivers could resort to closing their eyes or even sleeping during the trip to counteract the effects of motion sickness or can become sleepy or bored due to not being able to engage in non-driving tasks, thus potentially switching to other types of impaired states.

The exact effects of motion sickness on automation monitoring and resumption of control have been relatively unexplored. The sheer variety of symptoms and factors that may contribute to motion sickness can complicate studies that directly examine the effects of motion sickness on takeover performance, as different symptoms may have different effects on performance or manifest differently in different individuals, making it difficult to obtain definitive results or develop comprehensive theories. However, it is a valid assumption that many of the symptoms induced by motion sickness would impair monitoring and takeover performance.

9.5.2 DETECTION

9.5.2.1 Self-Report Measures

Subjective questionnaires, most prominently the Pensacola Motion Sickness Questionnaire (Kennedy & Graybiel, 1965) and the Motion Sickness Susceptibility Questionnaire (Reason & Brand, 1975), can be used to assess whether drivers are susceptible to or are experiencing symptoms associated with motion sickness. However, such questionnaires can only provide singular measures that are taken before or after a drive, rather than measures that can be used to provide some form of assistance to alleviate the symptoms in real time. It could be useful, though, if a vehicle's intelligent or automated system can learn in advance that a driver is susceptible to motion sickness and disable some features (e.g., flexible seating, non-driving tasks that require a driver to look away from the road) that can trigger or exacerbate symptoms. However, in a review of simulator sickness, Stoner, Fisher, and Mollenhauer (2011) argue that priming simulator study participants with a questionnaire or instructions related to simulator sickness before a drive can potentially induce or exacerbate symptoms, an effect that may also be observed in an automated vehicle on the road. Drivers can also be given the option to report whether they are experiencing any discomfort or sickness, and the severity of those symptoms, intermittently throughout the drive, so that remediation strategies can be prompted. Such methods should be used with caution, however, as they might cause further distraction to the driver or even worsen symptoms of motion sickness (for example, if they are presented visually).

9.5.2.2   Other Measures

Physiological measures, such as skin conductance or skin temperature, can be used in real time, as cold sweating is one of the symptoms associated with motion sickness (e.g., Benson, 2002). Video analysis of the driver's face can also be used to detect whether the driver has turned pale, another possible symptom of motion sickness (e.g., Benson, 2002). Measures of brain activity, such as EEG, have also been used to predict the incidence of motion sickness (Lin, Tsai, & Ko, 2013). However, because of the wide variety of motion sickness symptoms that can manifest in different individuals, it may be difficult to train detection algorithms to identify motion sickness in drivers of automated vehicles; algorithms may have to be trained for the particular individual. Stoner et al. (2011) state that in the case of simulator sickness, relatively limited studies have found a correlation between simulator sickness and physiological measures, thus further research is needed to assess the efficacy of physiological measures for detecting motion sickness in automated vehicles.

Body movement measures, obtained using video cameras or motion sensors for example, can be used to detect whether drivers are in positions that could induce motion sickness. However, video cameras may raise privacy concerns and the signal quality may degrade with ambient light. Physiological measures have similar issues in addition to the potential intrusiveness of the sensors. Further research is needed to evaluate physiological and body movement measures of motion sickness, and to develop methods of obtaining such measures that are practical for use in commercial vehicles.

Vehicle-based measures can be used to detect, for example, levels of acceleration or types of road geometry that are associated with inducing motion sickness and can support motion sickness detection when used in combination with other measures. For example, Jones et al. (2018) used a combination of vehicle-based, self-report, and head tilt measures in an ongoing effort to develop a platform to quantify motion sickness. However, further research is needed to evaluate the efficacy of vehicle kinematics in supporting other measures in detecting motion sickness.

9.5.3 REMEDIATION

Although research is limited, some remediation strategies have been proposed for motion sickness in automated vehicles. Such strategies can include behavioral and postural adjustments such as avoiding certain head tilt angles or seating positions, use of medication, as well as interface and vehicle adjustments. For example, it has been shown that certain head tilt angles are associated with an increased likelihood of motion sickness (e.g. Wada, Fujisawa, & Doi, 2018), thus drivers can be guided to adjust their head tilt accordingly. Drivers who are susceptible to motion sickness can also be encouraged to refrain from non-driving activities and avoid rearward facing positions for vehicles that provide flexible seating. Drugs that can alleviate motion sickness symptoms can also be used; however, such drugs might cause side effects like drowsiness and thus can still impair performance of the automated driving task (Diels & Bos, 2016).

Diels and Bos (2016) outline different techniques and design considerations that can help remediate the symptoms of motion sickness and increase driver's comfort. Such techniques include enhancing visual cues of the driving environment and the driver's ability to anticipate the vehicle's trajectory, for example, by increasing the window surface area or by using augmented reality displays. Moreover, non-driving tasks could be designed in a way that does not require drivers to move their heads to positions that can induce or exacerbate motion sickness symptoms, for example, by repositioning the non-driving task interface or presenting the task using head-up or augmented reality displays that do not fully compromise the driver's ability to view the driving environment and vehicle's trajectory. An alternative approach is to redesign vehicle suspension systems to reduce jerk (e.g., Ekchian et al., 2016; Giovanardi et al., 2018).

9.6 CONCLUSION

This chapter presented an overview of the four different impairments to automated driving, namely, distraction, sleepiness, impairment from AODs, and motion sickness. We explored the sources of these impairments (Figure 9.1) as well as their effects on automated driving performance. We have also briefly discussed possible detection and remediation techniques; a more detailed consideration of driver state monitoring approaches is presented in Chapter 11. In general, with increasing advances in automated and intelligent vehicle technologies, further research is needed in defining and quantifying the effects of different types of impairment, including the interaction between them, as well as on developing and evaluating techniques for detection and remediation.

A noteworthy consequence of increasingly automated driving environments is—ironically—periods of increased risk. This seems counterintuitive at first blush given automation's promise to protect safety by reducing risk. And automation does look to do just that *dans l'ensemble*; overall, risk reduces. But there exists a meaningful *redistribution of risk* that disproportionately increases safety risk during critical transition junctures. Especially relevant are relatively brief but safety critical situations such as a driver's decision to resume control while supposedly acting in a supervisory capacity (SAE Level 2), or sudden, unexpected resumption of driver control via automated request (SAE Level 3).

Each scenario presents challenging risks, particularly prior to, during, and immediately following a takeover transition. Such risk could be further compounded by impairment factors acting independently or in combination. For example, a sleepy driver may have the foggy notion that something is not quite right with the vehicle or the road itself and hazily recognize the need to quickly resume vehicle control (SAE Level 2). When experiencing diminished vigilance related to sleepiness prior to the event, the driver may not be quite sure what to suddenly do, especially if they have historically overly relied on the technology to self-correct prior to recognizing the near-term need to resume control. In essence, the driver had relinquished the SAE Level 2 role of supervisory control. Now sleepy, and perhaps distracted as well by sudden noises or auxiliary incoming visual stimuli, the driver can easily over- or under-correct the vehicle and contribute to a single, multi-vehicle, or pedestrian crash depending on the surroundings.

Similarly, drivers not sustaining or necessarily even intending to sustain active vigilance given SAE Level 3 driving could literally be in microsleep when an alarm sounds in automated recognition of conditions requiring near-immediate driver intervention and control. Sleepiness combined with motion sickness from a sudden jolt—or residual effects perhaps unrecognized from a few too many afternoon beers at the game hours earlier—could combine to create yet another crash, much more likely during such critical transition periods due to relatively increased risk. But fortunately, such an outcome is not *fait accompli* merely due to risk redistribution; rather, redistribution is itself a *call to action* to engage in remediation strategies as noted previously in the chapter and to be further developed as new research and risk-reducing innovations emerge to address this conundrum of automation. Across scenarios—including where there is potential for increased safety risk—automation need not harm.

ACKNOWLEDGMENTS

The authors would like to thank the members of the Human Factors and Applied Statistics (HFASt) laboratory at the University of Toronto, particularly Dengbo He, Chelsea DeGuzman, and Mehdi Hoseinzadeh Nooshabadi, for their valuable feedback on this chapter.

REFERENCES

Aghaei, A. S., Donmez, B., Liu, C. C., He, D., Liu, G., Plataniotis, K. N., … Sojoudi, Z. (2016). Smart driver monitoring: When signal processing meets human factors. *IEEE Signal Processing Magazine, 33*(6), 35–48.

Åkerstedt, T. & Folkard, S. (1995). Validation of the S and C components of the three-process model of alertness regulation. *Sleep, 18*(1), 1–6.

Altmann, E. M. & Trafton, J. G. (2002). Memory for goals: An activation-based model. *Cognitive Science, 26*(1), 39–83.

Barr, A. & Mackinnon, D. P. (1998). Designated driving among college students. *Journal of Studies on Alcohol, 59*(5), 549–554.

Beck, O., Sandqvist, S., Dubbelboer, I., & Franck, J. (2011). Detection of Δ9-tetrahydrocannabinol in exhaled breath from cannabis users. *Journal of Analytical Toxicology*, *35*(8), 541–544.

Benloucif, M. A., Sentouh, C., Floris, J., Simon, P., & Popieul, J.-C. (2017). Online adaptation of the level of haptic authority in a lane keeping system considering the driver's state. *Transportation Research Part F: Traffic Psychology and Behaviour*, *61*, 107–119.

Benson, A. J. (2002). Motion sickness. In K. B. Pandolf & R. R. Burr (Eds.), *Medical Aspects of Harsh Environments* (pp. 1048–1083). Washington, DC: U.S. Department of the Army, Office of the Surgeon General.

Berning, A., Compton, R., & Wochinger, K. (2015). *Results of the 2013–2014 National Roadside Survey of Alcohol and Drug Use by Drivers* (DOT HS 812 118). Washington, DC: National Highway Traffic Safety Administration.

Blomberg, R. D., Peck, R. C., Moskowitz, H., Burns, M., & Fiorentino, D. (2005). *Crash Risk of Alcohol Involved Driving: A Case-Control Study*. Stamford, CT: Dunlap.

Borbély, A. A. (1982). A two-process model of sleep regulation. *Human Neurobiology*, *1*(3), 195–204.

Borbély, A. A., Daan, S., Wirz-Justice, A., & Deboer, T. (2016). The two-process model of sleep regulation: A reappraisal. *Journal of Sleep Research*, *25*(2), 131–143.

Borkenstein, R. F., Crowther, R. F., & Shumate, R. P. (1974). The role of the drinking driver in traffic accidents (the Grand Rapids study). *Blutalkohol*, *11*(Suppl), 1–131.

Borkenstein, R. F., Crowther, R. F., Shumate, R. P., Ziel, W. B., & Zylman, R. (1964). *The Role of the Drinking Driver in Traffic Accidents*. Bloomington, IN: Indiana University.

Boyle, L. N., Tippin, J., Paul, A., & Rizzo, M. (2008). Driver performance in the moments surrounding a microsleep. *Transportation Research Part F: Traffic Psychology and Behaviour*, *11*(2), 126–136.

Braunagel, C., Geisler, D., Rosenstiel, W., & Kasneci, E. (2017). Online recognition of driver-activity based on visual scanpath classification. *IEEE Intelligent Transportation Systems Magazine*, *9*(2), 23–36.

Bugarski, V., Bačkalić, T., & Kuzmanov, U. (2013). Fuzzy decision support system for ship lock control. *Expert Systems with Applications*, *40*(10), 3953–3960.

Cabrall, C. D. D., Janssen, N. M., & de Winter, J. C. F. (2018). Adaptive automation: Automatically (dis)engaging automation during visually distracted driving. *PeerJ Computer Science*, *4*, e166.

Caird, J. K., Johnston, K. A., Willness, C. R., Asbridge, M., & Steel, P. (2014). A meta-analysis of the effects of texting on driving. *Accident Analysis & Prevention*, *71*, 311–318.

Caird, J. K., Simmons, S. M., Wiley, K., Johnston, K. A., & Horrey, W. J. (2018). Does talking on a cell phone, with a passenger, or dialing affect driving performance? An updated systematic review and meta-analysis of experimental studies. *Human Factors*, *60*(1), 101–133.

Cammisa, M., Ferguson, S., & Wells, J. (1996). *Laboratory Evaluation of PAS III Sensor with New Pump Design*. Arlington, VA: Insurance Institute for Highway Safety.

Carsten, O., Lai, F. C. H., Barnard, Y., Jamson, A. H., & Merat, N. (2012). Control task substitution in semiautomated driving: Does it matter what aspects are automated? *Human Factors*, *54*(5), 747–761.

Chen, H. Y. W. & Donmez, B. (2016). What drives technology-based distractions? A structural equation model on social-psychological factors of technology-based driver distraction engagement. *Accident Analysis and Prevention*, *91*, 166–174.

Chen, H. Y. W., Hoekstra-Atwood, L., & Donmez, B. (2018). Voluntary- and involuntary-distraction engagement: An exploratory study of individual differences. *Human Factors*, *60*(4), 575–588.

Compton, R. P. (2017). *Marijuana-Impaired Driving - A Report to Congress* (DOT HS 812 440). Washington, DC: National Highway Traffic Safety Administration.

Compton, R. P. & Berning, A. (2015). *Drug and Alcohol Crash Risk* (DOT HS 812 117). Washington, DC: National Highway Traffic Safety Administration.

Costa, G., Gaffuri, E., Ghirlanda, G., Minors, D. S., & Waterhouse, J. M. (1995). Psychophysical conditions and hormonal secretion in nurses on a rapidly rotating shift schedule and exposed to bright light during night work. *Work & Stress, 9*(2–3), 148–157.

Couper, F. J. & Logan, B. K. (2014). *Drugs and Human Performance Fact Sheets* (DOT HS 809 725). Washington, DC: National Highway Traffic Safety Administration.

Cuddy, J. S., Sol, J. A., Hailes, W. S., & Ruby, B. C. (2015). Work patterns dictate energy demands and thermal strain during wildland firefighting. *Wilderness and Environmental Medicine, 26*, 221–226.

Cunningham, M. L. & Regan, M. A. (2018). Driver distraction and inattention in the realm of automated driving. *IET Intelligent Transport Systems, 12*(6), 407–413.

Dawson, D. & Reid, K. (1997). Fatigue, alcohol and performance impairment. *Nature, 388*(6639), 235.

de la Torre, R. & Pichini, S. (2004). Usefulness of sweat testing for the detection of cannabis smoke. *Clinical Chemistry, 50*(11), 1961–1962.

de Winter, J. C. F. F., Happee, R., Martens, M. H., & Stanton, N. A. (2014). Effects of adaptive cruise control and highly automated driving on workload and situation awareness: A review of the empirical evidence. *Transportation Research Part F: Traffic Psychology and Behaviour, 27*(PB), 196–217.

Diels, C. (2014). Will autonomous vehicles make us sick? In S. Sharples & S. Shorrock (Eds.), *Contemporary Ergonomics and Human Factors* (pp. 301–307). Boca Raton, FL: CRC Press.

Diels, C. & Bos, J. E. (2016). Self-driving carsickness. *Applied Ergonomics, 53*, 374–382.

Diels, C., Bos, J. E., Hottelart, K., & Reilhac, P. (2016). Motion sickness in automated vehicles: The elephant in the room. In G. Meyer & S. Beiker (Eds.), *Road Vehicle Automation 3. Lecture Notes in Mobility* (pp. 121–129). Berlin: Springer.

Dijk, D.-J. & Czeisler, C. A. (1994). Paradoxical timing of the circadian rhythm of sleep propensity serves to consolidate sleep and wakefulness in humans. *Neuroscience Letters, 166*(1), 63–68.

Dijk, D.-J. & Larkin, W. (2004). Fatigue and performance models: General background and commentary on the circadian alertness simulator for fatigue risk assessment in transportation. *Aviation, Space, and Environmental Medicine, 75*(3, Suppl.), A119–A121.

Dinges, D. F. (1995). An overview of sleepiness and accidents. *Journal of Sleep Research, 4*(S2), 4–14.

Dinges, D. F., Pack, F., Williams, K., Gillen, K. A., Powell, J. W., Ott, G. E., … Pack, A. I. (1997). Cumulative sleepiness, mood disturbance, and psychomotor vigilance performance decrements during a week of sleep restricted to 4–5 hours per night. *Sleep, 20*(4), 267–277.

Dingus, T. A., Guo, F., Lee, S., Antin, J. F., Perez, M., Buchanan-King, M., & Hankey, J. (2016). Driver crash risk factors and prevalence evaluation using naturalistic driving data. *Proceedings of the National Academy of Sciences, 113*(10), 2636–2641.

Ditter, S. M., Elder, R. W., Shults, R. A., Sleet, D. A., Compton, R., & Nichols, J. L. (2005). Effectiveness of designated driver programs for reducing alcohol-impaired driving: A systematic review. *American Journal of Preventive Medicine, 28*(5S), 280–287.

Dogan, E., Honnêt, V., Masfrand, S., & Guillaume, A. (2019). Effects of non-driving-related tasks on takeover performance in different takeover situations in conditionally automated driving. *Transportation Research Part F: Traffic Psychology and Behaviour, 62*, 494–504.

Dong, Y., Hu, Z., Uchimura, K., & Murayama, N. (2011). Driver inattention monitoring system for intelligent vehicles: A review. *IEEE Transactions on Intelligent Transportation Systems, 12*(2), 596–614.

Donmez, B., Boyle, L., & Lee, J. D. (2008a). Designing feedback to mitigate distraction. In M. A. Regan, J. D. Lee, & K. L. Young (Eds.), *Driver Distraction: Theory, Effects, and Mitigation* (1st ed., pp. 519–531). Boca Raton, FL: CRC Press.

Donmez, B., Boyle, L., & Lee, J. D. (2003). Taxonomy of mitigation strategies for driver distraction. *Proceedings of the Human Factors and Ergonomics Society Annual Meeting*, 1865–1869.

Donmez, B., Boyle, L. N., & Lee, J. D. (2006). The impact of distraction mitigation strategies on driving performance. *Human Factors, 48*(4), 785–804.

Donmez, B., Boyle, L. N., & Lee, J. D. (2007). Safety implications of providing real-time feedback to distracted drivers. *Accident Analysis and Prevention, 39*(3), 581–590.

Donmez, B., Boyle, L. N., & Lee, J. D. (2008b). Mitigating driver distraction with retrospective and concurrent feedback. *Accident Analysis & Prevention, 40*(2), 776–786.

Ekchian, J., Graves, W., Anderson, Z., Giovanardi, M., Godwin, O., Kaplan, J., … DiZio, P. (2016). *A High-Bandwidth Active Suspension for Motion Sickness Mitigation in Autonomous Vehicles*. SAE Technical Paper Series (Vol. 1). Warrendale, PA: Society for Automotive Engineers.

Endsley, M. R. (1995). Toward a theory of situation awareness in dynamic systems. *Human Factors, 37*(1), 32–64.

Endsley, M. R. & Kiris, E. O. (1995). The out-of-the-loop performance problem and level of control in automation. *Human Factors, 37*(2), 381–394.

Fawcett, J. M., Risko, E. F., & Kingstone, A. (2015). *The Handbook of Attention*. Cambridge, MA: MIT Press.

Ferguson, S. A., Traube, E., Zaouk, A., & Strassburger, R. (2009). Driver Alcohol Detection System For Safety (DADSS) – A non-regulatory approach in the development and deployment of vehicle safety technology to reduce alcohol-impaired driving. *Proceedings of the 21st International Technical Conference on the Enhanced Safety of Vehicles*. Stuttgart, Germany: ESV.

Fisher, D. L. & Strayer, D. L. (2014). Modeling situation awareness and crash risk. *Annals of Advances in Automotive Medicine, 58*, 33–39.

Gaspar, J. G., Schwarz, C., Kashef, O., Schmitt, R., & Shull, E. (2018). *Using Driver State Detection in Automated Vehicles*. Iowa City, IA: SAFER-SIM University Transportation Center.

General Motors. (2018). *2019 Cadillac CT6 Owner's Manual*. Detroit, MI: General Motors LLC.

Giovanardi, M., Graves, W., Ekchian, J., DiZio, P., Ventura, J., Lackner, J. R., … Anderson, Z. (2018). An active suspension system for mitigating motion sickness and enabling reading in a car. *Aerospace Medicine and Human Performance, 89*(9), 822–829.

Gold, C., Damböck, D., Lorenz, L., & Bengler, K. (2013). "Take over!" How long does it take to get the driver back into the loop? *Proceedings of the Human Factors and Ergonomics Society Annual Meeting* (pp. 1938–1942). SAGE Publications, Los Angeles, CA.

Gold, C., Körber, M., Lechner, D., & Bengler, K. (2016). Taking over control from highly automated vehicles in complex traffic situations: The role of traffic density. *Human Factors, 58*(4), 642–652.

Hamid, M., Samuel, S., Borowsky, A., Horrey, W. J., & Fisher, D. L. (2016). Evaluation of training interventions to mitigate effects of fatigue and sleepiness on driving performance. *Transportation Research Record, 2584*, 30–38.

Haque, M. M. & Washington, S. (2014). A parametric duration model of the reaction times of drivers distracted by mobile phone conversations. *Accident Analysis and Prevention, 62*, 42–53.

He, D. & Donmez, B. (2018). The effect of distraction on anticipatory driving. *Proceedings of the Human Factors and Ergonomics Society Annual Meeting*, 1960–1964.

He, D. & Donmez, B. (2019). Influence of driving experience on distraction engagement in automated vehicles. *Transportation Research Record, 2673*(9), 142–151.

He, D., Kanaan, D., & Donmez, B. (2019). A taxonomy of strategies for supporting time-sharing with non-driving tasks in automated driving. *Proceedings of the Human Factors and Ergonomics Society Annual Meeting* (pp. 2088–2092). Santa Monica, CA: HFES.

He, D., Risteska, M., Donmez, B., & Chen, K. (in press). Driver cognitive load classification based on physiological data. In *Introduction to Digital Signal Processing and Machine Learning for Interactive Systems Developers* (pp. 1–19). New York: ACM Press.

Heaton, K., Browning, S., & Anderson, D. (2008). Identifying variables that predict falling asleep at the wheel among long-haul truck drivers. *American Association of Occupational Health Nurses*, 56(9), 379–385.

Horrey, W. J. & Wickens, C. D. (2006). Examining the impact of cell phone conversations on driving using meta-analytic techniques. *Human Factors*, 48(1), 196–205.

Huestis, M. A., Scheidweiler, K. B., Saito, T., Fortner, N., Abraham, T., Gustafson, R. A., & Smith, M. L. (2008). Excretion of D 9-tetrahydrocannabinol in sweat. *Forensic Science International*, 174(2–3), 173–177.

Jacobé De Naurois, C., Bourdin, C., Stratulat, A., Diaz, E., & Vercher, J.-L. (2019). Detection and prediction of driver drowsiness using artificial neural network models. *Accident Analysis and Prevention*, 126, 95–104.

Jamson, A. H., Merat, N., Carsten, O. M. J., & Lai, F. C. H. (2013). Behavioural changes in drivers experiencing highly-automated vehicle control in varying traffic conditions. *Transportation Research Part C: Emerging Technologies*, 30, 116–125.

Jones, M. L. H., Sienko, K., Ebert-Hamilton, S., Kinnaird, C., Miller, C., Lin, B., … Sayer, J. (2018). Development of a Vehicle-Based Experimental Platform for Quantifying Passenger Motion Sickness during Test Track Operations. SAE Technical Paper Series (Vol. 1). Warrendale, PA: Society for Automotive Engineers.

Junaedi, S. & Akbar, H. (2018). Driver drowsiness detection based on face feature and PERCLOS. *Journal of Physics: Conference Series*, 1090.

Kanaan, D., Ayas, S., Donmez, B., Risteska, M., & Chakraborty, J. (2019). Using naturalistic vehicle-based data to predict distraction and environmental demand. *International Journal of Mobile Human Computer Interaction*, 11(3), 59–70.

Kelley-Baker, T., Berning, A., Ramirez, A., Lacey, J. H., Carr, K., Waehrer, G., … Compton, R. P. (2017). *2013–2014 National Roadside Study of Alcohol and Drug Use by Drivers: Drug Results* (DOT HS 812 411). Washington, DC: National Highway Traffic Safety Administration.

Kelley-Baker, T., Waehrer, G., & Pollini, R. A. (2017). Prevalence of self-reported prescription drug use in a national sample of U.S. drivers. *Journal of Studies on Alcohol and Drugs*, 78(1), 30–38.

Kennedy, R. S. & Graybiel, A. (1965). *The Dial Test: A Standardized Procedure for the Experimental Production of Canal Sickness Symptomatology in a Rotating Environment* (NSAM-930). Pensacola, FL: Naval School of Aviation Medicine.

Kim, J., Jeerapan, I., Imani, S., Cho, T. N., Bandodkar, A., Cinti, S., … Wang, J. (2016). Noninvasive alcohol monitoring using a wearable tattoo-based iontophoretic-biosensing system. *Sensors*, 1(8), 1011–1019.

Kircher, K. & Ahlstrom, C. (2018). Evaluation of methods for the assessment of attention while driving. *Accident Analysis & Prevention*, 114, 40–47.

Kircher, K., Kircher, A., & Ahlström, C. (2009). *Results of a Field Study on a Driver Distraction Warning System*. Linköping, Sweden: Swedish National Road and Transport Research Institute.

Klauer, S. G., Dingus, T. A., Neale, V. L., Sudweeks, J. D., & Ramsey, D. J. (2006). *The Impact of Driver Inattention on Near-Crash/Crash Risk: An Analysis Using the 100-Car Naturalistic Driving Study Data* (DOT HS 810 594). Washington, DC: National Highway Traffic Safety Administration.

Köhn, T., Gottlieb, M., Schermann, M., & Krcmar, H. (2019). Improving take-over quality in automated driving by interrupting non-driving tasks. *24th International Conference on Intelligent User Interfaces (IUI'19)* (pp. 510–517). New York: ACM.

Lacey, J. H., Kelley-Baker, T., Berning, A., Romano, E., Ramirez, A., Yao, J., … Compton, R. (2016). *Drug and Alcohol Crash Risk: A Case-Control Study* (DOT HS 812 630). Washington, DC: National Highway Traffic Safety Administration.

Lackner, J. R. (2014). Motion sickness: More than nausea and vomiting. *Experimental Brain Research, 232,* 2493–2510.

Lee, J. D. (2014). Dynamics of driver distraction: The process of engaging and disengaging. *Annals of Advances in Automotive Medicine, 58,* 24–32.

Lee, J. D., Young, K. L., & Regan, M. A. (2008). Defining driver distraction. In M. A. Regan, J. D. Lee, & K. L. Young (Eds.), *Driver Distraction: Theory, Effects and Mitigation* (pp. 31–40). Boca Raton, FL: CRC Press.

Leffingwell, T. R., Cooney, N. J., Murphy, J. G., Luczak, S., Rosen, G., Dougherty, D. M., & Barnett, N. P. (2013). Continuous objective monitoring of alcohol use: Twenty-first century measurement using transdermal sensors. *Alcoholism: Clinical and Experimental Research, 37*(1), 16–22.

Lehrer, A. M. (2015). A systems-based framework to measure, predict, and manage fatigue. *Reviews of Human Factors and Ergonomics, 10*(1), 194–252.

Lehrer, A. M. & Popkin, S. M. (2014). *Current and Next Generation Fatigue Models* (DOT-VNTSCOST-14-01). Cambridge, MA: Volpe Center.

Levitin, D. J. (2014). *The Organized Mind: Thinking Straight in the Age of Information Overload.* New York: Plume/Penguin Books.

Li, G., Brady, J. E., & Chen, Q. (2013). Drug use and fatal motor vehicle crashes: A case-control study. *Accident Analysis and Prevention, 60,* 205–210.

Li, Z., Bao, S., Kolmanovsky, I. V., & Yin, X. (2018). Visual manual distraction detection using driving performance indicators with naturalistic driving data. *IEEE Transactions on Intelligent Transportation Systems, 19*(8), 2528–2535.

Li, Z., Chen, L., Peng, J., & Wu, Y. (2017). Automatic detection of driver fatigue using driving operation information for transportation safety. *Sensors, 17*(6), 1212.

Liang, Y. & Lee, J. D. (2014). A hybrid Bayesian Network approach to detect driver cognitive distraction. *Transportation Research Part C, 38,* 146–155.

Lim, J. & Dinges, D. F. (2008). Sleep deprivation and vigilant attention. *Annals of the New York Academy of Sciences, 1129*(1), 305–322.

Lin, C. T., Tsai, S. F., & Ko, L. W. (2013). EEG-based learning system for online motion sickness level estimation in a dynamic vehicle environment. *IEEE Transactions on Neural Networks and Learning Systems, 24*(10), 1689–1700.

Ljungblad, J., Hök, B., Allalou, A., & Pettersson, H. (2017). Passive in-vehicle driver breath alcohol detection using advanced sensor signal acquisition and fusion. *Traffic Injury Prevention, 18*(Suppl. 1), S31–S36.

Louw, T., Markkula, G., Boer, E., Madigan, R., Carsten, O., & Merat, N. (2017). Coming back into the loop: Drivers' perceptual-motor performance in critical events after automated driving. *Accident Analysis & Prevention, 108,* 9–18.

Louw, T. & Merat, N. (2017). Are you in the loop? Using gaze dispersion to understand driver visual attention during vehicle automation. *Transportation Research Part C: Emerging Technologies, 76,* 35–50.

Lund, A. K. & Jones, I. S. (1987). Detection of impaired drivers with a passive alcohol sensor. *Proceedings of the 10th Conference on Alcohol, Drugs, and Traffic Safety* (pp. 379–382). Amsterdam: ICADTS.

Magnusson, P., Jakobsson, L., & Hultman, S. (2011). Alcohol interlock systems in Sweden: 10 years of systematic work. *American Journal of Preventative Medicine, 40*(3), 378–379.

McCartt, A. T., Rohrbaugh, J. W., Hammer, M. C., & Fuller, S. Z. (2000). Factors associated with falling asleep at the wheel among long-distance truck drivers. *Accident Analysis and Prevention, 32*, 493–504.

McDonald, A. D., Alambeigi, H., Engström, J., Markkula, G., Vogelpohl, T., Dunne, J., & Yuma, N. (2019). Towards computational simulations of behavior during automated driving takeovers: A review of the empirical and modeling literatures. *Human Factors, 61*(4), 642–688.

McDonald, A. D., Ferris, T. K., & Wiener, T. A. (2019). Classification of driver distraction: A comprehensive analysis of feature generation, machine learning, and input measures. *Human Factors*.

McFarlane, D. & Latorella, K. A. (2002). The scope and importance of human interruption in human-computer interaction design. *Human-Computer Interaction, 17*(1), 1–61.

Merat, N., Seppelt, B., Louw, T., Engström, J., Lee, J. D., Johansson, E., ... Keinath, A. (2019). The "Out-of-the-Loop" concept in automated driving: Proposed definition, measures and implications. *Cognition, Technology & Work, 21*(1), 87–98.

Metz, B., Schömig, N., & Krüger, H.-P. P. (2011). Attention during visual secondary tasks in driving: Adaptation to the demands of the driving task. *Transportation Research Part F: Traffic Psychology and Behaviour, 14*(5), 369–380.

Michon, J. A. (1985). A critical view of driver behavior models: What do we know, what should we do? In L. Evans & R. C. Schwing (Eds.), *Human Behavior and Traffic Safety* (pp. 485–520). New York, NY: Plenus.

Miller, D., Sun, A., Johns, M., Ive, H., Sirkin, D., Aich, S., & Ju, W. (2015). Distraction becomes engagement in automated driving. *Proceedings of the Human Factors and Ergonomics Society Annual Meeting, 59*(1), 1676–1680.

Mitler, M. M., Carskadon, M. A., Czeisier, C. A., Dement, W. C., Dinges, D. F., & Graeber, R. C. (1988). Catastrophes, sleep, and public policy: Consensus report. *Sleep, 11*(1), 100–109.

Moore-Ede, M. C., Sulzman, F. M., & Fuller, C. A. (1982). *The Clocks that Time Us: Physiology of the Circadian Timing System*. Cambridge, MA: Harvard University Press.

National Center for Statistics and Analysis. (2018). *Alcohol-Impaired Driving: 2017 Data*. Washington, D.C.: NCSA.

National Highway Traffic Safety Administration. (2013). Visual-manual NHTSA driver distraction guidelines for in-vehicle electronic devices. *Federal Register, 78*(81), 24818–24890. (Washington, DC: National Highway Traffic Safety Administration.)

National Highway Traffic Safety Administration. (2016). Visual-manual NHTSA driver distraction guidelines for portable and aftermarket devices. *Federal Register, 81*(233), 87656–87683. (Washington, DC: National Highway Traffic Safety Administration.)

National Highway Traffic Safety Administration. (2017). *Research Note: Distracted Driving 2015*. Washington, DC: National Highway Traffic Safety Administration.

National Highway Traffic Safety Administration. (2019). *Research Note: Driver electronic device use in 2017*. Washington, DC: National Highway Traffic Safety Administration.

National Safety Council. (2019). *Drowsy Driving Is Impaired Driving*. Retrieved from www.nsc.org/road-safety/safety-topics/fatigued-driving

National Sleep Foundation. (2019). *Sleepiness, Medication & Drugs: Why Your OTC Medications and Prescription Drugs Might Make You Tired*. Retrieved from www.sleepfoundation.org/excessive-sleepiness/causes/sleepiness-medication-drugs-why-your-otc-medications-and-prescription

Naujoks, F., Höfling, S., Purucker, C., & Zeeb, K. (2018). From partial and high automation to manual driving: Relationship between non-driving related tasks, drowsiness and takeover performance. *Accident Analysis & Prevention, 121*, 28–42.

Naujoks, F., Wiedemann, K., & Schömig, N. (2017). The importance of interruption management for usefulness and acceptance of automated driving. *Proceedings of the 9th ACM International Conference on Automotive User Interfaces and Interactive Vehicular Applications (AutomotiveUI '17)* (pp. 254–263). New York: ACM.

Ogden, E. & Moskowitz, H. (2004). Effects of alcohol and other drugs on driver performance. *Traffic Injury Prevention, 5*(3), 185–198.

Oviedo-Trespalacios, O., Haque, M. M., King, M., & Washington, S. (2017). Self-regulation of driving speed among distracted drivers: An application of driver behavioral adaptation theory. *Traffic Injury Prevention, 18*(6), 599–605.

Owens, J., Gruber, R., Brown, T., Corkum, P., Cortese, S., O'Brien, L., ... Weiss, M. (2013). Future research directions in sleep and ADHD: Report of a consensus working group. *Journal of Attention Disorders, 17*(7), 550–564.

Parasuraman, R. & Nestor, P. G. (1991). Attention and driving skills in aging and Alzheimer's disease. *Human Factors, 33*(5), 539–557.

Pech, T., Enhuber, S., Wandtner, B., Schmidt, G., & Wanielik, G. (2018). Real time recognition of non-driving related tasks in the context of highly automated driving. In J. Dubbert, B. Müller, & G. Meyer (Eds.), *Advanced Microsystems for Automotive Applications 2018. AMAA 2018. Lecture Notes in Mobility* (pp. 43–55). Berlin: Springer.

Platten, F., Milicic, N., Schwalm, M., & Krems, J. (2013). Using an infotainment system while driving - A continuous analysis of behavior adaptations. *Transportation Research Part F: Traffic Psychology and Behaviour, 21*, 103–112.

Politis, I., Brewster, S., & Pollick, F. (2017). Using multimodal displays to signify critical handovers of control to distracted autonomous car drivers. *International Journal of Mobile Human Computer Interaction, 9*(3), 1–16.

Proctor, R. W. & Zandt, T. Van. (2018). *Human Factors in Simple and Complex Systems* (3rd ed.). Boca Raton, FL: CRC Press.

Ranney, T. A., Garrott, W. R., & Goodman, M. J. (2000). *NHTSA Driver Distraction Research: Past, Present, and Future* (No. 2001–06–0177). Warrendale, PA: SAE.

Reason, J. T. (1978). Motion sickness adaptation: A neural mismatch model. *Journal of the Royal Society of Medicine, 71*, 819–829.

Reason, J. T. & Brand, J. J. (1975). *Motion Sickness*. Oxford: Academic Press.

Regan, M. A., Hallett, C., & Gordon, C. P. (2011). Driver distraction and driver inattention: Definition, relationship and taxonomy. *Accident Analysis and Prevention, 43*(5), 1771–1781.

Regan, M. A., Lee, J. D., & Young, K. (2008). *Driver Distraction: Theory, Effects and Mitigation*. Boca Raton, FL: CRC Press.

Reimer, B., Donmez, B., Lavallière, M., Mehler, B., Coughlin, J. F., & Teasdale, N. (2013). Impact of age and cognitive demand on lane choice and changing under actual highway conditions. *Accident Analysis and Prevention, 52*, 125–132.

Roche, F., Somieski, A., & Brandenburg, S. (2019). Behavioral changes to repeated takeovers in highly automated driving: Effects of the takeover-request design and the nondriving-related task modality. *Human Factors, 61*(5), 839–849.

Roehrs, T. & Roth, T. (2001). Sleep, sleepiness, and alcohol use. *Alcohol Research and Health, 25*(2), 101–109.

Romano, E. & Voas, R. B. (2011). Drug and alcohol involvement in four types of fatal crashes. *Journal of Studies on Alcohol and Drugs, 72*(4), 567–576.

Rosa, R. R., Bonnet, M. H., Bootzin, R. R., Eastman, C. I., Monk, T., Penn, P. E., ... Walsh, J. K. (1990). Intervention factors for promoting adjustment to nightwork and shiftwork. *Occupational Medicine: State of the Art Reviews, 5*(2), 391–415.

SAE International. (2018). *Taxonomy and Definitions for Terms Related to Driving Automation Systems for On-Road Motor Vehicles* (J3016). Warrendale, PA: Society for Automotive Engineers.

Sarter, N. B. & Woods, D. D. (1995). How in the world did we ever get into that mode? Mode error and awareness in supervisory control. *Human Factors*, *37*(1), 5–19.

Saxby, D. J., Matthews, G., & Neubauer, C. (2017). The relationship between cell phone use and management of driver fatigue: It's complicated. *Journal of Safety Research*, *61*, 129–140.

Schömig, N., Hargutt, V., Neukum, A., Petermann-Stock, I., & Othersen, I. (2015). The interaction between highly automated driving and the development of drowsiness. *Procedia Manufacturing*, *3*, 6652–6659.

Schroeter, R., Oxtoby, J., & Johnson, D. (2014). AR and gamification concepts to reduce driver boredom and risk taking behaviours. *Proceedings of the 6th International Conference on Automotive User Interfaces and Interactive Vehicular Applications - AutomotiveUI '14* (pp. 1–8). Seattle, WA: ACM Press.

Schwarz, C., Brown, T., Lee, J., Gaspar, J., & Kang, J. (2016). *The Detection of Visual Distraction Using Vehicle and Driver Based Sensors*. Warrendale, PA: SAE International.

Seppelt, B. D. & Lee, J. D. (2019). Keeping the driver in the loop: Dynamic feedback to support appropriate use of imperfect vehicle control automation. *International Journal of Human-Computer Studies*, *125*, 66–80.

Shen, S. & Neyens, D. M. (2017). Assessing drivers' response during automated driver support system failures with non-driving tasks. *Journal of Safety Research*, *61*, 149–155.

Sibi, S., Ayaz, H., Kuhns, D. P., Sirkin, D. M., & Ju, W. (2016). Monitoring driver cognitive load using functional near infrared spectroscopy in partially autonomous cars. *IEEE Intelligent Vehicles Symposium (IV)* (pp. 419–425). Gotenburg, Sweden: IEEE.

Simons, J. S., Wills, T. A., Emery, N. N., & Marks, R. M. (2015). Quantifying alcohol consumption: Self-report, transdermal assessment, and prediction of dependence symptoms. *Addictive Behaviors*, *50*, 205–212.

Steinberger, F., Schroeter, R., Foth, M., & Johnson, D. (2017). Designing gamified applications that make safe driving more engaging. *CHI 2017: Conference on Human Factors in Computing Systems, ACM SIGCHI* (pp. 2826–2839). Denver, CO: ACM Press.

Stockert, S., Richardson, N. T., & Lienkamp, M. (2015). Driving in an increasingly automated world – Approaches to improve the driver-automation interaction. *Procedia Manufacturing*, *3*, 2889–2896.

Stoner, H. A., Fisher, D. L., & Mollenhauer, M., Jr. (2011). Simulator and scenario factors influencing simulator sickness. In D. L. Fisher, M. Rizzo, J. K. Caird, & J. D. Lee (Eds.), *Handbook of Driving Simulation for Engineering, Medicine, and Psychology* (pp. 14:1–14:24). Boca Raton, FL: CRC Press.

Swift, R. M., Martin, C. S., Swette, L., Laconti, A., & Kackley, N. (1992). Studies on a wearable, electronic, transdermal alcohol sensor. *Alcoholism Clinical and Experimental Research*, *16*(4), 721–725.

Tassi, P. & Muzet, A. (2000). Sleep inertia. *Sleep Medicine Reviews*, *4*(4), 341–353.

Tepas, D. I. & Carvalhais, A. B. (1990). Sleep patterns of shiftworkers. *Occupational Medicine*, *5*(2), 199–208.

Tesla. (2019). *Model X Owner's Manual*. Palo Alto, CA: Tesla.

Thomas, G. R., Raslear, T. G., & Kuehn, G. I. (1997). *The Effects of Work Schedule on Train Handling Performance and Sleep of Locomotive Engineers: A Simulator Study* (DOT/FRA/ORD-97-09). Washington, DC: Federal Railroad Administration.

Tijerina, L., Gleckler, M., Stoltzfus, D., Johnston, S., Goodman, M. J., & Wierwille, W. W. (1998). *A Preliminary Assessment of Algorithms for Drowsy Driver and Inattentive Driver Detection on the Road* (HS-808 905). Washington, DC: National Highway Traffic Safety Administration.

Traffic Injury Research Foundation. (2018). *Distraction-Related Fatal Collisions, 2000–2015*. Ottawa, Canada: TRIF.

Turner, M. & Grifffin, M. J. (1999). Motion sickness in public road transport: The relative importance of motion, vision, and individual differences. *British Journal of Psychology, 90,* 519–530.

Van Dongen, H. P. A. (2006). Shift work and inter-individual differences in sleep and sleepiness. *Chronobiology International, 23*(6), 1139–1147.

Ver Steeg, B., Treese, D., Adelante, R., Kraintz, A., Laaksonen, B., Ridder, T., … Cox, D. (2017). Development of a solid state, non-invasive, human touch based blood alcohol sensor. *Proceedings of the 25th International Technical Conference on the Enhanced Safety of Vehicles.* Detroit, MI: ESV.

Victor, T. W. (2010). *The Victor and Larsson (2010) Distraction Detection Algorithm and Warning Strategy.* Gothenburg, Sweden: Volvo Technology.

Victor, T. W., Engström, J., & Harbluk, J. L. (2008). Distraction assessment methods based on visual behaviour and event detection. In M. A. Regan, J. D. Lee, & K. L. Young (Eds.), *Driver Distraction: Theory, Effects and Mitigation* (pp. 135–165). Boca Raton, FL: CRC Press.

Vogelpohl, T., Kühn, M., Hummel, T., Gehlert, T., & Vollrath, M. (2018). Transitioning to manual driving requires additional time after automation deactivation. *Transportation Research Part F: Traffic Psychology and Behaviour, 55,* 464–482.

Voss, A., Witt, K., Kaschowitz, T., Poitz, W., Ebert, A., Roser, P., & Bär, K.-J. (2014). Detecting cannabis use on the human skin surface via an electronic nose system. *Sensors, 14*(7), 13256–13272.

Wada, T., Fujisawa, S., & Doi, S. (2018). Analysis of driver's head tilt using a mathematical model of motion sickness. *International Journal of Industrial Ergonomics, 63,* 89–97.

Wandtner, B., Schömig, N., & Schmidt, G. (2018a). Effects of non-driving related task modalities on takeover performance in highly automated driving. *Human Factors, 60*(6), 870–881.

Wandtner, B., Schömig, N., & Schmidt, G. (2018b). Secondary task engagement and disengagement in the context of highly automated driving. *Transportation Research Part F: Traffic Psychology and Behaviour, 58,* 253–263.

Wang, J., Fang, H., Carreiro, S., Wang, H., & Boyer, E. (2017). A new mining method to detect real time substance use events from wearable biosensor data stream. *2017 International Conference on Computing, Networking and Communications (ICNC)* (pp. 465–470). Silicon Valley, CA: ICNC.

Watson, A. & Zhou, G. (2016). Microsleep prediction using an EKG capable heart rate monitor. *2016 IEEE First International Conference on Connected Health: Applications, Systems and Engineering Technologies (CHASE)* (pp. 328–329).

Wertz, A. T., Ronda, J. M., Czeisler, C. A., & Wright, K. P. (2006). Effects of sleep inertia on cognition. *Journal of the American Medical Association, 295*(2), 163–164.

Wesensten, N. J., Belenky, G., & Balkin, T. J. (2005). Cognitive readiness in network-centric operations. *Parameters: US Army War College Quarterly, 35*(1), 94–105.

Wickelgren, W. A. (1977). Speed-accuracy tradeoff and information processing dynamics. *Acta Psychologica, 41,* 67–85.

Wickens, C. D. (2008). Multiple resources and mental workload. *Human Factors, 50*(3), 449–455.

Wierwille, W. W., Wreggit, S. S., Kirn, C. L., Ellsworth, L. A., & Fairbanks, R. J. (1994). *Research on Vehicle-Based Driver Status/Performance Monitoring; Development, Validation, and Refinement of Algorithms for Detection of Driver Drowsiness.* Final Report (HS-808 247). Washington, DC: National Highway Traffic Safety Administration.

Wulf, F., Rimini-Doring, M., Arnon, M., & Gauterin, F. (2015). Recommendations supporting situation awareness in partially automated driver assistance systems. *IEEE Transactions on Intelligent Transportation Systems, 16*(4), 2290–2296.

Xie, J. Y., Chen, H.-Y. W., & Donmez, B. (2016). Gaming to safety: Exploring feedback gamification for mitigating driver distraction. *Proceedings of the Human Factors and Ergonomics Society Annual Meeting,* 60(1) 1884–1888.

Yoon, S. H. & Ji, Y. G. (2019). Non-driving-related tasks, workload, and takeover performance in highly automated driving contexts. *Transportation Research Part F: Traffic Psychology and Behaviour, 60,* 620–631.

Young, K. L., Regan, M. A., & Lee, J. D. (2008). Measuring the effects of driver distraction: Direct driving performance methods and measures. In M. A. Regan, J. D. Lee, & K. L. Young (Eds.), *Driver Distraction: Theory, Effects and Mitigation* (pp. 85–106). Boca Raton, FL: CRC Press.

Zador, P. L., Krawchuk, S. A., & Voas, R. B. (2000). Alcohol-related relative risk of driver fatalities and driver involvement in fatal crashes in relation to driver age and gender: An update using 1996 data. *Journal of Studies on Alcohol, 61*(3), 387–395.

Zeeb, K., Buchner, A., & Schrauf, M. (2015). What determines the take-over time? An integrated model approach of driver take-over after automated driving. *Accident Analysis & Prevention, 78,* 212–221.

Zeeb, K., Buchner, A., & Schrauf, M. (2016). Is take-over time all that matters? The impact of visual-cognitive load on driver take-over quality after conditionally automated driving. *Accident Analysis & Prevention, 92,* 230–239.

Zhang, B., de Winter, J., Varotto, S., Happee, R., & Martens, M. (2019). Determinants of take-over time from automated driving: A meta-analysis of 129 studies. *Transportation Research Part F: Traffic Psychology and Behaviour, 64,* 285–307.

Zijlstra, F. R. H., Cropley, M., & Rydstedt, L. W. (2014). From recovery to regulation: An attempt to reconceptualize "recovery from work." *Stress and Health, 30,* 244–252.

10 Driver Capabilities in the Resumption of Control

Sherrilene Classen
University of Florida

Liliana Alvarez
University of Western Ontario

CONTENTS

Key Points 218
10.1 Introduction 218
10.2 Michon's Model 219
10.3 Medically-at-Risk Conditions.. 219
    10.3.1 Deskilling: Implications of the Aging Process 219
        10.3.1.1 Age-Related Deskilling 219
        10.3.1.2 Functional Performance Deficits 220
        10.3.1.3 Effect on Driving Behaviors 220
        10.3.1.4 AV Technology to Compensate for Functional
                    Driving Impairments by SAE Level 221
    10.3.2 Low Vision.. 222
        10.3.2.1 Cataracts 223
        10.3.2.2 Age-Related Macular Degeneration 224
        10.3.2.3 Glaucoma 226
        10.3.2.4 Diabetic Retinopathy (DR) 228
        10.3.2.5 AV Technology to Compensate for Functional
                    Driving Impairments by SAE Level 229
        10.3.2.6 Case Study: Zane, an Older Adult with a Diagnosis of
                    Glaucoma 229
    10.3.3 Neurological and Neurodegenerative Disorders............................. 231
        10.3.3.1 Introduction.. 231
        10.3.3.2 ASD and ADHD 231
        10.3.3.3 Parkinson's Disease 232
        10.3.3.4 Strategic, Tactical, and Operational Deficits for
                    Those with ASD, ADHD, and/or PD 233
        10.3.3.5 Case Study: Elizabeth, a Client with PD 238
10.4 Conclusion 241
Acknowledgments... 241
References... 241

KEY POINTS

- Medical conditions, and their associated characteristics, affect the functional abilities (visual, cognitive, motor, and other sensory) and therefore the performance of drivers.
- These functional abilities are critical to perform the strategic, tactical, and operational tasks associated with driving.
- Driver support features such as in-vehicle information systems (any of the SAE levels, although sometimes identified synonymously with SAE Level 0), lane-centering assist, and braking/acceleration assist (SAE Level 1 and 2) may provide benefits but also challenges for drivers who are medically-at-risk.
- Automated vehicle technologies hold great potential to facilitate fitness-to-drive abilities in drivers wanting to resume control of the vehicle
- Empirical testing of the potential benefits of automated vehicle technologies is mission critical

10.1 INTRODUCTION

Because it is outside the scope of this chapter to focus on all medical conditions that may impact fitness to drive, the authors describe selected medical conditions grouped in four distinct categories, i.e., deskilling, visual disorders, neurological disorders, and neurodegenerative disorders. From these categories, the authors indicate the core clinical characteristics and explicate how those characteristics relate to functional performance deficits. Driving behaviors, an indicator of fitness to drive, are discussed within the structure of Michon's model (Michon, 1985)—which classifies driving behaviors on the strategic, tactical, and operational level.

Rehabilitation scientists and professionals are concerned with empowering clients to overcome deficient driving behaviors. Therefore, this chapter expounds how automation is providing exciting possibilities to address this challenge of deficient driving behaviors. The authors demonstrate the benefits of automated vehicle technologies (SAE Levels 0–2) to enable the driver to resume control of his/her fitness-to-drive abilities (Society of Automotive Engineers International, 2016). SAE Level 3 of automation may yield more risks than benefits for the medically-at-risk driver and will as such not be further discussed in this chapter. Levels 4 and 5 of automation may yield multiple benefits related to transportation equity, especially for the disadvantaged, medically-at-risk, and disabled populations. However, because the driver will not "resume control" when engaged with these levels of automation, but rather be an "operator" (SAE Level 4) or a "passenger" (SAE Level 5), the authors do not further discuss those levels of automation.

The authors present two case studies, to address the issues related to an aging adult with a visual disorder and an adult with a neurodegenerative condition, to tie the previously discussed concepts together.

10.2 MICHON'S MODEL

Michon's model of driving behaviors (Michon, 1985) is widely used and accepted in the driving literature, and it acts as a conduit to communicate driving behaviors in a way that is understandable to transportation engineers, psychologists, road traffic safety officials, and driver rehabilitation specialists. Michon's model categorizes aspects of the driver and the environment into three hierarchical levels. First, the *strategic* level requires the highest cognitive processing for the driver, and involves high-level cognitive skills such as decision-making, planning, and problem-solving to discern where, with whom, how, and how much to drive. The strategic level also incorporates discerning the level of risk, i.e., anticipating risks, such as skidding, sliding, or being unable to come to a stop when encountering icy roads. Usually these decisions are made prior to the driving task over a period of minutes, hours, or days—depending on the complexity of the trip and the environment. The *tactical* level requires intermittent behaviors when maneuvering the vehicle to travel from one destination to another. These behaviors include, but are not limited to, handling the vehicle, avoiding obstacles, accepting gaps, making a turn, backing up, or overtaking another vehicle. Such behaviors occur during the driving task and may last for minutes to hours, depending on the length of the route. The *operational* level demands the driver's skills related to motor coordination, reaction time, visual scanning, and spatial perception and orientation. These behaviors are critical when carrying out distinct tasks, such as braking in time when a child runs across the road or swerving around a distracted pedestrian. Usually such behaviors occur in seconds. Therefore, Michon's model (1985) emphasizes how the driver interacts with the environment when taking the driver's behaviors–influenced by attention, judgment, working memory, cognitive processing, and sensory-motor abilities—into consideration.

10.3 MEDICALLY-AT-RISK CONDITIONS

10.3.1 Deskilling: Implications of the Aging Process

Driving deskilling is a potential consequence of SAE Level 3 of automation, which requires the driver to cede control of all safety-critical functions of the vehicle, under certain conditions, but expects the driver to resume control if such conditions change (Trosterer et al., 2016). As noted earlier in this chapter, SAE Level 3 of automation is not further discussed, as its use may yield more risks than benefits for at-risk drivers. However, changes in the functional skills of drivers can also result in deskilling. For the remainder of this section, we will discuss the normal aging process as an example of natural deskilling and its implications for vehicle automation technologies.

10.3.1.1 Age-Related Deskilling

In the United States alone, in 2016, there were approximately 46 million adults 65 years of age or older—15% of the entire population, and this number is expected to almost double by 2060 when older adults will make up a quarter of the American population (Mather, Jacobsen, & Pollard, 2015). In addition to this expected demographic shift, older adults are also working longer, and prefer to age in place (Mather et al., 2015). The number of licensed older drivers across the United States reflects these demographic trends. In 2015, the Federal Highway Administration estimated that 18% of licensed drivers were 65 years of age and older, and that drivers over the age of 85 were the fastest growing demographic of road users (Federal Highway Administration, 2015). Driving enables the independence and community mobility of older adults, and supports aging in place. However, the functional performance deficits that characterize the normal aging process can result in driver deskilling.

10.3.1.2   Functional Performance Deficits

Increased age is associated with sensory, cognitive, and motor functional performance deficits. Age-related sensory performance deficits include decreases in visual acuity (which are commonly compensated for with glasses), contrast sensitivity, and peripheral vision, as well as increased glare sensitivity (Haegerstrom-Portnoy, Schneck, & Brabyn, 1999; Karthaus & Falkenstein, 2016). In addition, older adults commonly experience hearing impairments (Lin, Thorpe, Gordon-Salant, & Ferrucci, 2011). The normal aging process also impacts cognitive functions including visual processing speed, divided and selective attention, set shifting, inhibition of irrelevant information, and performance self-monitoring (Karthaus & Falkenstein, 2016). Finally, impaired motor functions include decreased muscle strength, speed of movement, flexibility, motor coordination, and limited range of motion (Chen, Xu, Lin, & Radwin, 2015; Vieluf, Godde, Reuter, Temprado, & Voelcker-Rehage, 2015). Together, these functional performance deficits rather than age alone are factors that decisively contribute to crash involvement in older adults (Papa et al., 2014).

10.3.1.3   Effect on Driving Behaviors

The aging trajectory of older adults will vary according to factors such as access to health and social services, type and severity of comorbidities, socioeconomic status, housing, and other social determinants of health (Sadana, Blas, Budhwani, Koller, & Paraje, 2016). These determinants contribute to the severity of functional impairments and thus play a role in the degree of driving deskilling an individual older adult might experience. Overall, aging results in driving deskilling at the strategic, tactical, and operational levels (Karthaus & Falkenstein, 2016). Examples of each are listed below.

Strategic. Difficulty in navigating busy intersections; negotiating high traffic environments; and planning, executing, and adapting route plans, especially in unfamiliar environments.

Tactical. Difficulty in identifying appropriate gaps in traffic, with lane maintenance, initiating emergency maneuvers in response to roadway hazards, and when merging into traffic.

Operational. Prolongation of braking time, difficulty detecting relevant traffic stimuli in the periphery, insufficient blind spot checking, difficulties with alternative and appropriate use of gas and brake pedals, difficulty maintaining steering wheel control.

Together, impairments at all three of these levels increase older adults' involvement in motor vehicle collisions, particularly due to difficulties with giving the right of way, negotiating turns, driving backwards during parking maneuvers, and navigating complex and unfamiliar roadways (Karthaus & Falkenstein, 2016). In fact, older adults over the age of 65 experience an increase in motor vehicle collisions per mile driven (Davis, Casteel, Hamann, & Peek-Asa, 2018). In addition, older adults can overestimate their driving skills. In a study conducted by Ross and colleagues (2012), 85% of older adults ($N = 350$) rated themselves as either good or excellent drivers, in spite of previous crash or citation rates. However, driving is the primary means of community mobility for older adults, particularly in Western countries. Restrictions in community mobility lead to social isolation, as well as cognitive and mental health decline (Fonda, Wallace, & Herzog, 2001). Furthermore, driving cessation can increase the need for long-term care in older adults compared with active drivers of the same age, independent of health status (Freeman, Gange, Munoz, & West, 2006). Therefore, autonomous vehicle (AV) technologies may provide opportunities for older adults to remain on the road when possible, while compensating for specific functional impairments.

10.3.1.4 AV Technology to Compensate for Functional Driving Impairments by SAE Level

The following are the common functional impairments in the older adult population and the potential AV technologies that can mitigate their impact on the fitness to drive. The mitigating factors are based upon the overview of in-vehicle information systems (IVIS) and advanced driver assistance systems (ADAS) by Alvarez and Classen (2017).

1. Vision: Older drivers experience declines in visual acuity, which impacts their ability to see objects in their path clearly or accurately estimate distances.
   a. *Parking distance control (SAE Level 0):* Provides the driver with information regarding the distance from the vehicle to any outstanding object or obstacle.
2. Vision: Older drivers experience decreased contrast sensitivity and increased sensitivity to glare, which can impair their driving especially at nighttime.
   a. *Automotive night vision (SAE Level 0):* This technology utilizes a thermographic camera to allow the driver to see in darkness or poor weather conditions, beyond the scope of the vehicle's headlights.

3. Vision: Older drivers can experience restrictions in their peripheral field of view which can increase risk for lane changes.
   a. *Lane departure warning (SAE Level 0):* Provides the driver with a visual or auditory warning when the vehicle is drifting from the lane.
4. Cognition: Decreased processing speed can result in difficulties adjusting to the position of the lead vehicle.
   a. *Front collision warning (SAE Level 0):* Monitors the distance to the lead vehicle and warns the driver when such distance becomes critical.
5. Cognition: Older drivers can experience decreased divided and selective attention, as well as set shifting.
   a. *Adaptive cruise control (SAE Level 1):* Maintains a certain speed and distance to the lead vehicle, as set by the driver.
6. Cognition: Older drivers can have difficulty inhibiting irrelevant information and self-monitoring their driving behaviors.
   a. *Lane-keeping assist (SAE Level 1):* Monitors and implements corrective action if the vehicle drifts towards adjacent lane.
   b. *Lane-centering control (SAE Level 1):* Keeps the vehicle continuously centered in the lane.
7. Motor: The normal aging process results in decreased muscle strength, speed of movement, and coordination.
   a. *Adaptive cruise control with steering assist (SAE Level 2):* Maintains the vehicle speed and position in the lane when activated.
8. Motor: Older drivers also experience decreased flexibility and range of motion which can impact complex maneuvers such as parking.
   a. *Parking assist (SAE Level 1):* Assists the driver with the steering of the wheel into a parking space, while the driver remains in control of the accelerator and brake.

10.3.2 LOW VISION

Approximately 1.3 billion people around the world experience some form of visual impairment (World Health Organization, 2018). According to the World Health Organization (2018), the five leading causes of visual impairment around the world are uncorrected refractive errors, cataract, age-related macular degeneration (AMD), glaucoma, and diabetic retinopathy (DR). Furthermore, adults over the age of 50 represent 65% and 82% of those who are visually impaired and blind, respectively (Pascolini & Mariotti, 2012). As such, many of these conditions are considered age related.

Refractive errors (e.g., nearsightedness, farsightedness, astigmatism, presbyopia) occur when the shape of the eye prevents it from focusing light directly on the retina, and uncorrected translates to the better eye presenting a visual acuity of 20/50 or poorer which improves with refraction (National Eye Institute, 2010; Willis, Jefferys, Vitale, & Ramulu, 2012). Driving licensing agencies and jurisdictions have established minimum visual acuity requirements that must be met in order for an individual to be eligible for a driver's license. For example, the Canadian Medical Association (2017) requires drivers to have a corrected visual acuity of at least 20/50 with both eyes opened and examined together. In the United States, all but three states have a minimum best corrected visual acuity requirement of 20/40 (Steinkuller, 2010). Refractive errors are commonly corrected through the use of prescription glasses, contact lenses, and refractive surgeries, and as such, global efforts are improving access to such services and interventions (World Health Organization, 2013). These corrections allow individuals to achieve the visual acuity requirements of their jurisdictions. Because correction is the goal of interventions for refractive errors, and drivers who do not meet the required standard are otherwise restricted or prevented from driving, the remainder of this section will focus on the subsequent four leading causes of visual impairment.

10.3.2.1   Cataracts

A cataract is a visual impairment that results from a clouding or opacity of the lens in the eye (National Eye Institute, 2013). The lens, located behind the iris and the pupil, focuses the light onto the retina. When the proteins that make up the lens accumulate, the lens becomes clouded, a phenomenon that can occur in one or both eyes (National Eye Institute, 2013).

10.3.2.1.1   Core Clinical Characteristics

A cataract reduces the sharpness of the image that reaches the retina. As the proteins continue to cluster together, the opacity of the lens can increase and cause blurred vision (National Eye Institute, 2013). Figure 10.1 illustrates a scene as viewed by a person with cataract.

10.3.2.1.2   Functional Performance Deficits

When first formed, a cataract can be small and the individual might not notice any significant changes in vision. As the cataract grows, however, the individual experiences gradual difficulty with glare, contrast sensitivity, poor night vision, increased blurry vision, and fading of color (National Eye Institute, 2013).

FIGURE 10.1   (See color insert.) Scene as viewed by a person with cataract. (Image from the National Eye Institute, National Institutes of Health.)

10.3.2.1.3   Effect on Driving Behaviors

Cataracts have a negative impact on driving performance. Drivers with cataracts are involved in 2.5 times more motor vehicle collisions than age-matched controls (Owsley, Stalvey, Wells, & Sloane, 1999). Specifically, cataracts significantly impair a driver as follows (Wood & Carberry, 2004):

> *Strategic.* Difficulty avoiding low-contrast hazards on the road, including high-way debris, speed bumps, or vulnerable road users; and driving in poor weather or under restricted visibility conditions including nighttime driving.
> *Tactical.* Negotiating traffic infrastructure; difficulty recognizing road signs, hazards, or roadway markings in low contrast situations and performing maneuvers to avoid them; and difficulty maintaining their position in the lane.
> *Operational.* Swerving, braking in response to hazards, and steering wheel control in response to lane markings.

10.3.2.1.4   AV Technology to Compensate for
Functional Driving Impairments

The following are the common functional impairments in individuals with cataract and the potential AV technologies that can mitigate their impact on fitness to drive. It is important to note that AV technologies could only support drivers that meet the legal visual requirements in their jurisdiction. The mitigating factors are based upon the overview of IVIS and ADAS by Alvarez and Classen (2017).

1. Decreased visual acuity
   a. *Parking distance control (SAE Level 0)*: See Section 10.3.1.4.
2. Low contrast and increased glare sensitivity:
   b. *Automotive night vision (SAE 0)*: See Section 10.3.1.4.
3. Difficulty identifying lane markings:
   c. *Lane departure warning (SAE Level 0)*: Can provide the driver with an auditory warning when the vehicle is drifting from the lane (also see Section 10.3.1.4).

10.3.2.2   Age-Related Macular Degeneration

AMD is a condition that causes progressive loss of central vision (Mitchell, Liew, Gopinath, & Wong, 2018). Approximately 170 million people globally have AMD, with 11 million individuals residing in the United States alone (Pennington & DeAngelis, 2016). Also, in the United States, the prevalence of AMD is similar to that of invasive cancer and almost twice that of Alzheimer's disease (Pennington & DeAngelis, 2016). Age is the primary risk factor for AMD. As the aging population increases, the prevalence of AMD is thus expected to increase proportionally.

10.3.2.2.1   Core Clinical Characteristics

AMD is a degenerative disease that primarily affects the macular—or central—region of the retina, causing blurriness in the center of vision while peripheral vision remains unaltered. As AMD progresses, the blurred area in the center of vision increases in size. Individuals with AMD may also see lines as wavy or distorted

FIGURE 10.2   (See color insert.) Scene as viewed by a person with AMD. (Image from the National Eye Institute, National Institutes of Health.)

and may see dark spots blanking out their central vision (Canadian Association of Optometrists, n.d.). AMD is broadly classified into two types. The dry form is the most common and presents as a gradual degeneration of the tissue in the macula with symptoms developing more slowly. In contrast, the wet form is more severe and results from the bleeding of weakened vessels under the macula which causes the symptoms to progress more rapidly (Mitchell, Liew, Gopinath, & Wong, 2018). Figure 10.2 illustrates a scene as viewed by an individual with AMD.

10.3.2.2.2   Functional Performance Deficits

Persons with AMD primarily experience blurred or blocked central vision. In addition, spatiotemporal contrast sensitivity is also a reliable indicator of functional impairment in individuals with AMD (Midena, Degli Angeli, Blarzino, Valenti, & Segato, 1997). Although in early stages of the disease visual acuity can remain adequate (20/40), individuals with AMD can experience other early symptoms including visual difficulties under reduced illumination and delays adapting to the dark (Owsley & McGwin, 2008).

10.3.2.2.3   Effect on Driving Behaviors

Driving is a primary concern for individuals with AMD. Owsley and colleagues (2006) conducted focus groups with individuals diagnosed with AMD, to discuss their vision. The most commonly cited difficulty by participants was driving. Although the literature exploring the driving performance of drivers with AMD is limited—many studies group visual impairments together, making it difficult to separate out drivers with AMD (Owsley, McGwin, Scilley, & Kallies, 2006)—the extant literature revealed poorer driving performance among individuals with AMD when compared with controls. In fact, drivers with AMD have slower braking response times, slower speed, more lane crossing, and collisions on a driving simulator than controls (Szlyk et al., 1995).

10.3.2.2.4   AV Technology to Compensate for Functional Driving Impairments
The following are common functional impairments in drivers with AMD and the potential AV technologies that can mitigate their impact on fitness to drive (Alvarez & Classen, 2017). It is important to note that these technologies can be helpful to drivers in the early stages of the condition when visual requirements are met. As such, these technologies cannot compensate for severe visual deficits; they can only provide support to fit drivers.

1. Decreased visual acuity and central processing speed:
    a. *Front collision warning (SAE Level 0)*: Monitors distance to the lead vehicle straight ahead of them (central vision) and warns the driver when such distance becomes critical (also see Section 10.3.1.4).
2. Decreased contrast and limited vision in low illumination settings:
    b. *Automotive night vision (SAE 0)*: See Section 10.3.1.4.

10.3.2.3   Glaucoma
The term glaucoma refers to a group of diseases characterized by progressive damage to the optic nerve. In the absence of early detection and treatment, glaucoma can eventually lead to blindness (National Eye Institute, 2015b). In 2013, it was estimated that approximately 64 million people around the world lived with glaucoma, a number that is expected to almost double by 2040 (Tham et al., 2014).

10.3.2.3.1   Core Clinical Characteristics
Glaucoma refers to a group of degenerative eye diseases that compromise the integrity of the optic nerve, often associated with elevated intraocular pressure (Cohen & Pasquale, 2014). The most common type of glaucoma is open-angle glaucoma, which accounts for nearly 90% of all glaucoma cases (Glaucoma Research Foundation, n.d.). Open-angle glaucoma is a result of a progressive clogging of the eye's drainage canals, which in turn increases pressure on the optic nerve. This clogging develops in spite of an adequately wide and open angle where the cornea and the iris meet, and the symptoms emerge only after the disease advances (Cohen & Pasquale, 2014). Angle-closure glaucoma, on the other hand, develops when the drainage canals are blocked due to the angle closure between the iris and the cornea. Angle-closure glaucoma develops rapidly and leads to severe noticeable symptoms (See: Aquino et al., 2011). Other types of glaucoma include normal-tension glaucoma, where the optic nerve is damaged in spite of relatively normal intra ocular pressure—the cause of which remains poorly understood (Anderson, 2011), and secondary glaucoma, which results from an injury around the eye that increases pressure on the optic nerve (Papadopoulos, Loh, & Fenerty, 2015). Figure 10.3 illustrates a scene as viewed by a person diagnosed with glaucoma.

10.3.2.3.2   Functional Performance Deficits
The onset of symptoms varies according to the type of glaucoma. For open-angle glaucoma, functional performance deficits can develop painlessly and gradually and

FIGURE 10.3   (See color insert.) Scene as viewed by a person with glaucoma. (Image from the National Eye Institute, National Institutes of Health.)

can go undetected until the disease has advanced considerably (Cohen & Pasquale, 2014). Once present, performance impairments are most often characterized by a loss of peripheral vision. As the disease progresses, it might compromise an individual's central vision and even cause blindness (National Eye Institute, 2015b).

10.3.2.3.3   Effect on Driving Behaviors

Glaucoma can impact the operational and tactical driving levels. These impairments result in increased crash risk for drivers with glaucoma when compared with controls (Szlyk, Mahler, Seiple, Edward, & Wilensky, 2005). In addition, glaucoma is a significant determinant of self-reported crashes (Tanabe et al., 2011). Specifically, drivers with glaucoma make more critical driving errors than controls including lane maintenance and gap acceptance, particularly at traffic lights and yield/give-way intersections (Wood, Black, Mallon, Thomas, & Owsley, 2016).

10.3.2.3.4   AV Technology to Compensate for Functional Driving Impairments

In addition to visual acuity requirements, driving licensing jurisdictions have established visual field requirements. For example, Canadian drivers must have a 120° continuous visual field in the horizontal meridian, and 15° above and below fixation with both eyes open and examined together (Canadian Medical Association, 2017). In the United States, similar requirements have been established by individual states, ranging from 110° to 140° minimum continuous visual field (Bron et al., 2010). Thus, the following are the potential AV technologies for drivers with glaucoma that meet their jurisdiction visual field requirements, in relation to visual field restrictions—the main characteristic of glaucoma.

1. Restricted visual field leading to lane maintenance errors:
   a. *Blind spot detector (SAE Level 0)*: Monitors and warns the driver if there are vehicles in the driver's blind spot.
2. Restricted visual field leading to gap acceptance errors:
   a. *Intersection assistant (SAE Level 0; requires V2V communications)*: Monitors traffic at an intersection or junction and prompts the driver to initiate braking if the gap becomes unsafe.

10.3.2.4 Diabetic Retinopathy (DR)

Approximately 285 million people around the world currently live with diabetes mellitus, a third of whom experience symptoms of DR (Lee, Wong, & Sabanayagam, 2015). DR is the most common form of vision loss among people with diabetes, and as the aging population increases, a rise in DR is also expected (Lee, Wong, & Sabanayagam, 2015).

10.3.2.4.1 Clinical Characteristics

DR is caused by damage to the small blood vessels in the retina, due to a high in-blood sugar concentration. As a result, the small vessels in the retina can hemorrhage and distort vision. As the disease progresses, new abnormal blood vessels can increase in number on the retina's surface, leading to scarring and damage to the retina (National Eye Institute, 2015a).

10.3.2.4.2 Functional Performance Deficits

In the early stages, DR can cause no noticeable symptoms. As the disease progresses, however, the hemorrhaging in the retina creates the appearance of "floating spots" in the person's field of view. Figure 10.4 illustrates a scene as viewed by an individual with DR.

FIGURE 10.4 (See color insert.) Scene as viewed by a person with DR. (Image from the National Eye Institute, National Institutes of Health.)

10.3.2.4.3 Effect on Driving Behaviors

Given the extent and trajectory of the condition, DR can result in severe visual loss below the legal jurisdictional visual requirements. However, drivers in earlier stages of the condition, or those treated for DR (Vernon, Bhagey, Boraik, & El-Defrawy, 2009), may experience visual impairments that, though not fully restricting their participation in driving, can impact their driving performance at the tactical and operational level including difficulty watching for hazards which results in slower brake response times and increased brake pressure; longer reaction and response times; and lane excursions and increased off-road times (Szlyk et al., 2004).

10.3.2.5 AV Technology to Compensate for Functional Driving Impairments by SAE Level

The following are the potential AV technologies that can support drivers with DR.

1. Decreased visual acuity and "spotting" in the field of view:
   a. *Parking assist (SAE Level 1)*: See Section 10.3.1.4.
2. Decreased contrast sensitivity:
   b. *Automotive night vision (SAE 0)*: See Section 10.3.1.4.

Finally, drivers diagnosed with all of the above-mentioned visual disorders experience difficulty watching for hazards. Pedestrian detection systems (SAE Level 0) can alert these drivers if there is a vulnerable road user (e.g., pedestrian) in the vehicle's path.

10.3.2.6 Case Study: Zane, an Older Adult with a Diagnosis of Glaucoma

Zane is a 72-year old male who immigrated to Canada in his 20s. Zane lives alone in a retirement condominium, where he moved five years ago after his wife passed away. He has two daughters, one who lives close to him and another who lives across the country. Zane is an accountant who worked in an accounting firm until his retirement, i.e., when he turned 65 years old. He enjoys reading and taking walks with his dog. Zane has a history of high blood pressure and a diagnosis of open-angle glaucoma. He underwent a laser trabeculoplasty—a laser surgery used to lower intraocular pressure—two years ago. He takes Acetazolamide to manage his glaucoma. He is experiencing worsening of his peripheral field of view even after surgery, and the optometrists have noticed some loss of visual acuity. He has been a licensed driver for over 50 years, and currently drives approximately three times a week in familiar environments and routes. He avoids driving at nights and in inclement weather. Table 10.1 summarizes Zane's strengths and challenges, which help guide the assessment process.

As part of his comprehensive driving evaluation, Zane underwent a visual assessment (details of these assessments are beyond the scope of this chapter). The findings included: 20/50 binocular visual acuity (20/50 is the jurisdictional visual requirement for driving); 125° continuous peripheral field of view (70° in the right eye; 55° in the left eye); intact contrast sensitivity, depth perception, color discrimination, and lateral and vertical phorias. No cognitive or motor impairments are noted.

TABLE 10.1
Strengths and Challenges

Strengths	Challenges
1. Has insight and has implemented self-restrictions on his driving.	1. Visual impairments are increasing even after surgical intervention.
2. Drives in familiar environments and routes while avoiding hazardous climates and nighttime driving.	
3. Has medical care and adequate follow-up from his circle of care.	2. Has high blood pressure which affects severity of his symptoms.
4. Has no known comorbidities beside high blood pressure.	
5. Has a family network of support.	

On-road assessment. The certified driving rehabilitation specialist (CDRS) conducted the on-road assessment noting the following:

- Zane makes narrow turns, potentially overcompensating for peripheral vision loss;
- he occasionally drifts from his lane, potentially as a result of difficulty seeing lane markings, but corrects lane positioning;
- he has difficulty navigating four-way stops, where he takes a long time to proceed even when having the right of way, potentially given the restricted peripheral field of view;
- he performs full head turns continuously throughout the drive, as a way of compensating for reduced peripheral field;
- he stops too close to the lead vehicle at intersections.

Recommendations. The CDRS recommended that Zane uses the two-second rule, leaving at least 2 seconds between him and the lead vehicle, as well as stopping where he can see the tires of the lead vehicle. She also suggests an alternative route to the supermarket that avoids four-way stops completely. She suggests that Zane works with her during upcoming sessions to practice this new route and build confidence. She recommends the use of a lane change assist and a blind spot detector. These technologies will warn Zane when he is drifting from the lane, as well as provide a warning if there are objects in his blind spot or in a position where a lane change would be unsafe. Other technologies that could support Zane include lane-centering control, adaptive cruise control, and a pedestrian/bicycle collision warning system so that when Zane is turning his head away from the forward view the forward roadway is being monitored by the AV technology to the extent possible to provide Level 2 assistance. However, given Zane's age and the resulting deskilling that is expected to emerge, integrating all these technologies could result in increased cognitive load. As such, the CDRS prioritizes these two.

Before recommending the technologies, however, she explores Zane's AV technology acceptance. Zane wants to be able to drive for as long as possible and is willing to purchase technologies that might be helpful. He heard one of his daughters talking about a new device in her car and he is actually excited to try one himself. The CRDS conducts education sessions with a vendor before any purchase, so that Zane can try the technology and use it appropriately.

10.3.3 Neurological and Neurodegenerative Disorders

10.3.3.1 Introduction

This section will present autism spectrum disorder (ASD) and attention deficit hyperactivity disorder (ADHD) as neurological disorders of concern, and Parkinson's disease (PD) as a neurodegenerative disorder of concern. The effect of each condition on the strategic, tactical, and operational components is discussed as a group in Section 10.4.4. The AV technologies that would benefit drivers with ASD, ADHD, and PD are discussed as a group later in Section 10.4.4.4.

10.3.3.2 ASD and ADHD

In the United States, ASD affects 1 in 59 children (Centers for Disease Control and Prevention, 2018). ADHD affects 1 in 11 of 4- to 17-year olds (Centers for Disease Control and Prevention, 2017; Visser, Bitsko, Danielson, Perou, & Blumberg, 2010) and approximately 13.2% of males and 5.6% of females under the age of 18. Should they meet the diagnostic criteria for both conditions, teens receive a dual diagnosis of ADHD/ASD (American Psychiatric Association, 2000, 2012; Clark, Feehan, Tinline, & Vostanis, 1999). The prevalence of those with a dual diagnosis is unknown.

10.3.3.2.1 Core Clinical Characteristics

ASD is characterized by deficits in executive functions (e.g., mental flexibility, planning, sequencing, and self-monitoring), visual-motor coordination, and motor coordination. Core ASD symptoms include repetitive behaviors, fixated interests, social interaction and communication deficits, and unusual response to sensory input—all of which may impact independent and safe driving. A survey of parents of teens (ages 15–18) with high-functioning ASD found that 12% of teens were involved in one or more motor vehicle crashes as the driver at fault—and another 12% received a citation for a moving violation in the prior year (Huang, Kao, Curry, & Durbin, 2012). ADHD symptoms are characterized by inattention, hyperactivity, and impulsivity (Centers for Disease Control and Prevention, 2017)—all of which may predict impaired fitness to drive.

10.3.3.2.2 Functional Performance Deficits

Functional performance, of those teens with ASD and/or ADHD, is impaired by deficits in *visual, cognitive, and motor* functions as described below.

Specifically, for *visual functions,* teens with ASD do not scan the driving environment as effectively as neurotypical peers, but instead maintain their visual focus on the driving horizon (Reimer et al., 2013). Teens with ASD have difficulty making eye contact (National Institute of Mental Health, 2017). Thus, they may fail to make eye contact with pedestrians at a crosswalk (Sheppard, Ropar, Underwood, & Van Loon, 2010), orient slower to driving hazards (Sheppard, Van Loon, Underwood, & Ropar, 2017), or perform worse in visual attention skills than neurotypical peers (Reimer et al., 2013). Thus, they may not notice potential hazards in their immediate path. Compared with neurotypical peers, teens with ADHD have impaired right visual acuity and impaired right peripheral field (Classen, Monahan, & Brown, 2013;

Classen et al., 2013); and teens with ADHD/ASD did worse on measures of visual performance (Classen, Monahan, & Brown, 2013; Classen et al., 2013).

For *cognitive functions,* teens with ASD have difficulty in *problem-solving* driving events such as approaching an emergency vehicle. During these driving maneuvers, when they experience increased demands on *working memory,* they make more steering and braking errors (Cox et al., 2016). Noticeably, as can be seen from Video 1, teens with ASD have difficulty carrying out the correct sequence when performing a turn.

Video 10.1: Teen with ASD Performing a Turn [UPLOAD VIDEO]

As such, they may not effectively sequence the adjustment of speed and rotation of the steering wheel to control a vehicle through a turn (Classen et al., 2013). Moreover, these teens divert their attention away from complex roadway situations that require increased cognitive demands (Reimer et al., 2013). Teens with ASD and/or ADHD performed worse than neurotypical peers on measures of cognition (Classen, Monahan, & Brown, 2013). Specifically, in the Classen et al. study, they responded late or not at all to traffic lights, regulatory signs, or pedestrians.

Reduced ability to estimate risk and impulsive tendencies of teens with ADHD impairs their judgment when driving (Barkley, 2004). Such impairments are evident in *misjudging* gaps in traffic and not *adjusting speed* for hazardous conditions. Compared with neurotypical peers, teens with ADHD also demonstrate impaired selective attention (Classen, Monahan, & Brown, 2013).

For *motor functions*, teens with ASD have difficulty with bilateral upper extremity motor coordination for turning the wheel when negotiating a turn in a high-fidelity driving simulator (Classen, Monahan, & Brown, 2013). Compared with neurotypical peers, teens with ADHD also showed impaired motor coordination during a simulator driving task (Classen, Monahan, Brown et al., 2013).

10.3.3.2.3   *Effect on Driving Behaviors*

Teens with ASD take longer to learn to drive and are older than their peers when obtaining a driver's license (Almberg et al., 2015). In one study, 33% of teens with ASD reported that they were driving, 12% were involved in crashes, and 12% had citations for moving violations (Huang et al., 2012). Compared with neurotypical drivers, drivers with ADHD had more self-reported motor vehicle crashes, more traffic citations, drove more without a driver's license, and drove more under the influence of alcohol (Fabiano et al., 2010; Jerome, Habinski, & Segal, 2006; Jerome, Segal, & Habinski, 2006). Teens with ASD and/or ADHD have more crashes in a driving simulator as compared with neurotypical peers (Classen et al., 2013).

10.3.3.3   Parkinson's Disease

In the United States, PD affects about 1 million Americans (Parkinson's Foundation, 2018). PD is most commonly diagnosed in people over the age of 60, with only 5% of all cases diagnosed before the age of 60. Men are 1.5 times more likely diagnosed with PD than women, and the incidence of PD increases with age (Parkinson's Foundation, 2018).

10.3.3.3.1 Core Clinical Characteristics

PD is an age-related, progressive, neurodegenerative disorder characterized by the development of four cardinal symptoms, i.e., resting tremor, rigidity, bradykinesia, and postural instability. In addition to motor difficulties, non-motor symptoms account for a significant source of disability, particularly as many non-motor symptoms often do not respond to dopaminergic medications. Non-motor symptoms may include visual deficits, cognitive impairment, depression, or other emotional impairments (e.g., apathy and disinhibition), sleep disorders, and autonomic dysfunction (Riggeal et al., 2007; Uc et al., 2005). These clinical features of PD may affect not only body functions but also functional performance or activities of such individuals (World Health Organization, 2001).

10.3.3.3.2 Functional Performance Deficits

Functional performance of those with PD is impaired by demographic factors, such as *age* (Crizzle et al., 2013), *disease duration,* and *disease severity* (Classen et al., 2011; Singh, Pentland, Hunter, & Provan, 2007; Worringham, Wood, Kerr, & Silburn, 2006). Other complicating factors include *daytime sleepiness* (Lachenmayer, 2000; Meindorfner et al., 2005), *medication use* (Meindorfner et al., 2005), as well as the accompanying *comorbidities* (Crizzle, Classen, & Uc, 2012).

Relationships exist between functional performance deficits resulting from PD and driving behaviors. Specifically in PD, such deficits pertain to: *binocular acuity* and *contrast sensitivity* (Uc et al., 2005; 2009; Worringham et al., 2006); *visual scanning and speed of processing* (Classen et al., 2011; Uc et al., 2006a; Worringham et al., 2006); *cognitive impairments* (Amick, Grace, & Ott, 2007; Dubinsky et al., 1991; Worringham et al., 2006); *set shifting and cognitive flexibility* (Amick et al., 2007; Uc et al., 2006a; 2006b; 2009); and *psychomotor speed,* including *reaction time, slowed walking, and fine motor movements* (Classen et al., 2011; Crizzle et al., 2012; Worringham et al., 2006). Such performance deficits can greatly impair functional driving ability and potentially lead to an elevated crash risk (Uc & Rizzo, 2008; Uitti, 2009), even in the early stages of PD (Crizzle et al., 2012; Devos, Ranchet, Akinwuntan, & Uc, 2015).

10.3.3.3.3 Effect on Driving Behaviors

Drivers with PD may have an increased risk of crashes per million miles traveled, specifically in those with greater disease severity (Dubinsky et al., 1991). In a small study ($N = 15$), researchers found that a third of the PD patients thought that their PD symptoms may have contributed to a recent crash (McLay, 1989), while another study suggests that PD drivers have no greater crash risk when compared with the rest of the population (Lings & Dupont, 1992). Yet, Meindorfner et al. (2005) found that 15% of over 6,000 PD respondents reported involvement in a motor vehicle crash, with 11% of the 15% reporting being at fault. However, whether crash rates are truly higher in PD is still unclear due to the lack of well-controlled studies.

10.3.3.4 Strategic, Tactical, and Operational Deficits for Those with ASD, ADHD, and/or PD

Taken together, the evidence indicates that the functional performance deficits experienced by drivers with ASD and/or ADHD, or PD result in impairments in driving behaviors on the strategic, tactical, and operational levels (Michon, 1985). Examples of such impairments on each of the levels of driving behavior are indicated below:

10.3.3.4.1 Strategic Impairment

Strategic impairments may include decisions that are made a priori, before starting to drive and incorporate decision-making, problem-solving, reasoning, judgment, and/or initation of strategies to overcome potential driving difficulties. Examples of impairments include experiencing

- Challenges with decisions related to trip planning, prior to the actual drive, which result in getting lost during travel, or being disoriented in time and space;
- Difficulty with judgment and decision-making, during the actual drive, when the driver with a cognitive impairment needs to negotiate a complex travel route; and
- Increased cognitive load when driving during peak traffic hours (vs. non-peak hours) which result in deterioration of driving performance.

10.3.3.4.2 Tactical Impairment

These impairments are evident during the driving task with routine functions being deficient. Such functions include steering, braking, accelerating, stopping, or controlling the vehicle in the driving environment, while adhering to the rules of the road and the traffic regulations. Examples include:

- Speeds too high or too low when merging on a highway;
- Difficulty with parking, backing up, pulling away, changing lanes, merging into traffic, or maintaining a safe headway distance;
- Challenges with anticipating and adjusting to traffic stimuli, such as not being able to supress road side distractions while making a turn; and
- Inadequately negotiating the traffic infrastructure, such as difficulty judging and accepting an appropriate gap when turning against oncoming traffic.

10.3.3.4.3 Operational

Such impairments occur when a swift reaction of the driver is necessary to avoid a potential obstacle or adverse condition. Examples include:

- Inability to swerve to avoid an obstacle in the road that may cause harmful effects;
- Challenges to control the gas and/or brake pedals in an emergency situation, such as a child running across the road; and
- Limited ability to manipulate swift steering control when a potential adverse event is unfolding, such as a driver in a parked car opening a car door in front of one's moving vehicle.

As such, because driving behaviors seem to be impaired at all levels (strategic, tactical, and operational) in those with ASD, ADHD, and/or PD drivers, AV technology holds out the promise to mitigate at least some of the functional impairments.

10.3.3.4.4 AV Technology Compensate for Functional Driving Impairments by SAE Level

Table 10.2 indicates the main functional deficits of drivers with neurological and neurodegenerative disorders, the AV technology by type and SAE level, as well as the potential to offset the functional deficits. *Note*: Most of the operational definitions of the AV technology were obtained from the MyCarDoesWhat.org site (*https:// mycardoeswhat.org/*, 2018).

TABLE 10.2

Functional Deficits of Drivers with Neurological or Neurodegenerative Disorders, the AV Technology by Type and SAE Level, as Well as the Potential to Offset the Functional Deficits Caused by the Disorder

Source	Deficit	AV Technology
	Teens with ASD and/or ADHD	
Vision	Teens with ASD do not scan the environment as effectively as neurotypical peers and may not notice potential hazards in their immediate path.	*Pedestrian detection and avoidance system (SAE Level 0)*: Assist the driver to detect pedestrians in the path by alerting the driver to such stimuli, and braking the car, should the driver not adjust his/her response.
		Blind spot detection (SAE Level 0): See Section 10.3.4.4.
Visual perception	Teens with ADHD may misjudge gaps in traffic.	*ACC (SAE Level 1)*: ACC automatically adjusts the vehicle speed to maintain a safe distance from vehicles ahead.
Cognition	Teens with ASD make more steering and braking errors when experiencing increased demands on working memory.	*Intelligent speed adaptation (SAE Level 0–1)*: On-board camera scans for signs and reads the speed limit while the GPS satellite passes road speed limits (Level 0) directly to the car, and the on-board computer warns the driver and slows (Level 1) the car to match the speed limit information received.
Cognition	Teens with ASD have difficulty carrying out the correct sequence of actions when performing a turn. As such, they may not effectively sequence the adjustment of speed and rotation of the steering wheel to control a vehicle through a turn.	*Intersection assist (SAE Level 0 or 1)*: Monitors cross traffic in an intersection or road junction. When it detects a hazardous situation, it prompts the driver to start emergency braking by activating visual and/or auditory warnings or automatically engages the brakes (also see Section 10.3.4.4).

(Continued)

TABLE 10.2 (*Continued*)

Functional Deficits of Drivers with Neurological or Neurodegenerative Disorders, the AV Technology by Type and SAE Level, as Well as the Potential to Offset the Functional Deficits Caused by the Disorder

Source	Deficit	AV Technology
Cognition	Teens with ASD may have difficulty diverting from routines. Subsequently when their habitual driving route is interrupted with a detour, they show a decline in driving performance.	*ANS (SAE Level 0)*: The ANS helps to find the route for the vehicle via satellite navigation. When directions are needed, routing can be calculated in real time.
Cognition	Teens with ADHD may have difficulty with planning aspects of driving such as route selection and time management for punctual arrival.	*Automated GPS (SAE Level 0 or 1)*: Navigation (Level 0) is integrated with the steering and wayfinding abilities of the vehicle. It indicates the estimated time of arrival and will steer the vehicle (Level 1) in the shortest or most advantageous route to offset deficits in route planning and time management.
Motor	Teens with ASD have difficulty with bilateral upper extremity motor coordination when turning the steering wheel. Teens with ADHD have impaired motor coordination resulting in difficulty maneuvering the vehicle in a parking bay.	*Parking assist (SAE Level 1 or 2)*: The system uses sensors and cameras to assist in parking. Parking assist is at a Level 1 when the driver controls the speed, but Level 2 where both steering and braking/accelerating/decelerating are controlled by the vehicle (also see Section 10.3.5.4.)
Driving performance	Teens with ADHD have more crashes and citations compared with neurotypical peers. Teens with ASD and/or ADHD have more crashes in a driving simulator compared with neurotypical peers.	*Adaptive cruise control with steering assist (SAE Level 2)*: The front and rear collision avoidance systems are designed to help maintain safe distances between vehicles as well as to prevent and reduce front and rear-end crashes and accompanying injuries (see, Section 10.3.1.4.).

Drivers with PD

Source	Deficit	AV Technology
Vision	Contrast sensitivity is impaired in drivers with PD, and drivers may not be able to clearly distinguish dark objects from a dark background.	*Automotive night vision assist (SAE Level 0)*: see Section 10.3.1.4.
Vision	Binocular acuity is impaired in drivers with PD, which may impair functions to detect approaching vehicles.	*Blind spot detection (SAE Level 0)*: See Section 10.3.4.4.
Vision	Contrast sensitivity is impaired in drivers with PD and drivers may not be able to clearly distinguish dark objects from a dark background.	*Automotive night vision assist (SAE Level 0)*: Section 10.3.1.4.

(*Continued*)

TABLE 10.2 (*Continued*)

Functional Deficits of Drivers with Neurological or Neurodegenerative Disorders, the AV Technology by Type and SAE Level, as Well as the Potential to Offset the Functional Deficits Caused by the Disorder

Source	Deficit	AV Technology
Visual scanning	Visual scanning may be impaired in drivers with PD when multiple tasks are required, such as making a turn and scanning the environment for pedestrians.	*Pedestrian detection system (SAE Level 0)*: The system acts as an extra set of eyes for the driver, helping them to avoid potentially catastrophic collisions by delivering either a visual or audible alert, or both (also see above).
Visual construction	This ability may be impaired in drivers with PD to adequately perceive all of the complexities in the driving environment.	*Pedestrian and cyclist detection (SAE Level 0) and avoidance (SAE Level 0)*: The system uses advanced sensors to detect pedestrians and cyclists and to alert the driver to respond. If the driver does not respond, the system will apply the brakes, via AEB, to slow and stop the vehicle (see above).
Visual-motor	Visual-motor tracking may be impaired in drivers with PD, and as such keeping the vehicle in the center of the lane while looking at the other traffic may be problematic.	*Lane-keeping assist (SAE Level 1) and lane-centering (SAE Level 1)*: See Section 10.3.1.4.
Cognition	Speed of processing and attention may be slowed in PD drivers and compromises the driver's ability to make a turn at an unprotected intersection adequately and safely.	*Intersection assistant (SAE Level 0)*: This requires V2V connectivity along with AEB (Level 0). Monitors cross traffic in an intersection or road junction. When it detects a hazardous situation, it prompts the driver to start emergency braking by activating visual and/or auditory warnings or automatically engages the brakes (also see Section 10.3.4.4).
Cognition	Set shifting and cognitive flexibility may be impaired in drivers with PD, which means that the driver may drive under the speed limit to offset the demands of cognitive load.	*ACC (SAE Level 1)*: See above.
Cognition	Verbal memory deficits affect working memory in drivers with PD and may lead to wayfinding problems.	*Automated wayfinding systems (SAE Level 0)*: Wayfinding systems include informational, directional, and identification signs and maps that work together to help drivers find their way.
Motor	Disdiadochokinesis may affect hand over hand coordination in drivers with PD, which is necessary in turning the steering wheel for parking a vehicle.	*Parking assist (SAE Level 2)*: The system uses sensors and cameras to automatically park the vehicle in a parking bay (also see Section 10.3.5.4.)

(Continued)

TABLE 10.2 (*Continued*)

Functional Deficits of Drivers with Neurological or Neurodegenerative Disorders, the AV Technology by Type and SAE Level, as Well as the Potential to Offset the Functional Deficits Caused by the Disorder

Source	Deficit	AV Technology
Motor	Reaction time is slowed in drivers with PD, resulting in slow response time to brake for an oncoming vehicle making a turn at an intersection.	*Turning assistant (SAE Level 0)*: The system monitors opposing traffic when turning left at low speeds. In critical situations, it uses AEB, to brake the car.
Motor	Fine motor movement may be impaired in drivers with PD, impeding their ability to activate controls, e.g., ACC and lane-centering, on the steering wheel of the vehicle.	Because activation of the controls is integrated into the steering wheel of the vehicle, and requires fine motor control, drivers with PD may have difficulty in engaging these systems. Consider different designs to accommodate fine motor control issues for drives with PD in activating controls for the in-vehicle technologies. For more information, see Chapter 16.

10.3.3.5 Case Study: Elizabeth, a Client with PD

Elizabeth is a 74-year-old female with a diagnosis PD and glaucoma in the right eye (ten years). She also has comorbidities including high blood pressure (40 years), arthritis in hips and lower back (10 years), cataract in right eye (5 years), sciatica (45 years), clinical depression (2 years), and sleep disorder (10 years). She is taking therapeutically prescribed dosages of eye drops for glaucoma, hydrochlorothiazide, and lisinopril for blood pressure, multivitamins, and glucosamine chondroitin. She wears prescription lenses, is medically stable, and reports some side effects, i.e., lightheadedness, from the blood pressure medication. She has a doctoral degree, and lives alone in a single story condominium in a residential neighborhood. She still drives five days a week, drives mostly alone, and she maintains her vehicle well. She is active in her community by participating in a garden club, doing her own shopping, and volunteering for work at the local hospital. She avoids driving in rain, rush hour traffic, nighttime driving, and interstate and highway driving. She has had no crashes or citations in the past three years. She has taken a classroom-based driving refresher course, more than three years ago. Table 10.3 summarizes Elizabeth's strengths and challenges.

Based on clinical assessments (content not discussed in this chapter), Elizabeth shows impairment in depth perception, recall, visual-motor perceptual skills, gait speed, and neck range of motion.

On-road assessment. Elizabeth has completed an on-road assessment administered by a CDRS, with the following results:

TABLE 10.3
Strengths and Challenges for Elizabeth

Strengths	Challenges
1. Regulating her driving	1. Neurodegenerative nature of PD
2. Avoiding potential hazards	2. Chronic and degenerative visual conditions
3. Living alone—independent in Activities of Daily Living (ADLs) and possibly most Instrumental Activities of Daily Living (IADLs)	3. Side effects of blood pressure medication (lightheadedness, dizziness)
4. Having higher education	4. Limited support network
5. Wearing prescription lenses	5. Effects of the comorbidities, e.g. stiff joints, depression, potential for daytime sleepiness
6. Having a good driving record	

- Stops over stop lines at intersections (potentially due to depth perception impairment)
- Makes wide turns at curves and turns into the furthest lane (potentially due to impairments in depth perception, visual-motor perception, and set shifting)
- Not scanning adequately to the left and right before crossing intersections (potentially due to impaired neck range of motion)
- Not reacting appropriately to cars in her blind spot (potentially due to impaired set shifting and neck range of motion)
- Not maintaining lateral lane position as she drifts to the left (potentially due to impairments in visual-motor perception, and set shifting)
- Not using turn signal consistently (potentially due to impairments in recall and set shifting)

The CDRS Has Made the Following Recommendations
- Continue with driving utilizing self-identified restrictions
- Restrict driving: no driving in challenging environments (new areas, poor weather, city driving); no highway driving; avoid rush hour traffic;
- Provide client with education on
  - use of reference points, i.e., vehicle features that align with lane markings and stop line;
  - planning the driving routes to include protected left turns; and
  - consistent use of turn signal;
- Use of AV technology to mitigate impairments related to functional declines as described in Table 10.4.
- Expose the client to three visits, for one month, for 60–75 minutes. The visits will entail in-clinic client education and instruction; skill training on the use of AV technology in the parking lot; and in-car residential and city (light traffic) driving skill retraining.
- Follow-up with the client in one year to monitor the neurodegenerative effects of PD on her fitness-to-drive abilities.

Elizabeth's long-term goal is to enhance her fitness-to-drive skills and to competently perform routine driving tasks, mainly dependent on tactical skills, in the restricted driving environment, while compensating for the effects of her medical conditions, i.e., depression and PD that can cause functional declines. The CDRS's recommendations and employment of strategies will help Elizabeth to achieve these goals. The CDRS discharges Elizabeth after three successful sessions and recommends that she maintains her regular medical appointments, including visits to the ophthalmologist, and a follow-up visit with the CDRS in one year (due to the degenerative qualities of PD), or if her health condition changes sooner. The CDRS reports the outcome of the comprehensive driving evaluation, as well as the success of the interventions, and the discharge plan to Elizabeth's primary care physician.

TABLE 10.4

Use of AV Technology to Mitigate Impairments Related to Functional Declines

Impairment in Driving Behaviors (Strategic/ Tactical/Operational)	AV Technology to Offset the Driving Impairments	Benefits for Elizabeth
Makes wide turns at curves and turns into the furthest lane (tactical)	Lane-keeping assist	Lane departure warning systems use visual/ auditory or vibrational (haptic) stimuli to alert a driver that they are drifting out of the lane and the lane-keeping assist centers the vehicle to its lane to overcome the notion of making wide turns.
Not scanning adequately to the left and right before crossing intersections (tactical)	Intersection assistant	Scans the environment for oncoming traffic and warns drivers of vehicles approaching from the sides at intersections, highway exits, or car parks to overcome the inadequate scanning behaviors of the client.
Not reacting appropriately to cars in her blind spot (tactical)	Blind spot detection	Alerts the driver through auditory and visual cues of approaching vehicles in the blind spot to overcome the impairment related to blind spot detection.
Not maintaining lateral lane position as she drifts to the left (tactical)	Lane departure warning and correction system	As described above.
Not using turn signal consistently (tactical)	Lane departure warning and correction system	As described above.

10.4 CONCLUSION

In this chapter the authors discussed deskilling, visual disorders, and neurological and neurodegenerative disorders. Specifically, the authors focus on the core clinical characteristics, associated functional performance deficits, the effect on driving behaviors, and potential of automated vehicle technologies (SAE Levels 0–2) to mitigate the effects of the functional performance deficits, and to enable the driver to resume control. The content discussed is tied together through the illustration of two case studies.

Although the empirical literature supports most of the sections in the chapter, the last section (i.e., deployment of automated vehicle technologies to resume control) represents a conceptual amalgamation of collective thinking on the possibilities of vehicle automation. This section is based on conjecture, informed by the authors' collective best clinical reasoning, understanding the medically-at-risk driver's functional performance deficits, as well as the potential possibilities that automated vehicle technologies hold for the driver to resume control.

The authors hope that this chapter lays the conceptual foundation for engineers, clinicians, rehabilitation scientists, and other transportation professionals to recognize the vulnerabilities of medically-at-risk drivers, as well as the opportunities that automation holds to enable these populations to be independent and safe in their driving tasks. The authors also hope that this chapter will prompt scientists, to empirically test the assumptions related to automated vehicle technologies, for at-risk drivers.

ACKNOWLEDGMENTS

Sarah Krasniuk, PhD student in Rehabilitation Science, University of Western Ontario, London, Ontario, Canada for formatting, coordinating, and technical support.

REFERENCES

Almberg, M., Selander, H., Falkmer, M., Vaz, S., Ciccarelli, M., & Falkmer, T. (2015). Experiences of facilitators or barriers in driving education from learner and novice drivers with ADHD or ASD and their driving instructors. *Developmental Neurorehabilitation, 20*(2), 59–67. doi:10.3109/17518423.2015.1058299

Alvarez, L. & Classen, S. (2017). In-vehicle technology and driving simulation. In S. Classen (Ed.), *Driving Simulation for Assessment, Intervention, and Training: A Guide for Occupational Therapy and Health Care Practitioners* (pp. 265–278). Bethesda, MD: AOTA Press.

American Psychiatric Association. (2000). *Diagnostic and statistical manual of mental disorders: DSM-IV-TR.* Washington, DC: American Psychiatric Association.

American Psychiatric Association. (2012). DSM-5 proposed criteria for autism spectrum disorder designed to provide more accurate diagnosis and treatment. *News Release.* Retrieved from DSM-5: The Future of Psychiatric Diagnosis website: www.dsm5.org/Pages/Default.aspx

Amick, M. M., Grace, J., & Ott, B. R. (2007). Visual and cognitive predictors of driving safety in Parkinson's disease patients. *Archives of Clinical Neuropsychology, 22*(8), 957–967.

Anderson, D. R. (2011). Normal-tension glaucoma (low-tension glaucoma). *Indian Journal of Ophthalmology, 59*(Suppl 1), S97–S101. doi:10.4103/0301–4738.73695

Aquino, M. C. D., Tan, A. M., Loon, S. C., See, J., & Chew, P. T. (2011). A randomized comparative study of the safety and efficacy of conventional versus micropulse diode laser transscleral cyclophotocoagulation in refractory glaucoma. *Investigative Ophthalmology & Visual Science*, 52(14), 2609–2609.

Barkley, R. A. (2004). Driving impairments in teens and adults with attention deficit hyperactivity disorder. *Psychiatric Clinics of North America, 27*, 233–260. doi:10.1016/S0193–953X(03)00091–1

Boisgontier, M. P., Olivier, I., Chenu, O., & Nougier, V. (2012). Presbypropria: The effects of physiological ageing on proprioceptive control. *Age (Dordrecht, Netherlands), 34*(5), 1179–1194. doi:10.1007/s11357-011-9300-y

Bron, A. M., Viswanathan, A. C., Thelen, U., de Natale, R., Ferreras, A., Gundgaard, J., … Buchholz, P. (2010. International vision requirements for driver licensing and disability pensions: Using a milestone approach in characterization of progressive eye disease. *Clinical Ophthalmology, 4*, 1361–1369. doi:10.2147/OPTH.S15359

Canadian Association of Optometrists. (n.d.). *Age-Related Macular Degeneration*. Retrieved from https://opto.ca/health-library/amd%20and%20low%20vision

Canadian Medical Association. (2017). *Determining Medical Fitness to Operate Motor Vehicles: CMA Driver's Guide*. Toronto, ON: Joule Inc.

Centers for Disease Control and Prevention. (2017). Facts about ADHD. Retrieved from www.cdc.gov/ncbddd/adhd/facts.html

Centers for Disease Control and Prevention. (2018). Prevalence of Autism Spectrum Disorder Among Children Aged 8 Years—Autism and Developmental Disabilities Monitoring Network, 11 Sites, United States, 2014. *Morbidity and Mortality Weekly*. Retrieved from www.cdc.gov/mmwr/volumes/67/ss/ss6706a1.htm?s_cid=ss6706a1_w

Chen, K. B., Xu, X., Lin, J.-H., & Radwin, R. G. (2015). Evaluation of older driver head functional range of motion using portable immersive virtual reality. *Experimental Gerontology, 70*, 150–156. doi:10.1016/j.exger.2015.08.010

Clark, T., Feehan, C., Tinline, C., & Vostanis, P. (1999). Autistic symptoms in children with attention deficit-hyperactivity disorder. *European Child and Adolescent Psychiatry, 8*(1), 50–55. doi:10.1007/s007870050083

Classen, S., Monahan, M., & Brown, K. (2013). Indicators of simulated driving performance in teens with attention deficit hyperactivity disorder. *The Open Journal of Occupational Therapy, 2*(1), Art. 3.

Classen, S., Monahan, M., Brown, K. E., & Hernandez, S. (2013). Driving indicators in teens with attention deficit hyperactivity and/or autism spectrum disorder. *Canadian Journal of Occupational Therapy, 80*(5), 274–283. doi:10.1177/0008417413501072

Classen, S., Witter, D. P., Lanford, D. N., Okun, M. S., Rodriguez, R. L., Romrell, J., … Fernandez, H. H. (2011). The usefulness of screening tools for predicting driving performance in people with Parkinson's disease. *The American Journal of Occupational Therapy, 65*(5), 579–588. doi:10.5014/ajot.2011.001073

Cohen, L. P. & Pasquale, L. R. (2014). Clinical characteristics and current treatment of glaucoma. *Cold Spring Harbor Perspectives in Medicine, 4*(6), a017236. doi:10.1101/cshperspect.a017236

Cox, S. M., Cox, J. M., Kofler, M. J., Moncrief, M. A., Johnson, R. J., Lambert, A. E., … Reeve, R. E. (2016). Driving simulator performance in novice drivers with autism spectrum disorder: The role of executive functions and basic motor skills. *Journal of Autism and Developmental Disorders, 46*, 1379–1391. doi:10.1007/s10803-015-2677-1

Crizzle, A. M., Classen, S., Lanford, D. N., Malaty, I. I., Rodriguez, R. L., McFarland, N. R., & Okun, M. S. (2013). Driving performance and behaviors: A comparison of gender differences in drivers with Parkinson's disease. *Traffic Injury and Prevention, 14*(4), 340–345. doi:10.1080/15389588.2012.717730

Crizzle, A. M., Classen, S., & Uc, E. Y. (2012). Parkinson Disease and driving: An evidence-based review. *Neurology, 79*(20), 2067–2074.

Davis, J., Casteel, C., Hamann, C., & Peek-Asa, C. (2018). Risk of motor vehicle crash for older adults after receiving a traffic charge: A case-crossover study. *Traffic Injury Prevention*, 1–26. doi:10.1080/15389588.2018.1453608

Devos, H., Ranchet, M., Akinwuntan, A. E., & Uc, E. Y. (2015). Establishing an evidence-base framework for driving rehabilitation in Parkinson's disease: A systematic review of on-road driving studies. *NeuroRehabilitation, 37*(1), 35–52.

Devos, H., Ranchet, M., Bollinger, K., Conn, A., & Akinwuntan, A. E. (2018). Performance-based visual field testing for drivers with glaucoma: A pilot study. *Traffic Injury Prevention*, 1–7. doi:10.1080/15389588.2018.1508834

Dubinsky, R. M., Gray, C., Husted, D., Busenbark, K., Vetere-Overfield, B., Wiltfong, D., … Koller, W. C. (1991). Driving in Parkinson's disease. *Neurology, 41*, 517–520.

Fabiano, G. A., Hulme, K., Linke, S., Nelson-Tuttle, C., Pariseau, M., Gangloff, B., … Buck, M. (2010). The Supporting a Teen's Effective Entry to the Roadway (STEER) program: Feasibility and preliminary support for a psychosocial intervention for teenage drivers with ADHD. *Cognitive and Behavioral Practice, 18*(2), 267–280. doi:1077–7229/10/267–280$1.00/0

Federal Highway Administration. (2015). *Highway Statistics 2014*. Washington, DC. Retrieved from www.fhwa.dot.gov/policyinformation/statistics/2014/

Fonda, S., Wallace, R., & Herzog, A. (2001). Changes in driving patterns and worsening depressive symptoms among older adults. *Journals of Gerontology. Series B: Psychological Sciences and Social Sciences, 56B*(6), S343–351.

Freeman, E. E., Gange, S. J., Munoz, B., & West, S. K. (2006). Driving status and risk of entry into long-term care in older adults. *American Journal of Public Health, 92*(8), 1284–1289.

Glaucoma Research Foundation. (n.d.). *Types of Glaucoma*. Retrieved from www.glaucoma.org/glaucoma/types-of-glaucoma.php

Haegerstrom-Portnoy, G., Schneck, M. E., & Brabyn, J. A. (1999). Seeing into old age: Vision function beyond acuity. *Optometry and Vision Science, 76*(3), 141–158.

Huang, P., Kao, T., Curry, A., & Durbin, D. R. (2012). Factors associated with driving in teens with autism spectrum disorders. *Journal of Developmental Behavioral Pediatrics, 33*(1), 1–5. doi:10.1097/DBP.0b013e31823a43b7

Jerome, L., Habinski, L., & Segal, A. (2006). Attention-deficit/hyperactivity disorder (ADHD) and driving risk: A review of the literature and a methodological critique. *Current Psychiatry Reports, 8*(5), 416–426.

Jerome, L., Segal, A., & Habinski, L. (2006). What we know about ADHD and driving risk: A literature review, meta-analysis and critique. *Journal of the Canadian Academy of Child and Adolescent Psychiatry, 15*(3), 105–125.

Karthaus, M. & Falkenstein, M. (2016). Functional changes and driving performance in older drivers: Assessment and interventions. *Geriatrics, 1*(2), 12.

Lachenmayer, L. (2000). Parkinson's disease and the ability to drive. *Journal of Neurology, 247*(Suppl 4), 28–30.

Lee, R., Wong, T. Y., & Sabanayagam, C. (2015). Epidemiology of diabetic retinopathy, diabetic macular edema and related vision loss. *Eye and Vision, 2*, 17–17. doi:10.1186/s40662-015-0026-2

Lin, F. R., Thorpe, R., Gordon-Salant, S., & Ferrucci, L. (2011). Hearing loss prevalence and risk factors among older adults in the United States. *The Journals of Gerontology. Series A, Biological Sciences and Medical Sciences, 66*(5), 582–590. doi:10.1093/gerona/glr002

Lings, S. & Dupont, E. (1992). Driving with Parkinson's disease: A controlled laboratory investigation. *Acta Neurologica Scandinavica, 86*, 33–39.

Mather, M., Jacobsen, L., & Pollard, K. (2015). Aging in the United States. *Population Bulletin, 70*(2), 1–17.

McLay, P. (1989). The parkinsonian and driving. *International Disability Studies, 11*(1), 50–51.

Meindorfner, C., Korner, Y., Moller, J. C., Stiasny-Kolster, K., Oertel, W. H., & Kruger, H. P. (2005). Driving in Parkinson's disease: Mobility, accidents, and sudden onset of sleep at the wheel. *Movement Disorders, 20*(7), 832–842. doi:10.1002/mds.20412

Michon, J. A. (1985). A critical view of driver behavior models: What do we know, what should we do? In E. L. Evans & R. Schwing (Eds.), *Human Behavior and Traffic Safety* (pp. 485–520). New York: Plenum.

Midena, E., Degli Angeli, C., Blarzino, M. C., Valenti, M., & Segato, T. (1997). Macular function impairment in eyes with early age-related macular degeneration. *Investigative Ophthalmology & Visual Science, 38*(2), 469–477.

Mitchell, P., Liew, G., Gopinath, B., & Wong, T. Y. (2018). Age-related macular degeneration. *The Lancet, 392*(10153), 1147–1159. doi:10.1016/S0140–6736(18)31550–2

National Eye Institute. (2010, October, 2010). Facts about Refractive Errors. *Refractive Errors.* Retrieved from https://nei.nih.gov/health/errors/errors

National Eye Institute. (2013, September, 2015). Facts about Cataract. Retrieved from https://nei.nih.gov/health/cataract/cataract_facts

National Eye Institute. (2015a, September, 2015). Facts about Diabetic Eye Disease. Retrieved from https://nei.nih.gov/health/diabetic/retinopathy

National Eye Institute. (2015b, September, 2015). Facts about Glaucoma. Retrieved from https://nei.nih.gov/health/glaucoma/glaucoma_facts

National Institute of Mental Health. (2017). *What Is Autism Spectrum Disorder (ASD)?* Retrieved from www.nimh.nih.gov/health/publications/a-parents-guide-to-autism-spectrum-disorder/what-is-autism-spectrum-disorder-asd.shtml

Owsley, C. & McGwin, G., Jr. (2008). Driving and age-related macular degeneration. *Journal of Visual Impairment & Blindness, 102*(10), 621–635.

Owsley, C., McGwin, G., Jr., Scilley, K., & Kallies, K. (2006). Development of a questionnaire to assess vision problems under low luminance in age-related maculopathy. *Investigative Ophthalmology & Visual Science, 47*(2), 528–535. doi:10.1167/iovs.05–1222

Owsley, C., Stalvey, B. T., Wells, J., & Sloane, M. E. (1999). Older drivers and cataract: Driving habits and crash risk. *Journal of Gerontology: Series A: Biological Sciences & Medical Sciences, 54*(4), M203–M211.

Papa, M., Boccardi, V., Prestano, R., Angellotti, E., Desiderio, M., Marano, L., … Paolisso, G. (2014). Comorbidities and crash involvement among younger and older drivers. *PLoS One, 9*(4), e94564. doi:10.1371/journal.pone.0094564

Papadopoulos, M., Loh, A., & Fenerty, C. (2015). Secondary glaucoma: Glaucoma associated with acquired conditions. *The Ophtalmic New and Education Network, Diease Reviews.*

Parkinson's Foundation. (2018). Retrieved from www.parkinson.org/

Pascolini, D. & Mariotti, S. P. (2012). Global estimates of visual impairment: 2010. *British Journal of Ophthalmology, 96*(5), 614–618. doi:10.1136/bjophthalmol–2011–300539

Pennington, K. L. & DeAngelis, M. M. (2016). Epidemiology of age-related macular degeneration (AMD): Associations with cardiovascular disease phenotypes and lipid factors. *Eye and Vision, 3*, 34–34. doi:10.1186/s40662-016-0063-5

Reimer, B., Fried, R., Mehler, B., Joshi, G., Bolfek, A., Godfrey, K. M., … Biederman, J. (2013). Brief report: Examining driving behavior in young adults with high functioning autism spectrum disorders: A pilot study using a driving simulation paradigm. *Journal of Autism and Developmental Disorders, 43*, 2211–2217. doi:10.1007/s10803-013-1764-4

Riggeal, B. D., Crucian, G. P., Seignoural, P. S., Jacobson, C. E., Okun, M. S., Rodriguez, R. L., & Fernandez, H. H. (2007). Cognitive decline tracks motor progression and not disease duration in Parkinson disease. *Neuropsychiatric Disease and Treatment, 3*(6), 955–958.

Ross, L. A., Dodson, J. E., Edwards, J. D., Ackerman, M. L., & Ball, K. (2012). Self-rated driving and driving safety in older adults. *Accident Analysis Prevention, 48*, 523–527. doi:10.1016/j.aap.2012.02.015

Sadana, R., Blas, E., Budhwani, S., Koller, T., & Paraje, G. (2016). Healthy ageing: Raising awareness of inequalities, determinants, and what could be done to improve health equity. *The Gerontologist, 56*(Suppl 2), S178–S193. doi:10.1093/geront/gnw034

See, J. L. S., Aquino, M. C. D., Aduan, J., & Chew, P. T. K. (2011). Management of angle closure glaucoma. *Indian Journal of Ophthalmology, 59*(Suppl 1), S82–S87. doi:10.4103/0301-4738.73690

Sheppard, E., Ropar, D., Underwood, G., & Van Loon, E. (2010). Brief report: Driving hazard perception in autism. *Journal of Autism and Developmental Disorders, 40*, 504–508. doi:10.1007/s10803-009-0890-5

Sheppard, E., Van Loon, E., Underwood, G., & Ropar, D. (2017). Attentional differences in a driving hazard perception task in adults with autism spectrum disorders. *Journal of Autism and Developmental Disorders, 47*, 405–414. doi:10.1007/s10803-016-2965-4

Shrestha, G. S. & Kaiti, R. (2014). Visual functions and disability in diabetic retinopathy patients. Journal of Optometry, *7*(1), 37–43. doi:10.1016/j.optom.2013.03.003

Singh, R., Pentland, B., Hunter, J., & Provan, F. (2007). Parkinson's disease and driving ability. *Journal of Neurology, Neurosurgery, and Psychiatry, 78*(4), 363–366. doi:jnnp.2006.103440 [pii] 10.1136/jnnp.2006.103440

Society of Automotive Engineers International. (2016). *Taxonomy and Definitions for Terms Related to On-Road Motor Vehicle Automated Driving Systems* (J3016_201401).

Steinkuller, P. (2010). Legal vision requirements for drivers in the United States. Virtual Mentor, 12, 938–940.

Szlyk, J. P., Mahler, C. L., Seiple, W., Edward, D. P., & Wilensky, J. T. (2005). Driving performance of glaucoma patients correlates with peripheral visual field loss. *Journal of Glaucoma, 14*(2), 145–150.

Szlyk, J. P., Mahler, C. L., Seiple, W., Vajaranant, T. S., Blair, N. P., & Shahidi, M. (2004). Relationship of retinal structural and clinical vision parameters to driving performance of diabetic retinopathy patients. *Journal of Rehabilitation Research and Development, 41*(3a), 347–358.

Szlyk, J. P., Pizzimenti, C. E., Fishman, G. A., Kelsch, R., Wetzel, L. C., Kagan, S., & Ho, K. (1995). A comparison of driving in older subjects with and without age-related macular degeneration. *Archives of Ophthalmology, 113*(8), 1033–1040.

Tanabe, S., Yuki, K., Ozeki, N., Shiba, D., Abe, T., Kouyama, K., & Tsubota, K. (2011). The association between primary open-angle glaucoma and motor vehicle collisions. *Investigative Ophthalmology & Visual Science, 52*(7), 4177–4181. doi:10.1167/iovs.10-6264

Tham, Y.-C., Li, X., Wong, T. Y., Quigley, H. A., Aung, T., & Cheng, C.-Y. (2014). Global prevalence of glaucoma and projections of glaucoma burden through 2040. *Ophthalmology, 121*(11), 2081–2090. doi:10.1016/j.ophtha.2014.05.013

Trosterer, S., Gartner, M., Mirnig, A., Meschtscherjakov, A., McCall, R., Louveton, N., ... Engel, T. (2016). *You* Never forget how to drive: Driver skilling and deskilling in the advent of autonomous vehicles. *Paper Presented at the Proceedings of the 8th International Conference on Automotive User Interfaces and Interactive Vehicular Applications*, Ann Arbor, MI. http://delivery.acm.org/10.1145/3010000/3005462/p209-trosterer.pdf?ip=129.100.55.89&id=3005462&acc=ACTIVE%20SERVICE&key=FD0 067F557510FFB%2E26B1F5E6B598D80D%2E4D4702B0C3E38B35%2E4D4702B0 C3E38B35&__acm__=1541780895_6dd6dd198cde137adcc00eb8fd121e1d

Uc, E. Y., Rizzo, M., Anderson, S. W., Qian, S., Rodnitzky, R. L., & Dawson, J. D. (2005). Visual dysfunction in Parkinson disease without dementia. *Neurology, 65*(12), 1907–1913.

Uc, E. Y., Rizzo, M., Anderson, S. W., Sparks, J., Rodnitzky, R. L., & Dawson, J. D. (2006a). Impaired visual search in drivers with Parkinson's disease. *Annals of Neurology, 60*(4), 407–413.

Uc, E. Y., Rizzo, M., Anderson, S. W., Sparks, J. D., Rodnitzky, R. L., & Dawson, J. D. (2006b). Driving with distraction in Parkinson disease. *Neurology, 67*(10), 1774–1780.

Uc, E. Y., Rizzo, M., Johnson, A. M., Dastrup, E., Anderson, S. W., & Dawson, J. D. (2009). Road safety in drivers with Parkinson disease. *Neurology, 73*(24), 2112–2119. doi:10.1212/WNL.0b013e3181c67b77

Uitti, R. J. (2009). Parkinson's disease and issues related to driving. *Parkinsonism and Related Disorders, 15*(3), A122–125.

Vernon, S. A., Bhagey, J., Boraik, M., & El-Defrawy, H. (2009). Long-term review of driving potential following bilateral panretinal photocoagulation for proliferative diabetic retinopathy. *Diabetic Medicine, 26*(1), 97–99. doi:10.1111/j.1464–5491.2008.02623.x

Vieluf, S., Godde, B., Reuter, E. M., Temprado, J. J., & Voelcker-Rehage, C. (2015). Practice effects in bimanual force control: Does age matter? *Journal of Motor Behavior, 47*(1), 57–72. doi:10.1080/00222895.2014.981499

Visser, S. N., Bitsko, R. H., Danielson, M. L., Perou, R., & Blumberg, S. J. (2010). Increasing prevalence of parent-reported attention-deficit/hyperactivity disorder among children - United States, 2003 and 2007. *Morbidity & Mortality Weekly Report, 59*(44), 1439–1443.

Willis, J. R., Jefferys, J. L., Vitale, S., & Ramulu, P. Y. (2012). Visual impairment, uncorrected refractive error, and accelerometer-defined physical activity in the United States. *Archives of Ophthalmology, 130*(3), 329–335. doi:10.1001/archopthalmol.2011.1773

Wood, J. M., Black, A. A., Mallon, K., Thomas, R., & Owsley, C. (2016). Glaucoma and driving: On-road driving characteristics. *PLoS One, 11*(7), e0158318-e0158318. doi:10.1371/journal.pone.0158318

Wood, J. M. & Carberry, T. P. (2004). Older drivers and cataracts: Measures of driving performance before and after cataract surgery. *Transportation Research Record, 1865*, 7–13.

World Health Organization. (2001). *International Classification of Functioning, Disability and Health*. Geneva: World Health Organization.

World Health Organization. (2013). *Universal Eye Health: A Global Action Plan 2014–2019*. Retrieved from Geneva, Switzerland. Retrieved from www.who.int/blindness/AP2014_19_English.pdf?ua=1

World Health Organization. (2018). *Blindness and Vision Impairment*. Retrieved from www.who.int/en/news-room/fact-sheets/detail/blindness-and-visual-impairment

Worringham, C. J., Wood, J. M., Kerr, G. K., & Silburn, P. A. (2006). Predictors of driving assessment outcome in Parkinson's disease. *Movement Disorders, 21*(2), 230–235. doi:10.1002/mds.20709

11 Driver State Monitoring for Decreased Fitness to Drive

Michael G. Lenné, Trey Roady, and Jonny Kuo
Seeing Machines Ltd.

CONTENTS

Key Points 247
11.1 Introduction–Why Driver State Monitoring Is Being Widely Promoted
as a Vehicle Safety Technology 248
11.2 Current Approaches to DSM 249
    11.2.1 Distraction and Engagement..250
    11.2.2 Drowsiness...254
    11.2.3 Automation-Driven Changes to Driver State...255
11.3 Future Directions and Applications in DSM..256
11.4 Human Factors and Safety Considerations...257
    11.4.1 Identifying Emerging Risks..258
    11.4.2 Interfacing with the Driver 258
11.5 Conclusion 259
References..260

KEY POINTS

- Governments and the automotive industry have come to recognize the need for DSM as a central safety measure in the next wave of advanced driver assistance systems (ADAS) technologies.
- There are a number of different underlying measurement approaches, such as vehicle-based measures; however, most research and industry applications now pursue camera-based approaches focusing on eye, head, and facial features.
- Driver drowsiness and distraction are two of the most studied driver states, with some algorithms now published in the research literature.
- There are a number of human factors issues to work through, including ensuring insights from the field inform development of future solutions and ensuring that vehicle technologies interface with the driver in ways that support safe performance.

11.1 INTRODUCTION–WHY DRIVER STATE MONITORING IS BEING WIDELY PROMOTED AS A VEHICLE SAFETY TECHNOLOGY

It is sobering to reflect that over 1.3 million people die annually across the world's road transport systems (World Health Organization, 2018). The leading driver behavior risk factors have held constant over several decades: drunk driving, drowsy driving, unbelted occupants, and speed. Worldwide, new road safety strategies are tackling the risks associated with distracted driving and drugged driving. Doubtlessly, directly targeting behavior and improving road infrastructure and vehicle technology have reduced road trauma, and now there are calls for novel approaches to reduce behavioral risk factors. Specifically, with respect to vehicle technologies, there is renewed focus on reducing the outcomes from a safety event causing injury (e.g., lane departure warning systems) (Cicchino, 2018) or in stopping the behavior from becoming an event (e.g., a system that alerts a distracted driver before the vehicle leaves the lane) (NTSB, 2017).

This industry shift to deliver increasing vehicle autonomy has been a key driver of change. Vehicle automation will fundamentally alter what it means to safely operate a vehicle. Future vehicles will afford the driver opportunities to decrease their active control of the vehicle through taking their hands off the wheel and taking their eyes and mind off the road. Driver State Monitoring (DSM) can be based on various methods of input, including vehicle control measures, duration of driving, and many others; however, camera-based methods are recognized to afford the greatest level of specificity in identifying risky behaviors and driver states. In an automated driving environment, a camera-based approach becomes even more critical given that other measures, including some vehicle-based measures, take on reduced significance for driver state estimation given the reduced interactions the driver has with the steering wheel and pedals. For these reasons, this approach to DSM is being actively pursued by almost all automotive original equipment manufacturers (OEMs).

In parallel, there is a huge drive by governments, regulatory bodies, academia, and the automotive industry to broadly introduce new measures to improve vehicle and road user safety. Over the past five years, the European Commission (EC) has been actively assessing a wide range of candidate technologies to reduce injury. A significant step in identifying potential new solutions was taken in 2015 when an EC commissioned report provided an overview of the feasibility of 55 candidate measures for possible inclusion in the revised General Safety and Pedestrian Safety Regulations (Hynd et al., 2015). The output of the study included indicative cost benefits to differentiate those measures that were deemed very likely, moderately likely, or very unlikely to provide a benefit consistent with the cost of implementation. Within the category of "Driver Interface, Distraction and ITS," "driver distraction and drowsiness recognition" was one of only two measures recommended for legislation, with a projected benefit cost ratio greater than one (Hynd et al., 2015).

Following much discussion over several years, the European Parliament approved measures in February 2019 to mandate the introduction of a range of new vehicle technologies. DSM, or occupant status monitoring, is one of the new safety measures that will be introduced into vehicles as a primary/active safety measure. It is a recommended approach to managing the risks associated with distracted and drowsy driving in non-automated vehicles. There is ample crash data available from organizations in the EC and from the United States (e.g., National Highway Traffic Safety Administration, Federal Highway Administration), and other regions across the world, that highlights the role of distraction and drowsiness in road crashes and the need to consider new approaches to managing these growing risks (see also, this Handbook, Chapter 9). DSM also has a key safety role in automated driving. While there is comparatively very little crash data available for automated vehicles due to their early phase of introduction to the market, emerging research papers (e.g., Favarò, Nader, Eurich, Tripp, & Varadaraju, 2017) have highlighted how driver behavior changes with these technologies and the threats to safety that need to be managed. There is considerable research, much reviewed elsewhere in this book, that points to drivers becoming "out of the loop" during automated driving (see, e.g., this Handbook, Chapters 7, 21). While the crash and incident data do not yet exist in sufficient number, it is almost certain that monitoring of the driver state is an essential crash prevention measure.

In parallel to recommendations to the European Parliament supporting the introduction of new technologies, in September 2017, the European New Car Assessment Programme (NCAP) 2025 roadmap was released that outlines targets for when these technologies will be incorporated into crash assessment protocols (Euro NCAP, 2017). Driver monitoring is introduced under Primary Safety, where it is noted that "Euro NCAP envisages an incentive for driver monitoring systems that effectively detect impaired and distracted driving and give appropriate warning and take effective action" (p.7; Euro NCAP, 2017). The roadmap also notes that "Effective driver monitoring will also be a prerequisite for automated driving, to make sure that, where needed, control can be handed back to a driver who is fit and able to drive the vehicle."

11.2   CURRENT APPROACHES TO DSM

Monitoring driver state is a primary safety measure for all forms of vehicle use. Hynd et al. (2015) review a range of methods for measuring drowsiness and distraction. It includes the following: physiological measures (e.g., ocular metrics); physical (e.g., posture); behavioral (e.g., steering inputs); and biomathematical models. In reviewing the available evidence for these different classes section F5.8 of the report concludes

> Overall, the literature indicates that eye feature detection is the most established measure for drowsiness monitoring and has the strongest evidence base for real-time detection. It is non-invasive and the latest (aftermarket) systems seem to be overcoming limitations associated with operator compatibility such as different eyewear and problems with low light. It can be combined with wider face and head detection methods using the same camera technology for improved accuracy.
>
> Of the other methods, vehicle control measures are typically most suited to highway environments as they are often based on steering inputs and lane keeping behaviour. For drowsiness monitoring they are more reactive than predictive.

They perhaps have most relevance as part of a composite monitoring system that includes physiological measures. However, vehicle control measures are the most established method of drowsiness monitoring fitted by automotive manufacturers as original equipment. (p. 415)

While driver monitoring can potentially be done using different sensor inputs (e.g., Lenné & Jacobs, 2016), there is a consensus that camera-based driver monitoring is the best approach. This is particularly true for automated vehicles where traditional vehicle input measures are likely unavailable and vehicle metrics take on new meaning. Notwithstanding, even when vehicle-based metrics are available, there is an argument that these are more diagnostic of the driver state rather than impairment per se. For example, identifying that there is a lane departure is terrific from the safety viewpoint but provides more limited insight into the driver state that may lead to that safety event.

The following section discusses some measures and measurement approaches that are available in the literature for two of the more prominent driver states that will emerge during automated driving—distraction and drowsiness.

11.2.1 DISTRACTION AND ENGAGEMENT

Driver distraction is defined as a diversion of attention away from those activities required for safe driving (Regan, Lee, & Young, 2008). In the context of automated driving this definition has evolved to acknowledge that automated vehicle technologies do allow the driver to take their focus away from the forward roadway for defined periods of time (see this Handbook, Chapter 9). Automated driving functions are likely to change the way a driver interacts with the environment, including gaze behavior, and therefore what is critical is the level of driver attentiveness or engagement for the conditions rather than only the time spent with attention directly away from the roadway.

Much of human factors research into driver state during Level 2 (L2) and Level 3 (L3) automated driving introduces the concept of the driver being "out of the loop" (Merat et al., 2018), defined as being "not in physical control of the vehicle, and not monitoring the driving situation OR in physical control of the vehicle, but not monitoring the driving situation." Recent research documents both the circumstances and the implications for performance and safety.

When assessing driver engagement there are three general distraction classes: manual, visual, and cognitive—or simply "hands on wheel, eyes on road, mind on driving" (Vegega, Jones, & Monk, 2013). Interventions to reduce manual distraction do improve response time but they do not improve visual or cognitive distraction (Victor et al., 2018). Visual attention is measured via glance behavior, with glance defined as the "maintaining of visual gaze within an area of interest, bounded by the perimeter of the area of interest" (ISO 15007-1, 2014). Visual and cognitive attention are more closely linked: improving visual attention may improve cognitive attention, and visual overfocus is an indicator of cognitive inattention.

Table 11.1 illustrates some of the distraction states and their involvement with each type of distraction. Drivers are fully attentive when they are directing

TABLE 11.1

Engagement States and Types of Distraction

State	Manually Engaged	Visually Engaged	Cognitively Engaged
Fully Attentive	Yes	Yes	Yes
"On the loop" (Merat et al., 2018)	No	Yes	Yes
Divided attention (Parasuraman, 2000)	Optional	No	Yes
"Out of the Loop" (Merat et al., 2018)	Optional	Yes	No
Completely disengaged	Optional	No	No

sufficient effort to drive effectively and are completely disengaged when they fail to direct substantive attention to the task. Divided attention is the phenomenon where attention is directed to two tasks, simultaneously, resulting in visual time sharing (VTS). "On the loop" is a state of active awareness without direct mechanical control, which may result in slightly slowed takeover due to motor skills adapting to task.

For over a decade, human factors research has measured visual distraction with eye behaviors in laboratory and field studies. Glance metrics are typically used to assess the impact of different in-vehicle display designs on driver distraction and mobile phone use. Due to this widespread acceptance, many research groups have developed statistical models or algorithms to assess or classify driver state in a more sophisticated manner. Many gaze algorithms classify disengagement and represent variations and modifications of several distinct concepts:

1. *Single Off-Road Glance Threshold*: Off-road glances should be minimal; those which exceed a threshold value (usually 2 seconds) are classified as "Distracted."

2. *Attentive-Glance Threshold*: When drivers are considered as "Distracted," on-road glances should be long enough for the driver to substantively improve their awareness of the road (usually 1 second) before drivers are re-classified as "Attentive" (e.g. Kircher & Ahlström, 2009).

3. *Multiple Short-Term Glances*: Assessment of gaze patterns should also account for multiple glances, as there may be no single glance that exceeds the threshold, but attention is clearly directed away from the road.

4. *Overfocusing*: Finally, it is possible to overfocus, and too much gaze fixation in one central location can be an indicator of cognitive distraction and multi-tasking. One common method is via ratios of central-vs-peripheral vision, such as implemented in "percent road center" (e.g., Victor, 2010, as cited in Lee et al., 2013).

5. *Speed-Based System Availability*: Accounting for differences based on speed helps identify differences in road types. Existing algorithms generally are designed for highway driving, and setting a cutoff for the algorithm's operation based on speed helps exclude other driving contexts. Ideally, algorithms for multiple contexts could shift between modes.

6. *Differentiated Risk Regions*: Glances to different areas of the car carry different risks. Glances to the rear-view mirror are more relevant to driving than those to the driver's lap and should be weighted differently.
7. *Head-Tracking*: Gaze tracking requires a certain quality of video tracking the driver. When this is not available, a less-accurate approximation can be implemented from tracking the driver's head.
8. *Yaw Recalculated Road Center*: Gaze prioritizes the road center, which shifts and bends with the road's curves. Road center is shifted based on vehicle yaw.

The following are the four major algorithms that have incorporated one or more of the above concepts: AttenD (Kircher & Ahlstrom, 2009), AttenD as modified by Seppelt et al. (2017), the multi-distraction algorithm (Victor, 2010, as cited in Lee et al., 2013), and the multi-distraction algorithm as modified by Lee et al. (2013). The concepts that each of the algorithms incorporate are displayed in Table 11.2.

Instead of just classifying general "eyes off forward roadway" and "risky visual scanning patterns," the AttenD algorithm (Ahlström, Kircher, & Kircher, 2009; Kircher, Ahlström, & Kircher, 2009) compares eye movements to the world model of the vehicle, allowing for customization based on physical vehicle features. It distinguishes between three glance categories: forward roadway glances, safe driving glances (i.e., at the speedometer or mirrors), and glances not related to driving. Forward roadway glances (a visual angle of 90° side-to-side and the vehicle windows, excluding the mirrors) increment an attention buffer at 1 unit per second up

---

TABLE 11.2

Gaze Algorithm Feature Classification

	#1: Single Off-Road Glance	#2: Single Attentive Glance	#3: Multi-glance	#4: Overfocusing	#5: Speed Cutoffs	#6: Risk Regions	#7: Head-Tracking	#8: Yaw Center
AttenD: Kircher and Ahlström (2009)	x	x	x			x		
AttenD: Seppelt et al. (2017)	x	x	x	x				
Multi-Distraction: Victor (2010)	x		x	x	x			
Multi-Distraction: Lee et al. (2013)	x		x	x	x	x	x	x

to 2 seconds (after a 0.1 second latency), whereas glances to rearview and speedometer less than 1 second are buffer-neutral, and glances elsewhere decrement from the attention buffer at 1 unit per second.

Seppelt et al. (2017) made modifications to AttenD, and though the specifics remain proprietary, three updates are noted: changes to the increment rate of attentive glances (#2, Table 11.2), changes to latency delay effects prioritizing off-road glance region durations, and addition of an Overfocus component to detect cognitive distraction (#4, Table 11.2). Most notably, this suggests the importance of recognition of cognitive distraction, and that the time required to form a reliable model of the roadway is not 2 seconds (though it remains unclear whether it is more or less).

Victor (2010, as cited in Lee et al., 2013) developed the Multi-Distraction algorithm to identify both visual and cognitive distraction in real time through measurement of Percent Road Centre (PRC; the proportion of time focusing on a 10° radius circle, centered on the driver's most frequent point of gaze on the forward roadway). At speeds above 50 km/h, three major time windows are considered: a long single glance, a shorter PRC below 60%, and a longer PRC exceeding 92% (Figure 11.1). An additional PRC window is implemented to capture VTS, where drivers are classified distracted if PRC decreases to below 65% and subsequently increases above 75% within a 4 second window. This approach does not account for different off-road regions or differentiate between driving-related and non-driving-related glances.

Building on the original Multi-Distraction, Lee et al. (2013) made four modifications in developing their implementation: (1) if sensor quality degrades, tracking shifts from considering gaze, to head position, to posture; (2) the cognitive distraction PRC threshold is decreased to 83%; (3) speed thresholding is implemented to limit tracking to speeds over 47 km/h, with imposed hysteresis to prevent rapid switching across a single value; and (4) the center cone shifts based on vehicle yaw.

Lee et al. (2013) concluded that the modified Multi-Distraction algorithm demonstrated a comparative advantage (True Positive: 90+% vs. 70%–90%; False Positive: 40%–60% vs. 10%–30%). However, this assessment predated Seppelt et al.'s (2017) modifications. Also noteworthy are the six years of improvement in hardware and eye tracking that have occurred since publication.

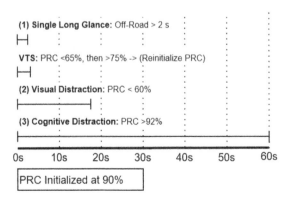

FIGURE 11.1 Representation of the multi-distraction algorithm. (Victor, 2010.)

11.2.2 Drowsiness

Alongside distraction, driver drowsiness remains a significant contributing factor to road crashes worldwide. The National Highway Traffic Safety Administration (2008) estimated that there are 56,000 crashes each year, in which drowsiness or fatigue was cited by police as a causal factor. These crashes lead to, on average, 40,000 non-fatal injuries and 1,550 fatalities per year. Driver drowsiness in the context of L2 and L3 driving remains a nascent field of research at the time of writing. However, drivers' tendency toward disengagement and "out-of-the-loop" states under automated driving, as previously discussed, would suggest the continued prevalence of and critical need to examine this form of impairment (see also, this Handbook, Chapter 9).

While electroencephalography (EEG) is often regarded as the gold standard for objectively quantifying discrete stages of sleep, existing research on its applicability to the transitional stage of drowsiness has yet to yield conclusive results. Significant variability exists in both how participants are observed as well as which candidate signals are proposed for analysis (Ahlström, Jansson, & Anund, 2017; Hung et al., 2013; Perrier et al., 2016). In contrast, greater consistency has been reported from research on ocular measures of drowsiness (e.g. Barua, Ahmed, Ahlström, & Begum, 2019). In practical terms, there is the additional advantage of using an unobtrusive, camera-based system for measuring ocular metrics (a method that, as of yet, has not been developed for measuring neural activity).

With respect to ocular measures, PERCLOS continues to be used as a drowsiness indicator (Jackson et al., 2016; McDonald, Lee, Schwarz, & Brown, 2014), though implementation and statistical approaches differ. PERCLOS is a measure of the proportion of time that the eyes are closed, or nearly closed, over a given period of time, typically between 1 and 20 minutes. The accepted use of the term PERCLOS refers to the proportion of time eyes are more than 80% closed (i.e., based on the degree of eyelid closure; Wierwille, Ellsworth, Wreggit, Fairbanks, & Kirn, 1994), although how this is determined does vary. Early studies used PERCLOS to assess drowsiness as established by performance on the Psychomotor Vigilance Task (PVT) and reported greater coherence in PERCLOS when longer time windows were used. The research by Dinges, Mallis, Maislin, and Powell (1998) found that PERCLOS was a more reliable indicator of drowsiness when considered minute-by-minute over a 20-minute window compared with shorter durations of 1 or 2 minutes, hence recommendations were made to use a 20-minute window. In these studies, PERCLOS is calculated through examination of video by manual annotators who make assessments on the degree of eye closure compared with a set of reference images. As discussed following, more recent studies have examined PERCLOS in driving studies predominantly conducted using driving simulators, for example, to use PERCLOS to either establish a level of drowsiness in a given sample or condition over a period of time.

In recent studies, there is greater interest in the use of real-time assessments of drowsiness to link with much more specific safety-related events, such as lane departure events, steering movements, and micro sleeps and other eye and facial metrics. By necessity, the 20-minute PERCLOS time window is reduced in these studies to increase the confidence that the potential drowsiness level determined through PERCLOS applies to a time at which the driving behavior of interest occurred.

When this approach is taken, the performance of PERCLOS is not as good as found with longer time windows. For example, McDonald et al. (2014) in a driving simulator used a 2-minute window and found poorer performance of PERCLOS compared with both previous research and also a novel steering-based algorithm for the detection of drowsiness associated with lane departure events. One potential explanation provided for the poor PERCLOS performance is that the 2-minute PERCLOS time window is indeed too short to accurately detect acute drowsiness associated with lane excursion events.

Related to PERCLOS, blink duration is another metric that has received considerable attention in drowsy driving research. In a study of real-world driving, Hallvig, Anund, Fors, Kecklund, and Akerstedt (2014) reported intra-individual mean blink duration to be a significant predictor of unintentional lane departures. The importance of warning latency cannot be overlooked in an operational real-world system and, in comparison to existing implementations of PERCLOS, measures of blink duration are generally able to perform closer to real time. An extension to this concept is the idea of a pre-drive assessment of fitness to drive. Using a camera-based driver monitoring system, Mulhall et al. (2018) demonstrated the significant predictive ability of mean blink duration measured before a drive on subsequent, real-world lane departure events.

As a caveat to the studies described above, it is important to note the critical distinction between DSM versus driver *eye tracking*. While the aforementioned metrics show significant promise in the real-time measurement of driver drowsiness, unidimensional ocular features form but one component of the multi-dimensional problem space that is driver state. A simple example of this can be seen in that of the drowsy driver who, with the vehicle safely stopped at a red light, voluntarily rests his/her eyes compared with the drowsy driver who suffers a microsleep event while the vehicle is in motion.

These issues are further confounded by interactions *between* driver states, where drivers have been shown to have an increased propensity toward distracting behavior following sleep deprivation (Anderson & Horne, 2013; Kuo et al., 2019). DSM systems based solely on canonical eye tracking metrics such as eyes-off-road time or eye closure are unlikely to provide the insights necessary to infer high-level driver state. This is especially pertinent in the context of autonomous driving where there is a fundamental shift in the nature of the driving task and objective risks of behaviors typically associated with safety critical events.

11.2.3 AUTOMATION-DRIVEN CHANGES TO DRIVER STATE

While DSM technologies are relevant for manual driving, their inclusion is driven by OEM autonomous technology development. However, increasing vehicle automation changes driver's behavior, their relationship with the vehicle, and the operating context. Drivers are accustomed to direct control and changing their role to a supervisor of automation can generate confusion and discomfort if they must drastically increase their vigilance (see also, this Handbook, Chapters 6, 7). Even in cases where individuals are skeptical of the car's ability to respond appropriately to a given event, their uncertainty slows reaction times and prevents resumption of control (Victor et al., 2018). As we require less-and-less of the driver there will be fewer indicators to determine their capability to intervene, and the indicators we do have may change. For instance, automated lane-keeping features change the requirement for drivers to visually monitor their lane position, and consequently change their scanning behavior. Shifting driving from motor-control to monitoring, driver visual behavior will change to match the task, widening the visual range that drivers scan (Louw & Merat, 2017). Further, steering is only useful as a measure of driver state when the driver is in control.

While there are vastly differing opinions on the rate at which different levels of automated driving features will be widely adopted on roads around the world, it is apparent now that the jump to mass-available autonomous vehicles is not coming all at once. Rather it is likely that waves of vehicles with Level 1 to Level 5 automation functionality will coexist on the road with widespread regional variation. It remains to be seen if dense regional centers will have rapid autonomy adoption, mirroring technologies such as cell phones and internet, as some anticipate (Litman, 2019; Corwin, Jameson, Pankratz, & Willigmann, 2016), but this will likely be determined by the degree of dependence on infrastructure improvements. For instance, GM Super Cruise automation has succeeded by implementing geofencing within regions that have been Light Detection and Ranging (LIDAR)-mapped in advance, an approach more beneficial for heavily traveled routes.

Currently, the age of the average vehicle on U.S. roads is 12 years, with 10% of vehicles being older than 20 years (Federal Highway Administration, 2017); if this pattern holds, half of the vehicles on the roads of 2030 are already here. This mixed fleet means that different levels of automation will have to interact with each other and human drivers who must anticipate behavior, changing and invalidating old patterns (see also, this Handbook, Chapter 19). New drivers will eventually lack familiarity with old modes of driving. Just as a 99.99% reliable anti-lock braking system (ABS) will result in new drivers with no concept of "pumping the brake," some drivers may only be truly qualified to "drive" in geo-fenced areas. With increasing automation, DSM must be able to identify a wide range of driver capabilities and determine an appropriate level of engagement for the specific automation and driving context.

11.3 FUTURE DIRECTIONS AND APPLICATIONS IN DSM

Rapidly evolving trends will change what drivers need and expect to meet their goals around safety, comfort, and convenience. The automotive industry has converged toward the development, implementation, and recommendation of camera-based DSM technologies, which will be rewarded in future crash assessment programs. DSM's early generations target both known and emerging risks to safety and will identify driver states including drowsiness, distraction, inattention, and disengagement from driving. Future applications may recognize mental workload and emotion, or other forms of impairment, such as medical conditions that result in incapacitation (see, e.g., this Handbook, Chapters 10, 17). Additionally, given the studies which show that drivers who have high levels of trust are especially likely to over-rely on their technologies, willing to let the vehicle travel outside of the operational design domain (Victor, 2010), algorithms are needed which can measure levels of driver trust using camera-based technologies. Progress has already been made in that regard (Hergeth, Lorenz, Vilimek, & Krems, 2016).

Drivers experiencing medical distress will be identifiable by their vehicles. Current detection of driver drowsiness and attention needs to expand to handle cases of unresponsive drivers, such as those experiencing cardiac arrest. Modeling advances may identify conditions such as obstructive sleep apnea, which is reportedly undiagnosed in 80% of cases, is more frequent in commercial fleets, and can cause excessive driver sleepiness (Bonsignore, 2017).

Emotionality is also of interest. While emotions do not *cause* action, they do motivate pursuit and avoidance of certain actions and increase or decrease risk aversion (Baumeister, Vohs, DeWall, & Zhang, 2007). Strong negative emotions (e.g., anger, sadness, frustration, etc.) require effort to manage and process, a form of cognitive distraction that can impact driving performance (Jeon, 2016).

Emotion recognition could also mitigate and model driver risk on an individual and aggregate level (e.g., identifying problem features within a city or vehicle fleet), help build more effective coping countermeasures, and even provide useful data fusion with existing gaze classification algorithms.

Emotional recognition has a variety of approaches: facial expression detection, voice analysis, brain scanning, physiological measures, body posture, and combinations thereof. Facial recognition is widely explored, with many applications of the Facial Action Coding System (FACS), which maps small movements in the face to Ekman et al.'s (1987) "six universal emotions" (disgust, sadness, happiness, fear, anger, and surprise). Voice is effective for interpersonal communication but is less relevant in driving. Brain scanning and physiological measures give a massive level of information on driver experience but interpreting the importance of any specific event requires a mature understanding of context as well as skillful and timely processing. Further, many powerful measures are also prohibitively invasive, making them useful only in research settings. Most of these methods are interesting, alone, but the real value lies in aggregation. For instance, body posture's strongest benefits occur when coupled with facial recognition (Aviezer, Trope, & Todorov, 2012). For a detailed methods review, we recommend Calvo & D'Mello (2010), Zeng, Pantic, Roisman, and Huang (2009), and Mauss and Robinson (2009).

Emotion tracking has its challenges. First, Ekman's six universal emotions vary in relevance to driving (Jeon & Walker, 2011) (e.g., disgust versus anger), and the claimed universality may be less reliable across cultures (Gendron, Roberson, van der Vyver, & Barrett, 2014; Russell, 1994). Further, most recognition algorithms are trained on posed emotions in ideal conditions, not the messily framed, naturally occurring emotions that are critical. Finally, to find social acceptance, emotion recognition must navigate difficult personal boundaries between observation of others' feelings and how polite it is to mention them.

11.4 HUMAN FACTORS AND SAFETY CONSIDERATIONS

Many issues must be addressed to move the industry from policy to practice, and to fully realize the benefits of driver monitoring technology. What should the minimum viable technology concepts be? For non-automated driver state measurement, the EC focuses on measuring distraction and drowsiness. These driver states can be defined and operationalized in different ways, and the implications for effectiveness are of critical concern. By necessity, there is a tension between OEMs accounting for cost and driver experience and regulators who must ensure public safety, with both priorities meriting consideration.

11.4.1 IDENTIFYING EMERGING RISKS

How do we ensure we learn what is happening in the real world and that these insights inform technology enhancements? Identifying the role of driver state in studies of crashes and near misses is going to be a challenge. It is likely that surrogate measures, such as takeover performance (however it may be defined) will be pivotal to assessing real-world risk until larger bodies of near-miss and incident data become available. To move forward and to continue to develop effective solutions to impairment, we must ensure we have the best understanding of the role of and types of driver states leading to crashes.

11.4.2 INTERFACING WITH THE DRIVER

To achieve the desired road safety improvements any driver monitoring system needs to not only effectively measure driver state but also determine the most effective time and method to communicate with the driver and/or the vehicle. According to Hynd et al. (2015), measurement thresholds and parameters are keys to DSM performance. Systems with low specificity may have high false positive rates. Further, industry stakeholders advised that warning thresholds and intrusiveness should not be irritating for drivers. An effective interface underpins the success of systems that interact with the driver and must be carefully considered (see also, this Handbook, Chapter 15).

This is particularly the case for drowsiness. Research with the Monash University Accident Research Centre has shown that multiple levels of driver feedback are required to maximally reduce the rate of drowsy driving events. Two feedback mechanisms, in-cab driver feedback and feedback to the company, were evaluated in an Australian fleet from 2011 to 2015 (Fitzharris et al., 2017). Relative to conditions where no feedback was provided, in-cab warnings to the driver resulted in a 66% reduction in drowsiness events, with a 94% reduction achieved with the provision of real-time feedback to the company in addition to in-cab warnings to the driver.

This second level of feedback has important implications for drowsiness-related human–machine interface (HMI) design in passenger and light commercial vehicles: it allows the employer to assess and manage drowsiness in real time and eliminates false positives. When an in-vehicle drowsiness alert is activated in a passenger vehicle the driver will need to make the right decision with this information available—which inevitably would involve an effective drowsiness mitigation that might include a break of some sort, ideally a power nap or longer sleep. According to research, this is unlikely to occur in a notable portion of passenger car drivers without an employer (or creative alternative) overseeing their driving. A call-back service or a smart vehicle assistant could interact with the driver after a warning, assess their drowsiness, guide them to an appropriate resting place, and ensure that an effective mitigation strategy is adopted. This function would also minimize false positives, directly improving driver acceptance—a critical issue given that even lifesaving systems, such as lane departure warnings, may be deactivated by drivers due to unacceptable annoyance levels.

Similar findings have also been reported in the context of driver distraction. Kujala, Karvonen, and Mäkelä (2016) tested the efficacy of a proactive smart phone-based distraction warning system that adjusted warning thresholds according to the expected visual demands of an upcoming driving situation, feeding information back to the driver in real time. The system detected visually high-demanding driving scenarios in which visual distraction would be particularly dangerous (based on factors such as the experience level of the driver and the proximity of intersections and pedestrian crossings ahead) and aimed to identify when the driver was looking at the phone.

The closed test track study design showed that the distraction warning system significantly increased glance time on road while multi-tasking with the mobile phone. However, no effects on individual in-car glance durations were evident, although, as noted by the authors, this may have reflected limitations associated with the gaze tracking system that was used. These results further highlight the codependency between accurate DSM and HMI in developing an effective system.

It seems inevitable that the future vehicle will modify its performance and/ or take over control when it senses driver impairment. If a driver is drowsy, the vehicle could take many measures, such as alerting the driver and providing a time window to find a rest stop before disabling the engine; limiting speed; increasing the sensitivity of lane departure and other ADAS systems; communicating with nearby vehicles to coordinate passing and following distances; enabling autonomous driving; engaging the driver in conversation; guiding them to a rest stop; and so on.

11.5 CONCLUSION

Many regions in the world are striving toward the Vision Zero philosophy first implemented in Swedish Parliament in 1997. Technologies such as driver monitoring are acknowledged in EC documents to hold promise in reducing road injury. The design and implementation of these technologies will determine whether they have minimum or maximum injury reduction benefits.

There is much excitement, energy, and skepticism surrounding talk about the autonomous future and separating reality from fiction. What the introduction of autonomous vehicles has done is to get camera-based DSM into a vehicle to maintain driver safety when operating a vehicle in autonomous mode. Automation notwithstanding, this affords future benefits in non-automated driving to address longstanding risks with distraction and drowsiness. The issues noted herein must be carefully considered, however, if DSM technology is to achieve maximum benefits in reducing road injury.

REFERENCES

Anderson, C. & Horne, J. (2013). Driving drowsy also worsens driver distraction. *Sleep Medicine, 14*(5), 466–468.

Ahlström, C., Jansson, S., & Anund, A. (2017). Local changes in wake electroencephalogram precedes lane departures. *Journal of Sleep Research, 26*(6), 816–819.

Ahlström, C., Kircher, K., & Kircher, A. (2009). Considerations when calculating percent road centre from eye movement data in driver distraction monitoring. *Fifth International Driving Symposium on Human Factors in Driver Assessment, Training and Vehicle Design* (pp. 132–139).Iowa City, IA: University of Iowa.

Aviezer, H., Trope, Y., & Todorov, A. (2012). Body cues, not facial expressions, discriminate between intense positive and negative emotions. *Science, 338*(6111), 1225. doi:10.1126/science.1224313

Barua, S., Ahmed, M. U., Ahlström, C., & Begum, S. (2019). Automatic driver sleepiness detection using EEG, EOG, and contextual information. *Expert Systems with Applications, 115*, 121–135.

Baumeister, R. F., Vohs, K. D., DeWall, N. C., & Zhang, L. (2007). How emotion shapes behavior: Feedback, anticipation, and reflection, rather than direct causation. *Personality and Social Psychology Review, 11*(2), 167–203. doi:10.1177/1088868307301033

Bonsignore, M. (2017). Sleep apnea and its role in transportation safety. *F1000 Research, 6*, 2168.

Calvo, R. A. & D'Mello, S. (2010). Affect detection: An interdisciplinary review of models, methods, and their applications. *IEEE Transactions on Affective Computing, 1*(1), 18–37. doi:10.1109/T-AFFC.2010.1

Cicchino, J. (2018). Effects of lane departure warning on police-reported crash rates. *Journal of Safety Research, 66*, 61–70.

Corwin, S., Jameson, N., Pankratz, D., & Willigmann, P. (2016). *The Future of Mobility: What's Next? Tomorrow's Mobility Ecosystem – and How to Succeed in It*. New York: Deloitte University Press.

Dinges, D. F., Mallis, M. M., Maislin, G., & Powell, J. W. (1998). *Evaluation of Techniques for Ocular Measurement as an Index of Fatigue and as the Basis for Alertness Management* (DOT-HS-808-762). Washington, DC: National Highway Traffic Safety Administration.

Ekman, P., Friesen, W. V., O'Sullivan, M., Chan, A., Diacoyanni-Tarlatzis, I., Heider, K., & Tzavaras, A. (1987). Universals and cultural differences in the judgments of facial expressions of emotion. *Journal of Personality and Social Psychology, 53*(4), 712–717. doi:10.1037/0022-3514.53.4.712

Euro NCAP. (2017). *Euro NCAP 2025 Roadmap: In Pursuit of Vision Zero*. Brussels, Belgium. Retrieved 30 April, from www.euroncap.com/en/for-engineers/technical-papers/

Favarò, F. M., Nader, N., Eurich, S. O., Tripp, M., & Varadaraju, N. (2017). Examining accident reports involving autonomous vehicles in California. *PLoS ONE, 12*(9). doi:10.1371/journal.pone.0184952

Federal Highway Administration. (2017). *National Household Travel Survey*. Washington, DC: FHWA.

Fitzharris, M., Liu, S., Stephens, A. N., & Lenné, M. G. (2017). The relative importance of real-time in-cab and external feedback in managing fatigue in real-world commercial transport operations. *Traffic Injury Prevention, 81*(S1), S71–78.

Gendron, M., Roberson, D., van der Vyver, J. M., & Barrett, L. F. (2014). Cultural relativity in perceiving emotion from vocalizations. *Psychological Science, 25*(4), 911–920. doi:10.1177/0956797613517239

Hallvig, D., Anund, A., Fors, C., Kecklund, G., & Akerstedt, T. (2014). Real driving at night – Predicting lane departures from physiological and subjective sleepiness. *Biological Psychology, 191*, 18–23.

Hergeth, S., Lorenz, L., Vilimek, R., & Krems, J. F., 2016. Keep your scanners peeled: Gaze behavior as a measure of automation trust during highly automated driving. *Human Factors, 58*, 509–519. doi:10.1177/0018720815625744

Hung, C. S., Sarasso, S., Ferrarelli, F., Riedner, B., Ghilardi, F., Cirelli, C., & Tononi, G. (2013). Local experience-dependent changes in the wake EEG after prolonged wakefulness. *Sleep, 36*(1), 59–72

Hynd, D., McCarthy, M., Carroll, J., Seidl, M., Edwards, M., Visvikis, C., ... Stevens, A. (2015). *Benefit and Feasibility of a Range of New Technologies and Unregulated Measures in the Fields of Vehicle Occupant Safety and Protection of Vulnerable Road Users: Final Report*. Report Prepared for the European Commission. Retrieved 30 April 2019, from htttps://publications.europa.eu/en/publication-detail/-/publication/47beb77e-b33e-44c8-b5ed-505acd6e76c0

ISO 15007-1. (2014). *Road Vehicles — Measurement of Driver Visual Behaviour with Respect to Transport Information and Control Systems — Part 1: Definitions and Parameters*. Retrieved November 19, 2018, from www.iso.org/obp/ui/#iso:std:iso:15007:-1:ed-2:v1:en

Jackson, M. L., Kennedy, G. A., Clarke, C., Gullo, M., Swann, P., Downey, L. A., ... Howard, M. E. (2016). The utility of automated measures of ocular metrics for detecting driver drowsiness during extended wakefulness. *Accident Analysis & Prevention, 87*, 127–133. doi:10.1016/j.aap.2015.11.033

Jeon, M. (2016). Don't cry while you're driving: Sad driving is as bad as angry driving. *International Journal of Human–Computer Interaction, 32*(10), 777–790. doi:10.1080/10447318.2016.1198524

Jeon, M. & Walker, B. N. (2011). What to detect? Analyzing factor structures of affect in driving contexts for an emotion detection and regulation system. *Proceedings of the Human Factors and Ergonomics Society Annual Meeting, 55*, 1889–1893. doi:10.1177/1071181311551393

Kircher, K. & Ahlström, C. (2009). Issues related to the driver distraction detection algorithm AttenD. *First International Conference on Driver Distraction and Inattention*. Gothenburg, Sweden: Chalmers.

Kircher, K., Ahlström, C., & Kircher, A. (2009). Comparison of two eye-gaze based real-time driver distraction detection algorithms in a small-scale field operational test. *Fifth International Driving Symposium on Human Factors in Driver Assessment, Training and Vehicle Design* (pp. 16–23).

Kujala, T., Karvonen, H., & Mäkelä, J. (2016). Context-sensitive distraction warnings – Effects on drivers' visual behavior and acceptance. *International Journal of Human-Computer Studies, 90*, 39–52. doi:10.1016/j.ijhcs.2016.03.003

Kuo, J., Lenne, M., Mulhall, M., Sletten, T., Anderson, C., Howard, M., ... Collins, A. (2019). Continuous monitoring of visual distraction and drowsiness in shift-workers during naturalistic driving. *Safety Science.* 119, 112–116.

Lee, J., Moeckli, J., Brown, T., Roberts, S., Victor, T., Marshall, D., ... Nadler, E. (2013). Detection of driver distraction using vision-based algorithms. *Proceedings of the 23rd Enhanced Vehicle Safety Conference*, Seoul, Korea: ESV.

Lenné, M. G. & Jacobs, E. M. (2016). Predicting drowsiness-related driving events: A review of recent research methods and future opportunities. *Theoretical Issues in Ergonomics Science, 17*, 533–553.

Litman, T. (2019). *Autonomous Vehicle Implementation Predictions*. Victoria, Canada: Victoria Transport Policy Institute.

Louw, T. & Merat, N. (2017). Are you in the loop? Using gaze dispersion to understand driver visual attention during vehicle automation. *Transportation Research Part C: Emerging Technologies, 76*, 35–50. doi:10.1016/j.trc.2017.01.001

Mauss, I. B. & Robinson, M. D. (2009). Measures of emotion: A review. *Cognition & Emotion, 23*(2), 209–237. doi:10.1080/02699930802204677

McDonald, A. D., Lee, J. D., Schwarz, C., & Brown, T. L. (2014). Steering in a random forest: Ensemble learning for detecting drowsiness-related lane departures. *Human Factors, 56*(5), 986–998. doi:10.1177/0018720813515272

Merat, N., Seppelt, B., Louw, T., Engström, J., Lee, J. D., Johansson, E., … Keinath, A. (2018). The "out-of-the-Loop" concept in automated driving: proposed definition, measures and implications. *Cognition, Technology & Work, 21*(2), 87–98. doi:10.1007/s10111-018-0525-8

Mulhall, M. D., Cori, J., Kuo, J., Magee, M., Collins, A., Anderson, C., … Howard, M. E. (2018). Pre-drive ocular assessment predicts driving performance in shift workers: A naturalistic driving study. *Journal of Sleep Research, 27*(S2).

National Highway Traffic Safety Administration. (2008). *National Motor Vehicle Crash Causation Survey* (HS 811 059). Washington, DC: National Highway Traffic Safety Administration.

National Transportation Safety Board. (2017). *Safety Recommendation H-17–042*. Washington, DC: National Transportation Safety Board.

Parasuraman, R. (2000). *The Attentive Brain*. Cambridge, MA: MIT Press.

Perrier, J., Jongen, S., Vuurman, E., Bocca, M. I., Ramaekers, J. G., & Vermeeren, A. (2016). Driving performance and EEG fluctuations during on-the-road driving following sleep deprivation. *Biological Psychology, 121*, 1–11.

Regan, M. A., Lee, J. D., & Young, K. L. (2008). *Driver Distraction: Theory, Effects and Mitigation*. Boca Raton, FL: CRC Press.

Russell, J. A. (1994). Is there universal recognition of emotion from facial expression? A review of the cross-cultural studies. *Psychological Bulletin, 115*(1), 102–141. doi:10.1037/0033–2909.115.1.102

Seppelt, B., Seaman, S., Lee, J., Angell, L., Mehler, B., & Reimer, B. (2017). Glass half-full: On-road glance metrics differentiate crashes from near-crashes in the 100-Car data. *Accident Analysis and Prevention, 107*, 48–62.

Vegega, M., Jones, B., & Monk, C. (2013). *Understanding the Effects of Distracted Driving and Developing Strategies to Reduce Resulting Deaths and Injuries: A Report to Congress* (DOT HS 812 053). Washington, DC: National Technical Information Service.

Victor, T. W. (2010). *The Victor and Larsson (2010) Distraction Detection Algorithm and Warning Strategy*. Gothenburg, Sweden: Volvo Technology.

Victor, T. W., Tivesten, E., Gustavsson, P., Johansson, J., Sangberg, F., & Ljung Aust, M. (2018). Automation expectation mismatch: Incorrect prediction despite eyes on threat and hands on wheel. *Human Factors, 60*(8), 1095–1116. doi:10.1177/0018720818788164

Wierwille, W. W., Ellsworth, L. A., Wreggit, S. S., Fairbanks, R. J., & Kirn, C. L. (1994). *Research on Vehicle-Based Driver Status/Performance Monitoring: Development, Validation and Refinement of Algorithms for Detections of Driver Drowsiness* (DOT HS 808 247). Washington, DC: National Highway Traffic Safety Administration.

World Health Organization. (2018). *Global Status Report on Road Safety 2018*. Geneva: World Health Organization.

Zeng, Z., Pantic, M., Roisman, G. I., & Huang, T. S. (2009). A survey of affect recognition methods: Audio, visual, and spontaneous expressions. *IEEE Transactions on Pattern Analysis and Machine Intelligence, 31*(1), 39–58. doi:10.1109/TPAMI.2008.52

12 Behavioral Adaptation

John M. Sullivan
University of Michigan Transportation Research Institute

CONTENTS

Key Points 263
12.1 Behavioral Adaptation and Automated, Connected, and
  Intelligent Vehicles 264
  12.1.1 Why Is BA a Concern? 265
    12.1.1.1 Negative BA 265
    12.1.1.2 Positive Adaptation 267
    12.1.1.3 Discussion 268
  12.1.2 Modeling Behavioral Adaptation 269
    12.1.2.1 How Might Behavior Change with Adaptation? 272
    12.1.2.2 What Specific Behaviors Might Change with Adaptation? ... 273
    12.1.2.3 When Does Behavioral Adaptation Occur? 274
12.2 Behavioral Adaptation to Advanced Driving Assistance Systems 275
  12.2.1 Adaptation to Automation of Longitudinal and Lateral Control 276
    12.2.1.1 Automation of Longitudinal and Lateral Control 276
    12.2.1.2 Changes in Driver Event Detection Related to
      Control Automation 277
    12.2.1.3 Control Automation and Changes in Driving
      Behavior—Control Management 281
    12.2.1.4 Control Automation and Driver Workload 282
    12.2.1.5 Control Automation and SA 283
  12.2.2 Adaptation to ADAS Warning Systems 283
    12.2.2.1 Adaptation to ISA 284
    12.2.2.2 Lane Departure Warning 284
    12.2.2.3 Forward Collision Warning 285
    12.2.2.4 Combination Discrete Warning Systems 285
12.3 Conclusion 286
References 287

KEY POINTS

- A driver's behavior is malleable and may change (or adapt) in response to alterations in the function and performance of their vehicle. This change has generally been called *behavioral adaptation* (BA).
- BA can occur for a variety of reasons and be manifest in a variety of ways. A major concern involves adaptations that diminish the expected safety improvements after a driving enhancement is introduced.

- Early models of BA attempted to explain a driver's motivation to alter his or her behavior, casting the problem in terms of managing, or balancing various assessments of risk.
- A driver's mental model of an automated system is an important issue because behavior that is under the voluntary control of the driver may be adjusted based on their understanding of the system.
- Understanding BA involves identifying when, where, and how a driver activates and deactivates systems, decides when to delegate and retake control, monitors system displays for status, heeds warnings that might indicate critical conditions, and takes appropriate action in response to such warnings.

12.1 BEHAVIORAL ADAPTATION AND AUTOMATED, CONNECTED, AND INTELLIGENT VEHICLES

It has long been suspected that improvements in technologies to enhance driver safety could result in a change in driver behavior that alters the actual benefit in a way that differs from the expected benefit. In their seminal work on automobile driving, Gibson and Crooks (1938) mention offhandedly in a footnote that, "...more efficient brakes on an automobile will not in themselves make driving the automobile any safer. Better brakes will reduce the absolute size of the minimum stopping zone, it is true, but the driver soon learns this new zone...." (p. 458). The authors go on to argue that drivers will adjust their braking to maintain the same safety margin as before. Thus, early on Gibson and Crooks recognized that a driver's behavior is malleable and may change (or adapt) in response to alterations in the function and performance of their vehicle. This change has generally been called *behavioral adaptation* (BA).

As such, BA effects are not simply restricted to responses to changes in the operational characteristics of a vehicle. Indeed, drivers have demonstrated sensitivity to environmental conditions associated with the roadway environment. For example, Kallberg (1993) found that the improved road guidance provided by reflector posts resulted in faster travel speeds at night along roads with relatively poor geometric standards. BA effects might also be more informally observed among drivers who reduce their speed in adverse weather or change their travel routes to cope with rush hour conditions. Likewise, BA effects are also observed in response to driver state. For example, as drivers age, they alter their exposure to difficult driving conditions (Ball et al., 1998), apparently mindful of their own limitations. Thus, BA can occur for a variety of reasons and be manifest in a variety of ways described in these examples. This review will focus specifically on how the introduction of automated, connected, and intelligent vehicles (ACIVs) is known to affect BA. It will not address factors directly related to the road environment or driver state, although we recognize that ACIVs may, in some cases, raise a driver's awareness of environmental conditions (e.g., congestion, icy road conditions, lane closures) or driver state (e.g., diminished alertness, poor lateral control) and may influence behavior.

In this chapter, we will discuss why BA is important to understand, introduce a few examples of the different kinds of driver adaptation reported in the literature, and provide a few classic examples of positive and negative adaptation. We will then discuss some of the theoretical frameworks used to describe the problem of

BA, review more recent thoughts about the mechanisms that drive BA, and finally discuss how BA is influenced by various forms of vehicle automation, including the kind of part-task automation provided by advanced driver assistance systems (ADAS), as well as other connected vehicle applications.

12.1.1   Why Is BA a Concern?

Advancements in driving technologies have offered a means to overcome inherent driver limitations that contribute to the global crash rate. Indeed, various driver errors have been implicated in about 94% of all crashes (National Highway Traffic Safety Administration, 2015). These errors include recognition errors (e.g., inattention, poor lookout, distraction), decision errors (e.g., excessive speed for conditions, gap misjudgment, illegal maneuvers), performance errors (e.g., oversteering, insufficient braking), and non-performance errors (e.g., asleep at the wheel). New technologies introduced into the vehicle hold the promise of shoring up these weak points in a driver's performance to improve overall driving performance and safety. Whether such technological remedies actually deliver on the promise of improved performance and safety cannot be determined without understanding how drivers may adjust their behavior to these new capabilities. Only after the behavioral changes that a driver may make are understood and accounted for, is it possible to obtain a full picture of the net benefits of these new technologies.

12.1.1.1   Negative BA

Perhaps the chief concern about BA involves adaptations that diminish the expected safety improvements after a driving enhancement is introduced. This is known as negative BA. Classic examples of this include the observation made by Gibson and Crooks (1938), whereby drivers equipped with more efficient brakes adjusted their safety margins to account for the new stopping capability. Other early examples include the introduction of studded tires (Rumar, Berggrund, Jernberg, & Ytterbom, 1976) to improve traction under icy conditions. In this study, drivers drove faster in icy conditions when equipped with studded tires. Although a greater net safety margin was observed, it was not as great as it would have been if the driver's speed behavior on icy roads had not changed. Another example is the introduction of anti-lock brakes (ABS). While earlier studies of drivers equipped with ABS (e.g., Aschenbrenner, Biehl, & Wurm, 1988; OECD, 1990) seemed equivocal—no differences were observed in the number of accident involvements, but faster driving was noted by observers of drivers in vehicles equipped with ABS—later work found evidence that vehicles equipped with ABS are driven with significantly shorter headways and lower rates of seat-belt use (Sagberg, Fosser, & Sætermo, 1997).

In each case, the expected improvement in driver safety appears to be offset by a change in behavior that increases crash risk. How much the behavioral change offsets the reduction in risk likely varies, but some researchers suggest that drivers adjust their driving to achieve an ideal target risk, completely negating the risk reduction produced by the intervention. This has been called *risk homeostasis* (Wilde, 1998). This is one of the several explanations of BA that will be discussed in more detail later.

12.1.1.1.1 Misuse

Other forms of negative BA may involve the use of a new technology in circumstances for which it is not appropriate (Parasuraman & Riley, 1997). This problem is specifically associated with different forms of vehicle automation in which a driver is provided the capability of delegating part of the driving task to an automated system, or supplementing his or her monitoring abilities with enhanced sensing capabilities. Unlike the previous examples, in which the driver might increase crash risk by altering his or her choice of speed, gap acceptance, or following distance, misuse involves decisions about when, where, and how one elects to employ these new capabilities. In the previous example, performance improvements in pre-existing vehicle functions resulted in the modulation of pre-existing driving behaviors. Automation introduces new tasks and capabilities to drivers, changing the repertoire of actions drivers perform to manage their vehicles.

For example, adaptive cruise control (ACC) is designed to be used on limited access roadways that are free of pedestrians, bicyclists, mopeds, and skateboarders. This is because most ACC radar has difficulty detecting small, stationary, or low-speed objects. ACC is also ignorant of traffic signage, signals, and other characteristics of the roadway environment. A misuse of ACC would involve activating it in an inappropriate environment like a surface street—a misuse of *where* it is activated. ACC is also limited in its capabilities in fog, snow, and heavy rain. Use of ACC in these conditions is also inappropriate—a misuse of *when* it is activated. ACC might be used by some drivers to maintain relatively shorter headways than they would during manual driving with the objective of reducing the incidence of forward cut-ins. This form of misuse is associated with *how* it is used.

Because studies of BA in which drivers are assisted by automation are generally focused on performance comparisons of behavior with and without the automation, behavioral differences have also been attributed to adaptation effects. However, in the early stages of use, it may be more accurate to attribute these behavioral differences to a lack of understanding about the technology's functions and limitations. Drivers may modify their driving behavior based on their (sometimes flawed) understanding of how a system operates, accepting the system to stand in for oneself (see e.g., this Handbook, Chapter 3). Many automation systems invite drivers to back out of the control role to allow the system to take over. After drivers oblige, they may be slow to recognize the need to retake control, resulting in a delayed intervention for a critical event (see also this Handbook, Chapter 7). Given the limited awareness some drivers may have about their technologies (e.g., Beggiato, Pereira, Petzoldt, & Krems, 2015; Dickie & Boyle, 2009; Jenness, Lerner, Mazor, Osberg, & Tefft, 2008), compromises to safety may be rooted in ignorance of these system limitations.

It is also important to recognize that use of these systems involve changes in how the vehicle is managed. For example, it would be hard to argue that refraining from use of a brake pedal during ACC activation is a result of BA—this is a behavioral change that is mandated by the ACC system to allow it to function properly. Stepping on the brake disengages ACC. Such a change should not be considered an instance of BA—it is a requirement to allow the ACC to perform. Likewise, other forms of ADAS may require some degree of display monitoring or performing an activation sequence that may introduce new driving behaviors. While this may differ from routine non-ADAS equipped behavior, it should not be considered an instance of BA.

12.1.1.1.2 Disuse

Failure to realize safety benefits from automation technologies may also be a consequence of drivers choosing to not employ automation when it may provide some assistance. For example, lane departure warning (LDW) technologies alert drivers when their vehicle is about to depart the lane, unless the lane change is signaled. These camera-based systems work best on roadways with clearly marked lanes, and less reliably in construction areas or when weather conditions obscure the lane lines. While they have the clear potential to assist drivers (Scanlon, Kusano, Sherony, & Gabler, 2015; Sternlund, Strandroth, Rizzi, Lie, & Tingvall, 2017), drivers often do not elect to activate them (e.g., Eichelberger & McCartt, 2014; Reagan & McCartt, 2016). This may not be so much an instance of BA—the driver limits their exposure to the system and thus eliminates the possibility of any change in driving behavior— but it does represent an example of a failure to realize a potential safety benefit as a consequence of the driver's non-use behavior.

12.1.1.2 Positive Adaptation

Positive BA occurs when drivers' behavioral changes improve the likely safety benefit in ways that exceed those originally anticipated. There are generally fewer examples of positive adaptations reported in the literature. Most often, the reported behavioral change is a consequence of adopting driving behaviors demonstrated (or encouraged) by the technology. For example, users of an LDW system that produced an audible warning whenever an unsignaled lane change was executed, reduced their frequency of unsignaled lane changes by 43% on freeways and 24% on surface roads (LeBlanc, Sayer, Winkler, Bogard, & Devonshire, 2007); a similar effect was observed in a later study in which a 66% reduction in unsignaled lane changes was observed (Sayer et al., 2010). Another example was reported by Ervin et al. (2005) in which ACC drivers executed 50% fewer passing maneuvers than they did when driving without ACC.

Other forms of positive BA have been reported that compare some characteristics of driving performance while assisted with ADAS to unassisted performance. For example, Bianchi Piccinini, Rodrigues, Leitão, and Simões (2014) report safer overall time headways when drivers are permitted use of an ACC compared with fully manual driving. In this study, the behavioral component of driving that is given focus is headway and speed management during ACC-supported and unsupported segments of a drive. While identified as a BA outcome, the overall performance improvement during ACC-assisted segments of the drive may actually be determined by the degree to which the driver delegates some of the longitudinal control to the ACC system, which is less variable than manual control. We also note that the performance measures (speed and headway management) in Bianchi Piccinini et al. (2014) are focused exclusively on those parts of the driving task that are directly related to the ACC function. Other characteristics of the driver's behavior that were not monitored in this study might not fare as well. For example, Rudin-Brown and Parker (2004) reported a decline in responsiveness to safety-relevant brake light activation in a forward vehicle. It is important to keep in mind that myriad driving behaviors might change in response to changes in the driving task.

12.1.1.3 Discussion

In the past, BA effects were generally categorized as either positive, negative, or none. Earlier perspectives of BA effects (e.g., Evans, 1985; OECD, 1990; Wilde, 1998) evaluated adaptation effects in terms of the projected benefits, based on an assumption of no change in driver behavior. Thus, if the observed decline in crash risk after the introduction of a technology (like ABS, for example) fell short of the projected decline, it was believed that some degree of negative BA of the driver was responsible for the shortfall. Early improvements in vehicle technologies were largely associated with the performance of the vehicle (e.g., braking, handling, steering, and acceleration) or with occupant protection (e.g., air bags, seat belts). The driver's basic task was generally unchanged throughout these improvements. Within these constraints, BA was evaluated in terms of changes in overall crash risk, operationalized as elevated travel speed, aggressive acceleration and braking, short following distances, small gap acceptances, and the execution of other aggressive maneuvers.

The introduction of ADAS and other automation technologies into the driving task has somewhat altered perspectives on driving such that the evaluation of BA effects is less focused on overall assessment of crash risk. BA effects are now more focused on component driving behaviors. Perhaps this is because ADAS technologies add new tasks for drivers to perform during driving: drivers must activate and deactivate systems, decide when to delegate and retake control, monitor system displays for status, heed warnings that might indicate critical conditions, and take appropriate action in response to such warnings. Understanding BA now also involves identifying when, where, and how a driver makes these choices. At the same time, many of these systems relieve the driver of the tedium of performing various control tasks, assist in the monitoring the roadway environment for conflicts, and intervene to avoid imminent crashes or to stabilize control. Perhaps because ADAS technologies are doing parts of the driving task (e.g., monitoring for lane excursion, maintaining a safe headway, regulating speed), the driving task is now conceived of as many component tasks. Some of these tasks can be undertaken by technology (e.g., headway management), some remain with the driver (e.g., tactical decisions), and some are new to the driver (i.e., activation and monitoring of ADAS systems). In this new perspective, behavioral changes resulting from ADAS use are evaluated with reference to a baseline driving condition in which the ADAS technology is not available. Often this involves identifying how much a specific activity that the driver would normally perform under manual driving has changed with automation. Do drivers use their rear-view mirrors less when blind zone detection is available? Do drivers glance less at the lane boundaries with lane-keeping assistance? Do drivers monitor forward headway less with ACC engaged?

12.1.2 Modeling Behavioral Adaptation

Early models of BA attempted to explain a driver's motivation to alter his or her behavior, casting the problem in terms of managing, or balancing various assessments of risk. While many of these models cite specific driving behaviors as evidence of risky behavior, their focus is primarily on behavioral change that results from maintaining a target risk level (Wilde, 1982), crossing a risk threshold (Näätänen & Summala, 1974), threat avoidance (Fuller, 1984), or as a general maintenance of task difficulty or behavioral responses to adjustments in target risk (Fuller, 2011).

These initial views of BA were influenced by a study by Taylor (1964) on galvanic skin response (GSR) changes during driving. He found that variation in road segments and driving conditions produced little effect on GSR. His interpretation was that GSR rate could be considered an index of subjective risk or anxiety and appeared to be independent of variation in road conditions. He argued that this is a result of drivers voluntarily adjusting their risk level by choosing "...to engage in more risk on one part of the road than on another" (p. 149). This idea proved influential in shaping theories of driver BA that followed, directing the theoretical focus on subjective risk (Carsten, 2013). Näätänen and Summala (1974) suggested that an internal subjective risk monitor continuously evaluates risk and either allows an action—when no subjective risk is detected—or inhibits behavior when subjective risk reaches a critical threshold. The theory was later reformulated as the "Zero-Risk" theory by Summala (1988).

Use of the risk construct was also employed by Wilde (1982) who suggested that drivers each have a target level of risk that they will accept. If the perceived level of risk falls below this target, drivers will act to increase it; if the perceived risk falls above this target, drivers will act to reduce it. Thus, BA occurs to maintain this target risk level, achieving a kind of equilibrium between perceived and target risks. Wilde called this *risk homeostasis theory* (RHT). It proved controversial because it suggested that, no matter what measure is taken to enhance traffic safety, it would be countered by behavioral changes that completely offset the enhancement and result in no net improvement in safety. Increased safety could only be achieved if a driver's target risk could be altered.

The development of models of BA is more completely described by Carsten (2013) and Lewis-Evans, de Waard, and Brookhuis (2013). The focus of later BA models shifted away from characterizations of behavioral change as a consequence of crossing risk thresholds or maintaining target risk levels and moved toward more subjective ideas related to feelings of risk and underlying subconscious processes. In particular, motivational theories suggest that the driver's response to a given situation lies along a continuum of perceived risk that comes from the driver understanding something about his or her own personal skill, and the demands of the present situation. When challenged, driving behavior is modified to reduce that challenge (e.g., slow down) so that the driver remains safe (Fuller, 2000; 2005; Kinnear, Stradling, & McVey, 2008; Lewis-Evans & Rothengatter, 2009; Summala, 2007; Vaa, 2007). These models of BA, however, have been critiqued by Carsten (2013) and Lewis-Evans et al. (2013) who suggest that such models are primarily descriptive and are limited in their ability to generate testable hypotheses or sufficient guidance to predict the character and kind of BA that might occur.

As touched on in the discussion of negative BA associated with automation technologies, the motivational models of BA do not explicitly identify details of the driving task that might be subject to BA effects. As automation technologies have been introduced into the vehicle, models of the driving task have involved partitioning it into those parts that can be supported with automation and those that are left to the driver. For example, conventional cruise control (CCC) originally assisted with longitudinal control by allowing the driver to set a fixed cruise speed that would be maintained by automation. The driver remained responsible to manage forward headway in a safe manner. Thus, components of longitudinal control involved speed and headway management. As technology advanced, both speed and headway management could be supported by automation, and the driver's task changed again. Although motivational models of BA explained the impetus to change driving behavior (e.g., to adjust perceived risk level or to adjust the level of skill challenge), later models of BA attempt to identify particular driving behaviors that may be subject to BA (e.g., forward road monitoring, mirror checks, turn signal use). Thus, driving is conceived as a coordinated agglomeration of complex tasks involving several cognitive, perceptual, and motor functions. Some of these functions can be supported by automation, others cannot (see e.g., this Handbook, Chapter 8); the driver is now responsible to understand which parts of the task remain his or her responsibility and which are handled by automation. Sometimes this can be a challenge—early versions of automated parking assist would steer, but leave the brake pedal application, transmission operation, and accelerator pedal application to the driver.

Even before much automation began appearing on vehicles, Michon (1985) critiqued motivational models, when he suggested that they were actually discussing "...the products of cognitive functions (beliefs, emotions, intentions) rather than such functions themselves." Michon (1979) offered a "cognitive" framework as an alternative to the existing behavioral approaches (see also, Janssen, 1979). The cognitive framework looked to production system architectures and human information processing approaches (e.g., Anderson, 1993; Anderson & Lebiere, 1998; Lindsay & Norman, 1977; Newell, 1990; Newell & Simon, 1972) that employed "cognitive" procedures to account for driver behavior. These models involved a far more granular analysis than before, identifying explicit inputs, processes, and outputs to explain driver behavior. They departed from traditional control-theoretic models (e.g., Reid, 1983) by incorporating processes like pattern matching, propositional logic, learning mechanisms, and goal-directed behavioral hierarchies to explain driving behavior. Thus, Michon casts the task of driving into a hierarchical framework that divided driving into strategic, tactical (or maneuvering), and control (or operational) tasks (Janssen, 1979; Michon, 1979). The strategic level formulates travel goals and plans, the tactical level governs deliberate maneuvers (like passing), and the control level covers automatic actions like lane tracking and speed control (see Figure 12.1). A similar hierarchy was suggested by Rasmussen (1983) in his description of the performance levels of skilled operators, dividing the operation levels into knowledge-, rule-, and skill-based behavior (similar to Michon's strategic, tactical, and control levels, respectively). Ranney (1994) endorsed this hierarchical approach, linking it

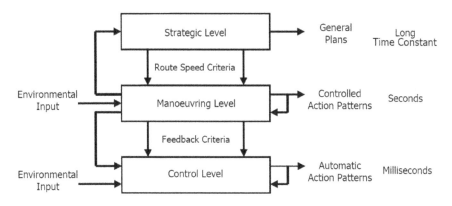

FIGURE 12.1 A hierarchical model of the task of driving. (From Michon, 1985.)

to cognitive theories of automatic and controlled processes (Schneider & Shiffrin, 1977; Shiffrin & Schneider, 1977), attention, and error, suggesting that it would lead to more fruitful and testable theories of driver behavior. Likewise, Smiley (2000) endorsed this view, suggesting that a driver's mental model of the safety system (or ADAS) would certainly influence the kind of adaptation effects observed.

To address adaptation involving ADAS, Rudin-Brown and Noy (2002) offered a qualitative model of BA that recasts the problem into a more cognitive/information-processing framework (Figure 12.2). Using a process diagram, they identified the flow of information between discrete component processes to account for factors that influence driving behavior. The model includes a driver component that models

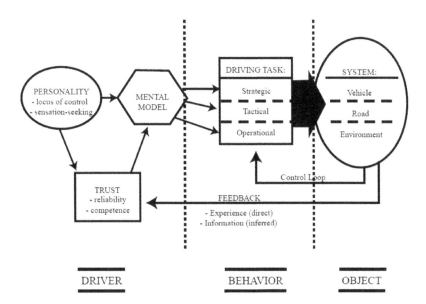

FIGURE 12.2 An early version of the qualitative model of behavioral adaptation. (Rudin-Brown & Noy, 2002.)

internal operations that affect driver behavior; a behavioral component that categorizes overt actions incorporating Michon's (1985) hierarchy; and an external world (Object) component that depicts the influence of the environment, roadway, and vehicle. On the driver side, components such as trust, the driver's mental model (of the ADAS and vehicle), as well as some personality factors such as locus-of-control and sensation seeking contribute to trigger BA (see Figure 12.2). Although modestly characterized as qualitative, the model has led to some testable hypotheses and results (e.g., Rudin-Brown & Noy, 2002; Rudin-Brown & Parker, 2004) that link personality factors such as locus-of-control and sensation seeking to BA effects. The model was later revised to include other driver factors such as gender, driver state, and age (Rudin-Brown, 2010).

12.1.2.1   How Might Behavior Change with Adaptation?

This more refined view of the driving task has several implications for understanding BA phenomena—BA effects might be observed as several behavioral changes at different levels in the driving behavioral hierarchy. For example, an adaptation might be observed at the skill or control level (identified as operational, in Figure 12.2), involving automatic action patterns that occur moment-to-moment (i.e., in milliseconds) perhaps without conscious awareness of the driver. Such automatic behavior includes regulating the vehicle's lateral and longitudinal position on the roadway. For example, when comparing drivers with and without automatic steering activated, a BA effect might be found in changes in a driver's standard deviation of lane position (SDLP) or variation in speed management when the automatic steering is not engaged (e.g., Merat, Jamson, Lai, Daly, & Carsten, 2014; Neubauer, Matthews, Langheim, & Saxby, 2012; Rudin-Brown, 2010; Sagberg et al., 1997). In this case, the adaptation effect is seen as a driver's decline in attending to lane position, resulting in more wandering in the lane.

Alternatively, adaptation might be observed at a tactical (or maneuvering) level involving controlled action patterns such as performing an overtaking maneuver, changing lanes, or preparing to exit the roadway. At this level, action selection is conscious and occurs in the temporal domain of seconds. Action patterns also include general lookout behaviors to monitor for potential conflicts in the roadway, traffic signals and signage, or gauges and indicators on the instrument cluster, as well as signaling when appropriate. For example, Ervin et al. (2005) report that drivers of vehicles equipped with ACC execute about 50% fewer passing maneuvers than drivers of vehicles under manual control. Examples of other behavioral impacts may include changes in a driver's road monitoring behavior, use of mirrors, delay in braking, or choice of the headway.

Finally, behavioral change may also occur at the strategic (or planning) level. This involves higher-level goal planning often prior to the initiation of the trip. This could include determination of the travel route and other associated goals of the trip (e.g., arrive in time for work). At this level, decisions are made at a longer temporal scale. An example of BA at this level might involve choosing a roadway that has particularly good support for vehicle automation and connectivity—e.g., clear lane marking, little construction activity, or full infrastructure connectivity.

12.1.2.2 What Specific Behaviors Might Change with Adaptation?

To determine where to look for BA phenomena after an enhancement is introduced, it is also important to know what the driver understands about how that enhancement supports the driving task. Thus, a recent focus of researches has been looking toward determining the kind of knowledge that drivers have about how these assistance systems support their driving. This knowledge is generally called the *mental model* of the system (Beggiato & Krems, 2013; Beggiato et al., 2015; Bianchi Piccinini, Rodrigues, Leitão, & Simões, 2015; de Winter, Happee, Martens, & Stanton, 2014; Lin, Ma, & Zhang, 2018; Piccinini, Simões, Rodrigues, & Leitão, 2012; Rudin-Brown & Noy, 2002; Saad, 2007; Smiley, 2000; this Handbook, Chapter 3).

We note that the mental model construct, while intuitively appealing, is somewhat imprecise and has meant different sorts of things in the psychological literature over time. The term was invoked by Johnson-Laird (1980) to explain lapses in syllogistic reasoning, and to differentiate between propositional representations and other mental representations that permit efficient derivation of other indirect physical relationships between components (e.g., Euler-circle mental representations and processes for manipulating them). It was later used to address an understanding of how software users come to understand the complex interactions with computer software (Carroll & Olson, 1988). They describe a mental model to include knowledge about how a system works, the system components, how the components are related, internal processes performed by the system, and how these processes affect the components. That is, a mental model is a representation of a physical system that can be "run" in the head to try out different actions and allow one to predict the results. Others have suggested mental models may involve some level of analogical thinking that allows one to map a familiar domain onto another domain (Gentner, 1983). We think it doubtful that drivers elaborate their representations of automation functions to this level of detail. Instead, we believe that in the context of driving and automation, the mental model construct seems to imply no more than some ability for the driver to forecast how the automation will behave, given a particular set of conditions or inputs (see also, this Handbook, Chapter 3).

The above view seems better aligned with a recent application of the predictive processing framework (Engström et al., 2018). The framework suggests that the brain is a statistical processor that seeks to minimize prediction error—it continuously predicts sensory input and seeks to minimize deviations between the predicted and actual input. The precision of these predictions improves over time; however, this time is likely affected by the variability of the sensory input and the amount of time the driver can observe this input. In the case of predictive processing of an ADAS system, the sensory input is likely affected by the variability in the conditions under which the ADAS performs and the amount of time the ADAS is operating. Rather than appeal to internal representations of physical systems, here the mental model may simply characterize the driver's ability to precisely predict what an ADAS will do, given a set of conditions. Indeed, Engström et al. (2018) suggest that the level of precision might be associated with trust. Perhaps once a driver achieves a level of predictive precision, a level of trust in the automation develops, which may then enable development of BA.

A driver's mental model of an automated system is an important issue because driving behavior that is under the voluntary control of the driver may be adjusted based on their understanding of the system (see also in this Handbook, Chapter 3). If a driver thinks that the automated system has a greater capability than it does, the driver may not quickly intervene to retake control when the system fails. Knowledge gaps about a system's competence can result in critical delays or unintentional misuse of the system. For example, Dickie and Boyle (2009) observed that some drivers of ACC systems had limited knowledge and were more likely to use it in inappropriate circumstances (e.g., curved roadways). Beggiato et al. (2015) observed that drivers of ACC systems became less aware of these limitations over time. If a driver believes that the ACC system is managing headway competently, perhaps they may be less inclined to monitor the forward direction as carefully as they might during manual driving. Similarly, misplaced trust (or *over-trust*) in automation to handle some driving tasks may result in the driver neglecting to monitor those tasks as they would normally monitor during manual driving (see also, this Handbook, Chapter 4). If drivers redirect their attentional resources in the driving task, based on their understanding of the support provided by the automation, this redirection will be reflected in their driving behavior and appear as a BA effect. This suggests that the behavior that is observed as a BA effect will be likely dependent on what the driver believes is supported by automation. That is, the observed BA effects depend on the driver's understanding of the functions provided by the ADAS, and a BA effect is observed when this understanding changes over time as with ACC.

12.1.2.3 When Does Behavioral Adaptation Occur?

The time course of BA phenomena is another question that is difficult to properly address. In many studies, a participant's exposure to automation can be relatively brief, lasting from a few short drives in a simulator (e.g., Stanton, Young, & McCaulder, 1997; Stanton, Young, Walker, Turner, & Randle, 2001; Young & Stanton, 2007a) to several weeks of use (Beggiato et al., 2015; Ervin et al., 2005; LeBlanc et al., 2006). Few studies observe BA beyond this interval. Consequently, few have observed adaptation effects over the longer term.

While some have suggested that proper use of ACC can be learned in as little as two weeks (Viti, Hoogendoorn, Alkim, & Bootsma, 2008; Weinberger, Winner, & Bubb, 2001), it is clear that even over longer periods, drivers may not have a completely accurate understanding of the system's capability (Beggiato et al., 2015). Several authors have suggested that adaptation occurs over several stages (Cacciabue & Saad, 2008; Manser, Creaser, & Boyle, 2013). Cacciabue and Saad (2008) distinguish a *learning and appropriation phase* in which the driver learns about the system, constructing a mental model of its competence and use, and developing confidence and trust in using it. This is followed by an *integration phase*, in which the driver more fully integrates the system's use into their driving. BA effects may not fully stabilize until after the integration phase has completed.

One problem that this raises is that the duration of *learning and appropriation phase* is likely to be related to the amount of exposure a driver has to the system's operation and the driver's ability to detect (or observe feedback from) its operation (Evans, 1985). Indeed, the Organisation for Economic Co-operation and

Development (OECD) expert group (OECD, 1990) distinguished three prerequisites for the adaptation to occur: (1) the driver must receive feedback from the system, (2) the driver must have the ability to alter behavior, and (3) the driver must be motivated to alter behavior.

Different driver assistance systems will operate with different degrees of obviousness to a driver. For example, automatic emergency braking (AEB) is likely to intervene on behalf of the driver less frequently than ACC. Consequently, the driver may receive much less feedback about AEB capabilities than they do about ACC and either never show any behavioral changes to AEB or change very slowly. The level of exposure to the operation of ACC may also differ substantially between two drivers if one driver primarily drives on uncongested roads or surface streets and another drives primarily in congested areas. The former driver is unlikely to learn about ACC as quickly as the latter. Lane-keeping support that continuously attempts to place the vehicle in the center of the lane is likely to intervene more frequently than lane departure prevention that only intervenes when the vehicle nears a lane edge. Consequently, it is arguably the case that drivers of vehicles equipped with lane centering systems are more likely to learn about them more quickly than drivers of lane departure prevention systems. A driver's ability to learn about the limitations of assistance systems is also likely to be limited if the opportunities to observe their limitations are rare. For example, it might take a long time for a driver to fully understand how limited the braking authority of an ACC might be.

Finally, each level in the driving skill hierarchy may involve different rates of adaptation. For example, behavior at the control level may adapt more quickly to changes than at the tactical or strategic level. Thus, alteration in management of headway or lane position might be observed before changes in a driver's choice of tactical maneuvers, which likely involves the integration phase.

Thus, unlike earlier investigations when BA phenomena were conceived as adjustments in the degree of risky driving actions to offset general reductions in crash risk, adaptation to systems that automate parts of the driving task (such as ADAS) may produce one or more adaptations at different levels in the driver's task hierarchy and affect different clusters of driving-related tasks over different periods of time. Such effects may differ significantly depending on the kind of automation introduced.

12.2 BEHAVIORAL ADAPTATION TO ADVANCED DRIVING ASSISTANCE SYSTEMS

In this section, we review the kinds of adaptation effects observed in the use of different types of ADAS. In the landscape of ACIVs, ADAS-equipped vehicles might be considered different forms of part-task automation of driving. Currently, many ADAS devices are different forms of Level 1 automation (e.g., they provide a control feature that automates steering or braking) and some provide Level 2 capabilities (e.g., they offer two forms of control support simultaneously). Some ADAS devices provide warnings and advice about potential threats and are not considered automation within the SAE J3016 levels of automation framework (SAE, 2018). For example, devices that warn drivers about potential collision conflicts could also be considered a form of lookout automation. They can enhance a driver's ability to detect conflicts or ensure that a conflict is not overlooked. BA to these capabilities may arise in the form of changes in the thoroughness with which a driver monitors the roadway environment around the vehicle. For example, Young, Fisher, Flynn, Gabree & Nadler (2018) show that drivers of Pedestrian Collision Warning Systems appear less likely to monitor the roadway for latent threats. However, warning systems are not widely thought of as automation.

Warning systems that extend the driver's ability to detect potential threats also share features with connected applications. For example, an in-vehicle night vision system that detects and warns drivers about the presence of pedestrians near the nighttime roadside in darkness might be considered similar to a connected application that warns about icy conditions ahead. In both cases, the driver is supplied with information about roadside conditions that cannot be detected directly. The only difference is that, in the first case, sensing is done by the vehicle and, in the second case, sensing is provided by external communication sources. Similar BA effects might be observed in both. For example, if drivers become too reliant that information about conflicts will be served up at appropriate times, the inclination to proceed cautiously (and slowly) through populated areas or in freezing conditions may decline.

In the next sub-sections we will discuss BA phenomena as it relates to control automation and warning systems.

12.2.1 ADAPTATION TO AUTOMATION OF LONGITUDINAL AND LATERAL CONTROL

By far the largest body of research on BA to ADAS has looked at the automation of longitudinal (i.e., speed and headway) and lateral (i.e., steering) control. In the next sections we first describe some details related to each form of automated control and then discuss some of the BA phenomena associated with the use of each of the forms of automated controls.

12.2.1.1 Automation of Longitudinal and Lateral Control

Longitudinal Control. Longitudinal control was first introduced as cruise control, a means of maintaining a selected speed over a long travel distance, relieving the driver of the task of monitoring speed while holding the foot in a fixed position on the accelerator pedal over a long period of time. This original form of longitudinal control is called CCC (conventional cruise control); an enhanced form that includes forward-looking sensors is called ACC (adaptive cruise control). ACC can detect and manage headway to a lead vehicle.

To control ACC, a driver selects both a target speed and a headway time that the vehicle will maintain in car-following situations. Headway to a lead vehicle is managed using forward sensors that detect distance and closing speed to a lead vehicle. Vehicle speed is regulated to maintain the selected headway or a target speed. Normally, headway time is set between 1 and 2 seconds—that is, for a given travel speed, the time it takes for a following vehicle to reach the current location of the end of the lead vehicle is the time headway. When not preceded by a lead vehicle, ACC operates like CCC, regulating speed alone.

Most ACC systems have limitations that could be worrisome for drivers. In particular, the sensor systems used for forward detection may not detect small forward objects such as pedestrians, bicycles, animals, or motorcycles. On curvy segments of road, forward objects may not be properly aligned with the radar such that the radar mistakes an object in an adjacent lane as a forward object. It might also fail to detect a forward object outside of the radar's field of view. Most high-speed ACC systems do not detect forward objects that are moving slowly or stopped. The stopping authority of most ACC systems is limited—many are incapable of braking above 0.3 g and will fail to avoid a forward collision in this situation. Finally, ACC performance deteriorates in snowy, rainy, or foggy weather, making the forward detection unreliable. These limitations are a concern, particularly if drivers are unaware of them and are responsible to intervene when they arise.

Lateral control. Active management of lane position is an outgrowth of LDW technologies that rely on lane markings to detect when a vehicle is crossing a lane or road boundary. Instead of simply warning the driver that a boundary is being approached, systems can now steer the vehicle back into the lane. For convenience, we will generically refer to them as active steering (AS) systems, although some intervene when the vehicle nears the lane boundary and others continuously center the vehicle in the lane. AS automation is another step toward fully autonomous vehicle control. When paired with ACC, it has been variously branded as Super Cruise (General Motors), Autopilot (Tesla), and Pilot Assist (Volvo). Most AS systems provide limited steering authority and cannot fully manage lateral position at high speed on high-curvature roadways.

Like ACC, AS systems are limited by the degree to which they can detect lane markings using video, laser, or infrared sensors. The system does one simple thing: it maintains the vehicle position between two lane detected boundary lines. If lane markings are obscured by snow, road wear, or other debris, the AS system will fail, returning lateral control to the driver. AS systems will not detect obstacles in the center of the roadway—no attempt will be made to avoid debris in the roadway. They detect lane lines, not objects. When lane boundaries are distorted by extreme roadway geometry (e.g., high curvature) or, if complicated line patterns are drawn in the road (as in a construction zone), the system's ability to maintain lane position deteriorates.

Most research on AS involves simulator studies. We include discussion of AS systems along with ACC because the two systems are frequently paired together in simulator studies that investigate progressively greater levels of automation (e.g., Carsten, Lai, Barnard, Jamson, & Merat, 2012; Jamson, Merat, Carsten, & Lai, 2001; Jamson, Merat, Carsten, & Lai, 2013; Merat, Jamson, Lai, & Carsten, 2012; Merat et al., 2014; Stanton & Young, 1998; Young & Stanton, 2002, 2007a; b).

12.2.1.2   Changes in Driver Event Detection Related to Control Automation

12.2.1.2.1   Short-Term Changes

The first studies of control automation began in the late 1990s with ACC. This work was specifically focused on how ACC use affected a driver's management of headway and travel speed, and its effect on the driver's preparedness to step in and take control of the vehicle when circumstances exceeded the ACC's performance boundaries. Using mostly simulator studies, ACC-equipped drivers were slower than manual drivers to react to critical traffic situations like the abrupt braking of a lead vehicle, unexpected cut-ins, the sudden appearance of a stationary vehicle in the travel path, or system failures (Bianchi Piccinini et al., 2015; de Winter et al., 2014; Hoedemaeker & Brookhuis, 1998; Larsson, Kircher, & Andersson Hultgren, 2014; Nilsson, 1995; Stanton et al., 1997; Stanton et al., 2001; Vollrath, Schleicher, & Gelau, 2011; Young & Stanton, 2007a). For example, Nilsson (1995) observed late braking among ACC-equipped drivers when approaching a stationary queue compared with manual drivers. Stanton et al. (1997) observed 4 of 12 drivers fail to retake control of their vehicle when the ACC system abruptly accelerated into a forward vehicle. Hoedemaeker and Brookhuis (1998) observed both larger brake force maximums and smaller minimum headway times for drivers of vehicles equipped with ACC. Larsson et al. (2014) observed longer brake reaction times (BRT) in response to cut-ins when using ACC, compared with manual driving. A similar pattern was also observed in a test-track study of Rudin-Brown and Parker (2004) where drivers with ACC took about 0.6–0.8 seconds longer to react to a lead vehicle's brake lights than the average 2.0 seconds when driving without ACC.

Comparable results were observed in studies of AS paired with ACC. Collectively, the pairing is referred to as highly automated driving (HAD) (e.g., de Winter et al., 2014; Merat et al., 2014). Strand, Nilsson, Karlsson, and Nilsson (2014) found more hard braking and collisions among HAD drivers than ACC-only drivers under conditions of automation failure. Merat and Jamson (2009) also found that drivers braked about 1.5 seconds later to respond to a forward vehicle braking with HAD compared with manual driving. In a meta-analysis, de Winter et al. (2014) report that most of the evidence suggests that HAD and ACC evoke "...long response times and an elevated rate of (near-) collisions in critical events as compared to manual driving" (p. 208).

It is unclear whether these effects are a consequence of BA, or simply the relative ignorance about the functional characteristics of an unfamiliar ADAS. As discussed earlier, BA is thought to stabilize after the *learning and appropriation phase,* during an *integration phase* (Cacciabue & Saad, 2008). This suggests that BA may take time to develop, however many of the above results are generated shortly after a driver is introduced to the system for the first time. That is, drivers are often relatively new to ACC and AS capabilities and the response to the critical event that is used to evaluate adaptation, occurs only after a brief period of exposure to these systems. In the driving simulator studies of Stanton and Young, trials with different levels of ACC lasted between 10 and 20 minutes and were preceded by about 5 minutes of practice (Stanton et al., 1997; 2001; Young & Stanton, 2007a); the simulator trials in Hoedemaeker and Brookhuis (1998) each lasted about 15 minutes. Although many later simulator studies use longer periods of ACC and AS exposure, they are not longer by much. For example, simulator studies by University of Leeds researchers employed about 45 minutes of simulator practice, followed by experimental trials that lasted 45 minutes each (Carsten et al., 2012; Jamson et al., 2001; Jamson et al., 2013; Merat et al., 2012; Merat et al., 2014). Based on track length and travel speed, other simulator experimental trials appear to exceed 30 minutes (Beggiato & Krems, 2013; Bianchi Piccinini et al., 2014; 2015; Vollrath et al., 2011).

Similar exposure levels were used in the test track study (Rudin-Brown & Parker, 2004), where exposure to the ACC system involved a briefing on ACC operation and a 30-minute warm-up session on the track. There were two 30-minute experimental trials with the ACC active.

If most studies of ACC and HAD involve drivers who are relatively unfamiliar with these systems, apart from an initial briefing and test drive, is it surprising that drivers are less prepared to respond to critical events that involve what could be characterized as different forms of ADAS failures? Perhaps participants' limited exposure to the ADAS in these studies reveal more about their *misunderstanding* about the system's limitations, than they do about ADAS BA effects. That is, should behavior that may be the result of a poor understanding of an ADAS be considered a BA effect? In the earlier theoretical discussion of BA mechanisms, the driver's mental model was identified as one of the several components that influence BA. While it is important to first have a mental model of an ADAS, it is also important to understand that both the objective accuracy of the model and the driver's level of confidence or trust in his or her model also play a role in adaptation.

12.2.1.2.2 Long-Term Changes

A driver's understanding of the functioning of ADAS has been explored by studying drivers with ACC experience (e.g., Bianchi Piccinini et al., 2015; Larsson et al., 2014). Comparing both experienced and novice ACC users, Larsson et al. (2014) found both groups had slower BRT with ACC automation, compared to full manual control, and the effect was smaller for experienced ACC users. A comparison of experienced and novice ACC users also revealed that experienced users were faster to respond than novices. These results suggest that experience with ACC can influence the degree to which drivers respond to unpredictable events.

Two studies directly examined a drivers' trust, acceptance, and mental models of ACC. Beggiato and Krems (2013) conducted a simulator study involving three separate drive sessions over a six-week period in which drivers were given different briefings about the ACC's functional capabilities. The *correct* group of drivers was given accurate information about the ACC that included details related to the system's difficulty detecting small vehicles, functioning in adverse weather conditions, and its management around narrow road bends. The *incomplete* group was provided with a basic functional overview of ACC, but was not advised of the ACC problem areas. An *incorrect* group was given the same information as the *correct* group, but was also given erroneous information that the system had problems with large vehicles and with white/silver cars. Later, mental models were assessed using questionnaires immediately following the ACC briefing, and after each experimental trial. Over time, the three groups' mental models converged. Notably, non-occurring, non-experienced failures originally called out in the descriptions (e.g., white/silver cars, large vehicles) were forgotten, while unexpected experienced failures (*incomplete* group) led to quick adjustments of the mental model toward the correct group. A follow-up on-road study (Beggiato et al., 2015) was also conducted among drivers with no prior ACC experience. Participants drove an ACC-equipped test vehicle in ten drives over a two-month period. They were initially given a complete description of the ACC function, which included specific details about ACC problem areas (i.e., detection of small vehicles, operation in adverse weather, detection of stationary objects, performance on curved roads). The drivers' mental models of the ACC were probed after drives 1, 2, 3, 5, and 10. Among the most interesting results is that an individual driver's awareness of ACC limitations drifted over time. Initially a strong disagreement with the statement "ACC detects stationary objects" drifted toward agreement by the third session and began to decline toward disagreement by the tenth session. This rise and fall effect is attributed to the driver's evolving experience with ACC—it might take time to directly encounter the ACC limitations called out in the owner's manual, before drivers adjust their mental model to include this. That is, learning about a system's capabilities is possibly mediated by a mechanism that minimizes predictive errors (e.g., Engstrom et al., 2018). Thus, only by observing deviations from an expected outcome will a driver amend their mental model of a system's function.

Lack of awareness about ADAS exceptions has also been noted in several survey studies of users of ACC. Jenness et al. (2008) and AAFTS (2008) report that drivers of ACC-equipped vehicles are generally unaware of the ACC's limitations and overestimate its ability to prevent collisions. There are some suggestions that driver awareness improves over time—for example, with prolonged use, drivers are less likely to be identified as unaware about their ACC function (Dickie & Boyle, 2009); prolonged use appears to be associated with greater awareness of system limitations (Larsson, 2012). It is also notable that when special measures are taken to inform drivers that an automation control has exceeded a bound, drivers take control more effectively (Lee, McGehee, Brown, & Marshall, 2007).

One interpretation of these results is that drivers are not so much slow to intervene because of BA to the ADAS, as they are mistaken about the ADAS's capabilities. Does it matter whether BA phenomena or a flawed mental model are responsible for undesirable behavior? Perhaps it does, if drivers act based on their mental model of ADAS operation, their confidence that their understanding is accurate, and their trust that the system will behave reliably, is presumably very strong. A faulty model, overconfidence in the model, or misplaced trust in the ADAS can result in a delayed response to assume control when the ADAS malfunctions or reaches a performance limit. It may be simpler to address a misunderstanding about a device, than a BA that optimizes a hidden goal. For example, a driver may configure an ACC to support overtaking maneuvers by selecting a higher-than-normal travel speed so that, when car following, the driver's vehicle speed matches the forward vehicle because it is regulated by headway. If the vehicle switches to an unoccupied lane, the speed of the vehicle will increase and the originally followed vehicle may be passed. This is an instance of tactical BA, based on the driver's knowledge of the ACC's function. The BA could adversely affect safety if the forward vehicle switches lanes removing the speed regulation, resulting in the vehicle accelerating to the high setting. Alternatively, a BA may also occur if a driver acts on a faulty understanding of the system's capability. For example, if the driver is confident that the ACC can detect all sized objects in the forward direction, forward vehicle headway may not be sufficiently monitored while following smaller vehicles that are only marginally detected by the ACC system. This could delay the driver's intervention if ACC fails to detect a forward target. The basis of the two kinds of BAs differs: one is based on clear knowledge of how ACC headway can act as a governor of travel speed; the other is based on incomplete awareness of the ACC's ability to detect small objects. It is unclear whether the accuracy of a driver's mental model should affect whether a behavioral adjustment is considered a BA. It seems BA results from the level of confidence (or trust) in the automation—regardless of whether that confidence (trust) is misplaced.

12.2.1.2.3 Driver Trust

Many studies of trust in automation have looked at the use of ACC. They suggest that both experienced and novice users over-trust automation, that trust increases with exposure, and that trust appears insensitive to failure (Itoh, 2012; Rudin-Brown & Parker, 2004). Similar results have been reported for inaccurate LDW systems (Rudin-Brown & Noy, 2002). When drivers are initially given many details about ACC operational exceptions, some of which are incorrect, their rated trust starts low; however, it increases over successive exposures to ACC (Beggiato & Krems, 2013). If drivers are given incomplete information about ACC function, trust starts high but declines with exposure to unanticipated boundary conditions—i.e., an unmistakable deviation from expectation that likely triggers an amendment to an internal model (Engstrom et al., 2018). Trust increases over sessions according to a power law (Beggiato et al., 2015) leveling out by about the fifth exposure session, although it would seem that this development may be affected by the variability in the predicted and observed system behavior (Engstrom et al., 2018). This suggests that trust in an ADAS will quickly grow with exposure alone and may become inappropriately high if the driver has limited or no experience with boundary conditions that reflect the system's performance variability over a wider operating range. When the operating range is limited, gaps or inconsistencies may develop in the driver's mental model.

12.2.1.3 Control Automation and Changes in Driving Behavior—Control Management

Early reports of BA to ACC suggested that drivers increased their travel speed and adopted shorter time headways to lead vehicles. Using a simulator, Hoedemaeker and Brookhuis (1998) observed driving speed and headway with and without ACC in two driving sessions. Average speed in the no-ACC condition was 107 kmph, and with ACC it increased to 115 kmph. Smaller minimum time headway was also observed with ACC. Similar results were reported in a test track study by Ward, Fairclough, and Humphreys (1995). Reduced minimum time headways were also observed in a simulator study of HAD by Merat and Jamson (2009). In contrast, Bianchi Piccinini et al. (2014) found longer headways and lower speeds among ACC users, especially for experienced ACC drivers. Stanton et al. (1997) reported no effects on either speed or headway. Hoedemaeker and Kopf (2001) reported lower speeds with ACC and no effects on headway in an ACC-enabled test vehicle. In a field test, Ervin et al. (2005) reported systematic reductions in the incidence of short headways (less than 1 seconds) with ACC use. Finally, in another field test, (Viti et al., 2008) reported increases in average headway, smaller headway variability, lower speeds, and smaller speed variability. Other than Stanton et al. (1997), no safety-compromising changes in speed or headway were found in either field tests (Ervin et al., 2005; Viti et al., 2008), studies of experienced users of ACC (Bianchi Piccinini et al., 2014), or in drives on public roads (Hoedemaeker & Kopf, 2001).

It is possible that the brief exposure times in the early studies may be responsible for the results, such that the effect is diminished with increased exposure.

Control automation may also affect a driver's tactical decisions, although there are few reports of this. Ervin et al. (2005) found that drivers using ACC linger behind forward vehicles twice as often as they do when ACC is inactive. Similarly, Jamson et al. (2001; 2013) report that in highly automated conditions, drivers refrain from behaviors that would require temporarily retaking manual control of the vehicle. Perhaps this occurs because of a disincentive to turn off automation once it is activated. For example, perhaps automation setup takes time and effort, or the workload reduction accompanying automation is attractive.

12.2.1.4 Control Automation and Driver Workload

Driver workload is principally measured in two ways: directly, using questionnaires which elicit subjective reports, and indirectly using secondary task performance which can be more objective. Subjective workload is most commonly measured using questionnaires like the NASA-Task Load Index (NASA-TLX) (Hart & Staveland, 1988) or the Rating Scale of Mental Effort (RSME) (e.g., Hoedemaeker & Brookhuis, 1998). Other customized scales have also been used (e.g., Ma & Kaber, 2005). Secondary task performance is used as an indirect measure of workload. It assumes that as the principal task workload lightens, the driver has greater mental resources which are available to apply to a secondary task (see also, this Handbook, Chapter 6). Consequently, secondary task performance increases as workload decreases.

In a comprehensive review of studies of ACC and HAD systems on driver workload and situation awareness (SA), de Winter et al. (2014) reported that subjective workload measures revealed small reductions were found comparing manual to ACC-equipped driving, and comparisons between manual and HAD driving found that workload was approximately half.

Secondary task performance measures show similar effects. In self-paced visual tasks (e.g., Rudin-Brown & Parker, 2004; Stanton et al., 1997; Young & Stanton, 2004; 2007a) secondary task performance improved when automation was introduced. In the meta-analysis of de Winter et al. (2014), self-paced secondary task performance was only marginally improved with ACC (+12%), but substantially improved with HAD systems (+152%). Control automation was less effective decreasing the workload associated with non-visual tasks. Little difference was found between more cognitively demanding tasks such as a twenty-question task (Merat et al., 2012) or an auditory/verbal task (Seppelt & Lee, 2007).

Assessments of workload using questionnaires do not always agree with measures of secondary task performance. Stanton et al. (1997) noted that while Nilsson (1995) found that drivers reported no difference in workload using ACC assessed using the NASA-TLX questionnaire, he found significantly better secondary task performance in the ACC condition. Stanton suggested that differences in each study's road geometry might be responsible (Stanton & Young, 1998). When roads are straight and longitudinal control is the driver's main control task, ACC provides significant assistance; when roads are curved and the lateral control task dominates control, ACC provides comparatively smaller reduction in workload compared to manual driving.

Because non-visual secondary tasks show little difference in performance in manual versus automated driving, it might be suggested that this is because the tasks do not rely on the redirection of the driver's visual attention off the roadway to perform the task. Reduction in road monitoring may be responsible for the observed declines in workload measures. Consistent with this, Carsten et al. (2012) noted that drivers redirected glances away from the roadway more under lateral control than under longitudinal control. Automation of lateral control produced dramatically greater workload reductions than longitudinal control alone (Carsten et al., 2012; Jamson et al., 2013; Stanton & Young, 2005; Young & Stanton, 2007a; b), perhaps because lateral control requires more regular monitoring of the roadway.

12.2.1.5 Control Automation and SA

Gugerty (2011) defines SA in the context of driving as, "The updated, meaningful knowledge of an unpredictably-changing, multifaceted situation that operators use to guide choice and action when engaged in real-time multitasking." Endsley (1995) distinguished three levels of SA: Level 1—Perception of Elements in the Environment; Level 2—Comprehension of the Current Situation; and Level 3—Projection of Future Status (see also this Handbook, Chapter 7). Simply put, SA means "knowing what is going on around you."

Drivers need to maintain a good level of awareness about the roadway environment: they must monitor for potential conflicts in many directions, pedestrians, cyclists, traffic signs, traffic signals, other vehicle signals, their vehicle's speed, lane position, headway, and other vehicle states displayed on the instrument cluster. When automation is introduced that supports other management functions like headway and lane position, it is possible for the driver to redistribute attention differently during driving. If SA measures target one of the areas to which the driver has redistributed attention, automation will appear to enhance SA. For example, pedestrian hazards may be better detected (Funke, Matthews, Warm, & Emo, 2007), and performance using SA Global Assessment Technique (SAGAT) queries shows improvement while supported by ACC. However, the drivers in the Funke et al. (2007) study were also aware that they could be asked about pedestrians and perhaps were motivated to devote some resources to monitor this.

When drivers are given less implicit direction about how to redistribute their attention, results are less positive. In the presence of both lateral and longitudinal control, drivers often take up non-driving-related secondary tasks, if available (Carsten et al., 2012; Llaneras, Salinger, & Green, 2013). The choice in how to redistribute their freed capacity could easily lead to drivers losing awareness of what is going on around the roadway and being slow to intervene when an unexpected situation arises.

12.2.2 ADAPTATION TO ADAS WARNING SYSTEMS

There are fewer studies of BA to ADAS that provide only warnings to drivers compared with those that provide active control support. Warning systems present a visual, audible, or haptic display that is triggered when a specific condition arises or limit is reached that the driver should be aware of to take appropriate action. For example, curve speed warnings (CSW) produce an audible alert when a driver approaches a curved road segment at a high rate of speed; LDWs are generated when a lane boundary is crossed, using either a haptic pulse or a rumble-strip sound; forward collision warnings (FCW) produce audible alerts when headway or time-to-collision to a forward vehicle reaches an established limit; and intelligent speed adaptation (ISA) systems can either provide warnings to the driver or intervene to control speed to conform to the posted speed limits. As mentioned earlier, some warning systems might be thought of as a form of connected vehicle application that relays supplemental information to the driver that may not be directly observable.

12.2.2.1 Adaptation to ISA

Several versions of ISA were developed in Europe in the late 1990s to address the problem of speeding (Brookhuis & de Waard, 1999). They vary primarily with respect to the degree of control they give a driver to modulate their function. Some simply advise a driver about the speed limit of the current roadway; others can be engaged by the driver to prevent exceedance of the current speed limit; others automatically engage and do not allow drivers to disengage; and finally, others may also incorporate information about weather and current road conditions to adjust speed to suit both posted speed and real-time conditions. Comte (2000) reports a simulator study on driver BA in three variations of ISA. Gap acceptance in turns and headway was generally reduced among drivers of ISA systems that constrained speed in a way that the driver could not override. This suggests that impatience or frustration with the system may have prompted drivers to tactically decrease their gap acceptance in turns and headway to make up for the perceived delay in travel speed. No other effects were observed in the performance of other driving maneuvers such as overtaking, incidence of traffic violations, or responses to surprise events.

In a field test of an ISA variant that provided speed regulation advice in the form of accelerator pedal counterforce against the driver's pedal application, speed reductions were observed without compensatory speed boosts around intersections (Varhelyi, Hjalmdahl, Hyden, & Draskoczy, 2004). Similar speed reductions have been observed in other field studies as well in which drivers are permitted to override the ISA (Lai & Carsten, 2012; Lai, Hjälmdahl, Chorlton, & Wiklund, 2010). ISA overriding appeared to increase with exposure. This was particularly true of the drivers who were likely to intentionally exceed the speed limit.

12.2.2.2 Lane Departure Warning

BA to LDWs was examined in both simulator and track studies by Rudin-Brown and Noy (2002). Their studies examined whether trust develops differently among drivers who differ in measures of locus-of-control and sensation seeking. Drivers used an accurate or inaccurate LDW (which generated occasional false-positive warnings and detection failures one-third of the time) while performing a secondary destination entry task on a navigation system. Performance was compared with a baseline in which the LDW was inactive. In a post-test session each group drove using the inaccurate LDW system. Trust increased with exposure for all groups, regardless of system accuracy. In the simulator study, drivers who completely departed the roadway when the LDW failed to report a lane departure were those who exhibited the highest levels of trust in the LDW, suggesting that they may have developed a greater reliance on the system to keep them informed than those who did not depart the roadway.

12.2.2.3   Forward Collision Warning

Few studies have examined BA in response to FCW, probably because FCW is usually bundled with ACC, and the adaptation effects of ACC would thus be confounded with those of FCW. While there is a general consensus that FCW provides a net benefit to drivers, FCW events are rare and unlikely to provide drivers with sufficient direct experience to produce any adaptation effects. Engagement in secondary tasks (measured as eye gaze behavior) has been used to assess warning-based BAs. Muhrer, Reinprecht, and Vollrath (2012) examined an enhanced FCW that also performed emergency braking. Although there were concerns that autonomous braking could lead to increased involvement in secondary tasks, no evidence of this was found. There has been a report of post-FCW negative adaptation by Wege, Will, and Victor (2013) using a brake-capacity FCW (B-FCW) system—a B-FCW system warns the driver of an impending collision if the situation exceeds the braking capacity of an ACC system. That is, it lets the driver know that footbrake intervention will be necessary to avoid a forward collision. In scenarios involving distracted drivers, they found that in the post-threat-recovery-period, after the B-FCW event, distracted drivers would glance from the road to the instrument cluster area where the warning was initiated. Perhaps this was an attempt to better understand the reason for the warning or to confirm that the threat is no longer present. Such behavior might be undesirable if drivers should continue to monitor the road in the aftermath of the braking event.

12.2.2.4   Combination Discrete Warning Systems

Adaptation effects have also been investigated in field studies of combination systems, including LDW systems in combination with other warning systems. LDW is often combined with FCW, CSW, and side-collision warnings like blind spot detection systems. A field test of the road departure crash warning (combination) system (RDCW) examined an LDW system in combination with CSW (LeBlanc et al., 2006; 2007; Wilson, Stearns, Koopman, & Yang, 2007). Behavioral changes observed with the system active included increased lane change signaling and reduced incidence of close proximity to the lane edge. Reduced lateral acceleration on ramps was also observed, although no effects were observed along other curved road segments. No change was observed in the driver's inclination to engage in secondary task activity.

The integrated vehicle-based safety system (IVBSS) field test also examined a combination of discrete warning systems that included FCW, LDW, lane change/merge warning (LCM), and CSW (Sayer et al., 2011; Nodine, Lam, Stevens, Razo, & Najm, 2011). Several performance comparisons were made to a baseline period when the system was inactive. Fewer lane departures (unintended deviations out of the lane), shorter lane departure durations, increased turn signal use, marginal reduction in lane offset, and increased lane changes (intentional deviations out of the lane) were observed when the system was activated. No changes were observed in drivers' engagement in secondary activities (e.g., eating, drinking, and cell phone use). Thus, in both field tests, generally positive BAs were observed.

A field test of another integrated warning system, called the Continuous Support system (Várhelyi, Kaufmann, & Persson, 2015), included devices that advised drivers about speed limit exceedance, CSW, forward collision, and warnings about vehicles in the blind zones. The on-road study involved two 45-minute drives along a prescribed route: once with the system turned on and once with the system off. There were few observed changes in driver behavior. No changes were found in speed or headway management, although curve speeds appeared to be lower when the system was active. Some negative adaptation outcomes were also observed—turn speeds through intersections were higher, and drivers appeared to come dangerously close to the sides of the road more frequently when the system is active.

12.3   CONCLUSION

This review of BA phenomena hints at several challenges human factors researchers face when looking for the impact of any change to the driving task on how a driver will adjust the way the vehicle is managed on the road. The challenges reviewed include identifying which of the many-component driving behaviors might be changed with the introduction of automation. Not only might there be many task components, but also they may occur across different levels of the driving task hierarchy. For each component, it will be important to characterize how the change is manifest, how long each takes to develop and stabilize, and what the ultimate consequence of this change is for safety. In some situations, the BA may occur at an executive level that may have downstream effects. For example, if a driver changes the way spatial attention is allocated during driving, we may find that the driver does not monitor the forward areas of the roadway as much as before, perhaps because it may not be deemed necessary since the automation is covering this. Perceived reduction in monitoring demand may enable a driver to redirect attention to other tasks that may be considered more important and easily managed. Some of these tasks may not be driving-related and ultimately result in diminished SA (see also, this Handbook, Chapters 7, 9). This diminished SA (or a distracted driver) might be considered an indirect byproduct of BA of attentional distribution.

In our view, BA is likely instigated by reaching an adequate level of trust (or confidence) in one's ability to predict how automation will respond in a variety of situations. How long it might take to reach this level is likely driven by direct experience using and observing the automation in action, as well as an individual's personal perception of what constitutes "adequate." Early studies of BA to vehicles equipped with airbags demonstrate that sometimes it is sufficient to simply know a capability is present without necessarily witnessing its operation, for a BA effect to materialize. In other cases, reaching the "adequate" level of trust may require direct experience of the system in operation over a longer period of time, especially if the system is more complex. As suggested by others, if the system is observed in a relatively narrow operational domain, trust may develop quickly (Engstrom et al., 2018). However, in this case, the driver may be left with little experience of the system performing at the edges of its capabilities. With high confidence and limited experience, BA effects may appear that look like misuse or over-reliance on the system.

With so many potentially interacting pieces to the BA puzzle, it will be a challenge to establish sufficient control over a driver's experience with automation to ensure that a coherent prediction can be made of what, when, and how BA will occur when different kinds of automation technologies are introduced into driving. And even when there is sufficient control, there are still differences of opinion as to whether it is behavior adaptation, or something else, that is the cause of a given behavior.

REFERENCES

AAAFTS. (2008). *Use of Advanced In-Vehicle Technology by Young and Older Early Adopters*. Washington, DC: AAA Foundation for Traffic Safety.

Anderson, J. R. (1993). *Rules of the Mind*. Hillsdale, NJ: Lawrence Earlbaum Associates.

Anderson, J. R. & Lebiere, C. (1998). *The Atomic Components of Thought*. Mahwah, NJ: Lawrence Earlbaum Associates.

Aschenbrenner, K. M., Biehl, B., & Wurm, G. W. (1988). *Is Traffic Safety Improved Through Better Engineering? Investigation of Risk Compensation with the Example of Antilock Brake Systems* [German]. Mannheim, Germany: BASt.

Ball, K., Owsley, C., Stalvey, B., Roenker, D. L., Sloane, M. E., & Graves, M. (1998). Driving avoidance and functional impairment in older drivers. *Accident Analysis and Prevention, 30*(3), 313–322.

Beggiato, M. & Krems, J. F. (2013). The evolution of mental model, trust and acceptance of adaptive cruise control in relation to initial information. *Transportation Research Part F, 18*, 47–57.

Beggiato, M., Pereira, M., Petzoldt, T., & Krems, J. (2015). Learning and development of trust, acceptance and the mental model of ACC. A longitudinal on-road study. *Transportation Research Part F, 35*, 75–84.

Bianchi Piccinini, G. F., Rodrigues, C. M., Leitão, M., & Simões, A. (2014). Driver's behavioral adaptation to adaptive cruise control (ACC): The case of speed and time headway. *Journal of Safety Research, 49*, 77.e71–84.

Bianchi Piccinini, G. F., Rodrigues, C. M., Leitão, M., & Simões, A. (2015). Reaction to a critical situation during driving with adaptive cruise control for users and non-users of the system. *Safety Science, 72*, 116–126.

Brookhuis, K. & de Waard, D. (1999). Limiting speed, towards an intelligent speed adapter (ISA). *Transportation Research Part F: Traffic Psychology and Behaviour, 2*(2), 81–90.

Cacciabue, P. C. & Saad, F. (2008). Behavioural adaptations to driver support systems: A modelling and road safety perspective. *Cognition, Technology & Work, 10*(1), 31–39.

Carroll, J. M. & Olson, J. R. (1988). Mental models in human-computer interaction. In M. Helander (Ed.), *Handbook of Human-Computer Interaction* (pp. 45–65). Amsterdam: North-Holland.

Carsten, O. (2013). Early theories of behavioural adaptation. In C. M. Rudin-Brown & S. L. Jamson (Eds.), *Behavioural Adaptation and Road Safety - Theory, Evidence and Action* (pp. 23–34). Boca Raton, FL: CRC Press.

Carsten, O., Lai, F. C. H., Barnard, Y., Jamson, A. H., & Merat, N. (2012). Control task substitution in semiautomated driving: Does it matter what aspects are automated? *Human Factors, 54*(5), 747–761.

Comte, S. L. (2000). New systems: New behaviour? *Transportation Research Part F: Traffic Psychology and Behaviour, 3*(2), 95–111.

de Winter, J. C. F., Happee, R., Martens, M. H., & Stanton, N. A. (2014). Effects of adaptive cruise control and highly automated driving on workload and situation awareness: A review of the empirical evidence. *Transportation Research Part F, 27*, 196–217.

Dickie, D. A. & Boyle, L. N. (2009). Drivers' understanding of adaptive cruise control limitations. *Proceedings of the Human Factors and Ergonomics Society Annual Meeting, 53*(23), 1806–1810.

Eichelberger, A. H. & McCartt, A. T. (2014). Volvo drivers' experiences with advanced crash avoidance and related technologies. *Traffic Injury Prevention, 15*(2), 187–195.

Endsley, M. R. (1995). Toward a theory of situation awareness in dynamic-systems. *Human Factors, 37*(1), 32–64.

Engström, J., Bargman, J., Nilsson, D., Seppelt, B., Markkula, G., Piccinini, G. B., & Victor, T. (2018). Great expectations: A predictive processing account of automobile driving. *Theoretical Issues in Ergonomics Science, 19*(2), 156–194.

Ervin, R. D., Sayer, J., LeBlanc, D., Bogard, S., Mefford, M., Hagan, M., … Winkler, C. (2005). *Automotive Collision Avoidance System Field Operational Test Methodology and Results, Volume 1: Technical Report* (DOT HS 809 900). Washington, DC: Department of Transportation.

Evans, L. (1985). Human-behavior feedback and traffic safety. *Human Factors, 27*(5), 555–576.

Fuller, R. (1984). A conceptualization of driving behavior as threat avoidance. *Ergonomics, 27*(11), 1139–1155.

Fuller, R. (2000). The task-capability interface model of the driving process. *Recherche - Transports - Sécurité, 66*, 47–57.

Fuller, R. (2005). Towards a general theory of driver behaviour. *Accident Analysis & Prevention, 37*(3), 461–472.

Fuller, R. (2011). Driver control theory: From task difficulty homeostasis to Risk Allostasis. In B. E. Porter (Ed.), *Handbook of Traffic Psychology* (pp. 13–26). San Diego, CA: Academic Press.

Funke, G., Matthews, G., Warm, J. S., & Emo, A. K. (2007). Vehicle automation: A remedy for driver stress? *Ergonomics, 50*(8), 1302–1323.

Gentner, D. (1983). Structure-mapping: A theoretical framework for analogy. *Cognitive Science, 7*(2), 155–170.

Gibson, J. J. & Crooks, L. E. (1938). A theoretical field-analysis of automobile-driving. *American Journal of Psychology, 51*, 453–471.

Gugerty, L. J. (2011). Situation awareness in driving. In J. Lee, M. Rizzo, D. L. Fisher, & J. Caird (Eds.), *Handbook for Driving Simulation in Engineering, Medicine and Psychology.* Boca Raton, FL: CRC Press.

Hart, S. G. & Staveland, L. E. (1988). Development of NASA-TLX (Task Load Index): Results of empirical and theoretical research. In P. A. Hancock & N. Meshlcati (Eds.), *Advances in Psychology: Human Mental Workload* (pp. 139–183): North Holland: Elsevier.

Hoedemaeker, M. & Brookhuis, K. A. (1998). Behavioural adaptation to driving with an adaptive cruise control (ACC). *Transportation Research Part F, 1*, 95–106.

Hoedemaeker, M. & Kopf, M. (2001). Visual sampling behaviour when driving with adaptive cruise control. *Ninth International Conference on Vision in Vehicles* (pp. 19–22). Loughborough, UK: Applied Vision Research Centre.

Itoh, M. (2012). Toward overtrust-free advanced driver assistance systems. *Cognition Technology & Work, 14*(1), 51–60.

Jamson, A. H., Merat, N., Carsten, O., & Lai, F. (2001). Fully-automated driving: The road to future vehicles. *6th International Driving Symposium on Human Factors in Driver Assessment, Training, and Vehicle Design.* Lake Tahoe, CA.

Jamson, A. H., Merat, N., Carsten, O., & Lai, F. (2013). Behavioural changes in drivers experiencing highly-automated vehicle control in varying traffic conditions. *Transportation Research Part C-Emerging Technologies, 30*, 116–125.

Janssen, W. (1979). *Routeplanning en geleiding: Een literatuurstudie* [Dutch] (IZF 1979 C-13). Soesterberg, The Netherlands: Institute for Perception, TNO.

Jenness, J. W., Lerner, N. D., Mazor, S., Osberg, J. S., & Tefft, B. C. (2008). Use of Advanced In-Vehicle Technology by Young and Older Early Adopters. *Survey Results on Adaptive Cruise Control Systems* (DOT HS 810–917). Washington, DC: National Highway Traffic Safety Administration.

Johnson-Laird, P. N. (1980). Mental models in cognitive science. *Cognitive Science, 4*(1), 71–115.

Kallberg, V.-P. (1993). Reflector posts - signs of danger? *Transportation Research Record, 1403*, 57–66.

Kinnear, N., Stradling, S., & McVey, C. (2008). Do we really drive by the seats of our pants? In L. Dorn (Ed.), *Driver Behavior and Traning, Vol III* (pp. 349–365). Andershot: Ashgate.

Lai, F. & Carsten, O. (2012). What benefit does Intelligent Speed Adaptation deliver: A close examination of its effect on vehicle speeds. *Accident Analysis and Prevention, 48*, 4–9.

Lai, F., Hjälmdahl, M., Chorlton, K., & Wiklund, M. (2010). The long-term effect of intelligent speed adaptation on driver behaviour. *Applied Ergonomics, 41*(2), 179–186.

Larsson, A. F. L. (2012). Driver usage and understanding of adaptive cruise control. *Applied Ergonomics, 43*(3), 501–506.

Larsson, A. F. L., Kircher, K., & Andersson Hultgren, J. (2014). Learning from experience: Familiarity with ACC and responding to a cut-in situation in automated driving. *Transportation Research Part F: Traffic Psychology and Behaviour, 27*, 229–237.

LeBlanc, D., Sayer, J., Winkler, C., Bogard, S., & Devonshire, J. (2007). Field test results of a road departure crash warning system: driver utilization and safety implications. *Proceedings of the Fourth International Driving Symposium on Human Factors in Driving Assessment, Training, and Vehicle Design*. Stephenson, WA.

LeBlanc, D., Sayer, J., Winkler, C., Ervin, R., Bogard, S., Devonshire, J. … Gordon, T. (2006). *Road Departure Crash Warning System Field Operational Test: Methodology and Results. Volume 1* (UMTRI-2006-9-1). Ann Arbor: University of Michigan Transportation Research Institute.

Lee, J. D., McGehee, D. V., Brown, T. L., & Marshall, D. (2007). Effects of adaptive cruise control and alert modality on driver performance. *Transportation Research Record, 1980*, 49–56.

Lewis-Evans, B., de Waard, D., & Brookhuis, K. (2013). Contemporary models of driver adaptation. In C. M. Rudin-Brown & S. Jamsom (Eds.), *Behavioural Adaptation and Road Safety* (pp. 35–59). Boca Raton, FL: CRC Press.

Lewis-Evans, B. & Rothengatter, T. (2009). Task difficulty, risk, effort and comfort in a simulated driving task—Implications for risk allostasis theory. *Accident Analysis & Prevention, 41*(5), 1053–1063.

Lin, R., Ma, L., & Zhang, W. (2018). An interview study exploring Tesla drivers' behavioural adaptation. *Applied Ergonomics, 72*, 37–47.

Lindsay, P. H. & Norman, D. A. (1977). *Human Information Processing: An Introduction to Psychology* (2d ed.). New York: Academic Press.

Llaneras, R. E., Salinger, J., & Green, C. A. (2013). Human factors issues associated with limited ability autonomous driving systems: Drivers' allocation of visual attention to the forward roadway. *Proceedings of the 7th International Driving Symposium on Human Factors in Driver Assessment, Training and Vehicle Design*. Iowa City, IA: University of Iowa.

Ma, R. & Kaber, D. B. (2005). Situation awareness and workload in driving while using adaptive cruise control and a cell phone. *International Journal of Industrial Ergonomics, 35*(10), 939–953.

Manser, M., Creaser, J., & Boyle, L. (2013). Behavioural adaptation: Methodological and measurement issues. In C. M. Rudin-Brown & S. Jamson (Eds.), *Behavioural Adaptation and Road Safety* (pp. 35–59). Boca Raton, FL: CRC Press.

Merat, N. & Jamson, A. H. (2009). How do drivers behave in a highly automated car? *Fifth International Driving Symposium on Human Factors in Driver Assessment, Training and Vehicle Design.* Iowa City, IA: University of Iowa.

Merat, N., Jamson, A. H., Lai, F. C. H., & Carsten, O. (2012). Highly automated driving, secondary task performance, and driver state. *Human Factors, 54*(5), 762–771.

Merat, N., Jamson, A. H., Lai, F. C. H., Daly, M., & Carsten, O. M. J. (2014). Transition to manual: Driver behaviour when resuming control from a highly automated vehicle. *Transportation Research Part F: Traffic Psychology and Behaviour, 27,* 274–282.

Michon, J. A. (1979). *Dealing with Danger* (VK 79-01). Groningen, The Netherlands: Traffic Research Center, University of Groningen.

Michon, J. A. (1985). A critical view of driver behavior models: What do we know, what should we do? In L. Evans & R. C. Schwing (Eds.), *Human Behavior and Traffic Safety* (pp. 485–520). New York: Plenum Press.

Muhrer, E., Reinprecht, K., & Vollrath, M. (2012). Driving with a partially autonomous forward collision warning system: How do drivers react? *Human Factors, 54*(5), 698–708.

Näätänen, R. & Summala, H. (1974). A model for the role of motivational factors in drivers' decision-making. *Accident Analysis & Prevention, 6*(3–4), 243–261.

National Highway Traffic Safety Administration. (2015). *Critical Reasons for Crashes Investigated in the National Motor Vehicle Crash Causation Survey* (DOT HS 812 115). Washington, DC: National Highway Traffic Safety Administration.

Neubauer, C., Matthews, G., Langheim, L., & Saxby, D. (2012). Fatigue and voluntary utilization of automation in simulated driving. *Human Factors, 54*(5), 734–746.

Newell, A. (1990). *Unified Theories of Cognition.* Cambridge, MA: Harvard University Press.

Newell, A. & Simon, H. A. (1972). *Human Problem Solving.* Englewood Cliffs, NJ: Prentice-Hall.

Nilsson, L. (1995). Safety effects of adaptive cruise control in critical traffic situations. *Second World Congress on Intelligent Transport Systems: Vol. 3.* Yokohama, Japan.

Nodine, E., Lam, A., Stevens, S., Razo, M., & Najm, W. G. (2011). *Integrated Vehicle-Based Safety Systems (IVBSS) Light Vehicle Field Operational Test: Independent Evaluation* (DOT HS 811 516). Washington, DC: National Highway Traffic Safety Adminstration.

OECD. (1990). *Behavioural Adaptations to Changes in the Road Transport System* (92-64-13389-5). Paris: Organisation for Economic Co-operation and Development.

Parasuraman, R. & Riley, V. (1997). Humans and Automation: Use, misuse, disuse, abuse. *Human Factors, 39*(2), 230–253.

Piccinini, G. F., Simões, A., Rodrigues, C. M., & Leitão, M. (2012). Assessing driver's mental representation of adaptive cruise control (ACC) and its possible effects on behavioural adaptations. *Work, 41,* 4396–4401.

Ranney, T. A. (1994). Models of driving behavior: A review of their evolution. *Accident Analysis & Prevention, 26*(6), 733–750.

Rasmussen, J. (1983). Skills, rules, and knowledge - signals, signs, and symbols, and other distinctions in human-performance models. *IEEE Transactions on Systems Man and Cybernetics, 13*(3), 257–266.

Reagan, I. J. & McCartt, A. T. (2016). Observed activation status of lane departure warning and forward collision warning of Honda vehicles at dealership service centers. *Traffic Injury Prevention, 17*(8), 827–832.

Reid, L. D. (1983). A survey of recent driver steering behavior models suited to accident studies. *Accident Analysis and Prevention, 15*(1), 23–40.

Rudin-Brown, C. M. (2010). 'Intelligent' in-vehicle intelligent transport systems: Limiting behavioural adaptation through adaptive design. *Intelligent Transport Systems, IET, 4*(4), 252–261.

Rudin-Brown, C. M. & Noy, Y. I. (2002). Investigation of behavioral adaptation to lane departure warnings. *Transportation Research Record, 1803,* 30–37.

Rudin-Brown, C. M. & Parker, H. A. (2004). Behavioural adaptation to adaptive cruise control (ACC): Implications for preventive strategies. *Transportation Research Part F-Traffic Psychology and Behaviour, 7*(2), 59–76.

Rumar, K., Berggrund, U., Jernberg, P., & Ytterbom, U. (1976). Driver reaction to a technical safety measure studded tires. *Human Factors, 18*(5), 443–454.

Saad, F. (2007). Dealing with behavioural adaptations to advanced driver support systems. In P. C. Cacciabue (Ed.), *Modelling Driver Behaviour in Automotive Environments* (pp. 147–161). London: Springer.

SAE. (2018). *Taxonomy and Definitions for Terms Related to Driving Automation Systems for On-Road Motor Vehicles* (J3016 Revised 2018-06). Warrendale, PA: SAE International.

Sagberg, F., Fosser, S., & Sætermo, I.-A. F. (1997). An investigation of behavioural adaptation to airbags and antilock brakes among taxi drivers. *Accident Analysis & Prevention, 29*(3), 293–302.

Sayer, J. R., Bogard, S. E., Buonarosa, M. L., LeBlanc, D. J., Funkhouser, D. S., Bao, S., & Blankespoor, A. (2011). *Integrated Vehicle-Based Safety Systems Light-Vehicle Field Operational Test: Key Findings* (DOT HS 811 416). Washington, DC: National Highway Traffic Safety Administration.

Sayer, J. R., Buonarosa, M. L., Bao, S., Bogard, S. E., LeBlanc, D. J., Blankespoor, A., ... Winkler, C. B. (2010). *Integrated vehicle-Based Safety Systems Light-Vehicle Field Operational Test, Methodology and Results Report* (UMTRI-2010-30). Ann Arbor, MI: The University of Michigan Transportation Research Institute.

Scanlon, J. M., Kusano, K. D., Sherony, R., & Gabler, H. C. (2015). Potential safety benefits of lane departure warning and prevention systems in the US vehicle fleet. *Paper presented at the 24th International Technical Conference on the Enhanced Safety of Vehicles (ESV)*, Gothenburg, Sweden: ESV.

Schneider, W. & Shiffrin, R. M. (1977). Controlled and automatic human information-processing: I. Detection, search, and attention. *Psychological Review, 84*(1), 1–66.

Seppelt, B. D. & Lee, J. D. (2007). Making adaptive cruise control (ACC) limits visible. *International Journal of Human-Computer Studies, 65*(3), 192–205.

Shiffrin, R. M. & Schneider, W. (1977). Controlled and automatic human information-processing: II. Perceptual learning, automatic attending, and a general theory. *Psychological Review, 84*(2), 127–190.

Smiley, A. (2000). Behavioral adaptation, safety, and intelligent transportation systems. *Transportation Research Record, 1724,* 47–51.

Stanton, N. A., Young, M., & McCaulder, B. (1997). Drive-by-wire: The case of driver workload and reclaiming control with adaptive cruise control. *Safety Science, 27*(2–3), 149–159.

Stanton, N. A., Young, M., Walker, G. H., Turner, H., & Randle, S. (2001). Automating the driver's control tasks. *International Journal of Cognitive Ergonomics, 5*(3), 221–236.

Stanton, N. A. & Young, M. S. (1998). Vehicle automation and driving performance. *Ergonomics, 41*(7), 1014–1028.

Stanton, N. A. & Young, M. S. (2005). Driver behaviour with adaptive cruise control. *Ergonomics, 48*(10), 1294–1313.

Sternlund, S., Strandroth, J., Rizzi, M., Lie, A., & Tingvall, C. (2017). The effectiveness of lane departure warning systems—A reduction in real-world passenger car injury crashes. *Traffic Injury Prevention, 18*(2), 225–229.

Strand, N., Nilsson, J., Karlsson, I. C. M., & Nilsson, L. (2014). Semi-automated versus highly automated driving in critical situations caused by automation failures. *Transportation Research Part F-Traffic Psychology and Behaviour, 27,* 218–228.

Summala, H. (1988). Risk control is not risk adjustment - The zero-risk theory of driver behavior and its implications. *Ergonomics, 31*(4), 491–506.

Summala, H. (2007). Towards understanding motivational and emotional factors in driver behavior: Comfort through satisficing. In C. Cacciabue (Ed.), *Modelling Driver Behavior in Automotive Environments* (pp. 187–207). London: Springer.

Taylor, D. H. (1964). Drivers galvanic skin-response and the risk of accident. *Ergonomics, 7*(1–4), 439–451.

Vaa, T. (2007). Modelling driver behaviour on basis of emotions and feelings: Intelligent transport systems and behavioural adaptations. In P. C. Cacciabue (Ed.), *Modelling Driver Behaviour in Automotive Environments* (pp. 208–232). London: Springer.

Varhelyi, A., Hjalmdahl, M., Hyden, C., & Draskoczy, M. (2004). Effects of an active accelerator pedal on driver behaviour and traffic safety after long-term use in urban areas. *Accident Analysis and Prevention, 36*(5), 729–737.

Várhelyi, A., Kaufmann, C., & Persson, A. (2015). User-related assessment of a driver assistance system for continuous support – A field trial. *Transportation Research Part F: Traffic Psychology and Behaviour, 30*, 128–144.

Viti, F., Hoogendoorn, S. P., Alkim, T. P., & Bootsma, G. (2008). Driving behavior interaction with ACC: Results from a Field Operational Test in the Netherlands. *2008 IEEE Intelligent Vehicles Symposium* (pp. 745–750). Piscataway, NJ: IEEE.

Vollrath, M., Schleicher, S., & Gelau, C. (2011). The influence of cruise control and adaptive cruise control on driving behaviour – A driving simulator study. *Accident Analysis & Prevention, 43*(3), 1134–1139.

Ward, N., Fairclough, S., & Humphreys, M. (1995). The effect of task automatisation in the automotive context: A field study of an Autonomous Intelligent Cruise Control system. *International Conference on Experimental Analysis and Measurement of Situation Awareness*. Daytona Beach, FL.

Wege, C., Will, S., & Victor, T. (2013). Eye movement and brake reactions to real world brake-capacity forward collision warnings—A naturalistic driving study. *Accident Analysis & Prevention, 58*, 259–270.

Weinberger, M., Winner, H., & Bubb, H. (2001). Adaptive cruise control field operational test—The learning phase. *JSAE Review, 22*(4), 487–494.

Wilde, G. J. S. (1982). The theory of risk homeostasis: Implications for safety and health. *Risk Analysis, 2*(4), 209–225.

Wilde, G. J. S. (1998). Risk homeostasis theory: An overview. *Injury Prevention, 4*(2), 89–91.

Wilson, B. H., Stearns, M. D., Koopman, J., & Yang, C. Y. D. (2007). *Evaluation of a Road-Departure Crash Warning System* (DOT HS 810 845). Washington, DC: National Highway Traffic Safety Administration.

Young, J., Fisher, D. L., Flynn, D. F., Gabree, S., & Nadler, E. (2018). Simulator study of pedestrian and bicycle collision warning systems: The effects of warning overreliance in latent threat scenarios. *Proceedings of the 97th Annual Meeting of the Transportation Research Board*. Washington, DC: Transportation Research Board.

Young, M., & Stanton, N. A. (2002). Malleable attentional resources theory: A new explanation for the effects of mental underload on performance. *Human Factors, 44*(3), 365–375.

Young, M. & Stanton, N. A. (2004). Taking the load off: Investigations of how adaptive cruise control affects mental workload. *Ergonomics, 47*(9), 1014–1035.

Young, M. & Stanton, N. A. (2007a). Back to the future: Brake reaction times for manual and automated vehicles. *Ergonomics, 50*(1), 46–58.

Young, M. & Stanton, N. A. (2007b). What's skill got to do with it? Vehicle automation and driver mental workload. *Ergonomics, 50*(8), 1324–1339.

FIGURE 2.2 Understanding of a real-world traffic scene in virtual space. (From Waymo.)

FIGURE 2.3 Assigning predictions to each object in the traffic scene. (From Waymo.)

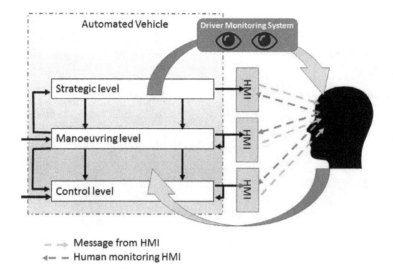

- - -▷ Message from HMI
◀- - - Human monitoring HMI

FIGURE 8.2 A proposed model showing the changing position and role of the human due to the introduction of automated functions in the driving task.

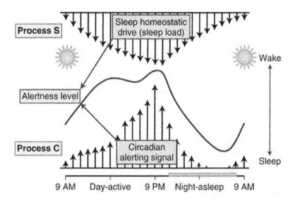

FIGURE 9.2 The two-process model of sleep–wake regulation (Borbély, 1982; Borbély, Daan, Wirz-Justice, & Deboer, 2016) depicting the interaction between homeostatic sleep drive (Process S) and sleep-independent circadian arousal drive (Process C) to produce S+C alertness level (solid line).

FIGURE 10.1 Scene as viewed by a person with cataract. (Image from the National Eye Institute, National Institutes of Health.)

FIGURE 10.2 Scene as viewed by a person with AMD. (Image from the National Eye Institute, National Institutes of Health.)

FIGURE 10.3 Scene as viewed by a person with glaucoma. (Image from the National Eye Institute, National Institutes of Health.)

FIGURE 10.4 Scene as viewed by a person with DR. (Image from the National Eye Institute, National Institutes of Health.)

FIGURE 13.1 Mill avenue on approach to the scene of Uber–Volvo collision. (*Source*: Google Maps.)

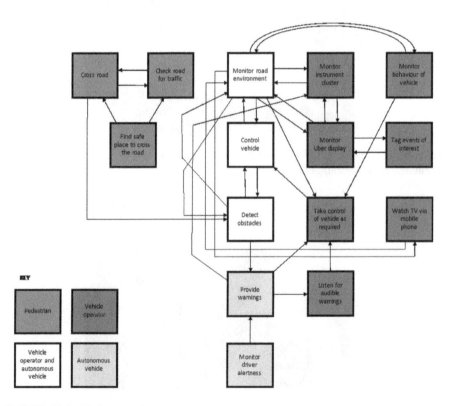

FIGURE 13.3 Task network.

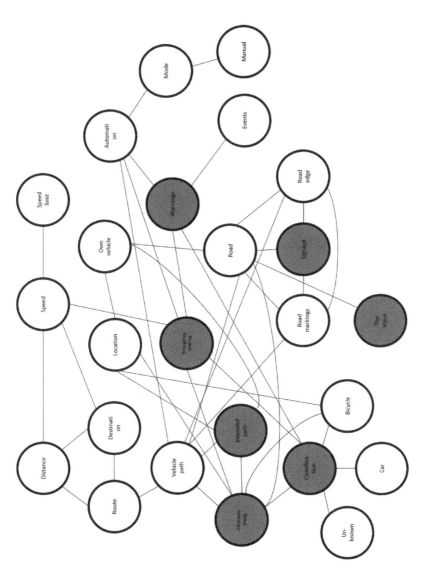

FIGURE 13.5 Information network. Nodes are shaded in red to show where agents did not have the information as required, where the information was understood but was incorrect, or where the information being used was inappropriate.

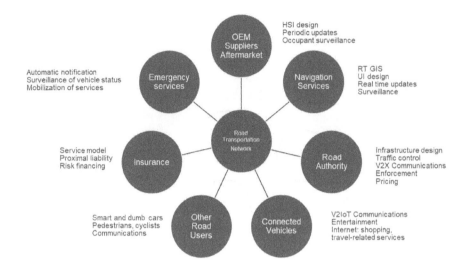

FIGURE 19.4　STS Star Model of an RTS. (Adapted from Noy et al., 2018.)

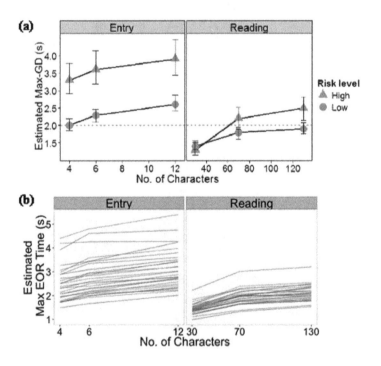

FIGURE 23.1　Data from a study on secondary task engagement. The data in (a) was reported in Peng et al (2013). A review of the data afterwards showed that the spread of maximum eyes-off-road time was greater in the text entry task when compared with the text reading task (b).

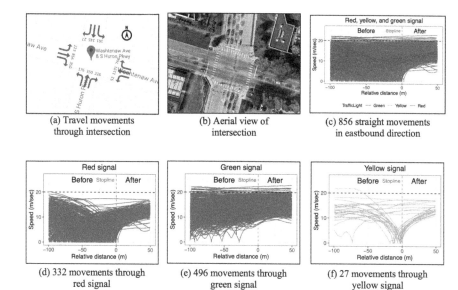

FIGURE 23.2 Washtenaw Ave & S Huron Pkwy, Ann Arbor, MI 48104 traffic patterns. (a) Travel movements through intersection, (b) Aerial view of intersection, (c) all movements in eastbound direction, and number of movements by (d) red signal (e) green signal, and (f) yellow signal.

13 Distributed Situation Awareness and Vehicle Automation
Case Study Analysis and Design Implications

Paul M. Salmon
University of the Sunshine Coast

Neville A. Stanton
University of Southampton

Guy H. Walker
Heriot-Watt University

CONTENTS

Key Points 294
13.1 Introduction 294
13.2 Situation Awareness... 296
    13.2.1 Individual SA... 296
    13.2.2 Team Models... 297
    13.2.3 System Models.. 297
13.3 SA on the Road 299
13.4 SA and Automated Vehicles 301
13.5 When DSA Breaks Down: Uber–Volvo Case Study 302
13.6 Uber–Volvo Incident Case Study.. 303
    13.6.1 Analysis of DSA in the Events Leading Up to the Collision........... 304
        13.6.1.1 Task Network 305
        13.6.1.2 Social Network.. 306
        13.6.1.3 Information Network 308
13.7 Implications for Automated Vehicle Design.. 308
13.8 Conclusions.. 311
References... 315

KEY POINTS

- The nature of automated vehicle systems is such that there is a need to move beyond simply considering the situation awareness (SA) needs of human road users to also focus on the SA needs of automated vehicles, infrastructure, and indeed the overall road transport system.
- The Distributed Situation Awareness (DSA) perspective has important ramifications for the design and implementation of automated vehicles and the road systems in which they will operate.
- Automated vehicles will have their own SA, they will be required to exchange SA with vehicle operators, and human agent SA (e.g., drivers, vehicle operators) and non-human SA (e.g., automated vehicles, infrastructure) will have to connect to support transportation.
- The Event Analysis of Systemic Teamwork (EAST) framework provides an integrated suite of methods for analyzing behavior, and specifically DSA, in complex systems.
- DSA frameworks could be usefully applied to the design, testing, and implementation of automated vehicles.

13.1   INTRODUCTION

Whilst all road collisions are caused by multiple interacting factors, one aspect is constant across them—almost always at least one of the road users involved is momentarily not aware of something important, be it other road users, the road conditions, hazards in the environment, or the safest way to negotiate a particular road situation (Salmon, Read, Walker, Lenne, & Stanton, 2018). Within human factors and safety science, the concept that we use to study and optimize awareness in complex and dynamic environments is known as "Situation Awareness" (SA; Endsley, 1995a; also see Chapter 7 in this Handbook). Based on over 30 years of applied research there are now various theoretical models and analysis methods that can be used to understand how humans, teams, organizations, and even entire systems develop an appropriate understanding of "what is going on" (Endsley, 1995a). SA has become an important lens through which to view and understand behavior and has been applied to support the design of tools, technologies, procedures, and environments aiming to optimize human performance in many areas (Salmon & Stanton, 2013; Wickens, 2008).

Whilst it may seem obvious that SA requirements should be a critical consideration during the design of automated and autonomous vehicles, and the road systems in which they operate, this is not always undertaken. Indeed, in relation to automated vehicles, specifically, there are concerns that SA is one in a long list of human factors concepts that may not be receiving the attention it warrants (Banks & Stanton, 2016; Hancock, 2019; Salmon et al., 2018). Despite the projected safety benefits of automated vehicles, it has been argued that the period between now and fully automated driving will be particularly problematic and could in fact lead to an increase in road crashes (Banks, Plant, & Stanton, 2018; Hancock, 2017, 2019; Salmon, 2019). There are various reasons for this (see Hancock, 2019), one of which is a failure to fully consider the SA requirements of road users, vehicles, and the road infrastructure (Salmon et al., 2018). These SA requirements include the information that relevant agents (e.g., drivers, vehicles, pedestrians, cyclists, road infrastructure) will need, as well as how SA should be exchanged between different agents (e.g., between an automated vehicle and its vehicle operator). Automated vehicles will, for example, need to be aware of the presence, location, and intended path of other road users, route and traffic-related information, and of their own location, speed, and intended path.

Similar oversights in the design of advanced automation have been seen in other areas, and the consequences can be catastrophic. The Air France crash of 2009 is one such instance where design-induced SA decrements resulted in the tragic loss of many lives (Salmon, Walker, & Stanton, 2016); but there are many more. In road transport specifically, recent incidents such as the Tesla Model S and Uber–Volvo fatal crashes provide a stark warning of the consequences that can arise when SA becomes sub-optimal. Both demonstrate that the introduction of highly automated vehicles will bring with it new emergent forms of collision, where it is the vehicle itself that is not aware of something (i.e., another road user or vehicle), where the human operator is not aware that they should be in control of the vehicle, or where SA is not adequately exchanged between the vehicle and the human operator. Such scenarios can be prevented through careful consideration of SA requirements during the design and testing of automated vehicles. To support this, however, an appropriate theoretical and methodological framework is required. In the field of SA, selecting one is not straightforward (Stanton, Salmon, Walker, Salas, & Hancock, 2017).

Since the 1990s there has been a constant debate regarding the utility and validity of different theoretical models and measures of SA (Endsley, 1995a; 2015; Salmon et al., 2008; Stanton, Salmon, Walker, & Jenkins, 2010; Stanton, Salmon, & Walker, 2015; Stanton et al., 2017; Smith & Hancock, 1995). Models either focus on SA at an individual (Endsley, 1995a), team (Salas, Prince, Baker, & Shrestha, 1995), or systems level (Stanton et al., 2006). Some argue that SA is exclusively a human property (Endsley, 1995a) whilst others argue that it spans both human and non-human agents (Stanton et al., 2006; 2010). Some argue that it can be shared between agents in the sense that they have the same understanding of situational elements (Endsley & Jones, 2001), whereas others argue that this level of shared SA is neither possible nor desirable (Stanton, Salmon, Walker, & Jenkins, 2009). As a result, it can be difficult for practitioners to select an appropriate model or measure of SA to support design activities. In a recent state-of-science review, Stanton et al. (2017) suggested a contingent approach whereby models and methods are matched to the problem space.

In this chapter, we argue that the nature of automated vehicle systems is such that there is a need to move beyond simply considering the SA needs of human road users to focus also on the SA needs of automated vehicles, infrastructure, and indeed the overall road transport system (for a complementary systems perspective, the reader may want to consider turning to Chapter 18 in this Handbook). In particular, emphasis on the exchange of SA between human and non-human agents is required. This aligns with models that view SA as a systems-level property that is distributed across human and non-human agents (e.g., Stanton et al., 2006; 2009; Salmon, Stanton, Walker, & Jenkins, 2009). This Distributed SA (DSA) perspective has important ramifications for the design and implementation of automated vehicles and the road systems in which they will operate. Accordingly, we provide an overview of the DSA model and discuss the implications for automated vehicle design. To demonstrate some of the core tenets of DSA we present an analysis of the recent Uber–Volvo collision, in which an automated test vehicle collided with a vulnerable road user, and close with a series of implications and future research requirements.

13.2 SITUATION AWARENESS

At its broadest level of description, SA refers to how agents, human or non-human, develop and maintain an understanding of "what is going on" around them (Endsley, 1995a; this Handbook, Chapter 7). Depending on the theoretical and methodological approaches employed, SA models and methods are used by researchers and practitioners to

- describe how individuals (Endsley, 1995a), teams (Salas et al., 1995), or systems (Salmon et al., 2009) develop and maintain appropriate levels of SA during task performance;
- make inferences on what SA comprises during different scenarios (i.e., what information is gathered and assimilated by individuals, teams, organizations, and systems in different scenarios: Stanton et al., 2010; 2015);
- inform the design of artefacts, technologies, procedures, training programs, and systems which support the development and maintenance of SA (Endsley, Bolte, & Jones, 2003);
- assess the quality of individual or team SA against a normative ideal (Endsley, 1995b); and
- identify how and why SA was "lost" or degraded during an adverse event of some sort (Salmon, Walker, & Stanton, 2015; 2016).

There are many definitions and models presented in the literature. These can be broadly categorized as those relating to the SA held by individuals, teams, and sociotechnical systems (STS). For a detailed review and comparison of models, the reader is referred to Salmon et al. (2008) and Stanton et al. (2017). A brief overview of popular definitions and models is given below.

13.2.1 INDIVIDUAL SA

Early definitions and models of SA focused on individual operators (e.g., drivers, pilots, control room operators) and the cognitive processes involved in developing and maintaining the awareness required to complete relevant tasks (e.g., driving). Mica Endsley, a pioneer in this area, introduced the most widely known and used definition, which describes SA as "the perception of the elements in the environment within a volume of time and space, the comprehension of their meaning and a projection of their status in the near future" (Endsley, 1988).

Endsley (1995a) outlined an information processing-based "three-level model" of SA (also see Chapter 7). This describes SA as an individual's understanding of the ongoing situation that encompasses three levels: Level 1, perception of the elements in the environment; Level 2, comprehension of their meaning; and Level 3, projection of future system states. The three-level model describes how SA is a central component of information processing that underpins decision-making and action. Within this model SA is influenced by various factors, including individual (e.g., mental models, workload), task (e.g., difficulty and complexity), and system factors (e.g., system complexity, interface design).

Level 1 SA involves perceiving the status, attributes, and dynamics of task-related elements in the surrounding environment (Endsley 1995a). Level 2 SA involves interpreting this data to understand its relevance in relation to one's goals. Level 3 SA involves anticipating the likely behavior of different elements in the environment. Level 1 and 2 SA is used along with mental models of similar situations to forecast likely events. Endsley describes how mental models play a critical role in SA, directing attention to pertinent elements in the environment (Level 1 SA), facilitating the integration of elements to aid comprehension (Level 2 SA), and supporting the generation of future states and behaviors (Level 3 SA; see also Chapter 7).

13.2.2 TEAM MODELS

Definitions and models of SA quickly expanded to encompass teams. Salas et al. (1995) defined team SA as "the shared understanding of a situation among team members at one point in time" (Salas et al., 1995, p. 131). Salas et al.'s model suggested that team SA comprises two primary components: individual SA and team processes. According to Salas et al. (1995) the perception of SA elements is influenced by the communication of mission objectives, individual tasks and roles, team capability, and other team performance factors. Comprehension of information is impacted by the interpretations made by other team members, and so individual SA is developed and then shared with other team members, which in turn leads to updates and modifications to each team member's SA. The model therefore incorporates a cyclical process of team members developing their own SA, sharing it with other team members, and then updating their own SA based on other team member's SA.

Salas et al. (1995) highlight the role that team processes, such as communication, assertiveness, and planning, play in the development and maintenance of team SA. Salas et al. also point to the key role of individual factors such as pre-existing knowledge and expectations and cognitive processing skills such as attention allocation, perception, data extraction, comprehension, and projection (Salas et al., 1995).

13.2.3 SYSTEM MODELS

In line with other concepts in human factors and safety science, there has been a significant shift over the last decade towards definitions and models that view SA as a systems-level property (e.g., Stanton et al., 2006; 2017; Salmon et al., 2009).

In the first article which initiated this paradigm shift, Neville Stanton, and colleagues defined distributed SA (DSA) as "activated knowledge for a specific task within a system.... [and] the use of appropriate knowledge (held by individuals, captured by devices, etc.) which relates to the state of the environment and changes as the situation develops" (Stanton et al., 2006, p. 1291).

Stanton et al. (2006) outlined a model of DSA, inspired by Hutchins' (1995a; b) seminal work on distributed cognition, which argues that SA is an emergent property that is held by the overall system and is built through interactions between "agents," both human (e.g., human operators) and non-human (e.g., tools, documents, displays). Whilst Hutchins' work on distributed cognition describes how information processing generally can be undertaken at a systems level by groups of individual agents and artifacts, Stanton et al.'s model focuses explicitly on SA, arguing similarly that SA can transcend individuals and be held by a system.

Stanton et al. (2017) recently updated the core tenets of the DSA model to include

1. *SA is an emergent property of STS*: Accordingly, the system represents the unit of analysis, rather than the individual agents working within it (see this Handbook, Chapter 19);
2. *SA is distributed across human and non-human agents*: Different agents have different views on the same situation. This draws on schema theory and the perceptual cycle model (Neisser, 1976), highlighting the key role of experience, memory, training, and goals;
3. *Systems have a dynamic network of information upon which different agents have their own unique view and contribution*: This is akin to a system's "hive mind" (Seeley et al., 2012). The compatibility between different agents' views is critical to support safe and efficient performance, with incompatibilities creating threats to performance and safety;
4. *DSA is maintained via transactions in awareness between agents*: These exchanges in awareness can be human-to-human, human-to-artefact, artefact-to-human, and/or artefact-to-artefact, and serve to maintain, expand, or degrade the network underpinning DSA;
5. *Compatible SA is required for systems to function effectively*: Rather than have shared awareness whereby all agents understand the situation in the same way, agents have their own unique views on the situation which connects to form the systems DSA. The compatibility between these different views is a key determinant of the quality of DSA, with high levels of compatibility found in optimally functioning systems (Salmon et al., 2009);
6. *Genotype and phenotype schemata play a key role in both transactions and compatibility of SA*: Schemata are mental templates of the world which guide perception and action (Neisser, 1976). Genotype schemata represent individuals' internally held mental templates that are developed based on their past experiences, and phenotype schemata represent task-activated genotype schemata that are triggered when individuals are engaged in relevant tasks. Human agents have genotype schemata for different situations

such as in the driving context, intersections, freeways, roundabouts, and driving in built-up areas. Once activated the genotype schemata become phenotype schemata, which direct and influence what agents attend to, how information is perceived and understood, and how the agent responds to the situation;

7. *DSA holds loosely coupled systems together*: Without this coupling, system performance may collapse. Dynamic changes in coupling may lead to associated changes in DSA; and

8. *One agent may compensate for degradation in SA in another agent*: Where one agent does not have optimal SA, other agents compensate, enabling the system to function as required. When other agents do not compensate, DSA breakdowns occur and performance is adversely impacted.

13.3   SA ON THE ROAD

The different definitions and models have important implications for how researchers and practitioners conceive SA to operate and indeed how they might propose to study and optimize it in automated vehicle systems. This can be demonstrated by using each perspective—individual, team, and distributed SA—as a lens to describe how SA works at road intersections (Salmon et al., 2018).

The individual perspective suggests that a driver develops SA of the current state of the intersection in relation to his or her driving goals. Elements in the intersection environment are perceived (Level 1), integrated, and comprehended in light of driving goals (Level 2). This information is used in combination with the driver's existing mental model of intersections to enable forecasting of future states (Level 3). For example, based on perceiving the status of the traffic lights, the surrounding traffic, and the weather and road conditions, a driver may anticipate that the lights are about to turn to red and that the vehicle in front is likely to brake sharply in response. This in turn directs decision-making and action, whereby the driver decides to slow down in anticipation of changing traffic lights and the vehicle in-front braking. Various pieces of information, or "SA elements," are critical, including the status of traffic lights, road signage, the driver's own vehicle speed and position, the positions and maneuvers of other vehicles, the behavior of other road users (e.g., pedestrians, cyclists), as well as roadway markings and weather conditions.

From a team SA viewpoint, SA at intersections can be examined at either the driving team level (e.g., a driver and their passengers) or the road user team level. The focus of analysis would initially be on how each team member acquires the SA required to negotiate the intersection. Beyond this, the team SA view would involve looking at how and what SA is shared and exchanged between team members. For example, sharing of SA might involve a passenger providing directions to a driver or telling the driver to stop as the traffic light is on amber. Alternatively, if the analysis was considering teams of road users, this might involve examining how the intentions of one road user are shared with another (e.g., via indicators or brake lights). Finally, how this sharing of SA updates each team members' own SA is also of importance. In the case of the passenger providing directions, the driver's SA is updated by the incoming communication from the other team member.

In contrast, the DSA model focuses on the SA of the overall intersection system as well as the role that each agent plays in SA development and maintenance (Salmon, Lenne, Walker, Stanton, & Filtness, 2014). According to DSA, the intersection's awareness comprises a network of information upon which different agents have distinct views and ownership—as mentioned above something that is akin to a hive mind of the system (Seeley et al., 2012). SA-related information is thus distributed across the human and non-human agents who operate within the intersection, including different road users (drivers, cyclists, motorcyclists, pedestrians), their vehicles, the road infrastructure (e.g., traffic lights, signage), and other elements of the intersection such as road markings (Salmon et al., 2018).

When agents are connected through "transactions," the intersection system can function effectively (Salmon et al., 2014). For example, a driver holds SA regarding their own goals, operational aspects of the driving task (e.g., position on road, intended path), and route information (e.g., directions required to achieve driving goals). Vehicle displays and exterior lights hold key SA-related information regarding speed and future maneuvers to be made (i.e., lane changes through indicators). The intersection infrastructure (i.e., traffic lights, pedestrian crossing lights, traffic light cameras) holds SA-related information regarding right of way through the intersection, traffic levels across the intersection, and road user infringements. DSA is an emergent property of the interactions between these agents.

None of the agents alone hold sufficient SA to allow safe operation of the intersection in its entirety. Drivers, for example, are not necessarily required to be aware of the level, location, and behavior of traffic coming through the intersection in the opposite direction. The intersection system as a whole, however, requires this information to be able to coordinate traffic signals and maintain an appropriate level of throughput and efficiency, and to enforce compliance. Likewise, drivers are not required to have the same understanding of the situation as other road users passing through the intersection; rather, they need to be able to exchange relevant pieces of SA with the other road users (Salmon et al., 2014). For example, a cyclist does not need to understand a driver's SA related to their goals, to vehicle control, and to the driver's SA of other vehicles in front of them; however, they do need to understand whether the driver has seen them and what the driver intends to do next. Compatible SA is achieved when these different views connect together to create an appropriate systems level of SA (Stanton et al., 2006; 2009).

Transactions, or exchanges of SA between agents, are also critical. Inappropriate transactions, or transaction failures, such as a driver failing to understand that a cyclist is about to enter the traffic flow in order to turn at the intersection, can lead to collisions. It is important to note that transactions represent more than just the communication of information; rather, they represent the exchange of SA where one agent interacts with another and both modify their SA as a result of the interaction. Agents interact with one another, exchange elements of their SA, and they are integrated with other information and acted on, and then passed on to other agents (Sorensen & Stanton, 2015; 2016).

13.4   SA AND AUTOMATED VEHICLES

A detailed investigation into driving revealed over 1,600 tasks, making it one of the most systemically complex activities that people perform in their everyday lives (Walker, Stanton, & Salmon, 2015). The introduction of automated vehicles will change the driving task, and in turn SA, considerably (see this Handbook, Chapters 7, 8). It is our view that the DSA model is the most appropriate perspective for automated vehicle system design and analysis. This is for various reasons, including the focus of DSA on both human and non-human agents, the emphasis placed on transactions in SA, and the notion of compatible SA (Salmon et al., 2017; 2018). These features fit neatly onto three key aspects of automated vehicle systems and the road systems in which they will operate. First, is the notion that automated vehicles will have their own SA. Second, is the notion that they will be required to exchange SA with vehicle operators. Third, and finally, is the notion that human agent SA (e.g., drivers, vehicle operators) and non-human SA (e.g., automated vehicles, infrastructure) will have to connect to support transportation.

All three aspects introduce critical design requirements. Automated vehicles will have their own goals, their own SA, and will be required to transact SA with their operators, other road users, other vehicles, and the road infrastructure (e.g., traffic lights). In relation to the example above, their SA will become an integral part of the intersection systems' SA; however, what it is, how it is acquired, and how it is used will be very different to the human road users within the intersection.

Given their advanced computational power, automated vehicles could acquire SA quicker than human road users. They may need less information to understand some elements of a situation (e.g., what a traffic light will do next), but more information to understand others (e.g., what a pedestrian will do next). They will also use and interpret information differently to other road users and other vehicles, depending on their goals and role within the system. For example, an automated vehicle might decide to stop based on an amber traffic light, whereas the same information may lead the driver of a car to decide to increase speed and pass through the intersection. Both actions may ultimately be safe; however, each agent is driven by a different understanding of the situation and its own internal rules for appropriate courses of action.

Given these subtle shifts in SA, it is critical that the SA requirements of different agents are considered, as well as how SA will be exchanged between agents. These SA requirements are summarized in Table 13.1.

Taking a broader view of the road system, there are other SA requirements that require consideration. These include those of other agents in the road transport system such as police, traffic management centers, insurance companies, vehicle manufacturers, licensing authorities, and road safety authorities (see this Handbook, Chapter 19). Given the capacity to do so via advanced technologies, these agents will likely wish to track and/or represent the road situation in order to maintain acceptable levels of performance and safety. Moreover, agents at the higher levels of road transport systems (e.g., road safety authorities) have to constantly monitor the state of the road transport system and use feedback on how it is operating to inform the development of road safety strategy, programs, and interventions. How automated vehicles exchange information with these broader road system agents also requires consideration.

TABLE 13.1

Automated Vehicle System Design SA Requirements

SA Requirement	Description	Examples
Human driver SA requirements	The information that the human driver needs to develop and maintain the SA required to safely drive the vehicle	• Current speed and speed limit • Route and directions • Location and actions of other road users • Hazards
Human operator SA requirements	The information that the human operator needs to supervise the vehicle in driverless mode and to understand when they are required to take over control of the vehicle	• Automation mode • Requirement to take over control of the vehicle • Location and actions of other road users • Hazards
Automated vehicle SA requirements	The information that the vehicle and its ADAS requires to operate automatically in a safe and efficient manner	• Current speed and speed limit • Route and directions • Location and actions of other road users • Hazards
Other road user SA requirements	The information that other road users require to understand what mode the automated vehicle is in, what it is doing, what it is aware of, and what it will do next	• Automation mode • Intended path • Awareness of surrounding road users
Other automated vehicle SA requirements	The information that other automated vehicles require to understand what mode the automated vehicle is in, what it is doing, what it is aware of, and what it will do next	• Automation mode • Intended path • Awareness of surrounding road users
Infrastructure SA requirements	The information that intelligent road infrastructure requires to understand what mode the automated vehicle is in, what it is doing, what it is aware of, and what it will do next	• Automation mode • Intended path • Awareness of surrounding road users

13.5 WHEN DSA BREAKS DOWN: UBER–VOLVO CASE STUDY

Recent incidents involving automated vehicles provide a clear indication of some of the issues that arise when SA requirements are not fully considered across all relevant agents. In the 2016 Tesla Model S crash which occurred in Florida, a driver was killed when the Tesla's Advanced Driver Assistance Systems (ADAS) failed to detect a crossing truck and collided with it whilst traveling at over 70 mph (Banks et al., 2018; NHTSA, 2016). The vehicle's ADAS was not aware of the truck's white trailer as it had failed to detect it against the bright sky. In addition, the driver had failed to place their hands on the steering wheel at previous points prior to the crash despite receiving repeated alerts to do so from the vehicle. Both the vehicle and the driver were ostensibly not aware of the truck and the risk of colliding with it, and the driver was seemingly not aware of the need to take over control of the vehicle. DSA provides a useful approach to respond to issues such as this by allowing designers to consider the SA needs of both the driver and the vehicle as well as how SA can be transacted between the driver and their vehicle and between vehicles (e.g., the truck and the Tesla).

More recently a Volvo fitted with Uber's self-driving system struck and killed a pedestrian in Tempe, Maricopa County, Arizona during operational testing (NTSB, 2018). Below we examine this incident through a DSA lens in order to articulate some critical DSA design requirements for automated vehicles. Although the full report was unavailable at the time of writing this chapter, there was sufficient information within the preliminary report (NTSB, 2018) to undertake a preliminary DSA-based analysis.

13.6   UBER–VOLVO INCIDENT CASE STUDY

At around 9:58 pm on Sunday 18 March 2018, a Volvo XC90 Sport Utility Vehicle (SUV) fitted with Uber's self-driving system struck and killed a pedestrian on Mill Avenue (see Figure 13.1), in Tempe, Maricopa County, Arizona (NTSB, 2018). The vehicle was being tested as part of Uber's Arizona testing program and was occupied by a vehicle operator at the time of the collision. The following description of the incident is adapted from the National Transportation Safety Bureau's (NTSB) preliminary investigation report (NTSB, 2018).

The test vehicle was traveling northbound along Mill Avenue, a two-lane road with a posted speed limit of 45 mph. The collision occurred when a 49-year-old female pedestrian, pushing a bicycle, attempted to cross Mill Avenue from a center median strip rather than via a pedestrian crosswalk, around 360 ft north of the collision site. Although the test vehicle detected an obstacle in the road ahead, it failed to

FIGURE 13.1   **(See color insert.)** Mill avenue on approach to the scene of Uber–Volvo collision. (*Source*: Google Maps.)

initiate an emergency braking maneuver, and the test vehicle operator failed to intervene until it was too late, only seeing the pedestrian immediately prior to impact.

The test vehicle was equipped with Uber's developmental self-driving system which comprised forward- and side-facing cameras, radars, Light Detection and Ranging (LIDAR), navigation sensors, and a computing and data storage unit (NTSB, 2018). Within the vehicle, a monitor located on the center console presented diagnostic information to the vehicle operator. The test vehicle had two control modes: computer and manual. It was also equipped with various Volvo driver assistance functions, including the City Safety™ system, which provides collision avoidance and automatic emergency braking. These functions, however, were disabled due to the vehicle being driven in the computer control mode and a desire to avoid an erratic ride in the vehicle, such as the vehicle braking in the event that objects in the vehicle's path were falsely detected.

At the time of the collision (approximately 9.58), the vehicle had been under computer control for around 19 minutes and was negotiating the second loop of an established test route. According to the NTSB report, the Uber system first registered radar and LIDAR observations of the pedestrian around 6 seconds prior to impact. The self-driving system initially classified the pedestrian as an unknown object and then as a bicycle, but could not identify an intended path. Around 1.3 seconds before impact, the self-driving system determined that an emergency braking maneuver was required in order to avoid a collision. Such a maneuver could not be initiated by the vehicle under computer control due to the City Safety™ system being disabled and so no braking action was initiated.

The vehicle operator only noticed the pedestrian and intervened less than a second before impact. Initially she engaged the steering wheel, but did not brake until after the vehicle hit the pedestrian. More recent accounts of the incident have suggested that the vehicle operator was watching the Hulu streaming service on her mobile phone (Stanton, Salmon, Walker, & Stanton, 2019a). The role of the vehicle operator was to observe the vehicle and to note events of interest on a central tablet. Vehicle operators were also supposed to monitor the environment for hazards and to regain control of the vehicle in the event of an emergency.

13.6.1 Analysis of DSA in the Events Leading Up to the Collision

The Event Analysis of Systemic Teamwork (EAST; Stanton, Salmon, & Walker, 2019b) was used to examine DSA leading up to the collision. As shown in Figure 13.2, EAST (Stanton et al., 2018) provides an integrated suite of methods for analyzing behavior, and specifically DSA, in complex systems. The framework supports this by providing methods to describe, analyze, and interrogate three network-based representations of activity: task, social, and information networks. Task networks are used to provide a summary of the activities performed within a system as well as the relationships between them. Social networks are used to analyze the organization of the system and the communications taking place between agents (both human and non-human). Information networks describe the information used to support task performance and how this information is distributed across different tasks and system agents.

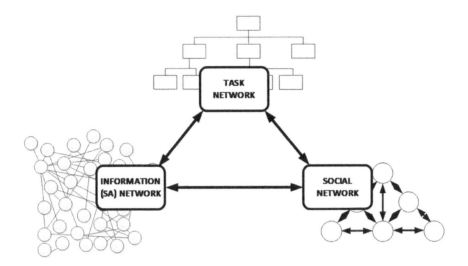

FIGURE 13.2 EAST network of networks approach.

EAST and its component methods have been used extensively to examine incidents involving DSA failures (Griffin, Young, & Stanton, 2010; 2015; Salmon et al., 2016). Using the NTSB preliminary reports, as well as other relevant documentation, we constructed task, social, and information networks for the events leading up to the Uber–Volvo collision. Specifically, the analysis focused on the behavior of the vehicle operator, the test vehicle, and the pedestrian. Construction of the networks involved working through the materials to identify tasks, interactions, and information relating to the behavior of the three core agents during the events leading up to the collision. This was undertaken initially by two of the authors (PS and NS), following which the third author reviewed the networks (GW) which were refined accordingly.

13.6.1.1 Task Network

The task network is presented in Figure 13.3, where the square nodes represent the tasks that were, and should have been, undertaken and the lines linking the nodes represent their interrelations where tasks were either undertaken together, sequentially, or influenced one another. Shading of the nodes is used to represent which agent was associated with each task.

Seemingly 7 of the 15 tasks within the task network were either not undertaken at all or were performed inadequately, leading to sub-optimal DSA and the subsequent collision. For the pedestrian, these included finding a safe place to cross and checking the road for traffic. For the vehicle operator, these included their monitoring of the road environment, detection of obstacles, and takeover of control of the vehicle. For the automated vehicle and its sub-systems, the inadequate tasks included the detection of obstacles, provision of warnings, and monitoring of driver alertness. Notably, many of the tasks that were not undertaken or were undertaken inadequately represent information gathering tasks or information communications tasks that are critical to DSA for this system. Further, the fact that the behaviors or omissions that

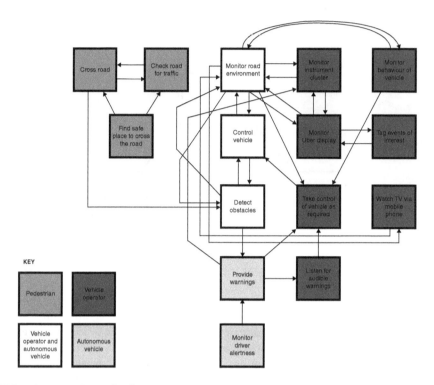

FIGURE 13.3   **(See color insert.)** Task network.

led to sub-optimal DSA were distributed across the vehicle, vehicle operator, and pedestrian/cyclist emphasizes the systemic nature of DSA. A final interesting feature of the task network is the "Watch TV via mobile phone" node. The TV show thus became an inappropriate part of DSA which had the effect of overriding and degrading other important aspects of DSA.

13.6.1.2   Social Network

Whilst the task network includes the three primary agents of the vehicle, the vehicle operator, and the pedestrian/cyclist, the social network presented in Figure 13.4 decomposes the system further to include 20 human and non-human agents. SA was thus distributed across these agents and was further exchanged between them throughout the incident. The lines connecting the agents show the communications pathways along which SA was exchanged between agents.

As shown in Figure 13.4, the relevant agents include the human road users (the vehicle operator and the pedestrian/cyclist), the test vehicle and associated sub-agents (e.g., radars, LIDAR, automatic emergency braking system), and road and road infrastructure agents (e.g., the road and road signage). According to the analysis, various failures in the transaction of SA between agents played a role in the incident. For example, whilst the vehicle's radar detected the pedestrian/cyclist (as represented by the link in Figure 13.4), the pedestrian/cyclist ostensibly did not detect the vehicle itself. These failed SA transactions are described in Table 13.2.

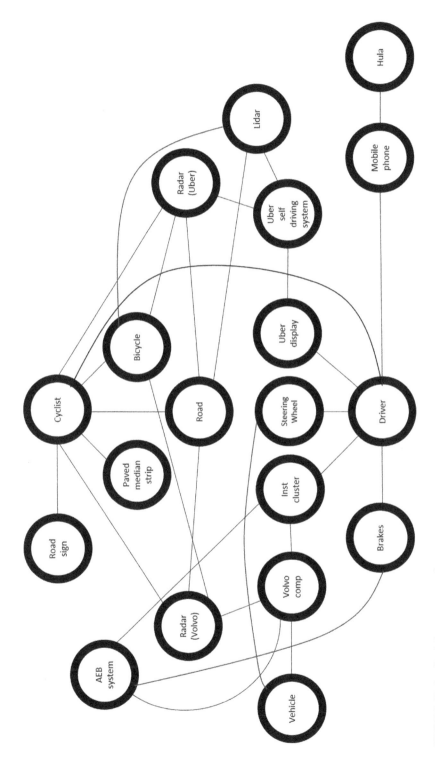

FIGURE 13.4 Social agent network.

TABLE 13.2
SA Transaction Failures

Transaction Failure Type	Transaction Required	Agents Involved
Absent transaction	Warning the pedestrian not to cross at chosen areas	Road signage, pedestrian
Absent transaction	Pedestrian to see approaching vehicle	Pedestrian, vehicle
Delayed transaction	Vehicle to detect and classify pedestrian	Pedestrian, vehicle
Absent transaction	Vehicle to determine pedestrians' intended path	Pedestrian, vehicle
Delayed transaction	Test operator to detect pedestrian	Pedestrian, test operator
Absent transaction	Vehicle to alert test operator of obstacle	Vehicle, test operator
Absent transaction	Vehicle to monitor test operator state	Vehicle, test operator
Inappropriate transaction	Mobile phone showing TV show to test operator	Mobile phone, test operator

13.6.1.3 Information Network

The information network is presented in Figure 13.5, where the nodes represent the information that underpinned DSA during the incident. Nodes are shaded in gray in the figure below (in red in the color insert) to show where DSA was ostensibly inadequate. These include instances where one or multiple agents did not have the information as required (e.g., vehicle operator not aware of the pedestrian as shown by the shaded "Obstacle (ped)" node), where the information was understood but was incorrect (e.g., the misclassification of the obstacle as shown by the shaded "Classification" node), or where the information being used was inappropriate (e.g., the voice TV show as shown by the shaded "The Voice" node). The other non-shaded nodes within the network represent information that was available and understood by the appropriate agent at the appropriate time.

Figure 13.5 shows that there were various DSA failures involved in the incident. Notably, these failures span the three agents involved in the incident. The pedestrian/cyclist was either unaware of the signage warning road users not to cross the road from the center median strip crosswalk or chose to ignore it. The test vehicle was initially not aware what type of road user the pedestrian/cyclist was and was then unable to identify an intended path. The vehicle operator was initially unaware that a pedestrian was occupying the road ahead, and did not receive any warnings from the vehicle once it had detected the pedestrian. Finally, the vehicle operator was allegedly watching a streaming television show on their mobile phone, which represents inappropriate information that was not required for tasks related to vehicle operation and monitoring. Notably, the vehicle was not aware of the vehicle operator's actions, resulting in lack of SA.

13.7 IMPLICATIONS FOR AUTOMATED VEHICLE DESIGN

The DSA model and Uber/Volvo collision analysis provide a series of important implications for automated vehicle design. These can be extracted via consideration of the DSA tenets outlined earlier. In Table 13.3, we present each tenet along with a relevant example from the incident and the resulting implications for automated vehicle design. Other design considerations are discussed in Chapters 15–17.

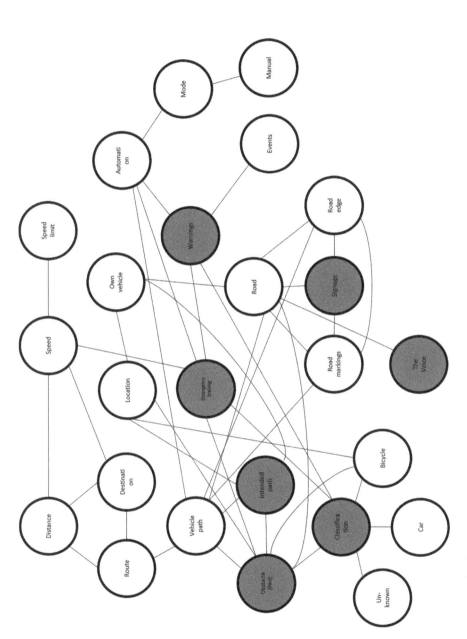

FIGURE 13.5 **(See color insert.)** Information network. Nodes are shaded in gray to show where agents did not have the information as required, where the information was understood but was incorrect, or where the information being used was inappropriate.

TABLE 13.3

Examples of DSA Tenets within Uber–Volvo Collision and Implications for Automated Vehicle Design

DSA Tenet	Example in Uber–Volvo Collision	Implications for Automated Vehicle Design
SA is an emergent property of STS	The SA required to ensure that there was a safe interaction between the vehicle operator, test vehicle, and pedestrian could only be achieved through interactions between 20 human and non-human agents (Social network, Figure 13.3).	• Designers should consider SA at a systems level and all human and non-human agents contributions to it. • Designers need to consider how information and SA are best exchanged between agents • Designers need to consider what information should be exchanged between agents, when it needs to be exchanged, and in what format
SA is distributed across human and non-human agents	SA was distributed across 20 human and non-human agents (Social network Figure 13.3).	• Designers should consider the SA needs of both human and non-human agents
Systems have a dynamic network of information on which each agent has their own view, and contribution to	The vehicle operator, test vehicle, and pedestrian were each using different combinations of information during the events leading up to the incident (Information network, Figure 13.4).	• Designers should develop predictive models of the overall systems' SA network and ensure that all of the information is available, in the right format, at appropriate times
DSA is maintained via transactions between agents	Various transactions in SA were made between 20 human and non-human agents (Social network, Figure 13.3), and many of the tasks required involved transactions in SA between these agents (Task network, Figure 13.2). A primary contributory factor was a failure of the vehicle to alert the vehicle operator to the presence of the pedestrian.	• Designers need to consider how SA is best transacted between agents, including what needs to be transacted, when, and in what format • Designers need to identify the SA requirements of all human and non-human agents
Compatible SA is required for systems to function effectively	The SA held by each of the three primary agents (vehicle operator, test vehicle, and pedestrian) was incompatible (Task, Social and Information networks, Figures 13.2–13.4).	• Designers need to ensure that the automated vehicle's SA is compatible with that of other road users, vehicles, and infrastructure in different road environments

(Continued)

TABLE 13.3 (*Continued*)

Examples of DSA Tenets within Uber–Volvo Collision and Implications for Automated Vehicle Design

DSA Tenet	Example in Uber–Volvo Collision	Implications for Automated Vehicle Design
Genotype and phenotype schemata play a key role in both transactions and compatibility of SA	The vehicle operator's schemata for this section of the route likely did not include pedestrians, which in turn meant that the operator was not keeping a constant look out (Task network, Figure 13.2)	• Designers should consider road users' existing and required schemata for different road environments when designing automated vehicles • Designers should consider the expectations of different road users in different road environments • Designers should consider the need to invoke appropriate driver schemata for different levels of vehicle automation • Automated vehicles should be designed to identify unexpected items in the road environment (e.g., cyclists, pedestrians) and inform vehicle operators of their presence
DSA holds loosely coupled systems together	Failures in DSA led to the system becoming uncoupled, resulting in fatal collision (Task, Social and Information networks, Figures 13.2–13.4)	• Designers should focus on the use of SA transactions to maintain constant coupling between agents
One agent may compensate for degradation in SA in another agent	Initially, the test vehicle compensated for the vehicle operator's degraded SA. With appropriate warning and/or an emergency braking system, the incident could have been prevented as a result (Task, Social and Information networks, Figures 13.2–13.4)	• Designers should use warnings and alerts to allow automated systems to update the SA of vehicle operators, other road users, and other automated vehicles • Designers should identify where agents can compensate for decrements in other agents' SA

13.8 CONCLUSIONS

The aim of this chapter was to provide an overview of the DSA model of SA and discuss some of the implications for automated vehicle design. In doing so, we presented a preliminary analysis of the recent Uber–Volvo collision and extracted relevant implications for automated vehicle design that relate the core tenets of the DSA model.

Automated vehicles will be highly disruptive. The next decade or so, therefore, represents a critical period for automated vehicles and road safety. Whilst there is an opportunity to create the safest and most efficient road transport systems of all time, there is also an opportunity to reverse decades of progress and create even more chaotic, unsafe, and congested road transport systems (Hancock, 2019; Salmon, 2019). SA, and specifically DSA, can play a key role in ensuring that we achieve the former and not the latter. A failure to consider DSA requirements and test DSA throughout the design life cycle will lead to automated vehicles, infrastructure, and road users that suffer losses of SA. In turn, this will create new forms of collision whereby the road transport system's DSA is insufficient to support safe interactions between road users. On the contrary, appropriate consideration and testing of DSA will help create efficient automated vehicles that are able to appropriately exchange SA with their operators, other road users, other vehicles, and the road infrastructure. A DSA-utopia on our roads is by no means out of reach. Of course, whether a DSA-utopia will optimize safety and efficiency in road transport systems is a question that also warrants further exploration.

How then can this be achieved? To conclude this chapter, we present a framework to support DSA-based design in road transport (Salmon et al., 2018, see Figure 13.6). This framework provides an overview of the kinds of analyses required to ensure that DSA can be understood and catered for during automated vehicle design life cycles. The framework includes the use of on-road naturalistic studies to examine DSA and road user behavior in existing road environments, the use of systems analysis methods such as Cognitive Work Analysis (CWA; Vicente, 1999) to identify key design requirements, the use of STS theory and an STS-Design Toolkit (STS-DT, Read, Beanland, Lenne, Stanton, & Salmon, 2017) to generate new design concepts, and finally the use of various evaluation approaches to evaluate design concepts.

It is recommended that the framework is used to support consideration of DSA in automated vehicle design. Although a significant body of driving automation research exists, knowledge gaps create a pressing need for further work. Previous studies have typically adopted one methodological approach (e.g., driving simulation) or have focused on one issue in isolation (e.g., automation failure, handover of control) when testing the safety risks associated with vehicle automation. For example, driving simulator projects have focused on the ability of drivers to regain manual control following automation failures (e.g., Stanton, Young, & McCaulder, 1997). Many key areas have been neglected, including the interaction of automated vehicles with vulnerable road users (e.g., cyclists and pedestrians), the interaction of automated vehicles with one another (e.g., vehicles designed by different manufacturers with differing algorithms and control philosophies), and the levels of SA held by automated vehicles in different scenarios. In addition, we do not know how vehicles with advanced levels of automation will interact with drivers operating vehicles without automation (and vice versa). This has prevented the full gamut of risks and emergent behaviors from being identified, which in turn means we do not currently understand the full potential impact of different levels of driving automation on behavior and safety.

It is our view that the framework presented in Figure 13.6 could be usefully applied to the design, testing, and implementation of automated vehicles (see also, this Handbook, Chapter 22). This could involve design and testing via modeling (e.g., with EAST and CWA) or testing via on-road studies involving both automated and non-automated vehicles. DSA-based research in this area will ensure that the

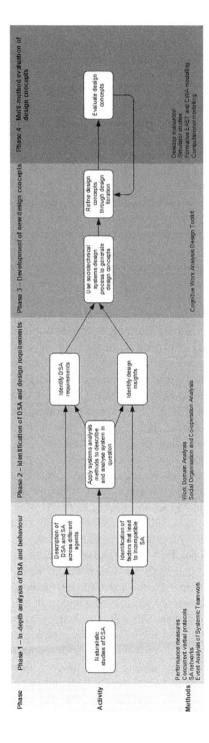

FIGURE 13.6 A framework for DSA-based design. (Salmon et al., 2018.)

SA needs of road users and automated vehicles are considered, and that appropriate analyses are undertaken to identify both design requirements and some of the emergent issues that might arise. Appropriate testing and revision of design concepts would then ensure that DSA requirements are met. Future crashes could be eradicated before they happen through more appropriate design. In the case of the recent Uber–Volvo collision, the SA requirements of the vehicle as well as the vehicle operator and other road users would be considered, meaning that the vehicle would provide adequate warning to the operator regarding its detection of the pedestrian. It is likely, for example, that a pro-active EAST modeling exercise would identify instances where the automated vehicle may detect hazards in the road environment, but where the vehicle operator may not. Based on this, appropriate design interventions could be made.

It is important to acknowledge that the consideration of DSA requirements in design should go beyond road users, automated vehicles, and the road environment. It is recommended that further study considers a larger component of road transport systems. For example, Salmon et al. (2016) recently developed a complex control structure model for the road transport system in Queensland, Australia, which included a description of all of the agents involved in road transportation, ranging from drivers and vehicles all the way up to government and international agencies. Each of these agents have DSA requirements relating to the safe operation of road transport systems. Indeed, the road transport systems' DSA incorporates the SA held by the police, traffic management centers, the media, road safety agencies, licensing authorities, insurance companies, government, etc. In particular, agents at the higher levels of road transport systems (e.g., road safety authorities) have to constantly monitor the state of the road transport system and use feedback on how it is operating to inform the development of road safety strategy, programs, and interventions. This system- level view could also be applied to other important cognitive mechanisms such as cognitive workload, mental models, and trust.

Banks et al. (2018) recently moved toward this form of systemic DSA analysis by applying EAST to examine DSA in future automated vehicle-based road transport systems. EAST is more useful in this context as it explicitly considers DSA, whereas Systems-Theoretic Accident Model and Processes (STAMP) examines control and feedback mechanisms (Leveson, 2004). Banks et al. (2018) concluded that, in future connected and autonomous vehicle (CAV)-based road transport systems, most of the agents within the system, including vehicles and infrastructure, will be connected. As a result, it is suggested that human factors issues will not be eliminated and, in fact, they will become more important. Banks et al. (2018) recommended EAST as a framework that can enable researchers to visualize the impact of automated vehicles on a much larger scale and to implement STS design principles to optimize performance.

It is also important to note that taking a systems perspective on DSA will enable wider system reforms that will contribute to the efficacy and safety of automated vehicles. By analyzing and understanding DSA at the overall road transport system level it will be possible to identify interventions beyond automated vehicle design. For example, modifications to road safety strategy, road rules and regulations, vehicle design guidelines and standards, enforcement tools and practices, and licensing and registration will all be required. Accordingly, we recommend further studies of DSA across current and future road transport systems. (Again, the reader is referred to a complementary chapter on the importance of systems analysis in this Handbook, Chapter 19).

Limitations of the DSA approach are worth noting. Attempting to identify the DSA requirements of multiple agents within a system is both difficult and resource intensive. In addition, EAST and the DSA model are not typically used to quantitatively assess the quality of DSA. As a result, other methods of assessing SA, such as the SA Global Assessment Technique (Endsley, 1995b), should also be used as part of a multi-methods approach.

REFERENCES

Banks, V. A., Plant, K. L., & Stanton, N. A. (2018). Driver error or designer error: Using the perceptual cycle model to explore the circumstances surrounding the fatal Tesla crash on 7th May 2016. *Safety Science, 108*, 278–285. doi:10.1016/j.ssci.2017.12.023

Banks, V. A. & Stanton, N. A. (2016). Keep the driver in control: Automating automobiles of the future. *Applied Ergonomics, 53*, 389–395.

Endsley, M. R. (1988). Situation awareness global assessment technique (SAGAT). *Proceedings of the National Aerospace and Electronics Conference (NAECON)* (pp. 789–795). New York: IEEE.

Endsley, M. R. (1995a). Towards a theory of situation awareness in dynamic systems. *Human Factors, 37*, 32–64.

Endsley, M. R. (1995b). Measurement of situation awareness in dynamic systems. *Human Factors, 37*, 65–84.

Endsley, M. R. (2015). Situation awareness misconceptions and misunderstandings. *Journal of Cognitive Engineering and Decision Making, 9*, 4–32.

Endsley, M. R., Bolte, B., & Jones, G. D. (2003). *Designing for Situation Awareness: An Approach to User-Centered Design.* Boca Raton, FL: CRC Press.

Endsley, M. R. & Jones, W. M. (2001). A model of inter- and intra-team situation awareness: Implications for design, training and measurement. In M. McNeese, E. Salas, & M. Endsley (Eds.), *New Trends in Cooperative Activities: Understanding System Dynamics in Complex Environments.* Santa Monica, CA: Human Factors and Ergonomics Society.

Hancock, P. A. (2017). Imposing limits on autonomous system. *Ergonomics, 60*, 284–291.

Hancock, P. A. (2019). Some pitfalls in the promises of automated and autonomous vehicles. *Ergonomics, 62*, 479–495.

Hutchins, E. (1995a). *Cognition in the Wild.* Cambridge, MA: MIT Press.

Hutchins, E. (1995b). How a cockpit remembers its speeds. *Cognitive Science, 19*, 265–288.

Griffin, T. G. C., Young, M. S., & Stanton, N. A. (2010). Investigating accident causation through information network modelling. *Ergonomics, 53*, 198–210.

Griffin, T. G. C., Young, M. S., & Stanton, N. A. (2015). *Human Factors Modelling in Aviation Accident Analysis and Prevention.* Aldershot, UK: Ashgate.

Leveson, N. G. (2004). A new accident model for engineering safer systems. *Safety Science, 42*(4), 237–270.

National Highway Traffic Safety Administration. (2016). *Office of Defects Investigation*: PE 16–007. Retrieved from https://static.nhtsa.gov/odi/inv/2016/INCLA-PE16007-7876.PDF

National Transportation Safety Board. (2018). *Preliminary Report Highway HWY18MH010.* Retrieved from www.ntsb.gov/investigations/AccidentReports/Reports/HWY18MH010-prelim.pdf

Neisser, U. (1976). *Cognition and Reality: Principles and Implications of Cognitive Psychology.* San Francisco, CA: Freeman.

Read, G. J. M., Beanland, V., Lenne, M. G., Stanton, N. A., & Salmon, P. M. (2017). *Integrating Human Factors Methods and Systems Thinking for Transport Analysis and Design*. Boca Raton, FL: CRC Press.

Salas, E., Prince, C., Baker, D. P., & Shrestha, L. (1995). Situation awareness in team performance: Implications for measurement and training. *Human Factors, 37*, 1123–1136.

Salmon, P. M. (2019). The horse has bolted! Why human factors and ergonomics has to catch up with autonomous vehicles (and other advanced forms of automation). *Ergonomics, 62*, 502–504.

Salmon, P. M., Lenne, M. G., Walker, G. H., Stanton, N. A., & Filtness, A. (2014). Exploring schema-driven differences in situation awareness across road users: An on-road study of driver, cyclist and motorcyclist situation awareness. *Ergonomics, 57*, 191–209.

Salmon, P. M., Read, G. J. M., Walker, G. H., Lenne, M. G., & Stanton, N. A. (2018). *Distributed Situation Awareness in Road Transport: Theory, Measurement, and Application to Intersection Design*. Boca Raton, FL: CRC Press.

Salmon, P. M. & Stanton, N. A. (2013). Situation awareness and safety: Contribution or confusion? *Safety Science*, 56, 1–5.

Salmon, P. M., Stanton, N. A., Walker, G. H., Baber, C., Jenkins, D. P., & McMaster, R. (2008). What really is going on? Review of situation awareness models for individuals and teams. *Theoretical Issues in Ergonomics Science, 9*, 297–323.

Salmon, P. M., Stanton, N. A., Walker, G. H., & Jenkins, D. P. (2009). *Distributed Situation Awareness: Advances in Theory, Measurement and Application to Teamwork*. Aldershot, UK: Ashgate.

Salmon, P. M., Walker, G. H., & Stanton, N. A. (2015). Broken components versus broken systems: Why it is systems not people that lose situation awareness. *Cognition, Technology and Work, 17*, 179–183.

Salmon, P. M., Walker, G. H., & Stanton, N. A. (2016). Pilot error versus sociotechnical systems failure? A distributed situation awareness analysis of Air France 447. *Theoretical Issues in Ergonomics Science, 17*, 64–79.

Seeley, T. D., Visscher, P. K., Schlegel, T., Hogan, P. M., Franks, N. R., & Marshall, J. A. (2012). Stop signals provide cross inhibition in collective decision-making by honeybee swarms. *Science, 335*(6064), 108–111.

Smith, K. & Hancock, P. A. (1995). Situation awareness is adaptive, externally directed consciousness. *Human Factors, 37*, 137–148.

Sorensen, L. J. & Stanton, N. A. (2015). Exploring compatible and incompatible transactions in teams. *Cognition, Technology and Work, 17*, 367–380.

Sorensen, L. J. & Stanton, N. A. (2016). Keeping it together: The role of transactional situation awareness in team performance. *International Journal of Industrial Ergonomics, 53*, 267–273.

Stanton, N. A., Salmon, P. M., & Walker, G. H. (2015). Let the reader decide: A paradigm shift for situation awareness in sociotechnical systems. *Journal of Cognitive Engineering and Decision Making, 9*, 44–50.

Stanton, N. A., Salmon, P. M., & Walker, G. H. (2018). *Systems Thinking in Practice: Applications of the Event Analysis of Systemic Teamwork Method*. Boca Raton, FL: CRC Press.

Stanton, N. A., Salmon, P. M., Walker, G. H., & Jenkins, D. P. (2009). Genotype and phenotype schema and their role in distributed situation awareness in collaborative systems. *Theoretical Issues in Ergonomics Science, 10*, 43–68.

Stanton, N. A., Salmon, P. M., Walker, G. H., & Jenkins, D. P. (2010). Is situation awareness all in the mind? *Theoretical Issues in Ergonomics Science, 11*, 29–40.

Stanton, N. A., Salmon, P. M., Walker, G. H., Salas, E., & Hancock, P. A. (2017). State-of-science: Situation awareness in individuals, teams and systems. *Ergonomics, 60*, 449–466.

Stanton, N. A., Salmon, P. M., Walker, G. H., & Stanton, M. (2019a). Models and methods for collision analysis: A comparison study based on the Uber collision with a pedestrian. *Safety Science, 120,* 117–128.

Stanton, N. A., Stewart, R., Harris, D., Houghton, R. J., Baber, C., McMaster, R., … Green, D. (2006). Distributed situation awareness in dynamic systems: Theoretical development and application of an ergonomics methodology. *Ergonomics, 49,* 1288–1311.

Stanton, N. A., Young, M., & McCaulder, B. (1997). Drive-by-wire: The case of driver workload and reclaiming control with adaptive cruise control. *Safety Science, 27,* 149–159.

Vicente, K. J. (1999). *Cognitive Work Analysis: Toward Safe, Productive, and Healthy Computer-Based Work.* Mahwah, NJ: Lawrence Erlbaum Associates.

Walker, G. H., Stanton, N. A., & Salmon, P. M. (2015). *Human Factors in Automotive Engineering and Technology.* Ashgate, UK: Aldershot.

Wickens, C. D. (2008). Situation awareness: Review of Mica Endsley's 1995 articles on situation awareness theory and measurement. *Human Factors, 50,* 397–403.

14 Human Factors Considerations in Preparing Policy and Regulation for Automated Vehicles

Marcus Burke
National Transport Commission Australia

CONTENTS

Key Points 320
14.1 Introduction 320
    14.1.1 Outline of the Chapter 320
    14.1.2 Why Is This Topic Important? 321
    14.1.3 How Do We Define Human Factors? 321
    14.1.4 What Are Automated Vehicles? 322
    14.1.5 What Is the Goal of the Automated Driving System? 323
14.2 Automated Driving Systems and Human Factors 323
    14.2.1 Human Factors and the Automated Driving System 323
    14.2.2 Specific Human Factor Risks That Could Impact Safety 325
    14.2.3 Which Parties Will Influence These Safety Risks? 327
    14.2.4 Human Factors Safety Risks and Automated Vehicles 327
14.3 Government, Transport, and Policy 328
    14.3.1 What Are the Roles of Government in Road Transport? 328
    14.3.2 Regulation and New Technology 328
    14.3.3 Quantifying the Safety Risks of Automated Vehicles 329
    14.3.4 What Policy and Regulatory Changes Do Automated Vehicles
        Require? 329
    14.3.5 Approaches for Government to Address Regulatory Barriers,
        Gaps, and Opportunities 331
14.4 How Does Policy and Regulation Address Human Factors Safety Risks? 332
    14.4.1 Government Regulation of Human Factors and Automation
        in Other Modes 332
    14.4.2 What Is the Role of Prescriptive Versus Principles-Based
        Regulation? 333

14.5 Conclusion—How Should Governments Regulate for Human
     Factors Safety Risks for Automated Vehicles?...334
Acknowledgments..335
References..335

KEY POINTS

- Automated vehicles will create new human factors-related safety risks, as these vehicles interact in new ways with humans in a variety of roles (pedestrian, passenger, driver, etc.).
- The human factors-related safety risks of automated vehicles are not well understood and cannot yet be fully quantified.
- Governments should seek to learn from other safety regimes that have already dealt with human factors-related safety risks and automation.
- Governments should consider outcome- or principles-based approaches to regulation.

14.1 INTRODUCTION

14.1.1 OUTLINE OF THE CHAPTER

This chapter examines the human factors issues in automated vehicle regulation. It seeks to answer the question "What human factors issues need to be considered in preparing policy and regulation for the deployment of automated vehicles?" The chapter is primarily aimed at those who are charged with developing policy for these vehicles—lawmakers, senior government executives, and policy officers. However, it will also likely be of interest to those who seek to influence that policy and by those who are affected by it—technology manufacturers, insurers, road safety groups, and, ultimately, the public.

I will first look at why this question is important and define automated vehicles, automated driving systems (ADS), and human factors. I will then examine the systems features of ADS. I will attempt to identify the individuals that will interact with automated vehicles and outline the safety risks in these interactions. I will go on to look at the role of government in general, and specifically for transport, before turning to different approaches to regulation. Finally, I will attempt to look at how regulation and other tools available to government might address the human safety risks of automated vehicles. Human factors issues relate ultimately to the design of systems, so the question becomes how much policy makers need to specify system design to account for human factors. I will largely use examples from Australian law as this is the author's major area of expertise; however, examples will also be drawn from other jurisdictions.

This chapter will use the terminology and taxonomy from the SAE Standard J3016 (SAE International, 2018). This includes terms such as "automated vehicle" and "ADS" along with the levels of automation for automated vehicles. The chapter will focus on regulation rather than broader policy (e.g., decisions around industry investment or labor policy). I will also largely focus on safety, rather than broader regulatory issues such as road pricing.

This chapter will focus on commercial deployment of ADS. There may be additional issues in relation to testing of these systems (see e.g., this Handbook, Chapter 22). These would likely mirror those of the commercial deployment but would be larger in severity; for example, even less standardization of technology and interfaces (more variability), less familiarity for other road users, and less mature technology. This may require additional mitigations during testing (e.g., safety drivers, additional monitoring, reduced speed).

14.1.2 WHY IS THIS TOPIC IMPORTANT?

Picture an everyday interaction on a public road: a car and a pedestrian at a pedestrian (or "zebra") crossing; that is, a crossing without traffic lights that gives right of way to pedestrians. A car approaches. At the same time a pedestrian is about to step out onto the crossing. The pedestrian looks at the car, to make sure it is slowing down (it is). The pedestrian makes eye contact with the car's driver. The pedestrian is now confident that the driver has seen her. The car driver nods or uses a hand gesture to wave the pedestrian through. The pedestrian walks across the crossing and the car proceeds on its way. Parts of this interaction are regulated (vehicle must give way to pedestrians), but others are not (making eye contact, gestures).

Replay the scene with one difference. The vehicle is automated. There may not be a person in the driver's seat. How does the pedestrian know that the vehicle has "seen" her? How does she have confidence that the vehicle will give her right of way? The passenger in the vehicle could wave them through, but would they be confident that the vehicle has seen the pedestrian? Are there safety risks here and, if so, what is the role of government (if any) in addressing those risks?

This is simply one example of many interactions between an ADS and a human—whether the human is a passenger in the vehicle, a driver handing over control, or another road user. Automated vehicles potentially offer significant safety benefits (Logan, Young, Allen, & Horberry, 2017; National Transport Commission, 2018), but they present potential new risks, including human factors risks (as discussed in more detail throughout this Handbook). Governments will need to decide how to set policy for these vehicles, likely with incomplete evidence about these risks. Poor decisions by governments could ultimately result in unnecessary deaths and injuries on our roads, through either allowing unsafe systems to operate or being too conservative and delaying or banning systems that could save lives. But to understand this topic, we must first explain what we mean by human factors and by automated vehicles.

14.1.3 HOW DO WE DEFINE HUMAN FACTORS?

Proctor and Van Zandt (2008) define human factors as:

> ...the study of those variables that influence the efficiency with which the human performer can interact with the inanimate components of a system to accomplish the system goals (p. 9).

Alternatively, it is defined by the industry as:

...the scientific discipline concerned with the understanding of interactions among humans and other elements of a system, and the profession that applies theory, principles, data, and other methods to design in order to optimize human well-being and overall system performance (Proctor & Van Zandt, 2008, p. 9).

The World Health Organization (2009) states that human factors

...examines the relationship between human beings and the systems with which they interact by focusing on improving efficiency, creativity, productivity and job satisfaction, with the goal of minimizing errors (p. 100).

The focus in all these definitions is on how humans interact with systems. To understand the human factors involved, therefore, we need to (1) identify the system and its goal (or goals), (2) identify the various individuals who will interact with that system, and (3) examine the specific safety risks of those interactions.

14.1.4 What Are Automated Vehicles?

As noted in earlier chapters, there are many different types and levels of automation (e.g., Chapters 1, 2), each with their own definition. Here, automated vehicles are defined as:

...vehicles that include an automated driving system (ADS) that is capable of monitoring the driving environment and controlling the dynamic driving task (steering, acceleration, and braking) with limited or no human input (National Transport Commission, 2018, p. 1).

This could include vehicles based on existing models, with automated functions, new vehicle types with automated functions, as well as aftermarket devices or software upgrades that add automated driving functions to existing vehicles.

The key feature of the automated vehicle is the ADS. The SAE Standard J3016 defines the ADS as:

The hardware and software that are collectively capable of performing the entire [dynamic driving task] on a sustained basis, regardless of whether it is limited to a specific operational design domain; this term is used specifically to describe a level 3, 4, or 5 driving automation system (SAE International, 2018, p. 3).

ADSs perform the entire dynamic driving task, though some will require a human in the vehicle ready to take back control when requested (Level 3 or "conditional" automation). These systems are no longer a driver assistance system—they are, in essence, the driver. Legally this creates a significant challenge—if a human is no longer doing the driving then who is responsible for the driving?

ADS is a new type of system designed to interact with humans in order to provide them with a service (driving). ADSs will drive passenger vehicles as well as freight vehicles.

ADSs will form part of a complex "system of systems," both in the vehicle, in some control or monitoring centers, and potentially in the road infrastructure (see also, this Handbook, Chapter 19). This could include "remote drivers" or "teleoperators," humans who exercise varying levels of control over the vehicle, but sit in an operations center rather than in the vehicle itself. Thus, the ADS will likely include:

- technologies within the vehicle (sensors, control systems, software and event data recorders);
- remote monitoring and cloud services provided by the vehicle manufacturer or technology provider, potentially including human supervisory systems and remote drivers;
- mapping systems and data to provide information to the ADS to choose appropriate routes and avoid hazards;
- commercial billing systems to charge passengers and users for services;
- links to government systems, either using government data (e.g., speed zone and roadworks information) or sending data (e.g. for tolling, which could also involve private road operators).

There are many unknowns in the deployment of ADSs, including:

- the exact limits of automation (e.g., will we ever have automated vehicles on dirt roads?);
- the timing of commercial deployments (how far away are we?);
- the applications that will be commercially successful;
- the mix of technologies that will be used;
- user acceptance of these systems, and;
- how they will change travel behaviors.

These unknowns limit the ability for governments to understand the interactions and safety risks. The data to quantify these risks, in many cases, simply do not yet exist.

Vehicle automation is already used in other environments, from trains to mining to aviation to ports. These applications, however, will generally not involve as wide a network of systems, and do not operate in a shared, public environment with such a range of other users. This complex operating environment creates a technology challenge as well as human factors and regulatory challenges.

14.1.5   What Is the Goal of the Automated Driving System?

ADSs will have different applications and purposes for which the technology is used. For example, "robotaxis" carrying people around cities have a different purpose from that of automated freight vehicles operating on freeways. But, broadly, the goal will remain the same—to drive the vehicle safely, efficiently, and legally to its destination.

14.2   AUTOMATED DRIVING SYSTEMS AND HUMAN FACTORS

14.2.1   Human Factors and the Automated Driving System

To achieve their goal, automated vehicles will need to continually interact with humans in many different roles. The exact list of roles that an automated vehicle will interact with will vary with the business model, application, and operational design domain of the technology (see also, this Handbook, Chapter 8). Roles of humans will also vary with different levels of automation—from a Level 3 vehicle to a Level 5 vehicle. However, persons who will likely interact with ADSs include:

1. Individuals inside the vehicles
   • Vehicle passengers
   • Vehicle drivers, who may hand over or take back control from the ADS, thereby becoming a passenger or fall-back ready user
   • Fall-back ready users, a role defined for Level 3 (Conditional) automation, who are able to operate the vehicle and must be receptive to requests by the ADS to intervene and be prepared for system failures (SAE International, 2018)
2. Individuals outside the vehicle that influence the control and movement of the vehicle:
   • Those calling up or directing a vehicle
   • Remote drivers
   • System operational managers, who monitor and supervise the day-to-day operation of the vehicles
3. Other individuals the ADS will encounter or interact with on public roads:
   • Drivers of other vehicles, including freight vehicles, passenger vehicles, motorcycles
   • Pedestrians
   • Cyclists
   • Enforcement officers, such as police
   • Emergency services workers, such as ambulance, police officers, and firefighters
   • Individuals directing traffic in public or private spaces (including workers at roadworks sites and individuals in private car parks)
   • Roadside workers

The latter list (#3) includes all the individuals who interact with human drivers today, but will also include those new roles that will come into existence with automated vehicles (such as remote drivers and fall-back ready users). This is because automated vehicles will also need to interact with *other* automated vehicles on the road. Some of these roles could be divided further—for example, interactions with pedestrians may differ when the pedestrian is very old or very young or visually impaired.

If we extend the idea of human factors to other parties who interact with the system, we could also include those who are programming, designing, and testing these vehicles as part of companies who are developing this technology, along with the senior executives of those companies. Testing phases will include test drivers monitoring the operation of the ADS.

There will be a wide range of interactions between the systems and individual roles. These could include an automated vehicle interacting with a pedestrian attempting to cross the road, an automated vehicle responding to a request from a human driver to hand over control, or a remote driver taking over control to assist an automated vehicle to negotiate a complex set of roadworks. These will be explored in more detail below.

In summary, we can state that interactions with ADSs will vary widely across a diverse range of users. These users will have varying levels of experience and understanding about how the technology works and what its limitations are. Many, if not most, will have no specific training in how the technology works, but will have experience interacting with human-driven vehicles. The risks are significant—getting an interaction wrong can result in injury or death. Major design problems could result in many deaths.

14.2.2 SPECIFIC HUMAN FACTOR RISKS THAT COULD IMPACT SAFETY

I now examine the specific human factor risks, based on some of the key roles set out above. These are largely framed as questions, as many are unknown. This is also not a comprehensive list; there will likely be other risks which emerge as trials and deployments continue. The risks should be considered in light of the research on human factors and advanced driver assistance systems (ADAS) which have identified human factors issues at lower levels of automation or in other modes of transport, including (amongst others) behavioral adaption (see also, this Handbook, Chapter 12); poor process monitoring (see also, this Handbook, Chapter 7); over-reliance and complacency (see also, this Handbook, Chapter 4; also van Wees & Brookhuis, 2005); and skill loss (see also, this Handbook, Chapter 21; Casner, Geven, Recker, & Schooler, 2014). Indeed, many drivers are confused by existing technology, let alone new or emerging technology.

I will now examine the key risks based on the categories of roles above in Section 14.2.1.

1. Human factors risks of individuals inside the vehicles

    Vehicles will have interfaces for those inside the vehicle, which could include on the one hand traditional vehicle controls or on the other hand screen-based interfaces in the dashboard or provided through devices. Safety risks include:
    • Vehicle passengers may interfere (inadvertently or deliberately) with the system
    • Fall-back ready users need to be awake, alert, sober, licensed, and ready to take back control. Each of these elements raises risks and creates a potential need for monitoring. Will the fall-back ready user be prepared to drive when called upon?
    • At Level 3, risks could be created in the handover back to a fully pre-pared fall-back ready user, including if the user fails to understand the request from the ADS or the specific risk that the vehicle is dealing with
    • At Level 4 safety risks could also occur in the handover between the ADS and the human driver
    • Is there a risk of deskilling and, if so, is there a need to change regulation for existing drivers? (see also, this Handbook, Chapter 10). Will there be a dip in performance after the human driver has taken back control or in the lead up to taking control?
    • How will drivers be educated about the risks/limitations of the technology? (see also, this Handbook, Chapter 18)

The point of handover between a human driver and an ADS will be a key safety risk. There may be different risks in handing control *from* the human driver to the ADS as opposed to handing control over from the ADS *to* the human driver. The latter may create greater risk.

2. Individuals outside the vehicle that influence the control and movement of the vehicle

   • Is there a way these users could send a vehicle to an unsafe location? Alternatively, could the vehicle put a passenger in an unsafe scenario (e.g., a vehicle that stops functioning in the middle of a freeway with a small child as the only passenger)?

   • Are there specific risks for remote drivers, given how removed they will be from the road environment? Are there risks that could impact how well these drivers will monitor the vehicles? Will they potentially supervise more than one vehicle? How do they need to be trained and licensed? What skills do they need to have? Will they have sufficient information to understand the environment the vehicle is operating in remotely?

   • What are the risks for system operational managers who are monitoring a network of vehicles? Will they react appropriately in the event of a risk? This could be analogous to air traffic controllers or controllers of complex facilities such as power stations.

3. Other individuals the ADS will encounter or interact with on public roads

   • Will pedestrians change their behavior if they perceive vehicles to be safer? Will they step straight out into the street in the way that people routinely wave their arms in front of the closing door of an elevator expecting that it will open for them?

   • Will other human drivers drive more aggressively?

   • Will other users seek to interfere with automated vehicles or to game them in some way?

   • Are there road user behaviors or interactions between road users that are not regulated today (such as the example at the beginning of the chapter on pedestrians making eye contact with a human driver, prior to crossing) that could create risks?

   • How will vehicles and other road users negotiate right of way in circumstances where they do so today? Will automated vehicles nudge out into traffic to signal intent in the way human drivers do?

   • Automated vehicles will drive more conservatively than many vehicles do today—in terms of speed, spacing, merging, etc. Will this create specific safety issues?

   • How will travel behaviors change with the opportunities provided by automated vehicles?

In addition to these questions, it is worth considering what are the human factors issues relating to automated vehicle manufacturers and technology providers. One of the key influences on safety will derive from the developers of the technology. As a result, it is important to understand the human factors issues that may influence decisions by individuals in these companies. What are they motivated by? How will they make decisions about what is acceptably safe? What will be the role of senior executives and how might they influence outcomes? Again, in all these roles, there may be different risks for different individuals, depending on age, experience, training, impairment, and understanding of the technology.

14.2.3 WHICH PARTIES WILL INFLUENCE THESE SAFETY RISKS?

Parties with a key influence on the safety of ADSs will include:

- Vehicle manufacturers
- Entities responsible for the development and ongoing performance of the ADS (where that is different to the vehicle manufacturer)
- Repair and maintenance providers
- Vehicle owners or operators
- Drivers
- Fall-back ready users
- Road managers
- Mapping providers
- Governments

An appropriate regulatory system will consider the safety risks but then go on to consider which parties have the greatest influence on those risks, how those parties are covered by existing regulation, and examine any gaps (National Transport Commission, 2018). There may be gaps for those parties that are new to the transport system. These include entities responsible for the development and running of the ADS, remote drivers, and fall-back ready users.

14.2.4 HUMAN FACTORS SAFETY RISKS AND AUTOMATED VEHICLES

We can conclude that there are a number of human factors risks in the deployment of ADSs. Today we are unable to quantify these risks. This raises the question of how to best address these risks and the roles of government, industry, and the public. What are the human–machine interface issues (see e.g., this Handbook, Chapters 15, 16)? Should these be left entirely to the market to resolve or is there a need for standardization? The automotive industry has recognized the need to develop verification and validation approaches for human–machine interactions (Wood et al., 2019), but we do not yet have detailed standards.

Some of these interactions are already governed by laws today in most countries, such as obligations to follow the direction of enforcement officers or to give way to other road users in certain circumstances. These laws vary in detail, but they were all originally designed exclusively for human drivers and so may be inadequate for interactions involving systems, and they do not cover the new roles that ADSs create. Other risks may be addressed by behavioral norms rather than laws. This raises the question for policy makers of whether laws should change for automated vehicles and, if so, how new laws would accommodate the human factors issues of ADSs?

14.3 GOVERNMENT, TRANSPORT, AND POLICY

Before we propose how governments should manage these human factors safety risks, we first need to examine what role governments have in transport and what policy tools they have at their disposal.

14.3.1 WHAT ARE THE ROLES OF GOVERNMENT IN ROAD TRANSPORT?

The broad roles of government include social regulation (including safety regulation), regulation of markets, and provision of services. In road transport, the focus of government is on:

- Standard setting (e.g., vehicle standards)
- Safety regulation and enforcement (which includes driver licensing and vehicle registration)
- Providing and managing infrastructure, such as the road network
- Funding and/or providing transport services (such as public transport)
- Regulation of certain transport markets (for example, taxi and rideshare markets).

The role of government in these areas will differ in different countries, as will the roles of different levels of government (national, state, local). Governments also have other tools available to encourage (or discourage) certain behaviors, including funding (e.g., for research) and public information campaigns.

14.3.2 REGULATION AND NEW TECHNOLOGY

What is the role of government with regard to new technology? New technology (transport or otherwise) could fall into three broad categories: (1) allowed under current laws, (2) banned under current laws, and (3) allowed with conditions or restrictions. The use of ADSs (vehicles at Level 3–5) is likely banned in most countries at the present time, due to laws assuming the presence of a human driver (National Transport Commission, 2016a; b). However, it appears that some countries, such as New Zealand, could support the deployment of automated vehicles under current laws and, in many, it may be that it is ambiguous—that is, it is unclear whether the laws would allow these vehicles or not.

An initial review of the regulatory barriers in Australia found over 700 regulatory barriers across state and federal legislation to the deployment of vehicles at Level 3 and above, which included regulations covering driving responsibilities, vehicle standards, interaction with enforcement officers, and interactions with other road users (National Transport Commission, 2016a). The question for governments that currently have barriers or where the law is unclear is: should they allow this new technology? Most governments appear to be moving towards allowing greater automation given the potential safety benefits. In the case of automated vehicles, however, governments need to weigh factors including whether: the technology offers significant benefits to the public; the technology also comes with significant potential risk; there is not yet clear research to understand and assess all of the risks; and, the technology is still in its infancy and is likely to evolve.

If these vehicles are allowed, the question becomes: should government place any restrictions or conditions on them, such as requiring that the technology meets basic safety requirements? Many countries take the view that regulation should only be introduced when there is a demonstrated market failure. This is difficult to demonstrate for new technologies where there are no historical data available.

Worldwide, over 1.2 million people die each year from road crashes. Driving is a dangerous activity. Heavy objects (cars and trucks) move at high speed in areas shared with other road users, including pedestrians. There is a clear argument for laws to manage safety, as governments do today. Automated vehicles could reduce current road tolls, but they could also introduce new road safety risks. Given the risks and the evident existing danger of roads, there would appear to be a clear case for regulation.

Governments currently regulate extensively in the area of road transport. This regulation tends to be highly prescriptive—how fast we can drive; how we should react to certain road signs and traffic lights; and to whom we should give way. Even so, these rules do not cover every potential situation on the road. In some situations, the rules may in fact be impossible to follow precisely.

14.3.3 QUANTIFYING THE SAFETY RISKS OF AUTOMATED VEHICLES

Part of the challenge is the lack of clear safety performance metrics for automated vehicles during the current testing phase. Due to the rarity of fatal crashes in terms of kilometers traveled (despite the large overall number of deaths), it is difficult to convincingly demonstrate that automated vehicles are safer without traveling a prohibitive distance. Some jurisdictions compel companies testing ADSs to report "disengagements"; that is, the frequency with which the ADS disengages and hands back control to the human safety driver. This is used by the media and the public as a proxy for the safety of the ADS. However, companies working in this area have indicated that this may not be an accurate indicator of safety. Disengagement rates could be "gamed" by companies driving only in the safest possible areas to improve their figures and could discourage safety drivers from taking back control in dangerous situations.

Governments must consider how they will gather research and evidence to determine the extent of risks. For example, at this point governments are not trying to define human–machine interfaces for manufacturers, but this does create risk; some will design better systems than others and the poorer ones could lead to crashes. There may be opportunities to develop guidance or high-level principles.

14.3.4 WHAT POLICY AND REGULATORY CHANGES DO AUTOMATED VEHICLES REQUIRE?

If governments decide to allow these vehicles—and again, most governments appear to be moving in this direction—governments must assess what regulatory changes are required. These will fall into three broad categories:

1. **Barriers**—remove existing barriers to enable the deployment of this technology. For example, rules related to driving that effectively require a human to be the driver are one barrier. This may include removing uncertainties in legislation.
2. **Gaps**—create new regulations to address new risks created by these vehicles. For example, regulation today does not set safety standards for automated vehicles.
3. **Opportunities**—are there areas where changing regulation may allow governments to take advantage to improve outcomes for the public that were not available previously? This could, for example, include governments using infrastructure more efficiently due the greater predictability and safety of these vehicles.

Gaps, barriers, and opportunities may differ in different jurisdictions. However, in most jurisdictions at least some changes are likely to be required in the areas of:

- Vehicle standards
- Road transport regulation—driving obligations or road rules, enforcement of these laws
- Vehicle and driver insurance requirements
- Heavy vehicle/commercial vehicle regulation—again including vehicle standards, regulation of hours of work, and other safety requirements
- Passenger transport regulation—including laws regulating buses, taxis, and rideshare
- Criminal law—such as dangerous driving and other on-road offences (Glassbrook, 2017; Channon, McCormick, & Noussia, 2019).

There are a range of other policy areas that may require regulatory changes that relate to other roles of government as provider of services and a regulator of markets. These include:

- Land use planning
- Financial regulation—e.g., regulation of private insurers impacted by motor accident injury insurance change
- Consumer law
- Education
- Labor relations
- Taxation and pricing
- Provision of infrastructure
- Infrastructure standards
- Competition/anti-trust regulation—in particular, if there are a limited number of suppliers, that these suppliers are running their own vehicles and that they potentially take over what are currently separate markets—e.g. taxis, repairers.

These will not be the focus of this chapter, but will need to be considered by governments to ensure a comprehensive response. Governments could address some of these issues after commercial deployment, once automated vehicles are an established part of the vehicle fleet. But government will need to address safety issues up front.

14.3.5 APPROACHES FOR GOVERNMENT TO ADDRESS REGULATORY BARRIERS, GAPS, AND OPPORTUNITIES

If we conclude that changes to regulation are required, there are several approaches that governments could take to regulate automated vehicles. One strategy is to simply "wait and see": do not make changes until there are detailed studies providing deep evidence of the safety benefits and the risks. However, this may not be feasible or desirable. The public may be keen to use technology that they see as providing benefits. Governments generally support beneficial technology, in particular for safety, even in the face of uncertainty. Policy is often made with imperfect information. Doing nothing in this case may prevent the use of technology that can save lives.

A second approach could be to open the market completely, without any regulation. Governments could rely on consumer choice, tort law, and the market to ensure safety without specific regulation. However, this may risk unsafe products entering the market that could cause fatalities and serious injuries. Lawsuits and market mechanisms may not respond quickly enough to remove these products. Crashes could undermine public confidence in the technology, which would ultimately hurt industry and decrease the uptake of even well-designed systems (see also, this Handbook, Chapter 5).

A third approach would be to develop detailed standards based on current expectations. This has the benefit of setting clear expectations to industry and the public. However, due to factors listed above in Section 14.3.2 (i.e., there is no clear research to understand all of the risks; the technology is still in its infancy and is likely to evolve) such an approach runs the risk of quickly becoming out of date as our understanding of the risks and the technology changes.

A fourth and final approach is to set forth clear outcomes in the form of principles. The focus would be the overall safety of the entire system, to guide the development of systems by industry, whilst allowing industry to innovate to meet broad outcomes. This comes with risks, including the uncertainty it leaves for the industry and the public. The benefit is in allowing innovation whilst holding industry responsible for the outcome—whether vehicles operate safely or not.

These approaches can also be combined or applied at different levels. Governments could also influence deployment through non-regulatory tools such as guidance, public education (to influence purchasing decisions and other behavior), and grants to encourage research. Eggers and Turley (2018) suggest five options for governments to consider for regulation of emerging technology generally, which align with some the approaches outlined above: (1) adaptive regulation; (2) regulatory sandboxes (lighter rules within limited areas or restricted operation); (3) outcome-based regulation; (4) risk-weighted regulation; and (5) collaborative regulation.

14.4   HOW DOES POLICY AND REGULATION
ADDRESS HUMAN FACTORS SAFETY RISKS?

Now that we have considered the role of human factors in ADSs and the role of government and regulation for these systems, we need to examine how they interact. How does government and regulation interact with human factors issues? If we consider human factors to be about the design of systems—that they are human-centered in design and take into account human fallibilities and mistakes, then the question becomes to what degree, and in what detail, governments should be involved in attempting to influence (1) how companies design systems for human users and (2) how humans interact with those systems?

Hedlund (2000) describes "three fundamental injury prevention strategies...

* Persuade person at risk to change their behavior
* Require behavior change by law or administrative rule
* Provide automatic protection through product and environmental design" (Hedlund, 2000, p. 82).

We need to be aware of the limitations of these approaches and how much we can expect people to change: "in designing a system, we have much more freedom in how we specify the operating characteristics of the machine than in how we specify the characteristics of the human operator. That is, we can redesign and improve the machine components, but we shouldn't expect to be able to (or be permitted to) redesign and improve the operator" (Proctor & Van Zandt, 2008, p.10).

Governments can attempt to exert control over human behavior in their interactions with technology through laws (e.g., laws to prevent drivers using mobile phones). Governments can also attempt to train and educate users, either formally (for example, requirements for driver licenses) or informally (such as through public education campaigns).

Even in highly controlled workplace environments, "we can carefully screen and extensively train our operators before placing them in our system but many limitations that characterize human performance cannot be overcome" (Proctor & Van Zandt, 2008, p. 10). In road transport, road users (drivers, pedestrians, cyclists, and other) will have a range of understanding of systems (and the law), and it would seem unrealistic to suggest that we can extensively screen and train all of them. Whilst we may improve public education, screening and training for the entire population (including children) appear completely unrealistic.

The alternative is to focus on influencing the design of the system. But before we consider this in detail, we should examine what we can learn from other modes.

14.4.1   GOVERNMENT REGULATION OF HUMAN FACTORS
AND AUTOMATION IN OTHER MODES

Road transport is the not the first area of regulation to deal with automation. We already have significant automation—from assisting to fully replacing human drivers—in rail, aviation, maritime, ports, mines, and other workplaces. These are covered by a variety of existing laws, from mode-specific laws (rail and aviation) to more general workplace health and safety laws.

To what degree do these other regulatory systems regulate human factors and what can we learn from them? Aviation regulation has a strong focus on human factors, including the risks of automation (see also, this Handbook, Chapter 21). Regulators provide detailed guidance on these issues for the design and operation of aircraft. However, airlines rely on a highly trained cohort of pilots to supervise automated systems.

Rail regulators have dealt with driverless trains for decades. These generally fall under a regime focused on broad safety duties for the rail operator, who must demonstrate how they are managing the safety risks. There are currently no specific regulations on automation in Australian rail law; the operation of these systems is addressed through the broader safety duties on operators and other parties (National Transport Commission, 2016a). That is, there is a focus on the overall safety of the system, a principles-based approach.

Maritime law already considers human factors, for example through the International Maritime Organization's Sub-Committee on Human Element, Training, and Watchkeeping. It is beginning to examine automation; however, it does not currently regulate automated ships.

Other workplaces, such as ports, come under general workplace health and safety regulation that provide overall obligations on employers to maintain a safe workplace. This, again is a principles-based approach.

These schemes generally provide broad safety duties to cover the risks, whilst allowing industry to develop the detail of systems and applications. More detailed standards are being developed over time, either by government or industry, depending on the different regime.

14.4.2 WHAT IS THE ROLE OF PRESCRIPTIVE VERSUS PRINCIPLES-BASED REGULATION?

We can conclude from the above sections that there will be human factors safety risks for ADSs, and these risks may not be addressed by the market, and at this stage, it is not possible to quantify fully these risks or even reliably identify all of the risks. Research is ongoing. Other regimes have sought to address similar risks in automation through general safety duties, coupled with industry standards.

Given the challenges identified above, governments will need to develop an appropriate approach to regulation. No detailed regulation is likely to address all of the risks already identified, let alone those that are currently unknown and may emerge as these vehicles are deployed.

International standards are likely to be developed over time, for example through the United Nations Economic Commission for Europe. However, these standards can take years to develop and have generally trailed the commercial release of driver assistance systems. According to Eastman (2016, p. 55) "[i]n the longer term, standardization of driver–vehicle interfaces and driver education and training programs will be necessary to facilitate wide-spread acceptance of autonomous vehicles." However, detailed standards (what is being referred to here as prescriptive-based regulation) at this stage of the technology's development may restrict industry innovation. Standards can only deal with known risks.

While dealing with known risks is an important starting point, there may be prescriptive obligations to address these risks, and this may still leave gaps. The automotive industry has recognized that there will be unknown risks in automated vehicles (Wood et al., 2019). A principles-based approach could better ensure risks are managed without restricting innovation. In particular, it could help ensure that system design accounts for human factors safety risks.

14.5 CONCLUSION—HOW SHOULD GOVERNMENTS REGULATE FOR HUMAN FACTORS SAFETY RISKS FOR AUTOMATED VEHICLES?

Policy makers considering how to regulate for automated vehicles should consider the following issues:

- *Start from the key safety risks*—can we attempt to categorize these risks and begin to assess their likelihood and impact? Government and researchers have begun to develop these classifications, with an initial focus on Level 2 vehicles (Russell et al., 2018). Significant additional research is required to better understand and quantify the safety risks (and safety benefits) of automated vehicles.
- *Examine the individuals involved*—including their motivations and existing behaviors.
- *Identify the parties that influence those safety risks*—including both companies and individuals.

Policy makers will need to examine whether there is a need for regulation and carefully consider the potential impacts on behavior, both positive and negative. Responsibility needs to be clear—if we are focused on the design of the system, then responsibility for system design needs to be clear, whether it is a single party or several parties.

If regulation is required, policy makers should consider how to ensure that regulation can be flexible and cover evolving technology. This could include: consideration of what is put into legislation versus guidelines; the choice of principles-based approaches or prescriptive rules; and consideration of how much discretion to provide to regulators or enforcement officers.

- *Monitor, research, and review*—As with humans, governments do not always monitor systems as closely as they could. There will be an opportunity with ADSs for governments to better monitor road safety, due to the data that these vehicles will provide. There will be a strong need for research and monitoring by governments to identify new risks, understand their likelihood and severity, and react accordingly. This will need to examine not just the human–machine interface between the vehicle and the passengers/

driver/fall-back ready user inside. It will also need to look at how other road users interact with these vehicles, how their behavior evolves over time, and what this might mean for safety.

Automated vehicles offer significant safety benefits. Government will need to ensure that regulation maximizes these benefits while minimizing the risks.

ACKNOWLEDGMENTS

This chapter is based on the work as part of the author's role at the National Transport Commission. I would like to thank all the members of the team, past and present, whose research, writing, and thinking contributed to the analysis in this chapter. The views expressed in this chapter are those of the author and do not represent the views of the National Transport Commission.

REFERENCES

Casner, S. M., Geven, R. W., Recker, M. P., & Schooler, J. W. (2014). The retention of manual flying skills in the automated cockpit. *Human Factors*, 56(8), 1506–1516.

Channon, M., McCormick, L., & Noussia, K. (2019). *The Law and Autonomous Vehicles.* Oxfordshire, UK: Informa Law Routledge.

Eastman, A. D. (2016). Self-driving vehicles: Can legal and regulatory change keep up with new technology? *American Journal of Business and Management*, 5(2), 53–56.

Eggers, W. D. & Turley, M. (2018). *The Future of Regulation: Principles for Regulating Emerging Technologies.* Deloitte Centre for Government Insights. Retrieved from www2.deloitte.com/us/en/insights/industry/public-sector/future-of-regulation/regulating-emerging-technology.html?id=us:2em:3pa:public-sector:eng:di:070718

Glassbrook, A. (2017). *The Law of Driverless Cars: An Introduction.* Somerset, UK: Law Brief Publishing.

Hedlund, J. (2000). Risk business: Safety regulations, risk compensation, and individual behavior. *Injury Prevention*, 6(2), 82–90.

Logan, D. B., Young, K., Allen, T., & Horberry, T. (2017). *Safety Benefits of Cooperative ITS and Automated Driving in Australia and New Zealand.* Sydney, Australia: Austroads.

National Transport Commission. (2016a). *Regulatory Barriers to More Automated Road and Rail Vehicles - Issues Paper.* Melbourne: National Transport Commission.

National Transport Commission. (2016b). *Regulatory Options for Automated Vehicles Discussion Paper.* Melbourne: National Transport Commission.

National Transport Commission. (2018). *Safety Assurance for Automated Driving Systems: Decision Regulation Impact Statement.* Melbourne: National Transport Commission.

Proctor, R. W. & Van Zandt, T. (2008). *Human Factors in Simple and Complex Systems.* Boca Raton, FL: CRC Press.

Russell, S. M., Blanco, M., Atwood, J., Schaudt, W. A., Fitchett, V., & Tidwell, S. (2018). *Naturalistic Study of Level 2 Driving Automation Functions* (DOT HS 812 642). Washington, DC: National Highway Traffic Safety Administration.

SAE International. (2018). *Taxonomy and Definitions for Terms Related to Driving Automation Systems for On-Road Motor Vehicles* (J3016). Warrendale, PA: Society of Automotive Engineers.

van Wees, K. & Brookhuis, K. (2005). Product liability for ADAS: Legal and human factors perspectives. *European Journal of Transport and Infrastructure Research, 5,* 357–372.

Wood, M., Robbel, P., Maass, M., Tebbens, R. D., Meijs, M., Harb, M., ... Schlicht, P. (2019). *Safety First for Automated Driving.* Retrieved from www.aptiv.com/docs/default-source/white-papers/safety-first-for-automated-driving-aptiv-white-paper.pdf

World Health Organization. (2009). *WHO Patient Safety Curriculum Guide for Medical Schools.* Geneva, Switzerland: World Health Organisation.

15 HMI Design for Automated, Connected, and Intelligent Vehicles

John L. Campbell
Exponent, Inc.

Vindhya Venkatraman, Liberty Hoekstra-Atwood, Joonbum Lee, and Christian Richard
Battelle

CONTENTS

Key Points 338
15.1 Introduction 338
15.2 Automated Vehicle HMI Design 339
   15.2.1 Communicating Information within a Given Mode 339
      15.2.1.1 Communicating AV System Status and Mode.................. 339
      15.2.1.2 HMI Guidelines for AV Warnings.................................... 340
   15.2.2 Conveying Information about the Transfer between the Driver
      and the ADS 343
      15.2.2.1 HMI Design Issues for TOC to and from the Driver......... 344
      15.2.2.2 HMI Solutions for SA Support: Improve SA for Both
            Normative and Time-Critical Driving Situations 345
15.3 Connected Vehicle HMI Design... 346
   15.3.1 Vehicle-to-Vehicle (V2V) Information.. 347
      15.3.1.1 HMI Design for Status of CVs.. 347
      15.3.1.2 HMI Principles for Presenting Warnings in CVs 348
   15.3.2 Vehicle-to-"X" (V2X) and "X"-to-Vehicle (X2V) Information....... 349
      15.3.2.1 V2X Information.. 350
      15.3.2.2 X2V Information: Prioritizing, Filtering, and Scheduling 350
      15.3.2.3 X2V Information: Multiple Displays 352
      15.3.2.4 X2V Information: Message Content 353
15.4 Intelligent Vehicle HMI Considerations 353
15.5 Conclusions.. 354
References.. 355

KEY POINTS

- The guidance that is available on the design of HMIs, plus various SAE and ISO documents, generally reflects pre-2015 research conducted on driver information/safety systems that provided little or no automated driving capability or connectivity;
- Insofar as the existing guidance is relevant to ACIVs, the basic driver information needs, HMI considerations for transitions of control alerts and warnings, and high-level principles of message management are well understood and have been documented in a variety of sources.
- The development of more comprehensive and effective HMI guidelines will require a better understanding of the changing nature of driving and of the implications of these changes for HMI design.

15.1 INTRODUCTION

A key design element in advanced vehicles is the human–machine interface (HMI).[1] The HMI refers to displays that present information to the driver and controls that facilitate the driver's interactions with the vehicle as a whole and indicate the status of various vehicle components and sub-systems (Campbell et al., 2016). In the context of vehicle safety systems in particular, the HMI should effectively communicate information while managing driver workload and minimizing distraction (Jerome, Monk, & Campbell, 2015).

HMI design requirements for automated, connected, and intelligent vehicles (ACIV) must be determined in the context of many considerations, including their influence on safety, public perception and perceived value, the mix and behaviors of legacy vs. connected vs. automated vehicles (AV) over time within the vehicle fleet, and the degree and type of automation associated with the HMI (see also Noy, Shinar, & Horrey, 2018). In general, safe and efficient operation of any motor vehicle requires that the HMI be designed in a manner that is consistent with driver needs, limitations, capabilities, and expectations—a continuing challenge is to identify just what these are amidst the changing and uncertain landscape of advanced vehicle technology.

Despite these challenges, our objective in this chapter is to summarize what we do know (or at least, what we think we know) regarding HMI design principles for ACIV. Many of these principles are aimed at the important issues raised in the preceding chapters; i.e., how can the design of the HMI be used to increase trust (Chapter 4), manage workload (Chapter 6), and improve situation awareness (SA) in ACIV (Chapters 7 and 13). All these goals support the broader goal of safety—the safety of the drivers and occupants of ACIV, as well as the safety of all road users.

---

[1] The literature generally uses the terms Human–Machine Interface (HMI) and Driver–Vehicle Interface (DVI) interchangeably; we will use HMI throughout this paper but view HMI and DVI to be synonymous for our purposes.

15.2   AUTOMATED VEHICLE HMI DESIGN

Considerable research has been conducted about how to design the HMI for vehicles with Level 0 and Level 1 automation (see Chapters 1 and 2 for a discussion of the levels of automation). Driver preferences, behaviors, and performance have been extensively studied using surveys, mock-ups, driving simulators, test tracks, and real-world driving; much of this research has been summarized and codified into design guidance (e.g., Campbell, Richard, Brown, & McCallum, 2007; Campbell et al., 2016), and into standards and best practices published by organizations such as the Society of Automotive Engineers (SAE) International and the International Standards Organization (ISO).[2] However, relatively few published research studies provide actionable insights into the questions surrounding how to design the HMI for vehicles with higher levels of automation. In many respects, this reflects the nascent level of maturity of the technology, but perhaps a broader challenge is the host of uncertainties surrounding the circumstances and scenarios in which automated driving system (ADS) HMIs will be fielded.[3] In particular, AV warnings may need to be richer and more carefully designed than the simpler hazard warnings that have been the focus of much of the published research. For example, driver warnings and even status information in AVs have the added long-term goals of aiding the driver to develop and maintain a functional mental model of the system, as well as supporting and increasing driver trust in the system. Information may also be frequently presented in situations where the driver is not fully engaged in the driving task and may be unaware of current conditions.

With these caveats in mind, we provide tentative design principles below for

- providing both basic status and mode information,
- identifying the key principles for the presentation of warning information,
- facilitating transfer of control (TOC), and
- supporting improved SA.

15.2.1   COMMUNICATING INFORMATION WITHIN A GIVEN MODE

When the driver is operating in a given level of automation (i.e., control is not being transferred between levels), the HMI must be designed appropriately for the level of automation.

15.2.1.1   Communicating AV System Status and Mode

Automation mode refers to the level and type of automation that is active at a particular time. This includes the specific driving functions (e.g., steering, speed maintenance, and/or braking) that are automated and other information that will aid the driver's understanding of the system's current operation. The status of automation refers to the information about the system overall, including mode, that is communicated to

---

[2] Jeong and Green (2013) as well as Campbell et al. (2016) provide extensive lists and summaries of documents published by SAE and ISO that are relevant to HMI design.

[3] Automated Driving Systems are used in this chapter to refer to automated vehicles with one or more driver support features (SAE Levels 0–2) and automated driving features (SAE Levels 3–5).

TABLE 15.1

Principles for Presenting System Status Information in AV

Type of Status Information	What Information to Provide	Why Information Is Provided
System activation or on/off status	A display indicating which automation feature/function/mode is currently active.	To support driver awareness of current automation mode when the driver seeks this information.
Mode transition status	A display indicating that a TOC is occurring or that one will occur in the near future.	Under normal operating conditions, this information is presented to help drivers maintain awareness of the driving tasks.
Confirmation of successful transfer from automated to manual control	A display or message confirming for the driver that control has been transferred to the driver as they would expect, or communication of a failed/incomplete TOC if the transfer is unsuccessful.	To indicate a successful TOC from the automation system to the driver.
System fault or failure	A display or message indicating that part of the system has failed, is not functioning correctly, or that the system has reached some operational limit.	To alert drivers that they must intervene and reclaim control of driving tasks that have previously been performed by automation, due to a system fault or failure.

the driver. Appropriate feedback about automation status and mode is important for (1) maintaining driver's SA, (2) communicating if the driver's requests (e.g., a request for a TOC) have been received by automation, (3) informing drivers if the system's actions are being performed properly, and (4) informing drivers if problems are occurring (Toffetti et al., 2009).

Overall, vehicles should display the information that drivers need to maintain an understanding of the current and impending automation status and modes. Table 15.1 (from Campbell et al., 2018) shows the types of status information that can be provided to the driver about the automation and design considerations for presenting this information.

15.2.1.2 HMI Guidelines for AV Warnings

Past research and guidelines are available to support the design of visual, auditory, and haptic warnings (see e.g., Campbell et al., 2007; 2016; 2018). A note of caution is warranted here: few research studies are available to support the range of warning situations and conditions associated with Level 2–4 automation. In general, the utility of a particular warning approach may vary with the level of automation, the manner in which the automation is implemented, and the driver's level of engagement with the driving situation and conditions. Thus, we will focus here on three design parameters of warnings that are both well understood and highly applicable to AVs: selecting warning modality, reducing false and nuisance warnings, and using a staged (or graded) approach to warnings.

15.2.1.2.1   Selecting Warning Modality

The modality of warning presentations can impact driver responses and behavior. The type of modality that is appropriate for a message depends on the driving environment (e.g., expected vehicle/cab noise and vibration, hazard scenario, etc.), the criticality of the message (e.g., hazard versus non-hazard situations), the location of the visual displays (assuming those locations cannot be changed), and other factors. Most of the relevant literature (e.g., Kiefer et al., 1999) suggest that performance can be improved by combining auditory and visual messages when presenting warnings. In general,

- *Auditory warnings* are capable of quickly capturing the driver's attention and can be used to present short, simple messages (e.g., simple or complex tones or speech messages) requiring quick or immediate action, including high-priority alerts and warnings (Lerner, Kotwal, Lyons, & Gardner-Bonneau, 1996). The auditory mode may be particularly effective in getting the driver's attention in situations where the driver is distracted or not looking at the roadway or the instrument panel. Especially with higher levels of automation, the presentation of auditory warnings may have to be integrated/coordinated with other sound sources (e.g., entertainment systems) to aid the driver's perception and understanding of the warning.
- *Visual messages* are best for presenting more complex information (Deatherage, 1972) that is non-safety-critical and does not call for immediate action, including continuous information (uninterrupted presentation of information over a trip segment, a trip, or even a longer period of time), lower-priority information such as navigation instructions, or cautionary information. In this regard, visual messages presented through the HMI may be used to help drivers recover their SA after a period of disengagement by presenting information about surrounding vehicles, locations of potential hazards, or upcoming turns. If the visual display is presented in a standard location inside the vehicle (e.g., the instrument panel or the center stack) and requires the driver to look away from the roadway to acquire the message, it can be distracting.
- *Haptic/tactile messages* (like auditory warnings) are capable of quickly capturing the driver's attention and can be used if an auditory message is unlikely to be effective. Two types of haptic interfaces are discussed in the literature: *vibrotactile* and *kinesthetic*; these types of haptic interfaces have fundamental differences that impact how well drivers detect and understand haptic messages. *Vibrotactile interfaces* provide information to the driver using vibrations and may be included in seat belts, seats, foot pedals, and the steering wheel. *Kinesthetic interfaces* provide information by causing limb or body motion. Some examples of this type of haptic interface are when counterforces are applied through the accelerator pedal to "push back" the driver's foot, or when brake pulse displays cause a sudden jerky motion, or when steering wheel rotations cause the drivers hands and arms to move.
- Haptic/tactile warnings may be useful, for example, when delivering a takeover request. A recent study (Petermeijer, de Winter, & Bengler, 2016)

indicated that vibrotactile displays have the potential to grab driver's attention when the automation reaches its functional limits. In this case, sufficiently salient stimuli should be used to grab driver's attention, thus either high amplitude and high frequency vibrations or vibrating a large area might be required. In general, haptic messages can serve a similar function as auditory messages and can be useful when drivers are engaged in secondary tasks (e.g., listening to music or watching a video at a high volume) or using portable devices not under the control or purview of the vehicle.

- Critically, all forms of haptic or tactile warnings require physical contact with the driver to deliver information (but also see Gupta, Morris, Patel, and Tan (2013) for recent interest in non-contact haptic interfaces). Depending on the level of automation associated with the vehicle and the way the automation is implemented, the driver may not be in consistent contact with pedals, the steering wheel, or even portions of the seat.
- *Multi-modal feedback* is recommended for takeover requests to minimize the likelihood of misses and provide redundancy gains (Prewett, Elliott, Walvoord, & Coovert, 2012). Some research has found that visual–vibrotactile feedback is more effective compared with visual-only feedback (Prewett et al., 2012). A survey of 1,692 people found that multi-modal takeover requests were the most preferred option in high-urgency scenarios, whereas auditory takeover requests were more preferred in low-urgency scenarios (Bazilinskyy, Petermeijer, Petrovych, Dodou, & de Winter, 2018).

15.2.1.2.2 Reducing False and Nuisance Warnings/Alarms

From Campbell et al. (2016), false alarms are alarms that indicate a threat when no threat exists. They should be avoided as they can cause driver distraction, lead to incorrect decisions and/or responses, and even increase driver's reaction time to true warnings. Nuisance alarms are alarms that correctly indicate a potential threat, but that the driver does not believe are warranted or needed, perhaps because the driver was already aware of the threat or believes that the threat will be resolved without driver intervention. Importantly, drivers may not necessarily make distinctions between false and nuisance alarms. Excessive false or nuisance warnings can increase workload and decrease the driver's trust in the AV system. Lerner et al. (1996) and Horowitz and Dingus (1992) provide some strategies for minimizing the frequency and impact of false/nuisance warnings, including

- Deactivate a warning device automatically when it is not needed during a particular driving situation (i.e., require the shift lever to be in reverse gear to place a backup warning device into the active mode).
- Allow the driver to reduce detection sensitivity to a restricted limit that minimizes false/nuisance warnings without significantly affecting the target detection capability of the device.
- Present a warning only after a target or critical situation has been detected as continuously present for some specified minimum time.

- Mitigate annoyance by allowing the driver to reduce warning intensity or volume.
- Change modality as the severity of the situation increases (e.g., warn first visually, then add auditory component as the severity increases).

15.2.1.2.3 Using Staged Warnings

A considerable body of research supports the value of staged warnings in the automotive environment for aiding both driver's comprehension of, and response to, a threat or hazard, as well as increased comprehension of system operation (e.g., Lerner et al., 1996; General Motors Corporation & Delphi-Delco Electronic Systems, 2002; Lee, Hoffman, & Hayes, 2004; Mendoza, Angelelli, & Lindgren, 2011). Staged (or graded) warnings may include two or more stages of cautionary information that increase in urgency proportionally (in terms of timing and perhaps modality) with the criticality of the hazard situation prior to the presentation of a warning indicating an imminent hazard. For example, in a collision warning application, a one-stage warning system may provide only an imminent collision warning (i.e., the warning requires immediate corrective action), while a two-stage system provides a cautionary collision warning (i.e., the warning requires immediate attention and possible corrective action) to cue the driver, followed by a separate imminent collision warning (see also Campbell et al., 2016).

The Cadillac Super Cruise™—a Level 2 vehicle—from General Motors provides a recent example of a staged warning.[4] If the system detects that the driver is not attending to the roadway, it provides a series of warning lights on the steering wheel, audible alerts, and/or a vibration in the seat. Only if the driver fails to respond appropriately to these alerts does the system apply the vehicle's brakes until it stops.

Two- or multi-stage warnings have the benefit of providing continuous information to the driver, provide more time for the driver to recognize and respond to an emerging threat, and may help drivers in developing a functional and coherent mental model and better awareness of the operation and limits of the automation system (Campbell et al., 2016).

15.2.2 Conveying Information about the Transfer between the Driver and the ADS

The HMI must also be designed to facilitate the TOC between the ADS and the driver; note that there are two types of TOC: (1) from the driver to the system and (2) rom the system to the driver. A recent naturalistic driving study collected 454 hours of autopilot use from Tesla drivers and found a total of 16,422 transfers of control (Reimer, 2017). In the study, the number of transfers from the human driver to the system was 8,211, and the number of transfers from the system to the human driver was 8,253. Furthermore, transfers from the system to the human driver split into two categories: (1) transfers initiated by the human drivers ($n = 8,211$) and (2) transfers initiated by the system ($n = 42$) (Reimer, 2017). Much of the literature in this area is

---

[4] https://www.consumerreports.org/autonomous-driving/cadillac-super-cruise-may-lead-to-safe-hands-free-driving/

concerned with a system-initiated TOC when the driver is unengaged or has low SA; many relevant scenarios here involve a potential hazard. However, TOC is an issue with AVs even when the driver is situation aware. Imagine that a driver is monitoring the dynamic driving task with their hands on the wheel and feet on the pedals. Automatic emergency braking could be too slow to avoid a child who suddenly runs out into the street, but a steering maneuver initiated by the driver may successfully avoid the hazard. Will the driver recognize this while in Level 2, as well as he or she does in Level 0 or 1? And, if not, how can the HMI be used to support the correct decision? If the driver is unengaged when the TOC request is made, this poses additional burdens on the HMI. Augmented reality head-up displays (HUDs) have been proposed as one solution to highlight the areas that pose potential threats or are of central concern to the immediate driving task (i.e., contain safety-critical traffic information). In general, the HMI needs to be designed to support all these situations; some general guidance for TOCs is presented below.[5]

15.2.2.1 HMI Design Issues for TOC to and from the Driver

15.2.2.1.1 TOC from the Driver to the System

This reflects a series of operations through which the driver transfers responsibility for performing part of or the entire driving task to the automated system. Providing appropriate information through the HMI is important during these transfers to maintain a driver's trust in the system and to help drivers maintain awareness of driving tasks and the broader driving situation. In general, the system should aid the transition from manual to automated driving by acknowledging a driver's request to engage the automation and providing information about the status of the TOC throughout the process. The following principles can be used to support this goal:

- The current system status should always be provided (Toffetti et al., 2009).
- Automation engagement requests should be acknowledged upon receipt to prevent duplicate or conflicting inputs from the driver and to prevent the driver from releasing control of the vehicle without the automation being activated.
- Feedback acknowledging a driver automation activation request should be provided within 250 ms of the driver's input (ISO 15005, 2002; AAM, 2006).
- If the transfer was successful, a notification should be provided to the driver along with an updated automation status display.
- If the transfer was unsuccessful, the driver should be provided with a notification as to the failure of the automation to engage and the reason why the automation did not engage (Tsao, Hall, & Shadlover, 1993; Merat & Jamson, 2009).

---

[5] Research on this topic is limited. These principles are perhaps most relevant to transitions occurring in lower levels of automation; i.e., Level 3 or below.

- The use of uni- or multi-modal notifications and messages should be based on the context of the situation.
- Distinctive messages should be used for successful and unsuccessful TOC events.

15.2.2.1.2   TOC from the System to the Driver

This reflects a series of operations through which the automated system transfers responsibility for performing part of or the entire driving task back to the driver. Importantly, this form of TOC can range from gradual and expected by the driver to immediate and completely unexpected. In general, the system should aid the transition from automated to manual driving by providing information about the need for the driver to take over vehicle control. The following principles can be used to support this goal:

- The driver should be provided with information on when they need to take control (Gold, Damböck, Lorenz, & Bengler, 2013; Blanco et al., 2015).
- The driver should be provided with information on how to take control if a specific control input is required (Toffetti et al., 2009).
- The driver should be provided with information on why the driver needs to take control. For time-critical situations, this may be a simplified "take control" message. For less time-critical situations, more information (e.g., upcoming system limits; Naujoks, Forster, Wiedemann, & Neukum, 2017) may be provided.[6]
- The current system status should always be provided, allowing the driver to validate the disengagement (Sheridan & Parasuraman, 2005).
- Notifications and messages related to a "take control" message should be multi-modal (Blanco et al., 2015; Toffetti et al., 2009; Brookhuis, van Driel, Hof, van Arem, & Hoedemaeker, 2008).

15.2.2.2   HMI Solutions for SA Support: Improve SA for Both Normative and Time-Critical Driving Situations

A consequence of automated driving may be a loss of SA by the driver (see also, this Handbook, Chapters 7, 13). Disengagement from the active control loop and/or a focus on non-driving activities while the vehicle is being controlled by automation may lower the driver's attention to and awareness of numerous aspects of the driving task, including traffic control devices, signs, other vehicles, and potential hazards. There may also be situations where the demands on the driver are too great to expect the driver to be able to assume control of the vehicle in the time available (see e.g., this Handbook, Chapter 6). As discussed in the previous sections, in Levels 2–3 systems, the system and the human driver may transfer control of the vehicle back and forth during driving as needed. When the system initiates the transition, the driver has to transition from being a passive monitor/supervisor in a vehicle being driven by the automated system to becoming an active controller of the vehicle. In this case, the driver must become fully engaged and takeover vehicle control in a timely manner.

---

[6] The driver's ability to take control may be faster and easier with lower levels of automation.

AV designers may assume drivers would perform their allocated roles (e.g., monitoring the driving situation and the system) and maintain their SA in automated driving, but recent fatal crashes (e.g., the self-driving Uber accident in Tempe, AZ) demonstrated the frailty of such assumptions (e.g., the driver watching a video instead of monitoring during automated driving). During the takeover phase, the human driver may have to identify the current situation, surroundings, and required actions in a very short period of time. If the human driver is totally disengaged and does not pay attention to the roadway environment (e.g., engaging in secondary tasks, sleeping, etc.) during automated operation, it will take a relatively longer time to build up SA to an appropriate level for intervention. In worst cases, drivers may not be able to take over within the available time or may contribute to other errors.

Proactive HMIs could help driver's attention management and takeover performance by incorporating information about driver state (e.g., overall readiness, glance history, secondary task activity, etc.; see also this Handbook, Chapter 11) and immediate tactical demands into an alerting approach that helps direct the driver's attention to time-critical information. Future vehicle designs may be able to use an advanced, proactive HMI that can help drivers re-engage with the driving task through information that is tailored to their specific information needs. For example, the HMI can be used to directly provide critical information to the driver or to help direct the driver's attention to critical information and/or information elements in the roadway environment. This could include timely presentation of attentional cues, alerts, or critical roadway information (e.g., missed guide signs or temporary roadside messages; see also principles for designing to support SA developed by Endsley, 2016). This could include design features such as an HUD with augmented reality capabilities that could be integrated with a driver state monitoring system to present such information. In the short term, the proactive HMIs can help drivers' takeover performance and decrease takeover time, and in the long term, this could improve driver trust and prevent potential misuse and disuse of AVs (Parasuraman & Riley, 1997).

15.3   CONNECTED VEHICLE HMI DESIGN

Connected vehicles (CV) wirelessly receive basic safety messages (BSM) from, and broadcast to, a range of sources and sinks in their environment including other vehicles and road users (Harding et al., 2014; ITS JPO, 2018). BSM packets contain information on position, heading, and speed of nearby connected road users (e.g., vehicles, motorcycles, pedestrians, etc.). The safety of the ego-vehicle movement is estimated using projections of others' movements as well as data on non-connected proximal objects using on-board short-range sensors. In addition, CV technologies allow for the linkage of nomadic devices (e.g., mobile phones, wearables, etc.) and cellular telematics (e.g., wireless updates of in-vehicle software), creating the potential for a veritable in-vehicle cockpit featuring various types of information, displays, and controls. Thus, while the central focus of the CV HMI is safety-relevant information, it is important to consider that it also serves as a medium for infotainment and entertainment services.

CV technologies can exist at any level of automation, with basic technologies already available in a substantial portion of the current fleet (e.g., Bluetooth connectivity to personal devices). The central design challenge of CV HMIs is the balance between the driver's access to information and the need for minimal distraction away from activities critical for safe vehicle control (Jerome et al., 2015; this Handbook, Chapters 6, 9). CV HMIs are also available to a variety of road users (pedestrians, bicyclists, and motorcyclists) via personal devices and roadway infrastructure (e.g., pedestrian flashing beacons). Each of these interfaces serves the safety purpose of alerting the user to potential path intrusions. However, the interface and information design considerations vary based on where and how they are being used (for related discussions in this Handbook, see Chapters 7, 13).

Currently, CV technologies are still in the developmental phase (Harding et al., 2014). The design features of CV HMIs, particularly the content and location of information provided to the vehicle occupants, will evolve with advancements in communication standards, infrastructure design, and vehicle technologies. Below, we discuss basic HMI design guidelines for the key configurations associated with CVs, with the expectation that these will change with related technological advancements.

15.3.1 VEHICLE-TO-VEHICLE (V2V) INFORMATION

The display real estate within the vehicle is limited, and safety critical information should be prioritized and salient. Much of this information is related to potential path intrusions by other road users. In this section, we will specify design requirements for displaying safety-relevant vehicle-to-vehicle (V2V) information. V2V information is primarily intended to provide an *omni-directional awareness of other vehicles* (NHTSA, 2017) to help drivers plan maneuvers (e.g., overtaking, merging, etc.), respond to unexpected incursions (e.g., sudden deceleration of lead vehicle), and monitor the movement of traffic.

15.3.1.1 HMI Design for Status of CVs

Drivers perform a variety of driving tasks while navigating in the midst of other vehicles, including lane changes, selecting and maintaining following distances, and overtaking other vehicles. Knowing the positions and movements of other vehicles can enhance the driver's SA and help in smooth execution of vehicle control. V2V technologies extend the amount of information available beyond the limits of the driver's perceptual capabilities and on-board sensors and cameras. This makes them particularly useful in warning the driver of hazards in difficult terrain and visibility, adverse weather conditions, and challenging traffic situations (NHTSA, 2017; Harding et al., 2014). CV HMIs can also provide traffic status information intended for use by the driver in preparation for upcoming maneuvers. Upcoming turns, distance to a destination, and relative positioning of the ego vehicle to a lead vehicle are some of the most commonly provided safety-relevant information. Status information is particularly safety-critical in situations when the driver is preparing to execute potentially demanding maneuvers and may want to survey the status of nearby traffic. Generally, continuous traffic status indications must be minimally intrusive during normal driving to prevent overloading the driver with non-critical information.

Two primary design considerations for provision of status information in CVs are listed below:

- Consider the provision of a display indicating the position and movements of the hazards in situations where the hazards may be hidden or hard to identify from the driver's viewpoint. This information is relevant in unexpected situations (e.g., pedestrian in a limited access highway) or in limited visibility. The driver must be given sufficient time to perceive and process the information on the display, without the risk of inopportune distraction.
- In certain situations, the CV HMI may provide integrated traffic status information (e.g., vehicle movements, traffic signal status, changes in traffic flow relative to planned maneuvers of the ego vehicle) to advise the driver of the safety risks of executing specific maneuvers; for example, see ISO 17387 (2008) and SAE (2010) for standards related to lane change coverage status information. One example is executing a lane change and checking for vehicles encroaching the blind spot. CV HMIs can provide a comprehensive display of the traffic along with advisory information on the safety of executing the lane change (Campbell et al., 2016).

Status information can also be useful in the case of heavy vehicles, such as automated truck platoons. The lead platoon vehicle communicates its movements to following vehicles through V2V communication; these followers can choose to enter or leave the platoon at will (Bergenhem, Hedin, & Skarin, 2012). HMIs in these vehicles may need to communicate different information and allow for different driver control inputs based on the position of the vehicle in the platoon (e.g., the leader, a follower behind an exiting vehicle, etc.) and the planned path (e.g., entering or exiting). A general consideration for heavy vehicle HMIs is to minimize the information units communicated to the driver due to the already dense control and display layouts in these vehicle interiors (Campbell et al., 2016).

15.3.1.2 HMI Principles for Presenting Warnings in CVs

Warnings about immediate hazards should be given priority over other information displayed on the CV HMI. Designers have a choice of providing staged warnings in these situations; however, single stage warnings may be preferred in manually controlled CVs to alert the driver of sudden path incursions and to minimize the incidence of false or nuisance alarms. A cautionary alert followed by an imminent collision alert can be useful for drivers of heavy vehicles to decrease the need for sudden hard braking or steering. Here, we present parameters and principles to consider when designing warnings in CVs (summarized in Campbell et al., 2016). Note that many of these principles apply to AVs (see also Section 15.3).

- Consider staged warnings based on the purpose of the warning and the criticality of the situation (Campbell et al., 2016, pp. 4–6 & 4–7). Single-stage warnings are useful in alerting drivers of imminent threats and minimizing the likelihood of false or nuisance alarms. Multi-stage warnings are useful when providing continuous information (e.g., visual display of decreasing proximity to a lead vehicle in a forward collision warning system).

- Choose warning timings based on driver and situation factors, including driver response times, types of driver responses relative to the movement and projected states of the hazards, and hazard visibility (ISO 15623, 2013; Campbell et al., 2016, pp. 4–9).

- In multi-stage warnings, provide meaningful timing between stages (Kiefer et al., 1999; Lee et al., 2004). The driver should not receive cautionary information so early that they perceive it to be a false alarm. Similarly, they should not receive the imminent collision warning too late to select and execute a suitable response. When it is not possible to provide sufficient time between warning stages, a single-stage imminent collision warning should be used.

- For every stage of a multi-stage warning, the modality for providing information can be varied, as long as it is selected to ensure perception, extraction, and comprehension of the information by the driver (Campbell et al., 2016, pp. 4–6 & 4–7). For example, a short visual message ("Stop signal ahead") can be provided as cautionary information when the driver approaches a traffic light hidden by a roadway curve. In the event that the driver does not slow down, imminent safety warnings can be provided using auditory alerts and/or haptic brake pedal pulses.

- Consider the complementary issues of rapid information perception vs. distraction when locating the CV HMI. Safety critical visual warnings are easily perceived if they are located near the central field of view of the driver. This might, for example, be implemented in an HUD located within ±5° relative to the driver's central line of sight (ISO, 1984; 2005). The instrument panel is also a popular choice, particularly for information that is minimally safety-critical. Care should be taken to not obscure the roadway view of the driver when mounting visual displays; this may be particularly relevant in information-dense heavy vehicle CV HMIs (Campbell et al., 2016, pp. 11–9). Visual warnings can be distracting in situations which require immediate speed or heading changes or difficult to perceive in conditions of high glare. Auditory warnings are generally preferred for collision avoidance information. Auditory and haptic warnings can also be localized to indicate the direction of the hazard (Campbell et al., 2016, pp. 7–12).

15.3.2 Vehicle-to-"X" (V2X) and "X"-to-Vehicle (X2V) Information

Options for CV configurations also include communication between vehicles (V) and other roadway users and infrastructure elements (X). Vehicle-to-"X" (V2X) HMIs can provide other road users with information about vehicles in the surrounding environment. "X"-to-Vehicle (X2V) HMIs give drivers access to information that has been transmitted from other non-vehicle data sources (e.g., pedestrians and cyclists, roadside equipment). From a broad systems implementation perspective, the source of V2X and X2V information is critical (e.g., where are the sensors located, is the data transfer time from the information source fast enough and sufficiently reliable to support the type of message being sent to the driver or other road user?). In this regard, Hartman (2015) provides resources on V2X and X2V implementations and system specifications. However, from the perspective of the HMI user (e.g., the driver or pedestrian) it often does not matter whether this information is collected via the roadway infrastructure, a pedestrian's mobile phone, or hardware in a bicycle, motorcycle, or other personal transportation device. Thus, many of the principles presented above for V2V guidance are relevant to a range of X2V scenarios using in-vehicle displays, but less relevant to HMI designs that involve displays that are not located within a vehicle.

Messages from X2V-supported HMIs are often intended to augment tasks already performed by the driver in order to make them easier, and to convey information that a driver may not be able to obtain—or obtain as promptly as the vehicle can—within the network. Not all information may be relevant or desired by the driver depending on the criticality of the information and the situation. For example, heavy vehicle drivers and bus drivers may have different information requirements than commuters, and a driver stopped at an intersection may have more information bandwidth available than they will once they initiate a turn through the intersection. Many X2V-specific issues and interface elements that have been evaluated within the literature address the task of managing this influx of information. Although the potential applications of X2V and V2X technologies are vast, the ecosystem is still developing, and there are a limited number of studies that have evaluated these HMIs (Lerner et al., 2014). The following sub-sections provide guidelines that apply to V2X/X2V systems and reflect issues unique to HMIs within the CV environment.

15.3.2.1 V2X Information

Much of the currently available V2X literature reflects technical demonstrations rather than formal HMI evaluations. An example of the demonstrations is proof-of-concept studies[7] for mobile phone applications that provide pedestrians or bicyclists with supplementary traffic information to use while crossing the street (e.g., the status of approaching vehicles). Another example is the design of warnings for motorcyclists. Song, McLaughlin, and Doerzaph (2017) evaluated rider acceptance of multi-modal CV collision warning applications for motorcycles in an on-road study. They found that haptic, auditory, and visual modalities were all viable for the collision warning message displays (as measured by rider acceptance), but that haptic and auditory messages were preferred because they do not represent visual distractions. Single-channel auditory messages were not recommended, since hearing tests are not required to legally operate a motorcycle, and loud engine noise can mask auditory messages. Beyond this evaluation, there has yet to be sufficient research on V2X HMI evaluation to support definitive guidance.

15.3.2.2 X2V Information: Prioritizing, Filtering, and Scheduling

In high workload scenarios, drivers may have trouble accessing relevant information in the driving environment if the HMI is burdening them with excessive or

---

[7] See for example: https://www.its.dot.gov/factsheets/pdf/CV_V2Pcomms.pdf

competing information. Messages may be managed through prioritizing, filtering, and scheduling (Campbell et al., 2016). Prioritization may be performed by only showing drivers the information that is most relevant to them. The existing X2V guidelines recommend using message urgency or criticality to derive relative priorities. Filtering can be performed by minimizing the dashboard complexity or creating lockouts for system notifications that may be irrelevant while the car is in motion. Scheduling system messages harnesses functionality afforded by CV capabilities to help pace the presentation of information to the driver and reduce the frequency of multiple warning scenarios (e.g., providing notifications in advance of conflict situations; Lerner et al., 2014)

15.3.2.2.1   Priority

Safety-critical warnings should be coded with sufficient urgency to prioritize the response (Ward et al., 2013). When managing simultaneous alerts, inferior messages may be suppressed based on this value system (Olaverri-Monreal & Jizba, 2016). Urgency of the driver response time is suggested as a way of categorizing alerts in X2V systems, and the highest levels of urgency may be reserved for situations where time-to-event is 5 seconds or less (Lerner et al., 2014). Lerner et al. (2014) also recommends limiting the number of warning categories so that drivers may easily discriminate the salience levels reserved for the highest stages of warning (e.g., "high threat, act now," "caution, measured action," and "no urgency, no action required"). Olaverri-Monreal and Jizba (2016) suggest using a standard priority index (ISO/TS 16951, 2004) to rank warning messages. This index derives the priority value from the response time for the driver to take an action and the potential resulting injuries or damages that may occur if no action is taken (see ISO/TS 16951, 2004 for the equation and categorization process).

When designing messages in an X2V context it is important to distinguish the format of non-safety-critical information from safety-critical information, and not design low-priority messages in a way that implies the driver is required to give an urgent response (Ward et al., 2013; Olaverri-Monreal & Jizba, 2016). This design may be approached in different ways (see Campbell et al., 2016), but consistency across X2V message design based on actual and perceived priority can facilitate fewer instances of perceived false/nuisance alarms, distraction, unnecessary workload, and distrust (Lerner et al., 2014).

15.3.2.2.2   Filtering

While it may be possible to present multiple non-speech auditory and visual X2V alerts concurrently without overloading the driver or negatively impacting performance, more effective driver responses will be elicited if the warning display interrupts and overrides all other messages (Lerner et al., 2014). Information lockouts may also be managed by the driver. In a test-track study where messages of varying relevance to the driving task could be presented, drivers tended to request that the system suppresses messages that aligned with content that the National Highway Traffic Safety Administration' s (NHTSA) visual manual distraction guidelines advise omitting (Holmes, Song, Neurauter, Doerzaph, & Britten, 2016; Olaverri-Monreal & Jizba, 2016).

15.3.2.2.3 Scheduling

To avoid simultaneous safety critical alerts, messages should be paced, if possible. If the X2V system has predictive capabilities, the preferred approach is to suppress non-safety-critical information within a time window preceding the onset of safety-critical messages (Ward et al., 2013). A safety-critical warning should continue until the driver responds appropriately, without provoking driver annoyance or suppressing consecutive safety-critical warnings. Non-safety-critical warnings should endure (unobtrusively) over a period of time that allows drivers to execute a self-paced response (Ward et al., 2013).

15.3.2.3 X2V Information: Multiple Displays

CV X2V technology allows for information that is directly relevant to a particular driver to be displayed in multiple ways within the vehicle as well as outside of the vehicle. Generally, the position of X2V displays (including displays inside and outside of the vehicle) should correspond directionally with key task-related external elements to cue rapid information extraction (Hoekstra-Atwood, Richard, & Venkatraman, 2019a; b; Richard, Philips, Divekar, Bacon-Abdelmoteleb, & Jerome, 2015a).

For messages within the vehicle, visual warnings conveyed simultaneously should only be presented on one physical display (Olaverri-Monreal & Jizba, 2016). Driver responses may be better if messages (even separate messages from separate devices) are presented on a single display rather than separate displays (Lerner & Boyd, 2005).

Driver's visual attention should not be directed towards in-vehicle displays when they need to be looking outside (Stevens, 2016; Svenson, Stevens, & Guglielmi, 2013; Richard et al., 2015a). In-vehicle displays that present warnings should be within the driver's visual field (see earlier sections for location considerations). When non-safety-critical information is presented inside the vehicle, it should be positioned near the periphery of the driver's field of view to be unobtrusive to the demands of the immediate driving task (Olaverri-Monreal & Jizba, 2016). Detailed guidance on HMI display location is provided in Campbell et al. (2016).

A display on roadway infrastructure, or a Driver-Infrastructure Interface (DII), may be part of the X2V system if this type of display provides context and facilitates simplified messaging along with reduced visual workload. See Richard et al. (2015b) for specific guidance on when DIIs may be appropriate. When positioning a DII, designers should consider the driving task, situation, and the proximal roadway environment (Hoekstra-Atwood, Richard, & Venkatraman, 2019a).

If a DVI and DII assess the same hazard, the information or instruction should be consistent and coordinated (Hoekstra-Atwood et al., 2019a; Richard et al., 2015a). However, safety-critical warning messages are not suitable for DIIs because they can be readily seen by road users that are not the intended recipient of the message. The ubiquitous visibility of infrastructure-based messages could have the unintended consequence of warning the wrong drivers and eliciting unnecessary evasive responses (Richard et al., 2015b). In this case, supplementing a cautionary DII message with an imminent collision warning on the in-vehicle HMI may facilitate the appropriate driver's crash-avoidance responses (Hoekstra-Atwood, Richard, & Venkatraman, 2019b).

15.3.2.4 X2V Information: Message Content

Even though X2V technologies afford an abundance of information to drivers, designers should seek to provide drivers with sufficient, but not excessive information and evaluate the levels of distraction and workload imposed by system displays (Olaverri-Monreal & Jizba, 2016). Olaverri-Monreal and Jizba (2016) recommend that while the car is moving, no more than 4 information units[8] should be presented in visual text messages by X2V systems (but no more than 2 information units for safety-relevant messages). Messages should elicit a binary response (single reaction) rather than a choice of responses to avoid imposing additional cognitive processing load (Ward et al., 2013). Greater amounts of information may not translate to greater trust in the information, and drivers could end up spending more time with eyes off the road to take in information with no added safety benefit (Inman, Jackson, & Chou, 2018). Designers must consider both the driving task and how drivers may interpret information differently on an X2V interface versus a road sign. A change in message context (e.g., transferring road sign message content directly to an in-vehicle interface) may change the accuracy of message interpretation by the driver (Chrysler, Finley, & Trout, 2018).

For specific application guidelines and information needs, see Richard et al. (2015b) and Hoekstra-Atwood et al. (2019a). These documents cover the following systems: stop sign assist, signalized left turn assist, red light violator warning, curve speed warning, spot-weather information warning–reduced speed, and pedestrian in crosswalk systems.

15.4 INTELLIGENT VEHICLE HMI CONSIDERATIONS

Intelligent vehicles (IVs) are described as being able to monitor driver behavior and state, and then reconfigure the HMI and vehicle control to improve safety (e.g., Cadillac Super Cruise[9] or "driver attention guard" in Tawari, Sivaraman, Trivedi, Shannon, & Tippelhofer, 2014). Thus, the "intelligent" part of IVs could support improved system performance in an AV or a CV. Configurable features may vary across automation capabilities, driver states, and the purpose of HMI reconfiguration. For example, an IV could communicate messages such as "Do Not Disturb While Driving" to prevent the driver from receiving messages while driving, or it could activate lane-keeping assistance and adaptive cruise control (ACC) when the system detects driver distraction. A central purpose of such intelligent systems is to optimize driver behaviors to improve safety.

There is very little published research in the area, and it is certainly not enough to support strong guidance. However, there are a number of general design principles that could serve as initial considerations. Drivers may be given control over at least parts of these reconfigurations, unless driver choices could result in safety deficits.

---

[8] Information units are a measure of information load. An information unit refers to key nouns and adjectives in the message that provide unique or clarifying information. For example, the phrase "Vehicle ahead. Merge to the right." contains the four information units underlined (Campbell et al., 2016).

[9] https://www.cadillac.com/world-of-cadillac/innovation/super-cruise

For example, L2 vehicles with intelligent features (e.g., Super Cruise) may monitor driver engagement levels, and slow down and eventually stop the vehicle if the driver remains disengaged or unresponsive. In this case, the IV's reconfiguration of safety maneuvers should not be disabled by drivers. At other times, the driver may have access to additional safety-relevant information that invalidates the need for system reconfiguration and should be able to provide control inputs.

IV HMIs are similar to proactive HMIs (discussed in the AV HMI design section), except that they may also reflect changes made to the function or level of automation as well as HMI based on estimations of driver state. Information that could be presented to the driver includes descriptive information about the current driver state (e.g., eyes are off the road, hand are off the wheel) and prescriptive information to help the driver maintain the desired level of automation (e.g., maintain eyes on the road, place hands on the wheel).

15.5 CONCLUSIONS

It should be clear from this chapter that there is much yet to learn about driver information needs and subsequent HMI design requirements for ACIVs. The rapid pace and changing nature of ACIV—combined with the relatively slow pace of research to support design—only adds to the challenges. The guidance that is available (e.g., guidance published by the National Highway Transportation Safety Administration (Campbell et al., 2016; 2018), plus various SAE and ISO documents, generally reflects pre-2015 research conducted on driver information/safety systems that provided little or no automated driving capability or connectivity. Perhaps the best that can be said about such guidance documents is that they provide provisionally useful design principles for ACIV supported by high-quality research. Basic driver information needs, HMI considerations for transitions of control alerts and warnings, and high-level principles of message management are well understood and have been documented in a variety of sources. The existing guidance can also serve as a roadmap for future research; i.e., holes or gaps in the topics covered by the available guidance may reflect areas where more research is needed.

The development of more comprehensive and effective HMI guidelines will require a better understanding of the changing nature of driving and of the implications of these changes for HMI design. Specifically, how will the range of ACIV functionality impact driver information needs, given the concurrent requirements to maintain driver trust, functional mental models, and SA? What new challenges are introduced through automation for which the HMI could serve as a solution? How could a broader focus on information management support driver engagement and SA? Could a proactive, flexible, and dynamic HMI address some of these challenges and, if so, how?

We are highly optimistic that answers to these and similar questions will be answered by the ACIV industry and broader research community. Even at this relatively early stage in the conceptualization and development of ACIV, many recent studies and analyses are serving to shed light on these and related topics and, we hope, will serve as a foundation for future HMI guidance; these include the changing role of the driver in AVs (Noy, Shinar, & Horrey, 2018); the definition and measurement of the "out-of-the-loop" concept (Merat et al., 2018; Biondi et al., 2018; this Handbook, Chapters 7, 21); driver engagement and conflict intervention performance (Victor et al., 2018); the challenges of partial automation (Endsley, 2017); and strategies for attention management in AVs (Llaneras, Cannon, & Green, 2017).

REFERENCES

Alliance of Automobile Manufacturers. (2006). *Statement of Principles, Criteria and Verification Procedures on Driver Interactions with Advanced In-Vehicle Information and Communication Systems*, including 2006 Updated Sections [Report of the Driver Focus-Telematics Working Group]. Retrieved from www.autoalliance.org/index. cfm?objectid=D6819130-B985-11E1-9E4C000C296BA163.

Bazilinskyy, P., Petermeijer, S. M., Petrovych, V., Dodou, D., & de Winter, J. C. (2018). Take-over requests in highly automated driving: A crowdsourcing survey on auditory, vibro-tactile, and visual displays. *Transportation Research Part F: Traffic Psychology and Behaviour, 56,* 82–98.

Bergenhem, C., Hedin, E., & Skarin, D. (2012). Vehicle-to-vehicle communication for a pla-tooning system. *Procedia-Social and Behavioral Sciences, 48,* 1222–1233.

Biondi, F. N., Lohani, M., Hopman, R., Mills, S., Cooper, J. M., & Strayer, D. L. (2018). 80 MPH and out-of-the-loop: Effects of real-world semi-automated driving on driver workload and arousal. *Proceedings of the Human Factors and Ergonomics Society Annual Meeting, 62*(1), 1878–1882.

Blanco, M., Atwood, J., Vasquez, H. M., Trimble, T. E., Fitchett, V. L., Radlbeck, J., … Morgan, J. F. (2015). *Human Factors Evaluation of Level 2 and Level 3 Automated Driving Concepts* (DOT HS 812 182). Washington, DC: National Highway Traffic Safety Administration.

Brookhuis, K. A., van Driel, C. J. G., Hof, T., van Arem, B., & Hoedemaeker, M. (2008). Driving with a congestion assistant: Mental workload and acceptance. *Applied Ergonomics, 40,* 1019–1025. doi:10.1016/j.apergo.2008.06.010

Campbell, J. L., Brown. J. L., Graving, J. S., Richard, C. M., Lichty, M. G., Sanquist, T., … Morgan, J. L. (2016). *Human Factors Design Guidance for Driver-Vehicle Interfaces* (Report No. DOT HS 812 360). Washington, DC: National Highway Traffic Safety Administration.

Campbell, J. L., Brown, J. L., Graving, J. S., Richard, C. M., Lichty, M. G., Bacon, L. P., … Sanquist, T. (2018). *Human Factors Design Guidance for Level 2 and Level 3 Automated Driving Concepts* (DOT HS 812 555). Washington, DC: National Highway Traffic Safety Administration.

Campbell, J. L., Richard, C. M., Brown, J. L., & McCallum, M. (2007). *Crash Warning System Interfaces: Human Factors Insights and Lessons Learned* (DOT HS 810 697). Washington, DC: National Highway Traffic Safety Administration.

Chrysler, S. T., Finley, M. D., & Trout, N. (2018). *Driver Information Needs for Wrong-Way Driving Incident Information in a Connected-Vehicle Environment.* Retrieved from https://trid.trb.org/view/1494783

Deatherage, B. H. (1972). Auditory and other sensory forms of information presentation. In H. P. Van Cott & R. G. Kinkade (Eds.), *Human Engineering Guide to Equipment Design* (Rev. ed.) (pp. 123–160). Washington, DC: U. S. Government Printing Office.

Endsley, M. R. (2016). *Designing for Situation Awareness: An Approach to User-Centered Design.* Boca Raton, FL: CRC Press.

Endsley, M. R. (2017). Autonomous driving systems: A preliminary naturalistic study of the Tesla Model S. *Journal of Cognitive Engineering and Decision Making, 11*(3), 225–238. doi:10.1177/1555343417695197

General Motors Corporation & Delphi-Delco Electronic Systems. (2002). *Automotive Collision Avoidance System Field Operation Test, Warning Cue Implementation Summary Report* (DOT HS 809 462). Washington, DC: National Highway Traffic Safety Administration.

Gold, C., Damböck, D., Lorenz, L., & Bengler, K. (2013). "Take over!" How long does it take to get the driver back into the loop? *Proceedings of the Human Factors and Ergonomics Society 57th Annual Meeting*, 1938–1942. Santa Monica, CA: HFES.

Gupta, S., Morris, D., Patel, S. N., & Tan, D. (2013). AirWave: Non-contact haptic feedback using air vortex rings. *Proceedings of the 2013 ACM International Joint Conference on Pervasive and Ubiquitous Computing* (pp. 419–428). New York: ACM.

Harding, J., Powell, G., Yoon, R., Fikentscher, J., Doyle, C., Sade, D., … Wang, J. (2014). *Vehicle-to-Vehicle Communications: Readiness of V2V Technology for Application* (DOT HS 812 014). Washington, DC: National Highway Traffic Safety Administration.

Hartman, K. K. (2015). *CV Pilot Deployment Program*. Retrieved from www.its.dot.gov/pilots/cv_pilot_apps.htm

Hoekstra-Atwood, L., Richard, C. M., & Venkatraman, V. (2019a). *Multiple Sources of Safety Information from V2V and V2I: Phase II Final Safety Message Report*. Seattle, WA: Battelle.

Hoekstra-Atwood, L., Richard, C. M., & Venkatraman, V. (2019b). *Multiple Sources of Safety Information from V2V and V2I: Redundancy, Decision-Making, and Trust Phase II P2V Applications*. Seattle, WA: Battelle.

Holmes, L., Song, M., Neurauter, L., Doerzaph, Z., & Britten, N. (2016). *Validations of Integrated DVI Configurations*. Retrieved from www.nhtsa.gov//DOT/NHTSA/NVS/Crash Avoidance/Technical Publications/2016/812310_ValidationsIntegratedDVIconfigurations.pdf

Horowitz, A. D. & Dingus, T. A. (1992). Warning signal design: A key human factors issue in an in-vehicle front-to-rear-end collision warning system. *Proceedings of the Human Factors Society 36th Annual Meeting*, 1011–1013. Santa Monica, CA: HFES.

Inman, V. W., Jackson, S., & Chou, P. (2018). *Driver Acceptance of Connected, Automation-Assisted Cruise Control—Experiment 1*. Retrieved from www.fhwa.dot.gov/publications/research/safety/18041/18041.pdf

Intelligent Transportation Systems Joint Program Office (ITS JPO). (2018). *How Will Connected Vehicles Be Used?* Retrieved from www.its.dot.gov/cv_basics/cv_basics_how_used.htm

ISO. (1984). *Development and Principles for Application of Public Information Symbols* (ISO/TR 7239). Geneva, Switzerland: International Organization for Standards.

ISO 15005. (2002). *Road Vehicles—Ergonomic Aspects of Transport Information and Control Systems—Dialogue Management Principles and Compliance Procedures*. Geneva, Switzerland: International Organization for Standards.

ISO/TS 16951. (2004). *Road Vehicles—Ergonomic Aspects of Transport Information and Control Systems (TICS)—Procedures for Determining Priority of On-Board Messages Presented to Drivers*. Retrieved from www.iso.org/iso/catalogue_detail.htm?csnumber=29024

ISO. (2005). *Road Vehicles—Ergonomic Aspects of Transport Information and Control Systems (TICS)—Warning Systems* (ISO/TR 16352). Geneva, Switzerland: International Organization for Standards.

ISO 17387. (2008). *Intelligent Transportation Systems—Lane Change Decision Aid Systems (LCDAS)—Performance Requirements and Test Procedures*. Geneva, Switzerland: International Organization for Standards.

ISO 15623. (2013). *Transport Information and Control Systems—Forward Collision Warning Systems—Performance Requirements and Test Procedures*. Geneva, Switzerland: International Organization for Standards.

Jerome, C., Monk, C., & Campbell, J. (2015). Driver vehicle interface design assistance for vehicle-to-vehicle technology applications. *24th International Technical Conference on the Enhanced Safety of Vehicles (ESV)*. Washington, D.C.: National Highway Traffic Safety Administration.

Kiefer, R., LeBlanc, D., Palmer, M., Salinger, J., Deering, R., & Shulman, M. (1999). *Development and Validation of Functional Definitions and Evaluation Procedures for Collision Warning/Avoidance Systems* (DOT HT 808 964). Washington, DC: National Highway Traffic Safety Administration.

Lee, J. D., Hoffman, J. D., & Hayes, E. (2004). Collision warning design to mitigate driver distraction. *Proceedings of the SIGCHI Conference on Human Factors in Computing Sciences* (pp. 65–72). Retrieved http://citeseerx.ist.psu.edu/viewdoc/download?doi=10.1.1.77.2168&rep=rep1&type=pdf

Lerner, N. & Boyd, S. (2005). *On-Road Study of Willingness to Engage in Distracting Tasks* (DOT HS 809 863). Washington, DC: National Highway Traffic Safety Administration.

Lerner, N., Kotwal, B. M., Lyons, R. D., & Gardner-Bonneau, D. J. (1996). *Preliminary Human Factors Guidelines for Crash Avoidance Warning Devices* (DOT HS 808 342). Washington, DC: National Highway Traffic Safety Administration.

Lerner, N., Robinson, E., Singer, J., Jenness, J., Huey, R., Baldwin, C., & Fitch, G. (2014). *Human Factors for Connected Vehicles: Effective Warning Interface Research Findings*. Retrieved from www.nhtsa.gov/DOT/NHTSA/NVS/Crash Avoidance/Technical Publications/2014/812068-HumanFactorsConnectedVehicles.pdf

Llaneras, R. E., Cannon, B. R., & Green, C. A. (2017). Strategies to assist drivers in remaining attentive while under partially automated driving. *Transportation Research Record, 2663*, 20–26. doi:10.3141/2663-03

Mendoza, P. A., Angelelli, A., & Lindgren, A. (2011). Ecological interface design inspired human machine interface for advanced driver assistance systems. *IET Intelligent Transport Systems, 5*(1), 53–59.

Merat, N. & Jamson, A. H. (2009). How do drivers behave in a highly automated car? *Proceedings of the International Driving Symposium on Human Factors in Driver Assessment, Training, and Vehicle Design, 5*, 514–521.

Merat, N., Seppelt, B., Louw, T., Engstron, J., Lee, J.D., Johannsson, E., Green, C.A., Katazaki, S., Monk, C., Itoh, M., McGehee, D., Sunda, T., Unoura, K., Victor, T., Schieben, A., & Keinath, A. (2018). The "Out-of-the-Loop" concept in automated driving: Proposed definition, measures and implications. *Cognition, Technology & Work, 21*(1), 87–98. doi:10.1007/s10111-018-0525-8

National Highway Traffic Safety Administration. (2017). *Vehicle to Vehicle Communication*. Retrieved from www.nhtsa.gov/technology-innovation/vehicle-vehicle-communication

Naujoks, F., Forster, Y., Wiedemann, K., & Neukum, A. (2017). A human-machine interface for cooperative highly automated driving. In N. A. Stanton, S. Landry, G. Di Bucchianico, & A. Vallicelli, *Advances in Human Aspects of Transportation* (pp. 585–595). Berlin: Springer.

Noy, I., Shinar, D., & Horrey, W. J. (2018). Automated driving: Safety blind spots. *Safety Science, 102*, 68–78.

Olaverri-Monreal, C. & Jizba, T. (2016). Human factors in the design of human–machine interaction: An overview emphasizing V2X communication. *IEEE Transactions on Intelligent Vehicles, 1*(4), 302–313. doi:10.1109/TIV.2017.2695891

Parasuraman, R. & Riley, V. (1997). Humans and automation: Use, misuse, disuse, abuse. *Human Factors, 39*(2), 230–253.

Petermeijer, S. M., de Winter, J. C., & Bengler, K. J. (2016). Vibrotactile displays: A survey with a view on highly automated driving. *IEEE Transactions on Intelligent Transportation Systems, 17*(4), 897–907.

Prewett, M. S., Elliott, L. R., Walvoord, A. G., & Coovert, M. D. (2012). A meta-analysis of vibrotactile and visual information displays for improving task performance. *IEEE Transactions on Systems, Man, and Cybernetics, Part C (Applications and Reviews), 42*(1), 123–132.

Reimer, B. (2017). Human Centered Vehicle Automation. *Presented at the European New Car Assessment Programme*, Antwerp, Belgium.

Richard, C. M., Philips, B. H., Divekar, G., Bacon-Abdelmoteleb, L. P., & Jerome, C., (2015a). Driver responses to simultaneous V2V and V2I safety critical Information in left-turn across path scenarios. *Proceedings of the 2015 Annual Meeting of the Human Factors and Ergonomics Society* (pp. 1626–1630). Santa Monica, CA: HFES.

Richard, C. M., Morgan, J. F., Bacon, L. P., Graving, J. S., Divekar, G., & Lichty, M. G. (2015b). *Multiple Sources of Safety Information from V2V and V2I: Redundancy, Decision Making, and Trust—Safety Message Design Report.* Seattle, WA: Battelle.

SAE J2802. (2010). *Blind Spot Monitoring System (BSMS): Operating Characteristics and User Interface.* Warrendale, PA: SAE International.

Sheridan, T. B. & Parasuraman, R. (2005). Human-automation interaction. *Reviews of Human Factors and Ergonomics, 1*, 89–129.

Song, M., McLaughlin, S., & Doerzaph, Z. (2017). An on-road evaluation of connected motorcycle crash warning interface with different motorcycle types. *Transportation Research Part C: Emerging Technologies, 74*, 34–50.

Stevens, S. (2016). *Driver Acceptance of Collision Warning Applications Based on Heavy-Truck V2V Technology.* Retrieved from www.nhtsa.gov//DOT/NHTSA/NVS/Crash Avoidance/Technical Publications/2016/812336_HeavyTruckDriverClinicAnalysis.pdf

Svenson, A. L., Stevens, S., & Guglielmi, J. (2013). Evaluating driver acceptance of heavy truck vehicle-to-vehicle safety applications. *23rd International Technical Conference on the Enhanced Safety of Vehicles.* Retrieved from www-esv.nhtsa.dot.gov/Proceedings/23/isv7/main.htm

Tawari, A., Sivaraman, S., Trivedi, M. M., Shannon, T., & Tippelhofer, M. (2014). Looking-in and looking-out vision for urban intelligent assistance: Estimation of driver attentive state and dynamic surround for safe merging and braking. *Intelligent Vehicles Symposium Proceedings* (pp. 115–120), IEEE.

Toffetti, A., Wilschut, E. S., Martens, M. H., Schieben, A., Rambaldini, A., Merat, N., & Flemisch, F. (2009). CityMobil: Human factor issues regarding highly automated vehicles on eLane. *Transportation Research Record: Journal of the Transportation Research Board, 2110*, 1–8. doi:10.3141/2110–01

Tsao, H.-S. J., Hall, R. W., & Shadlover, S. E. (1993). *Design Options for Operating Fully Automated Highway Systems.* Berkeley, CA: University of California PATH Institute of Transportation Studies.

Victor, T. W., Tivesten, E., Gustavsson, P., Johansson, J., Sangberg, F., & Ljung Aust, M. (2018). Automation expectation mismatch: Incorrect prediction despite eyes on threat and hands on wheel. *Human Factors, 60*(8), 1095–1116. doi:10.1177/0018720818788164

Ward, N., Velazquez, M., Mueller, J., & Ye, J. (2013). Response interference under near-concurrent presentation of safety and non-safety information. *Transportation Research Part F: Traffic Psychology and Behaviour, 21*, 253–266.

16 Human–Machine Interface Design for Fitness-Impaired Populations

John G. Gaspar
University of Iowa

CONTENTS

Key Points 359
16.1 Introduction 360
16.2 Adaptive Automation 361
    16.2.1 When to Adapt? 362
    16.2.2 How to Adapt? 363
    16.2.3 Invocation Authority 365
16.3 A Framework for AA for Impaired Drivers 366
    16.3.1 Distraction 368
    16.3.2 Drowsiness 370
    16.3.3 Alcohol and Other Drugs 372
16.4 Conclusions 372
Acknowledgments 373
References 373

KEY POINTS

- The human–machine interface provides the link between driver state detection and the human operator
- Using driver state information, adaptive automated systems could be designed to adjust their demands and/or the HMI based on the capacity of the driver
- Adaptive automation requires decisions about if, how, and when the automation, including both the vehicle systems and HMI, should adapt, and whether the automation or human has authority to invoke changes in the system
- Adaptive automation applied for driver impairment needs to consider the interaction between the state of the driver and the capability of automation

16.1   INTRODUCTION

The previous chapter (Chapter 15) discussed many important design considerations for the human–machine interface (HMI) for automated and connected vehicles. One additional and significant concern is how that design might be impacted by the ability to monitor driver state. Automation may indeed increase the incidence of drivers being unprepared or incapable of safely operating the vehicle (e.g., this Handbook, Chapter 9). For instance, recent research suggests that partial automation (i.e., Level 2) increases visual disengagement from driving, even in the absence of secondary tasks (Gaspar & Carney, 2019; See also, Russell et al., 2018). Similar research demonstrates an increase in the likelihood of fatigue and drowsiness with even moderately prolonged periods of automated driving (Vogelpohl, Kühn, Hummel, & Vollrath, 2019). Recent crashes involving partially automated vehicles highlight the potential consequences of driver impairment and disengagement from the dynamic driving task (e.g., NTSB, 2017).

Driver monitoring is often presented as the remedy to driver impairment in partially and highly automated vehicles (see e.g., this Handbook, Chapter 11). Indeed, in their report following the investigation of the Williston Tesla crash, the National Transportation Safety Board recommended that driver monitoring could provide a safeguard against driver disengagement and impairment in automated vehicles (NTSB, 2017). Previous chapters (Chapters 9, 11) discussed approaches to driver monitoring and their application in automated vehicles. However, simply knowing the state of the driver is not enough to improve safety. The vehicle must adapt in some fashion to account for the reduced capacity of the driver. This could be through modifying the HMI (e.g., providing feedback), adjusting the vehicle systems (e.g., tuning lane departure warnings), or some combination of the two.

This chapter builds on discussion of driver state monitoring by discussing how driver state information can be considered in HMI design in automated vehicles. Specifically, we consider how information about the state of the driver can be used to dynamically tailor the automation to the driver's capabilities on a moment-to-moment basis. This dynamic, state-based adaptation by the HMI is referred to as adaptive automation (AA). Unlike static automation, whose functionality remains constant when engaged, AA flexibly adjusts the HMI and level of automation based on information about the state of the human operator (Rouse, 1988). AA has been applied in a variety of complex tasks involving control distribution between human operators and automated aides, from monitoring air traffic control displays (Kaber, Perry, Segall, McClernon, & Prinzel III, 2006) to controlling unmanned vehicles (de Visser & Parasuraman, 2011).

This chapter is divided into two sections. First, we provide an overview of AA and the important design decisions that should be considered in its application to driving. We then present a framework for applying AA to driving, specifically driver impairment. The framework considers both the capability of the automation and capacity of the driver, as well as interactions between the two. This framework is considered across common modes of impairment, specifically distraction, drowsiness, and drugs and alcohol (also see this Handbook, Chapter 9).

16.2 ADAPTIVE AUTOMATION

AA refers to systems that dynamically adjust the level of automation or HMI based on the state of the operator (Hancock, Chignell, & Lowenthal, 1985; Rouse, 1988). This contrasts with static automation, which maintains the same level of automation independent of operator or environmental state. The goal of an adaptive system is to tailor the level of automation to meet the needs of the operator and maintain safe operation (Parasuraman, Bahri, Deaton, Morrison, & Barnes, 1992). With impairment (e.g., distraction, drowsiness, drugs) this involves dynamically adjusting the HMI, function of the automated systems, or both, in order to mitigate the detrimental effects of disengagement from the driving task. Note that, with respect to workload, in low-workload conditions, where automation complacency is likely, more control is shifted to the human operator to increase arousal (see also, this Handbook, Chapter 6).

Adaptive systems, as depicted in Figure 16.1, consist of two components, a state monitor and task manager. The state monitor detects and classifies the state of the human operator (and perhaps also the environment; this Handbook, Chapter 11). Operator state information is then fed forward to a task manager, whose role it is to adjust the allocation of tasks between the human operator and system automation (also see this Handbook, Chapter 8). The HMI serves as the link between the operator and automated system. For instance, an HMI might provide feedback to a distracted driver to return his gaze to the forward road. The outcome of behavior (e.g., lane-keeping) and the updated state of the driver are then fed back into the system.

A considerable body of research from different domains shows benefits of AA over static (i.e., non-adaptive) automation (see Scerbo, 2008). For example, Parasuraman, Mouloua, and Hilburn (1999) compared AA that provided adaptive aiding and adaptive task allocation against static automation in a simulated flight task. Adaptive aiding consisted of the system controlling one dimension of aircraft position in high-workload situations (i.e., takeoff and landing). Task control was also temporarily shifted back to the human operator in lower-workload conditions (i.e., the middle portion of the flight). Compared with a non-adaptive control group, AA improved tracking performance and reduced subjective workload.

While AA offers advantages over static automation across a number of tasks, HMI designers face several important questions in implementing AA in vehicles.

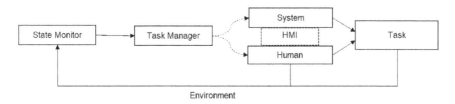

FIGURE 16.1 A framework for AA. The human–machine interface links the human operator to the automated system.

These include when the automation should adapt, what form that adaptation should take, and whether the human or automation has invocation authority. We next consider these questions in the context of vehicle automation.

16.2.1 WHEN TO ADAPT?

Adaptive systems first need to identify when to adjust the HMI. This is dependent on the method used to detect driver state. Systems can be classified based on whether they monitor driver state via driver-related or driving-related measures. The driver monitoring system must then establish a threshold for impairment, beyond which the system will adjust the HMI to manage driver state or provide automated support (see also, this Handbook, Chapter 11).

State monitoring data can come from two sources, the driver and the vehicle. Driver-based measures rely on either direct assessment of operator state through physiological measures or inputs to the vehicle controls. For example, Freeman, Mikulka, Prinzel III, and Scerbo (1999) used electroencephalography (EEG) to identify changes in workload during a monitoring and tracking task. Using this index of workload, they dynamically allocated control between the human and an automated system. Previous chapters of this volume described methods for direct evaluation of driver state through camera-based and other measures (Chapter 9, 11). Eyes-off-road is a common measure of visual distraction and can be used to trigger feedback to the driver (Donmez, Boyle, & Lee, 2008). State monitoring systems might also system input (e.g., steering wheel torque) to identify changes in driver state. Vehicle-based measures use vehicle sensors to detect changes in performance related to changes in driver state. For example, increased deviations in lateral vehicle position and increased lane departure rate can be used to identify likely increases in drowsiness (Schwarz et al., 2015).

These approaches each have advantages and potential drawbacks. Measuring changes to performance is often easier and is also a more direct evaluation of changes in safety. However, because changes in performance are the eventual manifestation of particular impairment states, relying on performance metrics as a state indicator may result in late detection of impairment. For instance, using run-off-road events to identify drowsiness may let drivers become drowsy past the point where an intervention could effectively mitigate impairment. Additionally, as automation assumes a greater share of vehicle control, performance measures will no longer represent the manifestation of driver state and will therefore prove ineffective for classifying impairment.

Physiological measures, on the other hand, provide a more direct evaluation of driver state. These approaches can therefore theoretically detect impairment earlier, perhaps even before impairment reaches dangerous levels. An adaptive system could thus intervene earlier, when more options might be available to preserve safety. Such sensitivity may, however, come with costs in that drivers may not yet be aware of changes in their own state at the early stages of impairment. If systems adapt in these instances, it could be perceived as a false alarm and decrease trust in the system (Parasuraman & Riley, 1997).

16.2.2 How to Adapt?

HMI designers also need to consider what form AA should take once the task manager calls for a change in system state. Sheridan (1992) provided a useful framework for considering how automation can be applied to the human–machine relationship (see Table 16.1). The framework consists of ten levels of automation, from full manual control to fully automated. Within this range, the distribution of control (and responsibility) between the human operator and automated system varies (see also, Chapter 8). A transition point occurs between Levels 6 and 7, where the human operator either does or does not have input on an automated decision before the action is executed. Inagaki and Furukawa (2004) therefore added an additional stage 6.5, where the automation simultaneously acts and informs the operator. The goal of such a stage is to combine the benefits of automated behavior (e.g., fast responding) while preventing automation surprises, where the operator is unsure why the automation behaved in a certain way. Limiting surprises is crucial to user acceptance of automation.

Inagaki and Furukawa (2004) considered how these levels might be applied in an adaptive cruise control system. In a Level 4 system (see Table 16.1, not to be confused with SAE levels), the system might provide a forward collision warning to the driver. In a Level 6 system, the system might give a forward collision warning and if the driver does not respond, initiate emergency braking (i.e., automatic emergency braking, AEB). In a Level 6.5 system, the vehicle applies emergency braking and provides a collision warning simultaneously. This has the advantage of initiating the response faster while still making the driver aware of the system's intentions. Finally, a Level 7 system might engage emergency braking and inform the driver after the fact that emergency braking was applied due to a forward collision situation. Such information, while seemingly unnecessary in most situations, can be potentially

---

TABLE 16.1
Levels of Automated Control

1. Full human control
2. Automation offers a set of action alternatives, and…
3. Narrows the selection, or
4. Suggests one, and
5. Executes the suggestion if the driver approves, or
6. Allows the human a restricted time to veto before automatic execution, or
7. Simultaneously executes automatically and tells the human what it is going to do, or
8. Executes automatically, then necessarily informs humans, or
9. Informs driver after execution only if asked, or
10. Informs driver after execution if automation decides to
11. Fully automate control

Note: After Inagaki and Furukawa (2004).

useful in teaching drivers about the edge cases that define the operational design domain of a system (see this Handbook, Chapter 18).

These degrees of automated control can be dynamically applied across different components of an information processing framework consisting of four stages: information acquisition, analysis, decision-making, and action execution (see Figure 16.2; Parasuraman, Sheridan, and Wickens, 2000; see also, this Handbook, Chapter 6). A distinction can be made between lower- and higher-order processes in this framework based on the degree to which information must be cognitively manipulated. Information acquisition and action execution are considered lower-order processes and analysis and decision-making, requiring greater cognitive processing, are considered higher-order functions.

Kaber, Wright, Prinzel III, and Clamann (2005) considered the potential implications of applying AA to each stage of the information processing framework in an air traffic control monitoring task. Participants were instructed to locate and "clear" aircraft on a control display before they reached a certain location. Participants could only move a portion of the display through a viewing portal and had to shift the portal to track multiple aircrafts. Operator state was evaluated via performance on a secondary gauge monitoring task and used to trigger AA. Participants experienced four automation conditions and a manual condition. Acquisition automation controlled movement of the viewing portal. Analysis automation provided a table of all active aircrafts. Decision-making automation prioritized aircraft to clear. Action implementation automation automatically cleared aircraft the operator had selected.

Kaber et al. (2005) found that AA applied to lower-order information processing stages (acquisition and execution) improved performance relative to manual control. However, applying AA to higher-order functions actually degraded performance. Operators had greater difficulty returning to manual control in the analysis and decision AA conditions. Kaber et al. (2005) suggest this effect may be due to the transparency of automation or how easy it is for the operator to assess the reliability of the AA (i.e., how well the automation is working at any point in time). With lower-level functions, such as automatically clearing selected aircraft, it is easy for operators to determine whether the automation is active and successful. With higher-order AA, additional processing is necessary to evaluate the automation's decisions against the mental model of the operator. Furthermore, if decision-making automation repeatedly makes and executes choices in a complex environment, it may be difficult for operators to maintain a clear understanding of the situation

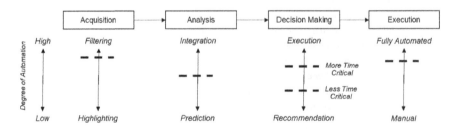

FIGURE 16.2 Automation applied to different information processing stages. Dashed lines represent the maximum degree of AA at each stage.

(Parasuraman et al., 2000). Similar costs of automation have been observed with high levels of automated information processing, such as display cueing (Yeh, Wickens, & Seagull, 1999).

The dashed lines in Figure 16.2 represent the extent to which a processing stage might be maximally automated, using the automation continuum from full manual control to full automation (see Table 16.1). Both lower-order processes can be highly automated (Levels 6.5–10), although, as noted earlier, insight into when and why automation performs a specific function might be helpful in improving driver understanding and awareness of automation functioning. Automation applied to higher-order processes, analysis and decision-making, is more likely to reduce situation awareness and take the driver out of the control loop (Scerbo, 2008). Thus high decision-making autonomy is only appropriate to the extent the driver can disengage from the driving task (see also, this Handbook, Chapters 7, 21). If drivers must remain aware of the driving situation (i.e., conditional automation), it is important that drivers at least have insight into the functions of the automation (Level 6.5 and below).

Parasuraman et al. (2000) outlined several other important considerations in how automation could be applied to the information processing framework. First, and most importantly, the resulting state of the joint driver–vehicle system should be safer with automation applied than if the driver was in full manual control. That is, the addition (or adaptation) of automation should increase safety and decrease the likelihood and severity of crashes. The goal of AA is to achieve a desired level of operator workload, not to obviate the dynamic driving task from the human operator in situations where doing so diminishes safety.

A designer must also consider the demands involved in a particular situation and the costs associated with a failure. In time-critical situations with insufficient time for the operator to respond, automated decision-making and action implementation may be ideal (Scerbo, 2008). Automatic emergency braking (AEB) is an example of such a situation, in that the vehicle can respond faster and with harder braking than a human driver possibly could. In less time-constrained high-risk situations, the extent of decision-making performed by the automation depends on the capability of the system and whether the driver is expected to intervene.

16.2.3 INVOCATION AUTHORITY

The third question designers need to consider is whether the system or human operator is responsible for adjusting the HMI. That is, should the system or the driver act as task manager? It is important to note the distinction between adaptive systems, where the adaptation is controlled by the system, and adaptable systems, where the human controls when and how the system adjusts (see also, this Handbook, Chapter 21). Research generally suggests that AA outperforms adaptable, human-initiated automation. Kaber and Riley (1999) compared mandated (i.e., adaptive) and elective (i.e., adaptable) automation in a radar monitoring task. Operator state was assessed via performance on a secondary task. Mandated automation adaptation resulted in significantly larger performance improvements relative to manual control compared with the elective system (see also, Bailey, Scerbo, Freeman, Mikulka, & Scott, 2006).

The major limitation of adaptable automation is that operators often lack insight into their own state. Humans are poor judges of their own mental and physical capacity and may therefore choose to invoke automation (or remain in manual control) at inappropriate times or fail to invoke automation when it is most needed, such as under high workload (Horrey, Lesch, Mitsopoulos-Rubens, & Lee, 2015; Morris & Rouse, 1986; Sarter & Woods, 1994). Indeed, Neubauer, Matthews, Langheim, and Saxby (2012) found that voluntary invocation of automation failed to reduce fatigue and stress in a sample of fatigued drivers. Humans are poor judges of the extent to which impairment states might negatively impact performance and safety. For example, Horrey, Lesch, and Garabet (2008) showed that drivers were poorly calibrated to the detrimental effects of distraction on closed-course driving. Therefore, it seems advantageous that an adaptive vehicle interface assume invocation authority in instances of driver impairment, particularly during safety-critical tasks.

Inagaki, Itoh, and Nagai (2007) proposed a situation-adaptive form of AA. The idea is that in certain situations, particularly those with high degrees of time-criticality where human operators may be incapable of responding fast enough, the automation should make decisions about when and how to respond (if it is capable of doing so). An important caveat of this idea is the importance of the automation informing the driver of its intentions, if a level of joint control is expected (see also, this Handbook, Chapter 8).

16.3 A FRAMEWORK FOR AA FOR IMPAIRED DRIVERS

The question then is how a vehicle equipped with driver monitoring technology should adapt based on different types and degrees of driver impairment. Impairment is defined by the extent to which the driver is capable of safely controlling the vehicle. As impairment increases, the capability of the human driver for safely performing aspects of the driving task decreases, shown in Figure 16.3, with shaded regions representing the transition in adaptive support provided by the impaired driver. At early stages of impairment, feedback via the HMI may be sufficient to alert the driver to a change in state and motivate corrective action. If impairment persists, the vehicle might provide adaptive support, such as tuning safety features like lane departure warning. At the highest degrees of impairment, it will be necessary for the automation to take full control of the vehicle because the operator is no longer capable of safely performing the requisite driving tasks.

Automation capability refers to the capacity of the system to assume various functions of the dynamic driving task. A key point in this discussion is that the capability of automation bounds the degree to which an adaptive system can support an impaired driver. The maximal extent to which the vehicle is capable of intervening is determined by the abilities and limitations of automation. In short, automation can only control aspects of the driving task it is capable of safely performing. Therefore, it is necessary that the driver state monitoring system intervenes before impairment exceeds the capabilities of automation.

The critical consideration in this framework thus becomes the joint human–automation capability, illustrated in Figure 16.4, where the bars represent the set of tasks that must be performed for safe driving. In certain situations, the capabilities of

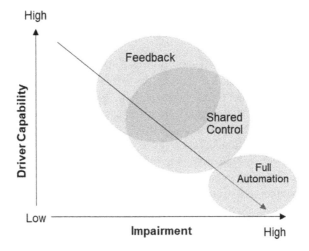

FIGURE 16.3 Relationship between capability, impairment, and vehicle adaptation, represented by the transition across shaded regions.

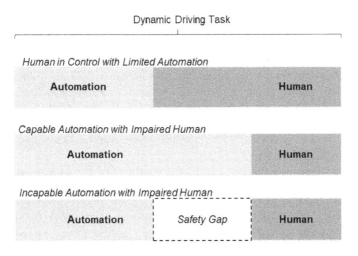

FIGURE 16.4 Relationship between human operator capacity, automation capability, and demands of the dynamic driving task.

the human operator may be limited by impairment. In such situations, the automation must intervene and control more of the driving task. If the automation fails or is incapable of intervening, this leaves a safety gap, a portion of the dynamic driving task not accounted for by either the automation or human operator. Yamani and Horrey (2018) applied this framework of shared control to the individual information processing stages (Figure 16.2). This model predicts how drivers might deploy attention across different levels of automation, considering varying levels of distributed control.

The goal of AA in automated vehicles is then to prevent the driver from reaching a level of impairment that exceeds the capability of the automated systems to control the driving task.

Consider two examples with a drowsy driver, first a vehicle with no automation and second a highly automated vehicle. In the first example, the vehicle is incapable of subsuming any portion of the driving task. The state detection system must therefore monitor the driver, and the task manager should intervene before the driver reaches a level of impairment, resulting in a safety gap between human and automated capabilities. For example, take a drowsy driver who is considering to starting a drive. A vehicle with low automation might warn the driver before the drive begins, because the automation is not capable of controlling the vehicle should the driver fall asleep. A highly automated vehicle, on the other hand, may be capable of performing the entire driving task. Such a vehicle might therefore allow the drowsy driver to disengage entirely (i.e., fall asleep), because the automation is capable of performing all tasks without leaving a gap in safety. Human input may in fact be harmful in such a situation, given how poorly drivers estimate their drowsiness levels (FHWA, 1998), and the responsibility of the AA would be to block the impaired driver from retaking control.

In specifying these impairment thresholds for adapting automation, the design must also strike a balance between safety and driver acceptance. Drivers must understand the correspondence between the impairment thresholds and changes in performance and safety. That is, they must trust the system to identify when driving is no longer safe. Changes in the HMI in situations the driver does not perceive as alarming are referred to as nuisance alerts (Kiefer et al., 1999). Nuisance warnings have a direct negative impact on trust and subsequent willingness to use a system (Bliss & Acton, 2003; Lee, Hoffman, & Hayes, 2004). Appropriate feedback about the state and function of automation is also critical to engender operator trust in both the state detection system and the automated vehicle control (Lee & See, 2004; this Handbook, Chapter 15).

In situations of joint human–automation control (i.e., partial automation), as in Figure 16.4, it is also important to consider the relationship between the expectations of the driver and the automated vehicle. For example, Inagaki et al. (2007) found that an action support system that automatically executed a response maneuver was effective at avoiding collisions. However, the system was, in general, not accepted by drivers. The authors posit that this was because the behaviors of the automation differed from the ways the driver expected the automation to behave, suggesting that the intentions of the adaptive system should match those of the driver. Similarly, in situations where the human operator is unable to detect unsafe levels of impairment or unwilling to alter unsafe behavior, automated intervention may be necessary (Saito, Itoh, & Inagaki, 2016).

16.3.1 DISTRACTION

Distraction can be defined as the diversion of attention away from the tasks necessary for safe driving (Lee, Young, & Regan, 2008). Much research in the last 20 years has explored the effects of distraction on driver performance and safety.

Distraction can be roughly classified as visual or cognitive, based on whether a secondary task diverts a driver's visual attention from the forward roadway (see also, this Handbook, Chapters 6, 9).

Visual distraction has a clear negative impact on safety. For manual driving, research indicates an appropriate threshold may be 2 seconds of eyes-off-road time. Off-road glances longer than 2 seconds increase crash risk and lane-keeping ability, as well as slowing response times to hazard events (Klauer, Dingus, Neale, Sudweeks, & Ramsey, 2006; Russell et al., 2018; Victor et al., 2015). Visual distraction can also be measured by the total eyes-off-road time associated with a task interaction. Research suggests that 12 or more seconds of total eyes-off-road time is an appropriate threshold for manual driving (NHTSA, 2012). Cognitive distraction also has the potential to disrupt safe driving, despite a driver's gaze never leaving the forward road (Strayer, Watson, & Drews, 2011), although the impact may be a risk somewhat lower than visual distraction (Dingus et al., 2016). It is worth noting that few tasks are purely cognitive in nature, and overall interactions with non-driving activities and devices such as cell phones increase crash risk (Dingus et al., 2016). Measuring cognitive distraction is more challenging than measuring visual distraction, particularly in production vehicle environments (e.g., Baldwin et al., 2017). In such cases, driving-based methods of impairment detection may be necessary to identify degraded state.

Research mostly on manual driving and distraction suggests adaptive feedback can reduce distraction and improve performance. Donmez, Boyle, and Lee (2007) provided real-time feedback based on off-road glances (using an eye tracker). A two-tiered feedback system provided alerts to the driver through a heads-up display or via the in-vehicle task display when gaze had been off road for two and two and a half seconds, respectively. Adaptive feedback reduced the amount of time drivers spent looking at the secondary task display, which theoretically increased the amount of time drivers spent attending to the forward road. Importantly, both real-time and retrospective feedback have shown the potential to improve driving performance and to reduce visual distraction (Donmez et al., 2008) (also see this Handbook, Chapter 9).

Beyond providing feedback, more capable automated systems can offer real-time assistance to drivers in high-workload or distracted states. Dijksterhuis, Stuiver, Mulder, Brookhuis, and de Waard (2012) compared a non-adaptive and adaptive lane-keeping support system with full manual control. The lane-keeping support consisted of feedback about lateral position via a heads-up display. In the non-adaptive condition, lane position feedback was continuous, whereas in the adaptive condition, feedback was triggered when performance thresholds were exceeded (e.g., time spent near lane edge). Compared with both non-AA and no support conditions, drivers showed improved lane-keeping performance and greater acceptance with the adaptive system. Dijksterhuis et al. (2012) posit that drivers in the adaptive condition may have used the adaptation as a form of feedback to know when a performance threshold was crossed.

A key question with higher levels of automation will be where the threshold for inattention should be drawn. As automation becomes more competent, drivers will conceivably be able to look away from the driving task for longer durations. In a small naturalistic study of partially automated driving, Gaspar and Carney (2019) found a significant percentage of individual glances and off-road interactions that exceeded established thresholds for manual driving (i.e., 2 and 12 seconds, respectively). Future research must establish how long is too long to look away in with different levels of automation capability.

16.3.2 DROWSINESS

Unlike distraction, which represents a relatively discrete disengagement from driving, drowsiness is a continuous and progressive state of impairment (also see this Handbook, Chapter 9). That is, over the course of a typical trip, drowsy drivers will become progressively drowsier (though see Williamson et al., 2011). Drowsiness can be simply defined as a state of reduced alertness associated with the inclination to fall asleep (Wierwille, 1995). At the early stages of drowsiness, behavioral changes such as increased reaction time manifest themselves (e.g., Kozak et al., 2005). In later stages, drivers actually begin to momentarily fall asleep, a phenomenon known as a microsleep event (Dinges, 1995). These events are marked by long (>500 ms) eyelid closures, resulting in clearly degraded driving performance, particularly the ability to maintain lateral vehicle control and heightened risk of roadway departures (Boyle, Tippin, Paul, & Rizzo, 2008).

Adaptive drowsiness countermeasures have mostly focused on providing feedback to drivers based on either physical state or driving performance (see this Handbook, Chapter 9, 11). Research suggests adaptive in-vehicle countermeasures can be effective in either mitigating or compensating for performance decrements related to drowsiness. In a simulator study, Gaspar et al. (2017) tested the effectiveness of several simple state-based drowsiness countermeasures, which were triggered based on the output of a steering-based drowsiness detection algorithm (see also, Atchley & Chan, 2011; Berka et al., 2005). The countermeasures consisted of either auditory-visual or haptic alerts and were either discrete (single stage) or staged (multiple stages of increasing urgency). These feedback warnings, and particularly warnings that escalated in severity with continued evidence of drowsiness, reduced the frequency of drowsy lane departures compared to a control group with no adaptive mitigation. The warnings in this study can be considered as a fairly straightforward type of feedback, similar to systems available in production vehicles.

Kozak and colleagues (2006) examined the effectiveness of different modalities of lane departure warnings for drowsy drivers. Warnings included steering wheel vibration, simulated rumble-strip sounds, and a heads-up display paired with steering wheel torque. Each of these warnings was effective for drowsy drivers, reducing response time to lane departures and decreasing the magnitude of lane excursions relative to baseline. May, Baldwin, and Parasuraman (2006) found that auditory forward collision warnings could reduce the chance of a head-on crash for drivers showing signs of active task-induced fatigue.

Higher levels of AA have also been employed successfully in the context of drowsy driving. Saito et al. (2016) studied the effectiveness of an adaptive lane-keeping system in a driving simulator. The adaptive system used changes in lane-keeping behavior as an index of drowsiness and implemented a staged adaptive system using corrective steering. At the first stage, which was triggered by a lane departure, automation provided corrective steering to prevent a severe lane departure (i.e., holding the vehicle in position). The driver then had time to perform a manual correction. If no manual correction was applied, the automation provided additional steering to re-center the vehicle. Saito et al. (2016) showed that this level of adaptive support was effective in preventing severe lane departures.

With drowsy driving, it is important to strike a balance between early detection of impairment and the need to avoid false alarms and nuisance alerts. As noted, the detrimental effects of drowsiness manifest early in the process of actually falling asleep. An adaptive system that can intervene at these stages may be able to keep a driver awake longer or motivate a driver to stop to rest while he is still capable of safely controlling the vehicle. However, drivers may be resistant to such systems in that they may not perceive these early signs of drowsiness as important for safety. As drowsiness detection technology becomes more sensitive to earlier symptoms of drowsiness, designers will have to consider the implications of potentially adapting automation earlier for drowsy drivers.

At the other end of the drowsiness continuum, a remaining question is what the vehicle should do in situations where a driver is too impaired to control the vehicle. In the adaptive system studied by Saito et al. (2016), if drivers repeatedly failed to provide a manual steering correction after the initial adaptive automated correction, the vehicle came to a stop in the current lane of travel. Another option would be for the automation (when capable) to pull the vehicle to the shoulder or, with more capable automation, for the vehicle to maneuver to a safe location for the driver to rest. Additional research is needed to consider the impact of such interventions and how partially and highly automated vehicles should behave when drivers are highly impaired. Another potential concern with adaptive systems is that drivers will over-rely on the automated assistance and continue driving longer than they would have without automation. For instance, Saito et al. (2016) reported instances of drowsy drivers continuing to drive with the adaptive lane-keeping system, presumably by relying on the automation to maintain safety (i.e., prevent severe lane departures) despite long eye closures. Saito et al. (2016) dealt with this situation by having the automation bring the vehicle to a stop if drivers repeatedly failed to provide a steering correction following the initial warning. Future research should focus on understanding the potential safety and long-term behavioral implications of strategies to yield appropriate reliance on AA.

Finally, nearly all the research on adaptive systems for drowsy drivers has focused on the efficacy of interventions over relatively short periods of driving (e.g., 1 hour). However, much of drowsy driving occurs during the course of long, multi-hour drives. The motivation tradeoffs in these situations become even more complicated, with drowsy drivers weighing the benefits of reaching a destination earlier against the potential safety costs of falling asleep at the wheel. Future research should address the impact of adaptive in-vehicle systems on driver behavior during longer trips.

16.3.3 ALCOHOL AND OTHER DRUGS

Alcohol impairment is more straightforward in that legal limits have been set defining thresholds for impairment (currently a breath alcohol concentration of 0.08 or greater). In-vehicle alcohol detection systems being developed as part of the Driver Alcohol Detection System for Safety (DADSS) program will be capable of assessing blood–alcohol content and locking the driver out when impairment above the legal limit is detected (Zaouk, Wills, Traube, & Strassburger, 2015; this Handbook, Chapter 9). As with drowsiness impairment, if alcohol thresholds exceed during the drive, the system would need to determine how best to bring the vehicle to a stop if it is incapable of fully taking control from the driver.

The situation with other drugs such as cannabis is considerably more complicated. One challenge is the lack of reliable and sensitive driver state evaluation technology (Compton, 2017). There currently exists no reliable roadside test to accurately evaluate concentrations of cannabis in the system. Instead, officers must rely on roadside impairment tests using behavior as an indicator of impairment. Furthermore, cannabis may remain in the body long after impairing effects have diminished, and chronic users may show significantly more muted impairment than novice users under similar dosages (Hall & Solowij, 1998). Similarly, prescription pain medications and other drugs have clear detrimental effects on driving performance (e.g., Brown, Milavetz, Gaffney, & Spurgin, 2018), yet measuring changes in driver state resulting from these drugs in the vehicle is a challenging task.

To this end, it may be more appropriate to use driving-based detection strategies to define driver impairment as a trigger for AA. Using performance-based vehicle adaptation has the advantage that instead of monitoring for intoxication directly, the system looks for degradation in driving performance. In most cases, there would likely be fluctuations in control of speed or lateral position (Brown et al., 2018). In such conditions, the system could provide automated support to the driver, such as the adaptive lane-keeping system employed by Saito et al. (2016) for drowsiness. Much research is needed to both identify methods for classifying and predicting impairment from drugs and to understand how an adaptive automated system might interact with a drug-impaired driver.

16.4 CONCLUSIONS

This chapter provides a framework for how adaptive automated systems can interact with impaired driving populations, using the relations among the type and degree of driver impairment, the information processing stage at which an intervention is needed, and the capability of the vehicle technology to identify the specific type of interaction that is needed. This framework considers the preceding body of research on AA in a number of tasks. In addition to the capability of the automation itself, designers of adaptive HMIs for automated vehicles need to consider how changes in automation will be invoked, when adaptation will occur, and to what degree various components of the driving task will be automated.

There are several key points that should be considered in this discussion as they relate to driver impairment. First, it is important that research helps explore how drivers will respond to these different interventions and how the joint human–vehicle system will behave under different task conditions. As Parasuraman et al. (2000) note, an automated system is only beneficial to the extent that the final human–machine relationship improves task performance and safety.

Second, it is important to consider the degree to which drivers are accepting and trusting different adaptive interfaces and behaviors. This will also require understanding the impact of factors like feedback and automation transparency in the design of adaptive interfaces. As this chapter shows, there are complex interactions between a number of factors that must be considered by HMI designers. These HMI design decisions have consequences for whether drivers will ultimately want to use particular systems.

AA has the potential to leverage exciting new developments in driver monitoring technology to make driving safer and more enjoyable for fitness-impaired populations. Yet driver state information is only useful to the extent it can be used to implement an HMI that will leverage the capability of automated vehicle systems to compliment or compensate for driver capacity.

ACKNOWLEDGMENTS

The author would like to thank William Horrey and Donald Fisher for their insightful and constructive feedback on earlier drafts of this chapter. The author would also like to thank several colleagues for discussion that led to the ideas outlined in this chapter, including Cher Carney for discussion of visual distraction in automation, Timothy Brown and Chris Schwarz for considering the role of driver monitoring in automation and investigating the efficacy of different countermeasures for drowsiness, and Daniel McGehee for discussions regarding trust and driver acceptance of driver monitoring technology in automated vehicles.

REFERENCES

Atchley, P. & Chan, M. (2011). Potential benefits and costs of concurrent task engagement to maintain vigilance: A driving simulator investigation. *Human Factors, 53*(1), 3–12.

Bailey, N. R., Scerbo, M. W., Freeman, F. G., Mikulka, P. J., & Scott, L. A. (2006). Comparison of a brain-based adaptive system and a manual adaptable system for invoking automation. *Human Factors, 48*(4), 693–709.

Baldwin, C. L., Roberts, D. M., Barragan, D., Lee, J. D., Lerner, N., & Higgins, J. S. (2017). Detecting and quantifying mind wandering during simulated driving. *Frontiers in Human Neuroscience, 11*, 406.

Berka, C., Levendowski, D., Westbrook, P., Davis, G., Lumicao, M. N., Ramsey, C., … Olmstead, R. E. (2005). Implementation of a closed-loop real-time EEG-based drowsiness detection system: Effects of feedback alarms on performance in a driving simulator. *1st International Conference on Augmented Cognition* (pp. 151–170), Las Vegas, NV.

Bliss, J. P. & Acton, S. A. (2003). Alarm mistrust in automobiles: How collision alarm reliability affects driving. *Applied Ergonomics, 34*(6), 499–509.

Boyle, L. N., Tippin, J., Paul, A., & Rizzo, M. (2008). Driver performance in the moments surrounding a microsleep. *Transportation Research Part F: Traffic Psychology and Behaviour, 11*(2), 126–136.

Brown, T. L., Milavetz, G., Gaffney, G., & Spurgin, A. (2018). Evaluating drugged driving: Effects of exemplar pain and anxiety medications. *Traffic Injury Prevention, 19*(suppl), S97–S103.

Compton, R. (2017). *Marijuana-Impaired Driving - A Report to Congress* (Report No. DOT HS-812-440). Washington, DC: National Highway Traffic Safety Administration.

de Visser, E. & Parasuraman, R. (2011). Adaptive aiding of human-robot teaming: Effects of imperfect automation on performance, trust, and workload. *Journal of Cognitive Engineering and Decision Making, 5*(2), 209–231.

Dijksterhuis, C., Stuiver, A., Mulder, B., Brookhuis, K. A., & de Waard, D. (2012). An adaptive driver support system: User experiences and driving performance in a simulator. *Human Factors, 54*(5), 772–785.

Dinges, D. F. (1995). An overview of sleepiness and accidents. *Journal of Sleep Research, 4*, 4–14.

Dingus, T. A., Guo, F., Lee, S., Antin, J. F., Perez, M., Buchanan-King, M., & Hankey, J. (2016). Driver crash risk factors and prevalence evaluation using naturalistic driving data. *Proceedings of the National Academy of Sciences, 113*(10), 2636–2641.

Donmez, B., Boyle, L. N., & Lee, J. D. (2007). Safety implications of providing real-time feedback to distracted drivers. *Accident Analysis & Prevention, 39*(3), 581–590.

Donmez, B., Boyle, L. N., & Lee, J. D. (2008). Mitigating driver distraction with retrospective and concurrent feedback. *Accident Analysis & Prevention, 40*(2), 776–786.

Federal Highway Administration. (1998). *The Driver Fatigue and Alertness Study*. Washington, DC: Federal Highway Administration.

Freeman, F. G., Mikulka, P. J., Prinzel, L. J., & Scerbo, M. W. (1999). Evaluation of an adaptive automation system using three EEG indices with a visual tracking task. *Biological Psychology, 50*(1), 61–76.

Gaspar, J. G., Brown, T. L., Schwarz, C. W., Lee, J. D., Kang, J., & Higgins, J. S. (2017). Evaluating driver drowsiness countermeasures. *Traffic Injury Prevention, 18*(supl), S58–S63.

Gaspar, J. & Carney, C. (2019). The effect of partial automation on driver attention: A naturalistic driving study. *Human Factors, 61*(8), 1261-1276. doi:/10.1177/0018720819836310.

Hall, W. & Solowij, N. (1998). Adverse effects of cannabis. *The Lancet, 352*(9140), 1611–1616.

Hancock, P. A., Chignell, M. H., & Lowenthal, A. (1985). An adaptive human-machine system. *Proceedings of the IEEE Conference on Systems, Man and Cybernetics, 15*, 627–629.

Horrey, W. J., Lesch, M. F., & Garabet, A. (2008). Assessing the awareness of performance decrements in distracted drivers. *Accident Analysis & Prevention, 40*(2), 675–682.

Horrey, W. J., Lesch, M. F., Mitsopoulos-Rubens, E., & Lee, J. D. (2015). Calibration of skill and judgment in driving: Development of a conceptual framework and the implications for road safety. *Accident Analysis & Prevention, 76*, 25–33.

Inagaki, T. & Furukawa, H. (2004). Computer simulation for the design of authority in the adaptive cruise control systems under possibility of driver's over-trust in automation. *Proceedings of the IEEE International Conference on Systems, Man and Cybernetics, 4*, 3932–3937.

Inagaki, T., Itoh, M., & Nagai, Y. (2007). Support by warning or by action: Which is appropriate under mismatches between driver intent and traffic conditions? *IEICE Transactions on Fundamentals of Electronics, Communications and Computer Sciences, 90*(11), 2540–2545.

Kaber, D. B., Perry, C. M., Segall, N., McClernon, C. K., & Prinzel III, L. J. (2006). Situation awareness implications of adaptive automation for information processing in an air traffic control-related task. *International Journal of Industrial Ergonomics, 36*(5), 447–462.

Kaber, D. B. & Riley, J. M. (1999). Adaptive automation of a dynamic control task based on secondary task workload measurement. *International Journal of Cognitive Ergonomics, 3*(3), 169–187.

Kaber, D. B., Wright, M. C., Prinzel III, L. J., & Clamann, M. P. (2005). Adaptive automation of human-machine system information-processing functions. *Human Factors, 47*(4), 730–741.

Kiefer, R. J., LeBlanc, D., Palmer, M. D., Salinger, J., Deering, R. K., & Shulman, M. (1999). *Development and Validation of Functional Definitions and Evaluation Procedures for Collision Warning/Avoidance Systems* (No. DOT-HS-808–964). Washington, D.C.: US Department of Transportation. National Highway Traffic Safety Administration.

Klauer, S. G., Dingus, T. A., Neale, V. L., Sudweeks, J. D., & Ramsey, D. J. (2006). *The Impact of Driver Inattention on Near-Crash/Crash Risk: An Analysis Using the 100-Car Naturalistic Driving Study Data* (Report No. DOT HS 810 594). Washington, DC: National Highway Traffic Safety Administration.

Kozak, H., Artz, B., Blommer, M., Cathey, L., Curry, R., & Greenberg, J. (2005). *Evaluation of HMI for Lane departure Warning Systems for Drowsy Drivers: A VIRTTEX Simulator Study.* Dearborn, MI: Ford Motor Company.

Kozak, K., Pohl, J., Birk, W., Greenberg, J., Artz, B., Blommer, M., ... Curry, R. (2006). Evaluation of lane departure warnings for drowsy drivers. *Proceedings of the Human Factors and Ergonomics Society Annual Meeting, 50*, 2400–2404.

Lee, J. D., Hoffman, J. D., & Hayes, E. (2004). Collision warning design to mitigate driver distraction. *Proceedings of the SIGCHI Conference on Human factors in Computing Systems* (pp. 65–72). New York: ACM.

Lee, J. D. & See, K. A. (2004). Trust in automation: Designing for appropriate reliance. *Human Factors, 46*(1), 50–80.

Lee, J. D., Young, K. L., & Regan, M. A. (2008). Defining driver distraction. In M. Regan, J.D. Lee, & K. Young (Eds.), *Driver Distraction: Theory, Effects, and Mitigation.* Boca Raton, FL: CRC Press.

May, J. F., Baldwin, C. L., & Parasuraman, R. (2006). Prevention of rear-end crashes in drivers with task-induced fatigue through the use of auditory collision avoidance warnings. *Proceedings of the Human Factors and Ergonomics Society Annual Meeting, 50*(22), 2409–2413. (Los Angeles, CA: Sage Publications.)

Morris, N. M. & Rouse, W. B. (1986). *Adaptive Aiding for Human-Computer Control: Experimental Studies of Dynamic Task Allocation* (No. TR-3). Burlington, MA: Alphatech Inc.

National Highway Traffic Safety Administration. (2012). *Visual-Manual NHTSA Driver Distraction Guidelines for In-Vehicle Electronic Devices.* Washington, DC: National Highway Traffic Safety Administration.

National Transportation Safety Board. (2017). *Collision between a Car Operating with Automated Vehicle Control Systems and a Tractor-Semitrailer Truck Near Williston, Florida* May 7, 2016 (Report No. NTSB/HAR-17/02). Washington, DC: National Transportation Safety Board.

Neubauer, C., Matthews, G., Langheim, L., & Saxby, D. (2012). Fatigue and voluntary utilization of automation in simulated driving. *Human Factors, 54*(5), 734–746.

Parasuraman, R., Bahri, T., Deaton, J., Morrison, J., & Barnes, M. (1992). *Theory and Design of Adaptive Automation in Aviation Systems* (Report No. NAWCADWAR-92033-60). Warminster, PA: Naval Air Warfare Center.

Parasuraman, R., Mouloua, M., & Hilburn, B. (1999). Adaptive aiding and adaptive task allocation enhance human-machine interaction. In M.W. Scerbo & M. Mouloua (Eds.), *Automation Technology and Human Performance: Current Research and Trends* (pp. 119–123). Mahwah, NJ: Lawrence Erlbaum.

Parasuraman, R. & Riley, V. (1997). Humans and automation: Use, misuse, disuse, abuse. *Human factors*, *39*(2), 230–253.

Parasuraman, R., Sheridan, T. B., & Wickens, C. D. (2000). A model for types and levels of human interaction with automation. *IEEE Transactions on Systems, Man, and Cybernetics-Part A: Systems and Humans*, *30*(3), 286–297.

Rouse, W. B. (1988). Adaptive aiding for human/computer control. *Human Factors, 30*(4), 431–443.

Russell, S., Blanco M., Atwood, J., Schaudt, W. A., Fitchett, V. L., & Tidwell, S. (2018). *Naturalistic Study of Level 2 Driving Automation Functions* (Report DOT HS 812 642). Washington, DC: National Highway Traffic Safety Administration.

Saito, Y., Itoh, M., & Inagaki, T. (2016). Driver assistance system with a dual control scheme: Effectiveness of identifying driver drowsiness and preventing lane departure accidents. *IEEE Transactions on Human-Machine Systems, 46*(5), 660–671.

Sarter, N. B. & Woods, D. D. (1994). Pilot interaction with cockpit automation II: An experimental study of pilots' model and awareness of the flight management system. *The International Journal of Aviation Psychology*, *4*(1), 1–28.

Scerbo, M. W. (2008). Adaptive automation. In R. Parasuraman & M. Rizzo (Eds.), *Neuroergonomics: The Brain at Work* (pp. 239–252). Oxford: Oxford University Press.

Schwarz, C., Brown, T. L., Gaspar, J., Marshall, D., Lee, J., Kitazaki, S., & Kang, J. (2015). Mitigating drowsiness: Linking detection to mitigation. *Proceedings of the 24th ESV Conference*, Gothenburg, Sweden.

Sheridan, T. B. (1992). *Telerobotics, Automation, and Human Supervisory Control*. MIT Press.

Strayer, D. L., Watson, J. M., & Drews, F. A. (2011). Cognitive distraction while multitasking in the automobile. *Psychology of Learning and Motivation*, *54*, 29–58.

Victor, T., Dozza, M., Bärgman, J., Boda, C. N., Engström, J., Flannagan, C., … Markkula, G. (2015). *Analysis of Naturalistic Driving Study Data: Safer Glances, Driver Inattention, and Crash Risk* (SHRP 2 Report S2-S08A-RW-1). Washington, DC: National Academy of Sciences.

Vogelpohl, T., Kühn, M., Hummel, T., & Vollrath, M. (2019). Asleep at the automated wheel—Sleepiness and fatigue during highly automated driving. *Accident Analysis & Prevention, 126*, 70–84.

Wierwille, W. W. (1995). Overview of research on driver drowsiness definition and driver drowsiness detection. *Proceedings: International Technical Conference on the Enhanced Safety of Vehicles* (Vol. 1995, pp. 462–468). Washington, D.C.: National Highway Traffic Safety Administration.

Williamson, A., Lombardi, D. A., Folkard, S., Stutts, J., Courtney, T. K., & Connor, J. L. (2011). The link between fatigue and safety. *Accident Analysis & Prevention, 43*(2), 498–515.

Yamani, Y. & Horrey, W. J. (2018). A theoretical model of human-automation interaction grounded in resource allocation policy during automated driving. *International Journal of Human Factors and Ergonomics, 5*(3), 225–239.

Yeh, M., Wickens, C. D., & Seagull, F. J. (1999). Target cuing in visual search: The effects of conformality and display location on the allocation of visual attention. *Human Factors, 41*(4), 524–542.

Zaouk, A. K., Wills, M., Traube, E., & Strassburger, R. (2015). Driver alcohol detection system for safety (DADSS)-A status update. *24th Enhanced Safety of Vehicles Conference*. Gothenburg, Sweden: ESV.

17 Automated Vehicle Design for People with Disabilities

Rebecca A. Grier
Independent

CONTENTS

Key Points 378
17.1 Introduction 378
17.2 Medical Model of Disabilities 379
17.3 Social Model of Disabilities 379
    17.3.1 Nature of the Task... 379
    17.3.2 Individual Differences 380
    17.3.3 Summary of the Social Model.. 381
17.4 Universal Design... 381
    17.4.1 Equitable Use.. 382
    17.4.2 Flexibility in Use 382
    17.4.3 Simple & Intuitive Use 382
    17.4.4 Perceptible Information 383
    17.4.5 Tolerance for Error 383
    17.4.6 Low Physical Effort 384
    17.4.7 Size and Space for Approach and/or Use 384
    17.4.8 Universal Design Summary.. 384
17.5 How Humans Will Interact with FAVs... 384
    17.5.1 Non-FAV Task Flow 384
    17.5.2 FAV Task Flow 385
    17.5.3 Negotiating Stopping Location for Pick-Up & Drop-Off................... 386
        17.5.3.1 People Who Use Wheelchairs.. 388
        17.5.3.1 Other Mobility Impairments.. 388
        17.5.3.2 Visual Impairments 389
        17.5.3.3 Final Thoughts on Stopping Locations for Pick-Up
                and Drop-Off 389
    17.5.4 Considerations in Unusual/Emergency Situations............................... 390
    17.5.5 Cabin Design Considerations 391
17.6 Conclusion 391
References.. 392

KEY POINTS

- Level 4/5 vehicles may remove the requirement of a driver's license, potentially allowing people with certain disabilities to travel in passenger vehicles independently.
- The principles of universal design should be considered for automated vehicles to potentially enhance the utility for people with disabilities.
- There are standards that exist for Human–Machine Interfaces (HMI) to be accessible to people with certain disabilities.
- Negotiating pick-up and drop-off locations are a factor for automated vehicles to be user friendly for people with disabilities.
- Considerations of how and what to communicate in emergency situations is an additional aspect to consider in enhancing automated vehicle accessibility to people with certain disabilities.

17.1   INTRODUCTION

Currently, to operate a vehicle, one must obtain a driver's license. The specific requirements for obtaining a license vary by jurisdiction. Generally speaking, one must have a certain level of vision acuity, pass a knowledge test, and pass a skills test. Due to these requirements, individuals with certain visual, cognitive, or motor impairments are not able to obtain a driver's license. As a result, they are required to rely on others to travel between locations. This situation creates an additional logistical burden for these individuals. This logistical burden makes it challenging to hold a job, attend medical appointments, and generally participate in commerce and society (Bureau of Transportation Statistics, 2003; World Health Organization, 2016). Fully Automated Vehicles (FAVs), in which a human is not expected to take over lateral (i.e., steering) or longitudinal (i.e., acceleration, speed, and braking) control of the vehicle at any time, have the potential of eliminating the need for a driver's license. For the purposes of this chapter, FAV indicates both SAE Level 4 and 5 (SAE, 2016) vehicles that do not have driver controls.

To be clear, it is likely that some SAE Level 4 and 5 vehicles for private ownership will have driver controls. However, it is unlikely that people with disabilities who cannot obtain a driver's license will be able to operate vehicles with driver controls. Similarly, it also appears that the first SAE Level 4 vehicles will be part of a ride hailing fleet rather than available for purchase (Walker, 2019). As such, this chapter focuses on the design of SAE Level 4 and 5 vehicles that do not have driver controls.

This chapter describes some considerations in developing FAVs to enhance the possibility that people with disabilities may be able to take advantage of the technology. Before discussing the specifics of vehicle design as it relates to people with disabilities, an overview of disabilities can be helpful (see also, this Handbook, Chapter 10). There are two distinct philosophical views of disabilities: the medical model and the social model (Family Voices, 2015). These philosophies may affect individuals with disabilities and the design of vehicles.

17.2 MEDICAL MODEL OF DISABILITIES

The medical model views individuals with disabilities as being defined by a deficiency that needs to be fixed by a medical or other specialized professional. The medical model defines the disability in relation to how biological function is different from normal. Per the medical model, a person's vision would be defined in terms of numerous dimensions, including (1) visual acuity in each eye and (2) size of visual field (which can be measured both horizontally and vertically as well as binocular or monocular for each eye). There are a plethora of other visual impairments (e.g., several versions of color blindness, etc.), but to obtain a driver's license, the above two are typically the primary considerations (Huggett, 2009).

17.3 SOCIAL MODEL OF DISABILITIES

Conversely, the social model takes the viewpoint that a disability is one of many neutral differences that make up an individual. In other words, a disability is no different from height, gender, handedness, or ethnicity. More specifically, within the social model, an individual's inability to perform a task is not a result of the disability, but rather an interaction of a certain set of characteristics and the world. The World Health Organization (2016) defines a disability as "the interaction between individuals with a health condition (e.g., cerebral palsy, Down syndrome, and depression) and personal and environmental factors (e.g., negative attitudes, inaccessible transportation and public buildings, and limited social supports)." That is, the world has certain barriers that make performing certain tasks difficult for some people and not for others.

For example, height has been a disqualifying characteristic for numerous activities. That is, one's height could be a barrier to certain jobs. In fact, two astronauts were retired from space flight because of their height when the United States retired the space shuttles. The Russian Soyuz was designed differently than the space shuttles, and Scott E. Parazynski (6 feet 2 inch) and Wendy Lawrence (5 feet 3 inch) were deemed too tall and too short to safely fly within the Soyuz capsule (NYT, 1995). Yet, no one considers their height disabilities; rather, it is recognized that the design of the Soyuz placed a barrier on their ability to fly in space.

Currently, the requirements to obtain a driver's license are a barrier to people with certain visual, motor, and cognitive disabilities to traveling without relying on others (Chapman, 2018). FAVs have the potential of removing the barrier of a driver's license and allowing for greater mobility. However, this is true only if the FAV is accessible to these individuals. It should be noted that what is accessible to one person with a given disability is not necessarily accessible to another individual with the same disability. The social model better explains this variability than the medical model. In particular, there are two aspects of the social model that support this understanding, which are ignored by the medical model: (1) the nature of the task and (2) individual differences.

17.3.1 NATURE OF THE TASK

As stated above, the World Health Organization (2016) defines a disability as "the interaction between individuals with a health condition (e.g., cerebral palsy, Down syndrome, and depression) and personal and environmental factors (e.g., negative attitudes, inaccessible transportation and public buildings, and limited social supports)." This definition emphasizes the importance of both the nature of the task and the specific medical diagnosis to the experience of the disability. Given the almost infinite number of medical diagnoses and tasks as well as the always-evolving nature of work and technology, there has not been an attempt to provide an exhaustive classification of disabilities. The classification schema for the Paralympics (International Paralympic Committee, n.d.) is a useful example of how much work would be required. There are two steps to classification: (1) impairment eligibility and (2) sport class classification.

For the Paralympics, there are ten different categories of impairment including impaired muscle power (e.g., paralysis), impaired passive range of movement, limb deficiency (e.g., amputees), leg length differences, short stature, hypertonia, ataxia, athetosis, and visual impairment. However, what defines an impairment depends on the sport. For example, the maximum height to be considered, short stature or the maximum amount of muscle power to be considered impaired, is different among the different sports (e.g., athletics, swimming, ...).

After a para-athlete is deemed eligible to compete in a specific sport, a classification panel determines what sport class the individual will compete in. These sport classes have been created to ensure that the events are competitive. Para-athletics (i.e., track and field) has the most number of sport classes at 52. This rather large number is because of the variety of activities. These sport classes may divide individuals based on the different levels of impairments (e.g., single versus double amputee for certain field events or races) or could combine different impairments (e.g., paralysis and amputation in wheelchair racing) into one sport class. In summary, what is an impairment depends on the sport in which one wants to compete.

This is all to say categorization of disabilities can be problematic if the to-be-performed tasks are not considered. For this reason, the universal design principles presented in Section 17.4 do not mention specific disabilities, but rather speak to the goals to be accomplished with the system design. Furthermore, this is what motivated the task analysis demonstrating the differences between traditional vehicles and FAVs that are presented in Section 17.5.

17.3.2 INDIVIDUAL DIFFERENCES

The second important design consideration highlighted by the social model is individual differences. That is, two individuals who have the same medical diagnosis may have different methods of interacting with the world and, as such, encounter different barriers. For example, an individual who is legally blind from birth has a very different experience than an individual who loses his/her sight suddenly as an adult. The individual who is blind from birth does not know anything other than interacting with the world without vision. The individual who loses his/her sight suddenly as an adult may go through a period of grief over the loss of sight. After this period, the person can usually learn strategies and continuously work to improve these strategies. Both of these individuals have a very different experience compared with an individual who loses vision over a period of years. The person who loses his/her sight gradually over the years is continuously learning new strategies to adapt to the worsening vision. As such, although these three individuals may ultimately have the same visual acuity, they interact with the world in very different ways. When conducting research with users with disabilities, the researchers should consider the important between-subjects variability of people with the same disabilities.

17.3.3 Summary of the Social Model

In summary, the medical model categorizes people based on their diagnosis, which does not fully account for the impact of individual differences in accomplishing tasks. In the social model, there is no one method of categorizing people with disabilities. The emphasis of the social model is on the interaction of an individual with the environment. That is, there are individual differences that may prevent some of us from accomplishing certain tasks. Thus, the social model is an excellent starting place for discussions of universal design (Vautier, 2014).

17.4 UNIVERSAL DESIGN

Universal design, at its simplest, is the utilization of the standard human factors approach to design in which people with disabilities are included in the user population. Earlier chapters have discussed the special design considerations for the interface to automated, connected, and intelligent vehicles (Chapter 15) and design considerations that need to be taken into account when the driver's fitness is impaired (e.g., the driver is distracted or impaired) (Chapter 16). The focus has been for the most part on the design of interface for vehicles with driver controls (SAE Levels 0–3). Here, the discussion focuses on the universal design of vehicles that do not have driver controls (SAE Levels 4–5).

There is evidence that universal design can result in designs that enhance the usability for nearly everyone. The classic example of such benefit is the curb cutout. In 1990 with the passage of the Americans with Disabilities Act, curb cut-outs came into wide usage within the United States. However, as early as 1945, the city of Kalamazoo, MI had installed curb cut-outs for injured veterans coming home from World War II (Brown, 1999). As more people experienced curb cut-outs, their utility to the general public increased. Curb cut-outs make sidewalks easier for people with luggage or strollers (i.e., prams). In fact, in Australia, they are sometimes referred to as pram ramps. This example illustrates how something designed for people with disability was beneficial to the general public. Another example is the typewriter, which was invented in the early 1800s to help a blind noble woman write legible letters to her friends (Niven, 2012). These are just two examples. The interested reader is encouraged to review other case studies compiled by Niven (2012) and the National Disability Authority (2014) of universal designs that had benefits beyond those for persons with disabilities.

The universal design process encourages product and system designers to involve people with a variety of disabilities and experiences in all stages of the design process. To maximize the gain from the involvement of people with disabilities, it is best to focus on the interactions that are unique to the system being built rather than on general interactions. To that end, North Carolina State University (NCSU), funded by the National Institute for Disability & Rehabilitation (NIDR) under the U.S. Department of Education, developed seven principles of Universal Design in 1997 (Connell et al., 1997). These principles are as follows:

1. Equitable use,
2. Flexibility in use,
3. Simple & intuitive use,
4. Perceptible information,
5. Tolerance for error,
6. Low physical effort, and
7. Size and shape for approach and use.

The next sections describe these principles in brief. Interested readers are encouraged to review the original documents. In the original publications, Connell et al. (1997), describe several guidelines to be used in the design process for each principle. What is presented next is merely a paraphrase of each principle and the associated design guidelines. In addition, an example of a design choice to illustrate a potential implementation of the principle is provided for some of the principles.

17.4.1 EQUITABLE USE

What is meant by *equitable use* is that the design should be appealing, safe, and secure for as many people as possible, regardless of abilities. Ideally, the means with which the user interacts with the technology is identical for all users, but when that is not possible equivalent means should be provided. If it is not possible to have one design for all, the equivalent means provided should be most appealing and least stigmatizing to utilize. For example, instead of having a label stating "settings for the visually impaired" it may be better to have labels like "screen magnifier" "select color scheme," and "change contrast."

17.4.2 FLEXIBILITY IN USE

What is meant by *flexibility in use* is that the design does not force the user to perform actions in one way. Rather, the design is accommodating to the widest user base, by considering a wide range of abilities or preferences. The design helps the user make accurate and precise responses where required. A user can perform tasks with either the right or the left hand. The design does not force the user to interact faster or slower than s/he prefers. The technology presents multiple ways of interacting with it. Ideally, these multiple methods of interaction utilize different control methods (e.g., voice and keyboard) or sensory modalities (e.g., auditory and visual).

17.4.3 SIMPLE & INTUITIVE USE

Systems that support *simple and intuitive use* are more likely to be comprehended by anyone regardless of their past experience, current knowledge, ability to concentrate on the task, or language skills. In the words of Steve Krug (2000) "Don't make me think!" Towards this end, information should be arranged consistently and in accordance with its importance. Prompts and feedback should be designed to maximize effectiveness for the user throughout his or her interaction with the technology. The interface should promote an accurate mental model of the technology without unnecessary complexity (see also, this Handbook, Chapter 3). In the words of the French novelist, Antoine de Saint-Exupery, "A designer knows he has achieved perfection not when there is nothing left to add, but when there is nothing left to take away."

17.4.4 PERCEPTIBLE INFORMATION

Systems that are designed in accordance with the principle of *perceptible information* recognize the different sensory capabilities amongst the population. In addition, the presentation of information considers the ambient conditions (e.g., background noise, sun glare) in which the system will be used. Adequate contrast between presented information and its surroundings should be offered, particularly when the information is essential. Just as in the flexibility of use principle, the technology should be able to present information via multi-modal interfaces (i.e., verbal, pictorial, tactile). In addition, interface elements should be designed such that they can be described in multiple ways that do not require a specific, single sensory modality. For example, a good instruction and design could be, "push the green square labeled go, which is the rightmost button in the top row." In response, a person with normal sight may look for the green button or the button labeled "'Go." Whereas a person with a visual impairment might feel for an array of buttons and then for the top right button within that array. Essential words or symbols should be legible to the widest audience within the context of use. Information shall be perceptible by people using assistive devices for sensory impairments. To elucidate, here are three examples: (1) Many accessibility devices such as portable braille displays can connect via standard methods to enable access to otherwise inaccessible user interfaces; (2) People with hearing impairments sometimes need things repeated verbatim—if an instruction is not repeated verbatim it can lead to even more confusion; (3) The ability to adjust contrast, brightness, color scheme, or font size/type is helpful to individuals with visual difficulties.

17.4.5 TOLERANCE FOR ERROR

The next principle is *tolerance for error*. The system's design should reduce the likelihood of the user making a mistake. If a mistake is made, the design should further minimize the negative consequences of that mistake. This is accomplished in part by ensuring that the elements that are most easily activated or controlled are the most commonly used elements. In addition, those controls that could potentially have negative results if activated accidently should be located remotely or shielded. A classic example of such a control is a fire alarm, which often has a two-step process to ensure that it is not accidently activated. The design should reduce the likelihood of automatic or unconscious actions in tasks that require alertness. In addition, there should be warnings an fail-safe features for hazards and errors.

17.4.6   Low Physical Effort

The principle of *low physical effort* means that the user can interact with the system without exerting a great deal of strength or energy. Moreover, the user should not be required to contort him/herself, and the design should not require excessive repetitive motions or actions.

17.4.7   Size and Space for Approach and/or Use

The final principle is *size and space for approach and/or use*. This means that the design should consider a variety of body sizes, postures, and mobility levels, as well as abilities to see, reach, and manipulate controls. Visual obstructions between the user and the interface should be minimized. There should be consideration for different hand sizes and different grip strengths. Finally, the design should ideally allow people to use assistive devices (e.g., such as pointers, artificial limbs) when interacting with the technology.

17.4.8   Universal Design Summary

These seven principles were developed in order to provide information and guidance regarding the breadth of disabilities and how to accommodate them (Connell et al., 1997). They do not supplant the need to conduct iterative design research with persons with disabilities. However, if these universal design principles are considered prior to the involvement of people with disabilities, the designer may be able to focus his/her attention on more technology-specific challenges.

17.5   HOW HUMANS WILL INTERACT WITH FAVs

The addition of automation does not always remove tasks from the human, but often changes the tasks that the human performs (Grier, 2015; this Handbook, Chapter 8). The following two sections describe the differences in task flow when using a non-FAV in 2018 and a hypothesized or potential task flow for an FAV in the future. For the purpose of this comparison, both vehicles are personally owned and the owner is traveling alone with only hand-carried personal items. Furthermore, the trip is routine in that there are no malfunctions or crashes. As FAVs are not yet commercially available, the task flow for the FAV is hypothetical. The goal of this comparison is to show the differences in actions that potentially need to be considered when designing FAVs.

17.5.1   Non-FAV Task Flow

For the purposes of this comparison, let us describe the interactions a human has with a vehicle without automation (i.e., SAE Level 0). The human likely travels to where s/he parked the vehicle. The human gains access to the vehicle by unlocking/opening the doors. The human then enters the vehicle. The human or the vehicle may or may not adjust seats and mirrors. The human fastens the seat belt. The human starts the engine. The human disengages the parking brake, if it was set. The human shifts the vehicle out of the park and into the appropriate gear. The human monitors the environment while maintaining lateral and longitudinal control of the vehicle. The human also is in charge of wayfinding with or without assistance from Global Positioning System (GPS) navigation. The human identifies a parking space near his/her destination and parks the vehicle. After the vehicle is parked, the human gathers his/her personal items and exits the vehicle. The human then closes/locks the doors of the vehicle and travels from the vehicle to his/her destination.

17.5.2 FAV TASK FLOW

The task flow for FAVs is not likely to be a simple reassignment of the monitoring of the driving environment, navigation, and lateral and longitudinal control from the human to the vehicle. FAVs may potentially have a very different task flow. As FAVs are still being designed and are not yet commercially available the task flow that is presented here is merely hypothetical and not a statement of fact. First, the human and vehicle negotiate where and when the pick-up will occur.

The vehicle may have to travel on its own to that location to arrive at the time negotiated. The human may also have to travel to this location. The human may have to identify the appropriate vehicle. The vehicle doors are opened by either the vehicle or the human. The human then enters the vehicle. The human potentially fastens the seat belts. The human potentially indicates that s/he is ready for the vehicle to begin travel. After the vehicle has begun its trip, it is assumed the vehicle will safely enter traffic. At some point, presumably before indicating s/he is ready for travel, the human indicates his/her travel intentions to the vehicle. These travel intentions could include negotiating the drop-off location and perhaps the purpose of the ride. The purpose of the ride could affect the path taken or the driving style of the vehicle. These intentions may change while in transit. The human may be able to monitor the vehicle's progress along the route and may be alerted when the route or estimated time of arrival is altered significantly. The human also may be alerted when approaching or upon arrival at the negotiated drop-off location. At the negotiated drop-off location, it is assumed that the human exits the vehicle. When exiting is complete, either the human dismisses the vehicle or the vehicle senses that it may depart based on some set of parameters. The vehicle then departs. Meanwhile, it is assumed the human navigates and travels between the drop-off location and his/her destination.

As with all introductions of automation throughout history, there are several differences between these task flows (see Table 17.1 for a summary). One difference that emerges is with regard to what information is passed between the human and the vehicle. It is anticipated that much of the interaction will be using human–machine interfaces (HMI) that are used in other domains (e.g., touchscreens, voice, physical controls like knobs, switches, and buttons, audio displays, and visual displays). Best practices for the design of many HMI to be accessible to people with disabilities are documented elsewhere [see for example Section 508 Amendment to the Rehabilitation Act of 1973 (www.section508.gov) and Web Content Accessibility Guidelines (WCAG) developed by the World Wide Web Consortium (W3C); www.w3.org/standards/webdesign/accessibility)]. The reader is encouraged to examine

TABLE 17.1

Summary of the Differences between the Two Task Flows (i.e., Non-FAVS and Potential FAVs)

Stage	Non-FAVs	Potential FAVs
Pre-Trip	Human travels to where s/he parked vehicle	Human negotiates with vehicle where and when s/he will be picked up.
Trip Initiation	Human starts engine, puts car into appropriate gear, and presses accelerator	Vehicle engine is already started. Human tells the vehicle s/he is ready to begin trip. Vehicle then safely pulls into traffic.
Traveling	Human controls longitudinal and/or lateral control over vehicle, monitors environment, and navigates with or without the aid of GPS.	Human indicates purpose of ride and negotiates drop-off location.[1] Vehicle controls longitudinal and lateral control to meet the human's purpose and arrive at the designated drop-off location. Vehicle has the ability to communicate route and progress to human.
Trip ending	Human parks vehicle, exits vehicle, and makes way to destination.	Vehicle arrives at drop-off location and alerts human. Human exits vehicle. Human indicates the vehicle can depart or vehicle senses it can depart. Human makes way to destination.

[1] This indication could occur at any time before this, but this is the latest stage at which this could occur.

these sources, which can provide more information than this chapter regarding selecting and designing HMIs that are usable by people with disabilities. Based on the hypothesized task flow, HMI is critical to accessibility. Another critical design consideration is related to vehicle entry and exit. Specifically, the human and the vehicle may have to negotiate where this will occur.

17.5.3 Negotiating Stopping Location for Pick-Up & Drop-Off

In a personally owned non-FAV, the driver (1) makes his/her way to the location the vehicle is parked, (2) drives the vehicle, (3) parks the vehicle somewhere near the destination, and (4) makes his/her way to the destination. With an FAV, the process may potentially be more similar to the process a passenger uses with a taxi or other hired driver. In this situation, the passenger typically negotiates the stopping location with the driver in two phases: (1a) the passenger provides driver/service a place of interest or street address; (1b) as necessary, the passenger provides additional information regarding location (i.e., specific entrance name/description such as airline or store); (2) driver and passenger consider current conditions (e.g., weather, other vehicles, infrastructure) and individual needs (e.g., disability, cargo) to determine the best available stopping location. This process happens easily for most humans today, because of a shared language and knowledge. However, without shared language and knowledge, the process could potentially be challenging. For example, some people with certain disabilities (e.g., communication disabilities) or those who do not speak the same language as the driver might find coordinating a stopping location challenging with a human driver.

For designers of FAVs, there will need to be careful consideration of the task flow and HMI for the negotiation of stopping location, as the quality of both are necessary to ensuring shared knowledge and adequate communication between the FAV and the passenger. If asked to describe where a person can expect an FAV to stop, people may say as close as possible to the origin/destination. However, the shortest distance between the origin/destination and the FAV may have barriers that pose challenges to an individual with a disability. For this reason, accessible parking spaces (i.e., parking spaces reserved for individuals with a disability permit) within the United States are required to be located on the shortest *accessible* path to an accessible entrance (United States Access Board, 2016). An accessible path is one that is free of barriers, most notably stairs and curbs, but also poles, trees, and other objects that can reduce the width of the path. The closest parking spaces to the door may not have an accessible path. If there is no acceptable path between the space and the entrance, then the spaces would not be considered accessible spaces. Moreover, if there are several parking spaces on different accessible paths, those that are on the shortest path are the ones that should be designated as accessible parking (United States Access Board, 2016). Collectively, this is not meant to indicate that FAVs should ONLY be able to pick-up and drop-off in accessible parking spaces. Rather, it is meant to illustrate the importance of a barrier-free path between the vehicle and origin/destination when selecting a stopping location (Disability Right Education and Defense Fund, 2018). What constitutes a barrier-free path varies by individual. The following sections describe different potential barriers for different disabilities (summarized in Table 17.2).

TABLE 17.2

Parameters Related to Vehicle Stopping Location Important to People with Disabilities

Distance from curb cut-outs

Distance from (accessible) entrance

Incline of road

Distance from curb

Space needed for lift/ramp

Space needed for wheelchair to approach/exit lift/ramp

Presence of physical obstacles (e.g., poles, vehicles, puddles, etc.)

Side of street of origin/destination

Bike lanes between vehicular traffic and sidewalks

17.5.3.1   People Who Use Wheelchairs

For people in wheelchairs, the accessible path described by the United States Access Board (2016) is the primary concern. Thus, rather than being closest to the main entrance (which may not be accessible), wheelchair users would prefer to be let out of a vehicle near a curb cut-out. All other things being equal, they would prefer this curb cut-out be as close as possible to the nearest accessible entrance. More specifically, the curb cut-out that is the closest to the end of any ramp access to the building. However, if the roadway has an incline, it may be helpful if the user could select a pick-up location that is downhill from their origin and a drop-off location that is uphill from the destination (SAE, 2019).

The above considerations generally apply to all people who use wheelchairs. That said, people who use wheelchairs vary greatly. For the purposes of negotiating drop-off and pick-up locations, people who use wheelchairs can be divided into two groups: (1) those that can transfer themselves between the wheelchair and a vehicle seat without assistance (see Schaupp et al., 2016, for a thorough task analysis of this process) and (2) those that need more assistance. There are several reasons why individuals who can transfer between their wheelchair and a vehicle seat do so. First, it is considered safer for an individual to sit in a vehicle's seat rather than in a wheelchair (Van Roosmalen, Ritchie Orton, & Schneider, 2013). Second, the ability to transfer to a vehicle seat allows an individual greater flexibility in terms of vehicle types they can utilize; not all vehicles can accommodate the seated height of an individual in a wheelchair for entry/exit.

These two groups of individuals have different needs as it relates to stopping location. We will assume these individuals are traveling alone and as such have no assistance. Individuals who remain in their wheelchair require a ramp or wheelchair lift to enter the vehicle. As such, it would be helpful if the stopping location considers the space needed for the ramp/lift (SAE, 2019). Similarly, it would be helpful if the space needed for the wheelchair to get into position to enter/exit the ramp/lift as well as the direction the person is coming from or going to be considered (SAE, 2019). For those that transfer themselves, it would be helpful if the vehicle stopped approximately 1 m away from the curb. This provides sufficient room for the wheelchair at the appropriate height to support transfer (SAE, 2019).

17.5.3.1   Other Mobility Impairments

For individuals with other mobility impairments who do not use wheelchairs (e.g., individuals who have hemiplegia, arthritis, cerebral palsy, and lupus), there are two concerns associated with stopping location: distance and stairs/curbs. These individuals may be limited in the distance that they can walk. That is, they would prefer the vehicle to stop as close to the origin and destination as possible. However, if that location has numerous stairs, they may prefer to be closer to the end of the ramp (SAE, in production). Similarly, they may find it helpful if the vehicle stops near enough to the curb, so that they can step directly on to the sidewalk when exiting the vehicle and directly into the vehicle from the sidewalk (SAE, 2019). For example, if the vehicle stops at a typical 6 to 12 inches from the curb, they may have difficulty trying to negotiate the curb with a small amount of space between themselves and the vehicle. Some individuals may prefer stopping near a curb cut-out depending upon its distance from the ramp or entrance (SAE, 2019).

17.5.3.2 Visual Impairments

For some individuals with visual impairments, moving or potentially moving objects are the barrier that may determine what is an accessible path (SAE, 2019). As such, it may be helpful if the FAV could allow these individuals to choose locations in which they do not have to cross a street or bike lane in transiting from/to the FAV. Similarly, they may prefer to avoid double parking (i.e., stopping next to an already parked vehicle; SAE, 2019).

Brinkley, Posadas, Woodward, and Gilbert (2017) and the Disability Right Education and Defense Fund (2018) specify some individuals with visual impairments have concerns related to their confidence in the vehicle location in relation to themselves for pick-up and in relation to their destination for drop-off. To address these concerns, it may be helpful to offer very specific information about the vehicle location that they can compare to their current location/destination (SAE, 2019). In addition, it may help to alleviate these concerns if the information provided could be used for wayfinding (Disability Right Education and Defense Fund, 2018; SAE, 2019).

In this context, wayfinding is the process of navigating and traveling safely between the vehicle and the origin/destination. This would include the location of the vehicle, the individual's current location (for pick-up) or the destination information (for drop-off), the presence of bike lanes, the presence of vehicles, the presence of infrastructure, etc. In addition, at drop-off this information would include what side of the vehicle is the safest for exiting (e.g., the curbside). With regard to the last of these, curbside is typically standard within a country, but there are exceptions such as one-way streets. For those exceptions, the vehicle could stop with traffic on either side of the vehicle (SAE, 2019). Per the universal design principle of perceptible information (#4, Section 17.4.4), the technology should be able to present the information via multi-modal interfaces (i.e., visual, auditory). Other HMI considerations that align with this universal design principle would be allowing the user to adjust visualization properties (e.g., zoom, color scheme, etc.) (Disability Right Education and Defense Fund, 2018).

17.5.3.3 Final Thoughts on Stopping Locations for Pick-Up and Drop-Off

In the above sections, several considerations associated with the negotiation of location for pick-up and drop-off were discussed (see Table 17.2 for a summary). Preferences regarding these considerations can be presented to the user for each ride, or they could be contained in a user profile similar to accessibility options for other technologies. Either way, it may be helpful for the user to have the ability to make the choices regarding pick-up and drop-off location that are best for him/herself (SAE, 2019).

Negotiating the stopping location is important, but it may not be sufficient. This is because it is possible that the vehicle arrives at the negotiated spot and there is an obstacle inhibiting ingress or egress. For this reason, it may be helpful if there was a way for the vehicle to move slightly or to re-negotiate the space if need be (SAE, 2019).

If neither of these results in a location that is acceptable for the passenger to ingress/egress, it may be helpful if the passenger had the option to request the vehicle wait/circle for an acceptable space to open.

17.5.4 CONSIDERATIONS IN UNUSUAL/EMERGENCY SITUATIONS

Another concern is related to situation awareness in unusual/emergency situations (see also, this Handbook, Chapter 7). Lastly, cabin design considerations will be described. The National Highway Traffic Safety Administration (NHTSA) has suggested that 94% of traffic crashes are caused by human errors (NHTSA, 2015). There are many that believe FAVs will reduce the number of crashes. However, there is still the potential that FAVs will be involved in crashes with wildlife (National Center for Rural Road Safety, 2016), human-driven vehicles (Marshall & Davies, 2018), and detritus that appears suddenly [e.g., rocks thrown; material dislodged from vehicles (e.g., AAA, 2017); rockslides (e.g., Abraham, 2011)]. In addition, riders may experience emergencies while in an FAV. As such, it may be helpful if vehicle- and passenger-initiated emergency procedures were designed into FAVs. To align with universal design principle of equitable use (#1, Section 17.4.1) there should be procedures available for the widest population possible.

For example, if an FAV needs to stop driving for some reason, it may be helpful for some individuals with cognitive or visual impairments if the FAV provides information to support the decision to exit or stay within the vehicle. If the FAV does provide such information, its presentation should ideally align with the universal design principle of perceptible information (#4, Section 17.4.4) (SAE, 2019). If it is safer for the person to exit the vehicle, it may be useful to individuals with visual impairments to have information about the roadway (e.g., number of lanes, speed limit, congestion, location of shoulders, sidewalks, or similar). Furthermore, it may be helpful if the presentation of the information took into consideration those individuals who may have never driven themselves.

In other circumstances, a passenger might want to initiate an emergency stop. To align with the universal design principle of equitable use (#1, Section 17.4.1), the design of such a feature should ideally accommodate the widest population possible. In terms of selection of the control device used to initiate an emergency stop, the designer should consider design options that balance the universal design principles of tolerance for error (#5, Section 17.4.5), simple and intuitive use (#3, Section 17.4.3), and size and shape for use (#7, Section 17.4.7). For example, if the HMI for such a feature was a physical button, it would be helpful if it were placed in a location that can be reached by either hand by a person with limited arm mobility while belted, while at the same time helping minimize inadvertent actuation. If the only location is one where it could be inadvertently activated, then one might consider a two-step process or a protective shield over the button.

Persons with disabilities may potentially have concerns regarding how they will be supported in an emergency or unusual situation. Two potential methods to mitigate the needs of people with disabilities in these situations may include the ability to name a contact person and indicate specialized information to first responders. With regard to emergency contact, this could be a person who is notified in unusual circumstances or who may potentially have the ability to monitor the FAV (Disability Right Education and Defense Fund, 2018). Also, these emergency contacts could potentially help to provide additional information to the first responders.

The second method is to allow individuals to voluntarily provide information with the known intent of being shared with a first responder in an emergency. This information could be useful for paramedics or for police. As an example, some people with special needs (e.g., some transplant recipients) have been told that they should only be treated at hospitals with certain specializations. It is important for the first responders to be aware of this information to ensure the person is taken to the appropriate hospital. Similarly, knowledge of disabilities, particularly in terms of communication impairments, autism, and mental health issues, may be useful to emergency responders (Autism Society, n.d.; Center for Development and Disability, n.d.; Perry & Carter-Long, 2016). Some may believe that in terms of these unusual scenarios, we might use the same methods used today. However, we need to consider that individuals today who cannot obtain a driver's license because of a disability are often traveling with other individuals who can communicate this information. In contrast, in an FAV these individuals may be alone. There have been incidents where individuals with disabilities were injured, because first responders were unaware of their special needs (Perry & Carter-Long, 2016).

17.5.5 CABIN DESIGN CONSIDERATIONS

The last category to be discussed within this chapter are considerations related to cabin design. Previously, the needs of wheelchair users as it relates to stopping location were discussed. The population was divided into two groups: those who remain in their wheelchairs; and those who transfer to the vehicle seat. Those who remain in their wheelchairs may benefit from securements for the wheelchair as well as chest restraints (Disability Right Education and Defense Fund, 2018; Bertocci, Hobson, & Digges, 2000). Standards exist for the design of vehicles to accommodate persons who remain in their wheelchairs (SAE, 1999a; b; c). Some individuals can transfer to a vehicle seat but cannot get their wheelchair/scooter into the vehicle without assistance. These individuals may currently use devices such as chair lifts to accomplish this (Disability Right Education and Defense Fund, 2018). Those individuals who can transfer themselves and their wheelchair may need a space within the cabin that allows them to keep the wheelchair within arm's reach throughout the journey (Disability Right Education and Defense Fund, 2018). Such a space may be useful for canes and walkers as well. Similarly, a wide variety of individuals rely on service/support animals, and space for these animals should be considered in the design. Lastly, the design of the cabin should consider the task analysis of the wheelchair to vehicle transfer process completed by Schaupp et al., 2016.

17.6 CONCLUSION

FAVs have the potential of making travel easier for people with disabilities (Chang & Gouse, 2017; see also, this Handbook, Chapter 10). However, designers of FAVs should consider a wide variety of disabilities (at the intersection of medical condition and task) and find solutions within the various constraints (Chang & Gouse, 2017). To that end, this chapter presented the framework of universal design and how it could potentially be applied to FAVs. The primary tenet of the universal design framework is that the designers of technology should consider people with disabilities as part of the user group just as the designers would consider other individual differences. Four areas related to FAVs were discussed in this chapter: HMI, on- and off-boarding, emergency situations, and cabin design. With regard to the first, there are numerous design standards created for other technologies that may potentially be leveraged to make the HMI within an FAV accessible to people with disabilities (also see Chapters 15 and 16 in this Handbook). Second, the on- and off-boarding process is one aspect that is potentially unique to FAVs. As such, considerations for various classes of disabilities related to this process were presented. Third, considerations for the development of responses to emergencies were discussed. Fourth the cabin should consider the needs of those who use wheelchairs, assistive mobility devices such as canes or walkers, and service animals. Finally, designers of FAVs are encouraged to consider disabilities as just another class of individual differences like height, nationality, gender, or handedness.

REFERENCES

American Automobile Association. (2017). *Prevent Road Debris*. Retrieved from https://exchange.aaa.com/prevent-road-debris/#.W8FA9LmWzIV

Abraham, H. (2011). *Rockslide Causes Accident on Route 28*. Retrieved from https://pittsburgh.cbslocal.com/2011/08/23/rockslide-causes-accident-on-route-28/

Autism Society. (n.d.). *Autism Information for Law Enforcement and Other First Responders*. Retrieved from www.autism-society.org/wp-content/uploads/2014/04/Law_Enforcement_and_Other_First_Responders.pdf

Bertocci, G., Hobson, D., & Digges, K. (2000). Development of a wheelchair occupant injury risk assessment method and its application in the investigation of wheelchair securement point influence on frontal crash safety. *IEEE Transactions on Rehabilitation Engineering, 8*(1), 126–139.

Brinkley, J., Posadas, B., Woodward, J., & Gilbert, J. (2017). Opinions and preferences of blind and low vision consumers regarding self-driving vehicles: Results of focus group discussions. *Proceedings of the 19th International ACM SIGACCESS Conference on Computers and Accessibility* (pp. 290–299). New York: ACM.

Brown, S. (1999). *The Curb Ramps of Kalamazoo: Discovering Our Unrecorded History*. Retrieved from www.independentliving.org/docs3/brown99a.html

Bureau of Transportation Statistics. (2003). *Transportation Difficulties Keep over Half a Million Disabled at Home*. BTS Issue Brief No. 3. Retrieved from www.bts.gov/sites/bts.dot.gov/files/legacy/publications/special_reports_and_issue_briefs/issue_briefs/number_

Center for Development and Disability. (n.d.). *Tips for First Responders*. Retrieved from https://srcity.org/DocumentCenter/View/2218/Tips-for-First-Responders-PDF?bidId

Chang, A. & Gouse, W. (2017). *Accessible Automated Driving System Dedicated Vehicles*. Retrieved from www.sae.org/standardsdev/news/mobility_benefits.htm

Chapman, L. (2018). *What Do Self-Driving Vehicles Mean for Disabled Travelers*. Retrieved from www.disabled-world.com/disability/transport/autonomous-vehicles.php

Connell, B., Jones, M., Mace, R., Mueller, J., Mullick, A., Oostroff, E., … Vanderheiden, G. (1997). *The Principles of Universal Design Version 2.0*. Retrieved from https://projects.ncsu.edu/design/cud/about_ud/udprinciplestext.htm

Disability Right Education and Defense Fund. (2018). *Fully Accessible Autonomous Vehicles Checklist: Working Draft.* Berkeley, CA: Disability Right Education and Defense Fund.

Family Voices Kids As Self Advocates. (2015). *Medical Model vs. Social Model.* Retrieved from www.fvkasa.org/resources/files/history-model.php

Grier, R. A. (2015). Situation awareness in command and control. In R. R. Hoffman, P. A. Hancock, M. W. Scerbo, R. Parasuraman, & J. L. Szalma, *The Cambridge Handbook of Applied Perception Research* (pp. 891–911). Cambridge, UK: Cambridge University Press.

Huggett, E. J. (2009). *Driving Safely.* Retrieved from http://lowvision.preventblindness. org/2009/08/25/driving-safely/

International Paralympic Committee. (n.d.). *Classification.* Retrieved from www.paralympic. org/classification

Krug, S. (2000). *Don't Make Me Think! A Common Sense Approach to Web Usability.* Indianapolis, IN: New Riders Publishing.

Marshall, A. & Davies, A. (2018). *Waymo's Self-Driving Car Crash in Arizona Revives Tough Questions.* Retrieved from www.wired.com/story/waymo-crash-self-driving-google-arizona/

National Center for Rural Road Safety. (2016). *Will Technology Bring an End to Wildlife Collisions?* Retrieved from https://ruralsafetycenter.org/uncategorized/2016/will-technology-bring-an-end-to-wildlife-collisions/

National Disability Authority. (2014). *3 Case Studies on UD.* Retrieved from http://universaldesign.ie/What-is-Universal-Design/Case-studies-and-examples/3-case-studies-on-UD/#oxo

National Highway Traffic Safety Administration. (2015). *Critical Reasons for Crashes Investigated in the National Motor Vehicle Crash Causation Survey* (DOT HS 812 115). Retrieved from https://crashstats.nhtsa.dot.gov/Api/Public/ViewPublication/812115

New York Times. (1995). *Americans Too Tall or Short for Russian Space Program.* Retrieved from www.nytimes.com/1995/11/10/us/americans-too-tall-or-short-for-russian-space-program.html

Niven, R. (2012). *How People with Disabilities Inspire Cutting Edge Technology.* Retrieved from www.channel4.com/news/gadgets-inspired-by-people-with-disabilities

Perry, D. M. & Carter-Long, L. (2016). *The Ruderman White Paper on Media Coverage of Law Enforcement Use of Force and Disability.* Boston, MA: Ruderman Family Foundation.

Schaupp, G., Seeanner, J., Jenkins, C., Manganelli, J., Henessy, S., Truesdail, C., … Brooks, J. (2016). *Wheelchair Users' Ingress/Egress Strategies While Transferring Into and Out of a Vehicle.* SAE. doi:10.4271/2016–01–1433

SAE. (1999a). *Structural Modification for Personally Licensed Vehicles to Meet the Transportation Needs of Persons with Disabilities* (J1725_199911). Warrendale, PA: SAE International.

SAE. (1999b). *Design Considerations for Wheelchair Lifts for Entry to or Exit from a Personally Licensed Vehicle* (J2093). Warrendale, PA: SAE International.

SAE. (1999c). *Wheelchair Tiedown and Occupant Restraint Systems for Use in Motor Vehicles* (J2249). Warrendale, PA: SAE International.

SAE. (2016). *Taxonomy and Definitions for Terms Related to Driving Automation Systems for On Road Motor Vehicles* (J3016 201609). Warrendale, PA: SAE International.

SAE. (2019). *Identifying Automated Driving Systems - Dedicated Vehicles (ADS-DV) Passenger Issues for Persons with Disabilities* (J3171). Warrendale, PA: SAE International.

United States Access Board. (2016). *Parking Spaces.* Retrieved from www.access-board.gov/guidelines-and-standards/buildings-and-sites/about-the-ada-standards/guide-to-the-ada-standards/chapter-5-parking

Van Roosmalen, L., Ritchie Orton, N., & Schneider, L. (2013). Safety, usability, and independence for wheelchair-seated drivers and front-row passengers of private vehicles: A qualitative research study. *Journal of Rehabilitation Research & Development, 50*(2), 239–252.

Vautier, L. (2014). Universal design: What is it and why does it matter? *LIANZA Conference.* Aukland, New Zealand. Retrieved from https://lianza.org.nz/sites/default/files/Vautier_L_Universal_Design.pdf

Walker, J. (2019). *The Self-Driving Car Timeline - Predictions from the Top 11 Global Automakers.* Retrieved from https://emerj.com/ai-adoption-timelines/self-driving-car-timeline-themselves-top-11-automakers/

World Health Organization. (2016). *Disability and Health.* Retrieved from www.who.int/mediacentre/factsheets/fs352/en/

18 Importance of Training for Automated, Connected, and Intelligent Vehicle Systems

Alexandria M. Noble and Sheila Garness Klauer
Virginia Tech Transportation Institute

Sahar Ghanipoor Machiani
San Diego State University

Michael P. Manser
Texas A&M Transportation Institute

CONTENTS

Key Points 396
18.1 Introduction 396
18.2 Training Overview...398
    18.2.1 Mental Models...399
    18.2.2 Consumer Protection 400
    18.2.3 Goals of Training for ACIV Systems 400
        18.2.3.1 Increased Improvement in Safety 401
        18.2.3.2 Appropriate Levels of Trust 401
        18.2.3.3 User Acceptance 402
18.3 Training Content for ACIV Systems 402
    18.3.1 System Purpose and Associated Risks of Use................................403
    18.3.2 Operational Design Domain 403
    18.3.3 Monitoring the Road and the System 404
18.4 Andragogical Considerations for ACIV Systems Training 404
    18.4.1 Driver-Related Factors...405
        18.4.1.1 Motivation 405
        18.4.1.2 Readiness to Learn 405
        18.4.1.3 Affect—Anxiety Toward Technology 406

18.4.2 Design of the Training 406
    18.4.2.1 Salient Relevance and Utility of Training Content...........406
    18.4.2.2 Framing of the Opportunity..407
    18.4.2.3 Opportunities to Practice and Apply Skills......................407
18.5 Training Protocols 408
  18.5.1 Trial and Error 408
  18.5.2 Vehicle Owner's Manual 409
  18.5.3 Demonstration...409
  18.5.4 Web-Based Content 410
  18.5.5 Future Training—Embedded Training for ACIV Systems 411
    18.5.5.1 Feedback (Knowledge of Results) 411
    18.5.5.2 When Does It End?... 412
18.6 Recommendations 412
18.7 Challenges and Conclusions 413
Acknowledgments... 414
References... 414

KEY POINTS

- Training is a critical component as the vehicle fleet begins its transition through the levels of automation and the driving task fundamentally changes with each technological advancement.
- When designing driver training for ACIV systems, the learner and the system must both be taken into consideration.
- Effective training must be supported by all stakeholders, including car companies, state and federal governments, academics, and traffic safety organizations.
- Intelligent HMI design and comprehensive and informative training should work as part of a coordinated effort to fully realize the safety benefits of ACIV systems.

18.1 INTRODUCTION

The mere idea of automated, connected, and intelligent vehicles (ACIVs) conjures up visions in which our vehicles cater to our every transportation need. A person walks out of their house and immediately steps into a waiting vehicle, which indicates their destination, and the vehicle quietly and effortlessly moves along freeways, suburban roadways, and local streets to deliver the person to work, a grocery store, or any desired destination. In the process, the vehicle exchanges information with other vehicles, the infrastructure, and the cloud, allowing for safe and efficient transportation. In this situation, the vehicle is, in fact, the driver. The person is only a passenger. A more realistic vision of ACIVs is one in which the vehicle and person form a partnership. In this approach, either the person or vehicle can be the "driver" that is essentially in control, but more commonly, both entities work together to sense the environment, control the vehicle, and avoid crashes. Given that humans are still an essential element of this current vision of ACIVs, it is easy to predict that safety will continue to be a significant concern. It was estimated that in 2016, approximately 94% of serious crashes were attributable to human error, including errors related to distraction, impairment, or drowsiness (National Highway Traffic Safety Administration, 2018). These statistics suggest that human-related factors will continue to require attention as the partnership between human and vehicle further develops. Indeed, the development and deployment of ACIV systems, such as advanced driver assistance systems (ADAS; sometimes referred to as driver support features—SAE Levels 0–2, automated driving features—Levels 3–5, and active safety features, e.g., automatic emergency braking [AEB]) that help to control vehicle acceleration, vehicle deceleration, and lane position, offer the potential to improve safety by relieving the human driver of tasks, particularly those which they are prone to performing with errors. It is in this situation of shared vehicle control that safety needs to be addressed.

For the last several decades, there has been an increasing focus on the technological development of ACIVs (Barfield & Dingus, 1997; Fancher et al., 1998; Tellis et al., 2016) in a shared control context to achieve the visions outlined earlier. Efforts in this area include the Automated Highway System projects in the 1990s (Congress, 1994); connected vehicle communications—vehicle to vehicle and vehicle to infrastructure—in the 2000s (Ahmed-Zaid et al., 2014; Bettisworth et al., 2015); and partially and fully automated vehicle technologies more recently (Blanco et al., 2015; Rau & Blanco, 2014; Tellis et al., 2016; Trimble, Bishop, Morgan, & Blanco, 2014). A basic premise within the human factors profession is that human–machine interface (HMI) systems should be intuitive and easy to use, both of which are a byproduct of good HMI design and standardization. In the absence of these factors, there is a critical need to provide training to drivers. However, there has been relatively little work examining the efficacy of training and learner-centered HMI design to positively impact the safety of ACIV systems, which is likely a critical key factor in promoting a *safe* person/vehicle partnership (see Chapter 15 for a more complete discussion of HMI design considerations for ACIV systems which are largely independent of learner-centered concerns).

A generally accepted definition of training, particularly relevant within the context of this work, is that it is a continuous and systematic process that teaches individuals a new skill or behavior to accomplish a specific task (Salas, Wilson, Priest, & Guthrie, 2006). This definition is particularly relevant to the domain of ACIVs because drivers must become proficient in a wide variety of tasks required to operate these technologies. However, the scope of the definition focuses solely on tasks and fails to acknowledge that the use of ACIV systems requires proficiency in tasks that rely on and interact with driving environments (e.g., lane markings that support lane-keeping assist [LKA] systems). Therefore, a more recent training definition may be more applicable to ACIVs. This definition describes training as the "systematic acquisition of knowledge, skills, and attitudes that together lead to improved performance in a specific environment" (Grossman & Salas, 2011). A successful training method will impart the necessary knowledge, skills, and attitudes related to the partnership between ACIVs and humans to promote improvements in safe driving.

This chapter will address the learner-centered design of training for ACIVs. We recognize the potential need for ACIV driver training across a wide range of topics, including sensor operation/capabilities and the limitations of the operational design domain (ODD); however, the state of driver training relative to ACIVs has not been examined extensively, so we have instead focused on training approaches which have received some research attention. This chapter summarizes the general topic of learner-centered training for ACIVs and provides information on specific training-related factors. The first section, Training Overview, will briefly discuss the importance of training, including how a person–vehicle partnership can be enhanced through training, while the second section addresses the critical question of what content should be included in training protocols for ACIV systems. The third section, titled Andragogical Considerations for ACIV Systems Training, will identify both driver and non-driver related key factors that should be considered in the development of training protocols. The fourth section provides a review of both current and future protocols that could be employed to train people on the use of ACIV systems. The chapter concludes with series of recommendations for ACIV systems training and training practices. This chapter will serve as a foundation for driver training stakeholders, technology developers, consumers, and legislatures to address the growing need to include relevant and effective training for ACIV systems as these technologies are developed and deployed.

18.2   TRAINING OVERVIEW

The prevalence of new vehicles equipped with ADAS has steadily increased over the last decade. Readers are directed to SAE J3016 (2018) for descriptions of ADAS technologies relationship to SAE automation Levels 0–5 (also, this Handbook, Chapters 1, 2). Some ADASs assist drivers in the lateral and longitudinal control of the vehicle within certain ODDs, while others provide warnings and information about the surrounding road environment. ADAS assistance in these fundamental tasks can help a driver by performing control functions and providing alerts about the presence of another vehicle or object. Despite the potential benefits associated with their use, many ADAS system capabilities and limitations are misunderstood by drivers (Larsson, 2012; McDonald, Carney, & McGehee, 2018; McDonald et al., 2015, 2016; McDonald, Reyes, Roe, & McGehee, 2017; Rudin-Brown & Parker, 2004).

As we extend the consideration of the state of the art with ADAS to the future with ACIV systems, we can see that many of these technologies share functional similarities among manufacturers; however, there are subtle differences in capabilities that are crucial for the users to understand. For example, a vehicle that has an LKA system will control the vehicle by oscillating between the lane lines on the right and left, through intermittent steering input, while a lane-centering system will attempt to center the vehicle in the lane through continuous steering input. Regardless of the fidelity of the lane-keeping system, or any Level 1 or 2 driver support feature, the driver is responsible for knowing and understanding the system's ODD. However, user's perceptions of the subtasks of the dynamic driving task that they are required to perform versus those that the driving automation system will perform are dependent on their perception and understanding of the system

(SAE, 2016). Moreover, there is little to no uniformity when it comes to naming for these technologies. Survey findings revealed the importance of vehicle technology naming and how the naming of an in-vehicle technology can influence the driver's mental model of system purpose, function, and operator responsibility (Abraham, Seppelt, Mehler, & Reimer, 2017; Casner & Hutchins, 2019).

18.2.1 MENTAL MODELS

If we think of product design as a game of Pictionary (Figure 18.1), designers are the ones drawing and users are the ones trying to interpret their creations. Designers want users to understand the conceptual models of their applications. The system image, according to Norman (2013), is the product, and this product is typically the only way that designers are able to communicate with their users. If the system image is not presented to the user in such a way that its purpose and function are adequately transparent and intuitive, then the image becomes abstract and is subsequently open to user interpretation, which leads, inevitably, to misinterpretations (see also, this Handbook, Chapter 3).

System users will look to the visible components, observe system behavior, and then make guesses about how the system works. Thus, users will create their own mental model, which may differ, often significantly, from the designer's intentions due to varying prior knowledge, individual variability, and different beliefs about the purpose and function of the system (Jonassen, 1995). Having an accurate mental model will appropriately guide user behavior and predictions about system behavior. Having an inaccurate mental model may lead to user errors and inaccurate predictions about system activity.

When drivers lack an appropriate understanding of ACIV systems, the consequences can be misused or underused, which undercut the potential benefits of these technologies (Parasuraman & Riley, 1997). This can also lead to gaps in knowledge

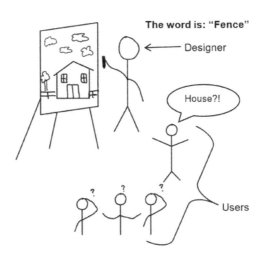

FIGURE 18.1 Example of a mental model misinterpretation.

that consumers will fill on their own, potentially with inaccurate information or beliefs derived from limited exposure or observation.

Mental models can be carefully formed and structured through training (Wickens, Hollands, Banbury, & Parasuraman, 2013). Using training combined with intuitive design practices can decrease variability in mental models among users. Furthermore, with increasing levels of automation and variability of system capabilities, increased levels of training are required (Sarter, Woods, & Billings, 1997). Additional training will improve the user's mental models to help ensure that vehicle operators are fully aware of their role and breadth of responsibilities while operating their vehicle on public roads.

18.2.2 CONSUMER PROTECTION

New challenges in litigation are reinforcing the importance of effectively training and educating drivers regarding vehicle operation. To minimize liability, a cornucopia of information regarding the operation of any vehicle that can be found in the manufacturer-provided operator's manual. However, there are a multitude of problems with this approach: many drivers don't read their owner's manual (Leonard, 2001), manuals are quite extensive (Brockman, 1992; Mehlenbacher, Wogalter, & Laughery, 2002), and studies suggest that when people believe that they are familiar with a product, they are less likely to read the documentation and warnings (Godfrey, Allender, Laughery, & Smith, 1983; Wogalter, Brelsford, Desaulniers, & Laughery, 1991; Wright, Creighton, & Threlfall, 1982). Furthermore, driver's knowledge of vehicle systems after reading the owner's manual has rarely translated to a demonstrated improvement in driving performance (Manser et al., 2019).

As the capabilities of driver support features progress to the higher levels of automation (SAE Levels 3–5), manufacturers may avoid potential legal claims by requiring consumers to undergo extensive training to ensure that they understand the various hazards and limitations of such systems (Glancy, Peterson, & Graham, 2015). At present, in the purview of SAE Levels 0–2, it is the driver's responsibility to use the system appropriately and understand system limitations. Some vehicle manufacturers are implementing systems to monitor driver behavior when automated features are activated, so the technology can be deactivated if the driver is not being attentive to system warnings or requests for intervention (Anderson et al., 2014; this Handbook, Chapter 14).

However, a recent National Transportation Safety Board report suggested that not all methods of driver monitoring are an effective means to ensure engagement in the driving task (National Transportation Safety Board, 2017a). Regardless of their effectiveness, recent research indicates that liability and control are among the top factors that drivers are most concerned about with regard to the use of vehicle automation (Howard & Dai, 2014; Kyriakidis, Happee, & de Winter, 2015; Schoettle & Sivak, 2014).

18.2.3 GOALS OF TRAINING FOR ACIV SYSTEMS

There are at least three overarching goals which training programs for ACIV Systems should aim to address. First, they should lead to improvements in safety.

Second, they should create appropriate levels of trust, and third, they should increase user acceptance of the benefits of the technology. These three goals are discussed in more detail below.

18.2.3.1 Increased Improvement in Safety

Training is one of the standard procedures used to aid people in acquiring safe behavior practices and remains the fundamental method for effecting self-protective behaviors (Cohen, Smith, & Anger, 1979). Current and future driver support features have the potential to reduce the number of crashes, injuries, and fatalities on public roadways, but only if they are accepted, used appropriately, and deployed responsibly. Benson, Tefft, Svancara, and Horrey (2018) estimated that technologies that include AEB, blind spot monitoring (BSM), forward collision warning (FCW), lane departure warning (LDW), and LKA, if installed on all vehicles, would have potentially mitigated roughly 40% of all crashes involving passenger vehicles, 37% of all injuries, and 29% of fatalities that occurred in those crashes in 2016.

The success of training depends on the use of positive approaches that stress learning safe behaviors rather than avoiding unsafe behaviors. Ensuring suitable conditions for practice that guarantee transferability of learned behaviors to real situations is critical (Cohen et al., 1979). Effective evaluation coupled with frequent feedback is also critical for reaching specified goals to mark progress.

18.2.3.2 Appropriate Levels of Trust

Mediating trust requires that ACIV system users have an accurate and realistic understanding of system capabilities, limitations, and responsibilities (see also, this Handbook, Chapter 4). Trust is defined by Lee and See (2004) as "the attitude that an agent will help achieve an individual's goals in a situation characterized by uncertainty and vulnerability." The formation of trust involves thinking and feeling, but emotions are the primary determinant of trusting behavior (Lee & See, 2004). The construct of trust is complicated and multi-faceted. However, the outcome of inappropriate trust in automated systems is fairly simple—when drivers do not possess appropriate trust in vehicle automation and understand the limitations and ODD of the ACIV systems, maximum safety benefits will not be realized due to users not using the systems (disuse) or using them inappropriately (misuse).

Research supporting this contention has shown that trust is a significant factor in the success of ADAS, as demonstrated by live road-, simulator-, and questionnaire-based studies (Banks, Eriksson, O'Donoghue, & Stanton, 2017; Eichelberger & McCartt, 2016; Rudin-Brown & Parker, 2004). Automation bias (or over trust) refers to the omission and commission of errors as a result of using automated cues as a heuristic replacement for vigilant information seeking and processing (Mosier, Skitka, Burdick, & Heers, 1996; Skitka, Mosier, & Burdick, 1999; 2000). In a study by Mosier, Skitka, Heers and Burdick (1998), automation bias played a significant factor in aircraft pilot interaction with automation aids. Pilots were not utilizing all available information when performing tasks and were making decisions with automated systems due to inappropriate trust in those systems. An example of misuse of systems was also found within the automotive domain by Victor et al. (2018), who found that, while system-based reminders influenced driver's eyes on-road and hands-on wheel behavior, prompts or explicit instructions regarding system limitations and supervision responsibilities were not able to prevent 28% of participants from colliding with an obstacle in the roadway despite seeing the hazard. This finding may suggest that participants were overly reliant on the vehicle system due to incorrect expectations of system capabilities.

18.2.3.3   User Acceptance

A critical concept relative to user acceptance is that user's attitudes and behavior have a bidirectional, multi-level relationship. Stated plainly, attitudes influence use and use influences attitudes (see also, this Handbook, Chapter 5). As drivers use ACIV systems, driving performance will likely improve as long as the systems are well-designed and the driver has a reasonable understanding of the system's ODD (see also, this Handbook, Chapter 12 for a discussion of behavioral adaptation to technologies).

Davis (1989, 1993) identified the lack of user acceptance as an impediment to the success of new information systems. The Technology Acceptance Model developed by Davis specifies the causal relationship between system design features, perceived usefulness, perceived ease of use, attitude toward usage, and actual usage behavior. Ghazizadeh, Lee, and Boyle (2012) extended that model to create the Automation Acceptance Model, which includes the explicit additions of trust and compatibility as well as a feedback loop that continues to inform user trust, perceived ease of use, compatibility, perceived usefulness, and behavioral intention to use. The Automation Acceptance Model captures the importance of experience and trial-and-error learning. Understanding the benefits of ACIV systems shapes the social norms regarding the use of these systems on the road. These norms may influence individual driver's perceptions of ACIVs and, ultimately, their use decisions (Ghazizadeh et al., 2012).

18.3   TRAINING CONTENT FOR ACIV SYSTEMS

Given that training is necessary, it is important to determine what is to be learned from that training. In the context of driver training for ACIVs, the question is: *what are the different skills, rules, and knowledge that a driver should be trained on to be able to safely operate a vehicle with automated features?* In light of the above, safely operating a motor vehicle is broadly to be interpreted as having appropriate trust in and acceptance of the technology as well as knowing the different skills, rules, and knowledge that are needed to successfully complete the driving task.

While drivers' familiarity with system activation, system deactivation, and the ability to perform both are essential components to system use, they are not the only criteria required for safe, efficient, and appropriate operation of ACIVs. Beyond these meager requirements, there exist higher-level knowledge-based skills that need to be learned, understood, and applied by operators of ADAS-equipped vehicles. The Federal Automated Vehicle Policy (U.S. DOT, 2016) recommends that training documents not only address the following topics for highly automated vehicles (e.g., SAE Levels 3–5) but also contain relevant elements across all levels of vehicle automation that could be applied to the ACIV framework. The recommended areas are as follows: highly automated vehicle system's intent, ODD, system capabilities and limitations, engagement/disengagement methods, HMI, emergency fallback scenarios, operational boundary responsibilities, and potential mechanisms that could change function behavior in service. Additionally, the policy recommends a hands-on experience program for consumers to ensure that they are fully aware of their vehicle's functions and capabilities. A report by the Pennsylvania Department of Transportation identified familiarity with assistance features and with required interactions between drivers and vehicles as areas that should be included in both knowledge and skill testing criteria (Hendrickson, Biehler, & Mashayekh, 2014). As of August 2016, seven states and the District of Columbia have enacted autonomous vehicle legislation, and one state has an executive order pertaining to the testing of autonomous vehicles on public roads (Council of State Governments, 2016). Training should educate consumers on the limitations and capabilities of HAVs, how to engage and disengage the system functions, risks of misuse, and how to deal with emergency situations related to the ACIV (American Association of Motor Vehicle Administrators, 2018). Furthermore, driver training should make every effort to ensure consumers will use the products within the established parameters.

18.3.1 SYSTEM PURPOSE AND ASSOCIATED RISKS OF USE

When a member of the motoring public elects to use an ACIV system, one of the required fundamental levels of knowledge includes the purpose of the system as well as any associated risks and benefits. Training drivers on this information could eliminate many negative consequences that result from the misuse of ACIV systems by appropriately modulating the user's expectations of the system's capabilities. Several factors influence drivers' decisions to use an ACIV system. As discussed above, user's attitude toward automated systems, the level of trust in the systems, mental workload, self-confidence, and the level of perceived risk are among the most important factors (Parasuraman & Riley, 1997). Drivers should be informed of the relationship between their motivations for utilizing ACIV systems and potential safety-critical outcomes if not used appropriately. For example, if the motivating factor is to mitigate fatigue, training should inform drivers that fatigue could lead to over-reliance and complacency, which could result in an inability to take action in emergency situations, such as system malfunction (Parasuraman & Riley, 1997).

18.3.2 OPERATIONAL DESIGN DOMAIN

Drivers must understand the capabilities and limitations of the ACIV system that they are using, a skill which, to date, many users have demonstrated deficiencies in through both knowledge-based tests and their own self-reported behavior while using these systems. Evidence of this situation was provided by McDonald et al. (2018), who found that only 21% of owners of vehicles with BSM systems correctly identified their inability to detect vehicles passing at very high speeds as a system limitation; the remainder expressed additional misconceptions about the BSM system's function and/or reported that they were unsure of the system's limitations. The authors also found that 33% of respondents with AEB systems did not realize that the system relied on cameras or sensors that could be blocked by dirt, ice, or snow.

Given that these are safety-critical limitations, users of these driver support and active safety features should have an understanding of when these systems can fail to operate and conditions under which it would be prudent to temporarily suspend system use. Better understanding could be achieved through the use of proper training.

The AAA Foundation for Traffic Safety in collaboration with the Massachusetts Institute of Technology AgeLab have developed a data-driven system to review and rate the effectiveness of new in-vehicle technologies that aim to improve safety (Mehler et al., 2014). This review focuses on legacy systems, such as Electronic Stability Control, and advanced features, such as Adaptive Cruise Control (ACC), Adaptive Headlights, Back-Up Cameras, FCW, Forward Collision Mitigation, and LDW. In addition to developing a rating system that considers the potential and demonstrated benefits offered by these technologies, the research team stated that while some systems require little or no familiarity with the technology to derive benefit, others have a steep learning curve (Mehler et al., 2014).

18.3.3 Monitoring the Road and the System

When a person's central concern is an individual task or decision, they are not likely to be interested in reading a comprehensive body of work on related matter. Instead, they will likely want the knowledge and skills that will be useful in dealing with the particular task or decision at hand (Tough, 1971). For this reason, it is important to highlight the responsibilities of the human operator and any system limitations. Twenty-nine percent of respondents in a study by McDonald et al. (2018) reported that they at least occasionally felt comfortable engaging in other activities while using ACC. Many respondents also reported not visually checking their blind spot when changing lanes in vehicles equipped with BSM (30%), and 25% of vehicle owners with rear cross-traffic alert systems reported not looking over their shoulder when backing up at least some of the time. These self-reported user behaviors further motivate the need for driver training just because such systems are fallible, and the most benefit is derived when the human operator is looking for potential threats as well as automation. In short, the training must be designed in a way that considers the driver as an active participant in the automation features, not just as a passive user.

18.4 ANDRAGOGICAL CONSIDERATIONS FOR ACIV SYSTEMS TRAINING

One of the standard pedagogical models of education assigns full responsibility to the instructor for making decisions about what will be learned, how it will be learned, when it will be learned, and whether it has been learned. The learner in this pedagogical model is a passive participant in their own education. Pedagogical methods are often improperly implemented for adult learners, whose intellectual aspirations are least likely to be aroused by the uncompromising requirements of authoritative, conventional institutions of learning (Lindeman, 1926). As individuals mature, their need and ability to self-direct, leverage experience, identify readiness to learn, and organize their learning around life problems increase (Knowles, Holton III, &

Swanson, 2005). Andragogical models bring into focus additional learner characteristics that should be considered in the development of training for ACIV systems (Knowles, 1979; Knowles et al., 2005). For vehicle automation, the training structure should focus on both near- and long-term improvements in driving performance as well as improved understanding of driver responsibilities and sustained understanding of vehicle system capabilities and limitations.

18.4.1 DRIVER-RELATED FACTORS

The characteristics of the target population may impact the design and delivery of instruction (Rothwell, Benscoter, King, & King, 2016). The training methods and content covered may need to be adjusted over time to allow users to gain the full benefit of their current driver support features based on individual differences. For example, drivers who are technology-averse may need a different level and type of training than their peers. Conversely, users who are technologically inclined may inadvertently place too much trust in technology. Additional factors that need to be considered in training design are discussed below.

18.4.1.1 Motivation

The motivational aspects of training design cover many theoretical concepts, including attribution theory, equity theory, locus of control, expectancy theory, need of achievement, and goal setting (Patrick, 1992). Trainee motivation can be influenced by individual characteristics as well as the characteristics of the training itself (Coultas, Grossman, & Salas, 2012). The temporal divisions of motivation were described by Quiñones (2003) as having an effect on (1) whether an individual decides to attend training in the first place, (2) the amount of effort exerted during the training session, and (3) the application of skills after training. Zhang, Hajiseyedjavadi, Wang, Samuel, and Qu (2018) found training transfer for hazard anticipation, and attention maintenance was observed only in drivers who were considered to be careful (e.g., low sensation seeking and aggressiveness). Due to its multi-faceted nature and the temporal inconsistencies within and between trainees, the consideration of factors affecting motivation requires significant attention when designing training programs.

18.4.1.2 Readiness to Learn

Learning can only happen when the learner is *ready* to learn. Readiness can be induced through exposure to models of superior performance, counseling, exercises, and other techniques to help trainees see value in the educational undertaking (Knowles et al., 2005). If a learner is deficient in a prerequisite skillset, readiness to learn a more advanced skillset may be futile. Gagné (1965) emphasized the importance of prerequisites in learning complex skills, as they are essentially subskills of the newer, more complex task. Consequently, trainees lacking necessary prerequisite skills will likely have to partake in remedial training prior to the onset of more advanced training in order to first achieve mastery (or at least satisfactory performance) of the remedial skill.

18.4.1.3 Affect—Anxiety Toward Technology

Affect, in this context, refers to the evaluation of an object or system as good or bad, evoked immediately and subconsciously within the individual (Finucane, Alhakami, Slovic, & Johnson, 2000; Slovic, Finucane, Peters, & Macgregor, 2006). Affect works both to focus a person's attention on what they perceive to be relevant details and to manage priorities of information processing. It is also believed to help people form complete mental models, as the cognitive complexity of decisions exceeds a human's ability to rationally evaluate a situation (Lee, 2006; Lee & See, 2004).

People's feelings toward a technology can predict their judgment of risk and benefit, regardless of the actual risk and benefit associated with the technology (Alhakami & Slovic, 1994). The affect heuristic indicates that there is a perceived inverse relationship between risk and benefit. Therefore, if a person has negative feelings about a technology, they are likely to view the technology as risky (Hughes, Rice, Trafimow, & Clayton, 2009). There are some decision-based studies suggesting that affect has an impact on subsequent information processing and judgments regarding that system or object (Cialdini, 2007). Merritt (2011; Merritt, Heimbaugh, LaChapell, & Lee, 2013) has studied the affective processes in human–automation interactions and found that user's implicit attitudes have important implications for trust in automation. Merritt found that user's decision-making processes may be less rational and more emotional than previously acknowledged, noting that implicit attitudes and positive affect may be used as a lever to effectively calibrate trust. This relationship between learner affective processes and technology may have an impact both on the prospective user's readiness to learn and on the transferability of training effects.

Biases induced by the affect heuristic may serve as a barrier to positive training transfer. A study by Smith-Jentsch, Jentsch, Payne, and Salas (1996) found that conceptually relevant negative events accounted for individual differences in learning and retention. Furthermore, the effect of [computer] anxiety in a meta-analysis by Colquitt, Lepine, and Noe (2000) demonstrated significant relationships with every training outcome examined (motivation to learn, post-training self-efficacy, declarative knowledge, skill acquisition, and reactions).

18.4.2 DESIGN OF THE TRAINING

18.4.2.1 Salient Relevance and Utility of Training Content

When training sessions are voluntary, adult learners may leave when the teaching or content fails to meet their interest or they no longer perceive value (Lindeman, 1926). The value or perceived utility in the knowledge or skill being taught needs to be salient to the learner, or their motivation to continue investing their time will plummet. To this end, simply stating learning objectives for adult learners will not be enough; rather, providing the learning objective with justification may provide enough transparency to sufficiently motivate further engagement. The link between the expected usefulness of training material and trainee motivation has been demonstrated in training and adult education literature time and time again (Burke & Hutchins, 2007; Noe, 1986; Tannenbaum, Mathieu, Salas, & Cannon-Bowers, 1991). Trainees who perceive that the new knowledge and skills are relevant and will improve performance will demonstrate positive training transfer (Alliger, Tannenbaum, Bennett Jr., Traver, & Shotland, 1997; Baumgartel, Reynolds, & Pathan, 1984; Rogers, 1951).

Another element of training content that has been shown to influence performance was found by Ivancic and Hesketh (2000), who investigated the effect of guided error training on driving skill and confidence on a driving simulator. In error training, learners are shown examples of errors that they themselves make in a test of their knowledge and skills and solutions for overcoming these errors. One group of participants was given feedback and training on their errors while driving (error training); another group was shown examples of errors in general, not their own errors, while driving (guided error training). The error training led to better performance in near and far transfer of training evaluations of performance. Moreover, error training reduced driver's self-confidence in their driving skill at the end of the training when compared with the group that received error-less training. This suggests, importantly, that training can improve performance without increasing confidence. However, error training may not be appropriate for all prospective users, especially those who demonstrate low confidence with using vehicle automation.

18.4.2.2 Framing of the Opportunity

In general, adults are motivated to learn as they experience needs and discover interests that learning new material will satisfy. Consequently, their involvement in this process should be considered voluntary—again, it is not possible to make someone learn when they do not want to. The way in which training programs are framed can enhance or diminish the effectiveness of training interventions (Quiñones, 1995). Adult learners need to feel a certain degree of autonomy in their educational endeavors. When presented with an activity labeled "education," "training," or anything synonymous, adult learners have a tendency to revert back to more dependent social roles, which is at odds with their innate desire to exercise their independence (Knowles et al., 2005).

Webster and Martocchio (1993) found that younger employees (under 40 years) who received training labeled as "play" demonstrated higher motivation to learn and performed better in an objective test of software knowledge than older employees (40 years or older). Interestingly, no differences were found between younger and older employees receiving training labeled as "work."

Games can provide models of good learning practices (Sandford & Williamson, 2005). Games are different from most other forms of learning, since players rarely want or need to read a manual before commencing play but instead they "learn by playing," which could be an enticing strategy for adult learners. The use of games to train drivers on ACIV technologies has not been examined to date but remains a potential alternative.

18.4.2.3 Opportunities to Practice and Apply Skills

Adults learn new knowledge, skills, values, and attitudes most effectively in the context of application to real-life situations (Knowles et al., 2005). This lends credence to the implementation of constructivist methods of instruction. Constructivism has roots in multiple psychological and philosophical domains and has implications for learning outcomes, such as reasoning, critical thinking, understanding and use of knowledge, self-regulation, and mindful reflection (Driscoll, 2000). Early contributors include Piaget (cognitive and developmental), Bruner and Vygotsky (interactional and cultural emphasis), as well as Dewey (1933), Goodman (1984), and Gibson (1977).

The first constructivist condition for learning is to embed learning in complex, realistic, and relevant environments. The rationale behind this condition is that "students cannot be expected to learn to deal with complexity unless they have the opportunity to experience complexity." Allowing learners to experience training through authentic activities will enhance the development of problem-solving capabilities and critical thinking skills. Problem-solving requires domain knowledge as well as structural knowledge (mental models), reasoning capabilities, and metacognitive skills.

Skills in complex tasks, including those with large social components, are usually taught best by a combination of training procedures involving both whole tasks and components or part tasks (Anderson, Reder, & Simon, 1996). Using situations that are relevant to the learner provides an opportunity for learners to engage in reflective thinking (Federal Aviation Administration, 2009; Shor, 1996). Incorporating a restrictive definition of authenticity in training design will result in learning environments that are authentic in a narrow context, thereby reducing the variability of skills and strategies acquired by the learner. Training should teach trainees to be innovative, creative, and adaptable so that they can deal with the demands of domains that are complex and ill-structured (Federation of American Scientists, 2005; Gee, 2003; Prensky, 2001).The inclusion of related cases in training can scaffold the learner's memory by providing representations of experiences that learners have not had (Jonassen, 1999), especially in ill-defined domains and for non-recurrent skills (Merrill, 2002; Schwartz, Lin, Brophy, & Bransford, 1999; van Merriënboer & Kirschner, 2017).

18.5 TRAINING PROTOCOLS

Current training methods for in-vehicle technologies show that while many of the strategies implemented by individual vehicle owners may suit some characteristics of adult learners (e.g., motivated to seek knowledge, task-oriented, need for self-direction), the current paradigm of consumer education does not provide sufficient motivation for the learner to search for a deeper understanding of the material. Several consumer-preferred methods for learning to use in-vehicle technologies were identified by Abraham, Reimer, Seppelt, Fitzgerald, and Coughlin (2017). These methods are discussed in the following sub-sections.

18.5.1 TRIAL AND ERROR

Learning through trial and error is a process that involves the forming of associations (connections) between the perception of stimuli and responses that manifest themselves behaviorally to those stimuli (Thorndike, 1913). A recent study found that 53% of drivers learned to use in-vehicle technology by means of trial and error at least some of the time (Abraham, Reimer, et al., 2017).

Generally, experience can improve performance, but learning through trial and error can waste time, may lead to a less-than-ideal solution, or may never result in a solution at all. Trial and error can also result in negative effects, such as increased frustration, reduced learner motivation, discontinued use of the system, or mental model recalibration through potentially dangerous experiences. Learning by doing is not a homogenous process. A study by Pereira, Beggiato, and Petzoldt (2015) found that mastering the use of ACC took different lengths of time. Furthermore, Larsson (2012) conducted a survey of 130 ACC users. The results indicated that drivers need to be especially attentive in those situations to which, during conventional driving, they would not be attentive. The system may not be self-explanatory enough for a strictly trial-and-error based approach. Larsson's survey results indicated that as drivers gained experience using the ACC system, they became more aware of the system's limitations. However, other studies have shown that safety-critical misunderstandings of system limitations are resilient and can persist over time (Kyriakidis et al., 2015; Llaneras, 2006). Tversky & Kahneman (1971) argued that users tended to place undue confidence in the stability of observed patterns, thus resulting in misunderstandings that are not corrected when relying solely on trial-and-error learning methods.

18.5.2 Vehicle Owner's Manual

Many vehicle owners learn to use their in-vehicle technologies by reading the owner's manual. In a survey conducted by Abraham, Reimer et al. (2017), 55% of respondents reported learning to use the in-vehicle technologies in their vehicles at least some of the time by reading the owners' manual.

Given the nature of this resource, learner engagement is highly dependent on the learner's perceived need of the information and their willingness to use the manual as a resource. As a completely self-guided method, learners are typically directed by a specific line of inquiry or a particular question they want answered. This observation is substantiated by Leonard (2001), who found that individuals are more likely to read specific portions of the operator's manual as opposed to the manual in its entirety. The majority of participants (62%) in the study by Leonard reported reading "special parts" of the operator's manual. Of the manual topics that were read, many pertained to equipment or maintenance. Approximately 2% of respondents reported not reading any safety information in the manual. Consequently, Leonard concluded that the limited use of the operator's manual is not a result of lack of availability and that while the manual is recognized as a source of information, its use is limited. If drivers are not motivated to read sections of the operator's manual, they will get no benefit from the information provided.

18.5.3 Demonstration

Demonstration, also known as behavior modeling, results in the display of a new pattern of behavior by a learner/trainee who observes a model (experienced user, professional), performing the task to be learned. The trainee is encouraged to rehearse and practice the model's behavior, and feedback is provided as the trainee refines their behavior to closer approximate that of the observed model. This training style is founded in social and developmental psychology centered on the research of Bandura (1977), who argued that by observing model behavior, people can develop a cognitive representation (mental model), which can then be used to guide future behavior. In theory, observational learning is a great strategy for situations in which early errors are viewed as problematic and visual guidance could provide a means to reduce the frequency and severity of errors. This makes behavior modeling an ideal way to learn how to perform simple tasks, such as system activation, which may result in performance deficiencies if learned using trial and error.

One limitation of the demonstration method is that it does nothing to improve knowledge of the system and it appears to result in fairly superficial learning. Findings from McDonald et al. (2017) show that ride-along demonstrations were ultimately no better than self-study of the owner's manual, as there were knowledge gains across all training types, of which none provided a statistically significant difference.

The behavioral model is also assumed to exhibit desirable behaviors and have accurate information. When demonstration takes place in automotive dealerships, sales people are assumed to have been trained on how to use ACIV systems and have accurate knowledge about their use. Unfortunately, this assumption is not always accurate. Abraham, McAnulty, Mehler, and Reimer (2017) conducted a study investigating sales employees from six vehicle dealerships in the Boston, MA area associated with six major vehicle manufacturers. They found that many sales people lacked a strong understanding of ACIV systems. It was also revealed that the training the employees received was meager, consisting mostly of web-based modules with very little (if any) hands-on experience. Furthermore, two of the sixteen employees with whom the researchers interacted gave explicitly wrong safety-critical information regarding the systems.

18.5.4 WEB-BASED CONTENT

Web-based content can include user-generated content or websites and materials generated by safety advocates, vehicle manufacturers, or other stakeholders in the automotive industry. The increased ubiquity and usability of technology has made it possible for a range of individuals and organizations with an interest in ACIVs to generate their own content for a variety of topics, including system performance, software updates, and maintenance.

Learning through user-generated content involves a large amount of independence, is question driven, and can result in additional extraneous load if the learner becomes too immersed in the material (Wickens et al., 2013). Additionally, less than desirable behavioral models can be found in user-generated content, such as attempts to "fool" or "hack" the ACIV system. Furthermore, information provided on the internet may not be accurate. Content may not be kept up-to-date with software updates and system capabilities, and those seeking knowledge about these systems will need to have a certain level of familiarity with their own vehicles to make the information useful.

18.5.5 Future Training—Embedded Training for ACIV Systems

Embedded training is described as a training program built into systems so that operational equipment can be switched over to a "training mode" during periods when it is not needed for operational use (Sanders & McCormick, 1993). Strategies mentioned in the previous section can be incorporated into this type of training to facilitate skill and knowledge acquisition at an appropriate level and enhance engagement by being appropriately difficult. Several important components of successful training practices that are missing from current knowledge acquisition methods for ACIV systems will be discussed.

The training needs of participants will change from their first exposure to the vehicle systems to their one hundredth exposure (Gagné, 1965). In their initial exposure, participants will need to be familiarized with the system interface, purpose, methods of activation and deactivation, and basic system signals. This could be done using the vehicle interface and multimedia methods so that auditory alerts are consistent, as are the placement, size, shape, and color of icons pertinent to the system. With increased exposure to the system, driver's perceived familiarity with the system will increase.

Godfrey, Allender, Laughery, and Smith (1983) conducted an evaluation of eight generic products and found that the more hazardous consumers perceived a product to be, the more likely they were to look for a warning. They also found that perceived hazardousness varied inversely with product familiarity, meaning the more familiar people thought they were with a product, the less hazardous they perceived it to be. This finding is particularly concerning, since studies have found that with limited exposure to vehicle automation, novice user's self-reported familiarity increased significantly (Manser et al., 2019). Over longer durations, drivers may become more aware of system limitations (Larsson, 2012), but safety-critical misunderstandings of system limitations have also been shown to persist over time (Kyriakidis et al., 2017; Llaneras, 2006).

18.5.5.1 Feedback (Knowledge of Results)

"Practice makes perfect" is a common saying; however, practice is of very little value when the results of an action are unknown or incorrect. Knowledge of results, or extrinsic feedback, is an important tool in the growth and development of learners because it provides an indication of discrepancy between actual and desired behavior (Patrick, 1992). Learning is promoted when learners are guided in their problem-solving by appropriate feedback mechanisms, which include error detection and correction (Gagné, 1965; Merrill, 2002). Feedback comes not only in the form of coaching, but also from the visual, auditory, and proprioceptive information associated with normal (correct) task execution. Consequently, Annett (1961) specified the difference between intrinsic feedback, which pertains to information concerning normal (non-training) task performance, and extrinsic feedback, which refers to additional knowledge supplied during training and not available during typical task performance.

Extrinsic feedback can be provided in real time during task performance or at some point after task completion (post hoc). Real-time and post hoc feedback for novice driver training has been shown to improve teen driving safety as well as reduce the frequency of risky driving behaviors (Klauer et al., 2017; Peek-Asa, Hamann, Reyes, & McGehee, 2016). Personalized feedback coupled with active practice was also shown to be superior to passive learning methods with no feedback when assessing older driver's scanning behaviors at intersections (Romoser, Pollatsek, Fisher, & Williams, 2013). Using real-time feedback for adults to assist them in learning/understanding ACIV systems could be a useful technique; however, additional research is needed on this topic.

If it can be determined that driver performance deficiencies are attributable to a lack of skill or knowledge, then an immediate training intervention after the first occurrence of the undesired behavior in situ may help to correct the behavior. However, if undesirable safety-related performance deficiencies cannot be attributed to a lack of skill or knowledge, then the solution does not lie in training, but in the application of salient differential consequences (Boldovici, 1992; Mager & Pipe, 1970). One example is the Tesla Autosteer system deactivation for non-compliance with hands-on wheel warnings (Telsa Inc., 2018).

18.5.5.2   When Does It End?

Giving learners complete control over when they may terminate learning invites overconfidence that a skill has been fully mastered. If the learner's metric for self-evaluation is heavily dominated by error-free performance, a highly salient measure, they may terminate their training too soon. Personal, unguided reflection on performance and understanding is a task people rarely perform well (Kruger & Dunning, 1999; Regehr & Eva, 2006). Fitts (1962) advocated that training should be continued beyond a minimum performance criterion and that skills acquired during training should be sufficiently versatile to withstand the change from a structured training situation to the less predictable application in the real world. Consequently, the criterion for trainee performance must be carefully defined and termination of training by trainee selection is not recommended.

According to Spitzer (1984), there is an unwritten law that training programs last a certain number of days, with lectures of a certain duration and breaks at specified times. This notion contributes to the erroneous idea that learning is a time-bound event that only occurs in certain defined time periods. Skills acquired during periods of training need to be sufficiently versatile to withstand the transition from the organized training context to the less predictable real-world domain. With merely "acceptable" performance, the fragile skill acquisition process will be easily disrupted unless further training is provided.

18.6   RECOMMENDATIONS

Training drivers is critical to the successful deployment of driver support features from low to high levels of automation. As discussed in this chapter, the high variability of drivers on our roadways and the competing strengths and limitations of current ACIV systems present challenges that transportation safety researchers must address. Given these challenges, there are several recommendations listed below.

First, the areas of training should remain dynamic as ACIV systems continue to develop, new data become available, and new skills become necessary. For example, new training requirements could arise from a driver's increased exposure (and familiarity) with particular ACIV systems, software updates, or the behavioral adaptation of non-system users.

Second, considerations for the training of system users should be included as a key point in the design cycle of these new systems. Training programs should be subject to proper evaluation and assessment to ensure that learning outcomes are achieved and that no unintended consequences are introduced by the program.

Third, in-vehicle driver monitoring systems may be an important option to consider for ACIV system training (see also, this Handbook, Chapter 11). Campbell et al. (2018) discussed the use of driver monitoring systems to avoid out-of-the-loop problems. Salinger (2018) discussed a driver monitoring system that presented multi-modal signals to capture drivers' attention and return focus back to the control or monitoring loop. Another approach to driver monitoring is to periodically provide a message to the driver. The National Transportation Safety Board has recommended implementing driver monitoring systems (National Transportation Safety Board, 2017b).

Fourth, traffic safety professionals need to develop effective training guidelines and procedures for ACIV systems. Currently, the California Department of Motor Vehicles requires that training programs for drivers who test such systems in public include familiarization with the automated driving system technology; basic technical training regarding the system concept, capabilities, and limitations; ride-along demonstrations by an experienced test driver; and subsequent behind-the-wheel training (Nowakowski, Shladover, Chan, & Tan, 2014). Perhaps another worthwhile endeavor in the near term would be to add some measure of advanced vehicle system components to future iterations of the basic knowledge test for the standard licensing requirement, similar to what was implemented in at least twenty states and the District of Columbia for distracted driving as of 2013 (Governors Highway Safety Association, 2013).

Finally, legislative action amending statutory and regulatory definitions of applicable terms (e.g. driver, vehicle, etc.) as well as reviewing and adapting existing rules regarding vehicle operation may be a persistent challenge until policy makers are well versed in the subject matter. Educating all entities on the need for acceptance and implementation of these universal terms and definitions will be an implementation challenge (American Association of Motor Vehicle Administrators, 2018). This is one reason why communication between researchers and legislators must be clear and concise so that, in the event legislation is required, it is based on science and not on other implicit or explicit biases. Furthermore, all stakeholders are encouraged to communicate with one another on the most effective ways to train novice and experienced drivers on ACIV systems. Educational materials that are developed should be proven effective and understood by the general motoring public.

18.7   CHALLENGES AND CONCLUSIONS

The true challenges that accompany the actualization of ACIV systems lie in the transition from fully manual to fully autonomous driving. As we transition through levels of automation, we are fundamentally changing the driving task primarily from the one of manual human control to automated vehicle system control. Additionally, we are changing the roles and responsibilities of driving from a process where the system supports the human driver to a process where the human driver supports the system.

Given this large change in the driving task, human factors professionals most likely will not be able to design their way to safety. The importance of intuitive and understandable HMI design is critical. However, we must also have effective and broadly available training for all users. All stakeholders will need to work together on this issue for ACIVs to deliver the safety benefits that will potentially save thousands of lives on our nation's roadways.

ACKNOWLEDGMENTS

This chapter is based in part on a Safe-D UTC report by the authors. Safety through Disruption (Safe-D) National UTC, a grant from the U.S. Department of Transportation's University Transportation Centers Program (Federal Grant Number: 69A3551747115).

REFERENCES

Abraham, H., McAnulty, H., Mehler, B., & Reimer, B. (2017). Case study of today's automotive dealerships: Introduction and delivery of advanced driver assistance systems. *Transportation Research Record*, *2660*, 7–14. doi:10.3141/2660–02

Abraham, H., Reimer, B., Seppelt, B., Fitzgerald, C., & Coughlin, J. F. (2017). *Consumer Interest in Automation: Preliminary Observations Exploring a Year's Change.* Cambridge, MA: MIT Agelab.

Abraham, H., Seppelt, B., Mehler, B., & Reimer, B. (2017). What's in a name: Vehicle technology branding and consumer expectations for automation. *Proceedings of the 9th International Conference on Automotive User Interfaces and Interactive Vehicular Applications - AutomotiveUI '17.* doi:10.1145/3122986.3123018

Ahmed-Zaid, F., Krishnan, H., Vladimerou, V., Brovold, S., Cunningham, A., Goudy, R., ... Viray, R. (2014). *Vehicle-to-Vehicle Safety System and Vehicle Build for Safety Pilot (V2V-SP) Final Report.* Washington, DC: National Highway Traffic Safety Administration.

Alhakami, A. S. & Slovic, P. (1994). A psychological study of the inverse relationship between perceived risk and perceived benefit. *Risk Analysis*, *14*(6), 1085–1096. doi:10.1111/j.1539-6924.1994.tb00080.x

Alliger, G. M., Tannenbaum, S. I., Bennett, W., Jr., Traver, H., & Shotland, A. (1997). A meta-analysis of the relations among training criteria. *Personnel Psychology*, *50*(2), 341–358. doi:10.1111/j.1744-6570.1997.tb00911.x

American Association of Motor Vehicle Administrators. (2018). *Jurisdictional Guidelines for the Safe Testing and Deployment of Highly Automated Vehicles.* Arlington, VA: AAMVA.

Anderson, J. M., Kalra, N., Stanley, K. D., Sorensen, P., Samaras, C., & Oluwatola, O. A. (2014). *Autonomous Vehicle Technology: A Guide for Policymakers.* Santa Monica, CA: Rand Corporation.

Anderson, J. R., Reder, L. M., & Simon, H. A. (1996). Situated learning and education. *Educational Researcher*, *25*(4), 5–11. doi:10.3102/0013189X025004005

Annett, J. (1961). *The Role of Knowledge of Results in Learning: A Survey.* Retrieved from https://apps.dtic.mil/docs/citations/AD0262937

Bandura, A. (1977). *Social Learning Theory* (1st ed.). Englewood Cliffs, NJ: Prentice-Hall.

Banks, V. A., Eriksson, A., O 'Donoghue, J., & Stanton, N. A. (2017). Is partially automated driving a bad idea? Observations from an on-road study. *Applied Ergonomics, 68,* 138–145. doi:10.1016/j.apergo.2017.11.010

Barfield, W. & Dingus, T. A. (1997). *Human Factors in Intelligent Transportation Systems.* New York: Psychology Press.

Baumgartel, H. J., Reynolds, J. I., & Pathan, R. Z. (1984). How personality and organisational climate variables moderate the effectiveness of management development programmes: A review and some recent research findings. *Management & Labour Studies, 9*(1), 1–16.

Benson, A. J., Tefft, B. C., Svancara, A. M., & Horrey, W. J. (2018). *Potential Reductions in Crashes, Injuries, and Deaths from Large-Scale Deployment of Advanced Driver Assistance Systems* (Research Brief). Washington, DC: AAA Foundation for Traffic Safety.

Bettisworth, C., Burt, M., Chachich, A., Harrington, R., Hassol, J., Kim, A., … Ritter, G. (2015). *Status of the Dedicated Short-Range Communications Technology and Applications: Report to Congress* (FHWA-JPO-15–218). Washington, DC: United States Department of Transportation.

Blanco, M., Atwood, J., Vasquez, H. M., Trimble, T. E., Fitchett, V. L., Radlbeck, J., … Morgan, J. F. (2015). *Human Factors Evaluation of Level 2 and Level 3 Automated Driving Concepts* (DOT HS 812 182). Washington, DC: National Highway Traffic Safety Administration.

Boldovici, J. A. (1992). *Toward a Theory of Adaptive Training.* Alexandria, VA: U.S. Army Research Institute for the Behavioral and Social Sciences.

Brockman, R. (1992). *Writing Better Computer User Documentation: From Paper to Online* (2nd ed.). New York: John Wiley & Sons.

Burke, L. A. & Hutchins, H. M. (2007). Training transfer. An integrative literature review. *Human Resource Development Review, 6*(3), 263–296. doi:10.1177/1534484307303035

Campbell, J. L., Brown, J. L., Graving, J. S., Richard, C. M., Lichty, M. G., Bacon, L. P., … Sanquist, T. (2018). *Human Factors Design Principles for Level 2 and Level 3 Automated Driving Concepts* (DOT HS 812 555). Washington, DC: National Highway Traffic Safety Administration.

Casner, S. M. & Hutchins, E. L. (2019). What do we tell the drivers? Toward minimum driver training standards for partially automated cars. *Journal of Cognitive Engineering and Decision Making, 13*(2), 55–66. doi:10.1177/1555343419830901

Cialdini, R. B. (2007). *Influence - The Psychology of Persuasion.* New York: Harper-Collins.

Cohen, A., Smith, M. J., & Anger, W. K. (1979). Self-protective measures against workplace hazards. *Journal of Safety Research, 11*(3), 121–131.

Colquitt, J. A., Lepine, J. A., & Noe, R. A. (2000). Toward an integrative theory of training motivation: A meta-analytic path analysis of 20 years of research. *Journal of Applied Psychology, 85*(5), 678–707. doi:10.1037//0021–9010.85.5.678

Congress, N. (1994). The automated highway system: An idea whose time has come. *Public Roads, 58*(1), 1–7.

Coultas, C. W., Grossman, R., & Salas, E. (2012). Design, delivery, evaluation, and transfer of training systems. In G. Salvendy (Ed.), *Handbook of Human Factors and Ergonomics* (4th ed., pp. 490–533). New York: John Wiley & Sons.

Council of State Governments. (2016). *State Laws on Autonomous Vehicles.* Retrieved from http://knowledgecenter.csg.org/kc/system/files/CR_automomous.pdf

Davis, F. D. (1989). Perceived usefulness, perceived ease of use, and user acceptance of information. *MIS Quarterly, 13.* Retrieved from www.jstor.org/stable/pdf/249008.pdf?refre qid=excelsior%3A59df6413a1047bd694b4c69a7eaf559e

Davis, F. D. (1993). User acceptance of information technology: System characteristics, user perceptions and behavioral impacts. *International Journal of Man-Machine Studies, 38*(3), 475–487. doi:10.1006/imms.1993.1022

Dewey, J. (1933). *How We Think: A Restatement of the Relation of Reflective Thinking to the Educative Process*. Boston: DC Heath.

Driscoll, M. P. (2000). *Psychology of Learning for Instruction*. Needham Heights, MA: Allyn & Bacon.

Eichelberger, A. H. & McCartt, A. T. (2016). Toyota drivers' experiences with Dynamic Radar Cruise Control, Pre-Collision System, and Lane-Keeping Assist. *Journal of Safety Research, 56*, 67–73. https://doi.org/10.1016/j.jsr.2015.12.002

Fancher, P., Ervin, R., Sayer, J., Hagan, M. R., Bogard, S., Bareket, Z., … Haugen, J. (1998). *Intelligent Cruise Control Field Operational Test* (DOT HS 808 849). Washington, D.C.: National Highway Traffic Safety Administration.

Federal Aviation Administration. (2009). *Aviation Instructor's Handbook*. Retrieved from www.faa.gov/regulations_policies/handbooks_manuals/aviation/aviation_instructors_handbook/media/faa-h-8083-9a.pdf

Federation of American Scientists. (2005). *Harnessing the Power of Video Games for Learning*. Retrieved from https://fas.org/programs/ltp/policy_and_publications/summit/Summit on Educational Games.pdf

Finucane, M., Alhakami, A., Slovic, P., & Johnson, S. M. (2000). The affect heuristic in judgements of risks and benefits. *Journal of Behavioral Decision Making, 13*, 1–17.

Fitts, P. M. (1962). Factors in complex skill training. In R. Glaser (Ed.), *Training Research and Education* (pp. 177–197). New York: Dover Publications.

Gagné, R. M. (1965). *The Conditions of Learning* (3rd ed.). New York: Holt, Rinehart and Winston.

Gee, J. P. (2003). What video games have to teach us about learning and literacy. *Computers in Entertainment, 1*(1), 20.

Ghazizadeh, M., Lee, J. D., & Boyle, L. N. (2012). Extending the technology acceptance model to assess automation. *Cognition, Technology & Work, 14*(1), 39–49. doi:10.1007/s10111-011-0194-3

Gibson, J. J. (1977). *The Theory of Affordances*. Hilldale, PA.

Glancy, D. J., Peterson, R. W., & Graham, K. F. (2015). *A Look at the Legal Environment for Driverless Vehicles*. Washington, DC: Transportation Research Board.

Godfrey, S. S., Allender, L., Laughery, K. R., & Smith, V. L. (1983). Warning messages: Will the consumer bother to look? *Proceedings of the Human Factors and Ergonomics Society Annual Meeting, 27*(11), 950–954.

Goodman, N. (1984). *Of Mind and Other Matters*. Cambridge, MA: Harvard University Press.

Governors Highway Safety Association. (2013). *2013 Distracted Driving: Survey of the States*. Washington, DC: GHSA.

Grossman, R. & Salas, E. (2011). The transfer of training: What really matters. *International Journal of Training and Development, 15*(2), 103–120. doi:10.1111/j.1468-2419.2011.00373.x

Hendrickson, B. C., Biehler, A., & Mashayekh, Y. (2014). *Connected and Autonomous Vehicles 2040 Vision*. Harrisburg, PA: Pennsylvania Department of Transportation.

Howard, D. & Dai, D. (2014). Public perceptions of self-driving cars: The case of Berkeley, California. *Proceedings of the Transportation Research Board Annual Meeting*. Washington, DC: Transportation Research Board.

Hughes, J. S., Rice, S., Trafimow, D., & Clayton, K. (2009). The automated cockpit: A comparison of attitudes towards human and automated pilots. *Transportation Research Part F: Traffic Psychology and Behaviour, 12*(5), 428–439. doi:10.1016/j.trf.2009.08.004

Ivancic, K. & Hesketh, B. (2000). Learning from errors in a driving simulation: Effects on driving skill and self-confidence. *Ergonomics, 43*(12), 1966–1984. doi:10.1080/00140130050201427

Jonassen, D. H. (1995). Operationalizing mental models: Strategies for assessing mental models to support meaningful learning and design-supportive learning environments. *The First International Conference on Computer Support for Collaborative Learning - CSCL '95*, 182–186. doi:10.3115/222020.222166

Jonassen, D. H. (1999). Designing constructivist learning environments. In C. M. Reigeluth & A. A. Carr-Chellman (Eds.), *Instructional-Design Theories and Models* (pp. 215–240). London: Taylor & Francis. doi:10.4324/9781410603784

Klauer, S. G., Ankem, G., Guo, F., Baynes, P., Fang, Y., Atkins, W., ... Dingus, T. (2017). *Driver Coach Study - Using Real-time and Post Hoc Feedback to Improve Teen Driving Habits*. Retrieved from https://vtechworks.lib.vt.edu/bitstream/handle/10919/81096/STSCE_DriverCoachFinalReport_Final.pdf?sequence=1&isAllowed=y

Knowles, M. S. (1979). The adult learner: A neglected species. *Educational Researcher, 8*(3), 20. doi:10.2307/1174362

Knowles, M. S., Holton III, E. F., & Swanson, R. A. (2005). *The Adult Learner: The Definitive Classic in Adult Education and Human Resource Development* (6th ed.). Oxford: Elsevier.

Kruger, J. & Dunning, D. (1999). Unskilled and unaware of it: How difficulties in recognizing one's own incompetence lead to inflated self-assessments. *Journal of Personality and Social Psychology, 77*, 1121–1134.

Kyriakidis, M., de Winter, J. C. F., Stanton, N., Bellet, T., van Arem, B., Brookhuis, K., ... Happee, R. (2017). A human factors perspective on automated driving. *Theoretical Issues in Ergonomics Science*, pp. 1–27. doi:10.1080/1463922X.2017.1293187

Kyriakidis, M., Happee, R., & de Winter, J. C. F. (2015). Public opinion on automated driving: Results of an international questionnaire among 5000 respondents. *Transportation Research Part F: Traffic Psychology and Behaviour, 32*, 127–140. doi:10.1016/j.trf.2015.04.014

Larsson, A. F. L. (2012). Driver usage and understanding of adaptive cruise control. *Applied Ergonomics, 43*(3), 501–506. doi:10.1016/j.apergo.2011.08.005

Lee, J. D. (2006). Affect, attention, and automation. In A. Kramer, D. Wiegmann, & A. Kirlik (Eds.), *Attention: From Theory to Practice* (pp. 73–89). Oxford: University Press.

Lee, J. D. & See, K. A. (2004). Trust in automation: Designing for appropriate reliance. *Human Factors, 46*(1), 50–80. doi:10.1518/hfes.46.1.50.30392

Leonard, D. S. (2001). Relation of owner's manuals to safety. *Proceedings of the First International Driving Symposium on Human Factors in Driver Assessment, Training and Vehicle Design* (pp. 125–130). Iowa City, IA: University of Iowa.

Lindeman, E. C. (1926). *The Meaning of Adult Education*. New York: New Republic Inc.

Llaneras, R. E. (2006). *Exploratory Study of Early Adopters, Safety Related Driving With Advanced Technologies. Draft Final Task 2 Report: In-Vehicle Systems Inventory, Recruitment Methods & Approaches, and Owner Interview Results* (DOT HS 809 972). Washington, DC: National Highway Traffic Safety Administration.

Mager, R. F. & Pipe, P. (1970). *Analyzing Performance Problems; or "You Really Oughta Wanna"*. Retrieved from https://eric.ed.gov/?id=ED050560

Manser, M. P., Noble, A. M., Machiani, S. G., Short, A., Klauer, S. G., Higgins, L., & Ahmadi, A. (2019). *Driver Training Research and Guidelines for Automated Vehicle Technology*. Retrieved from www.vtti.vt.edu/utc/safe-d/index.php/projects/driver-training-for-automated-vehicle-technology/

McDonald, A. B., Carney, C., & McGehee, D. V. (2018). *Vehicle Owners' Experiences with and Reactions to Advanced Driver Assistance Systems* (Technical Report). Washington, DC: AAA Foundation for Traffic Safety.

McDonald, A. B., McGehee, D. V., Chrysler, S. T., Askelson, N. M., Angell, L. S., & Seppelt, B. D. (2015). *National Consumer Survey of Driving Technologies*. Retrieved from http://ppc.uiowa.edu/sites/default/files/national_consumer_survey_technical_report_final_8.7.15.pdf

McDonald, A. B., McGehee, D. V., Chrysler, S. T., Askelson, N. M., Angell, L. S., & Seppelt, B. D. (2016). National survey identifying gaps in consumer knowledge of advanced vehicle safety systems. *Transportation Research Record, 2559*, 1–6.

McDonald, A. B., Reyes, M., Roe, C., & McGehee, D. V. (2017). Driver understanding of ADAS and evolving consumer education. *25th International Technical Conference on the Enhanced Safety of Vehicles.* Retrieved from http://indexsmart.mirasmart. com/25esv/PDFfiles/25ESV-000373.pdf

Mehlenbacher, B., Wogalter, M. S., & Laughery, K. R. (2002). On the reading of product owner's manuals: Perceptions and product complexity. *Proceedings of the Human Factors and Ergonomics Society Annual Meeting, 46*(6), 730–734.

Mehler, B., Reimer, B., Lavalliere, M., Dobres, J. J., Caughlin, J., Lavallière, M., … Coughlin, J. (2014). *Evaluating Technologies Relevant to the Enhancement of Driver Safety.* Washington, DC: AAA Foundation for Traffic Safety.

Merrill, M. D. (2002). First principles of instruction. *Educational Technology Research and Development, 50*(3), 43–59. doi:10.1007/BF02505024

Merritt, S. M. (2011). Affective processes in human–automation interactions. *Human Factors, 53*(4), 356–370. doi:10.1177/0018720811411912

Merritt, S. M., Heimbaugh, H., LaChapell, J., & Lee, D. (2013). I trust it, but I don't know why: Effects of implicit attitudes toward automation on trust in an automated system. *Human Factors, 55*(3), 520–534. doi:10.1177/0018720812465081

Mosier, K. L., Skitka, L. J., Burdick, M. D., & Heers, S. T. (1996). Automation bias, accountability, and verification behaviors. *Proceedings of the Human Factors and Ergonomics Society Annual Meeting, 40*(4), 204–208.

Mosier, K. L., Skitka, L. J., Heers, S., & Burdick, M. (1998). Automation bias: Decision making and performance in high-tech cockpits. *International Journal of Aviation Psychology, 8*(1), 47–63.

National Highway Traffic Safety Administration. (2018). *Traffic Safety Facts - 2016 Data.* Washington, DC: National Highway Traffic Safety Administration.

National Transportation Safety Board. (2017a). *Driver Errors, Overreliance on Automation, Lack of Safeguards, Led to Fatal Tesla Crash.* Retrieved from www.ntsb.gov/news/ press-releases/Pages/PR20170912.aspx

National Transportation Safety Board. (2017b). *Highway Accident Report Collision Between a Car Operating With Automated Vehicle Control Systems and a Tractor-Semitrailer Truck.* Retrieved from www.ntsb.gov/investigations/AccidentReports/Reports/ HAR1702.pdf

Noe, R. A. (1986). Trainees ' attributes and attitudes : Neglected influences on training effectiveness. *The Academy of Management Review, 11*(4), 736–749.

Norman, D. A. (2013). *The Design of Everyday Things.* New York: Basic Books.

Nowakowski, C., Shladover, S. E., Chan, C.-Y., & Tan, H.-S. (2014). Development of California regulations to govern testing and operation of automated driving systems. *Proceedings of the Transportation Research Board Annual Meeting.* Washington, DC: Transportation Research Board.

Parasuraman, R. & Riley, V. (1997). Humans and automation: Use, misuse, disuse, abuse. *Human Factors, 39*(2), 230–253. doi:10.1518/001872097778543886

Patrick, J. (1992). *Training : Research and Practice.* London: Academic Press.

Peek-Asa, C., Hamann, C., Reyes, M., & McGehee, D. (2016). A randomised trial to improve novice driving. *Injury Prevention, 22*(Suppl 2), A120.3-A121. doi:10.1136/ injuryprev-2016-042156.329

Pereira, M., Beggiato, M., & Petzoldt, T. (2015). Use of adaptive cruise control functions on motorways and urban roads: Changes over time in an on-road study. *Applied Ergonomics, 50*, 105–112. doi:10.1016/j.apergo.2015.03.002

Prensky, M. (2001). Digital game-based learning. *Computers in Entertainment, 1*(1), 21.

Quiñones, M. A. (1995). Pretraining context effects - Training assignment as feedback. *Journal of Applied Psychology, 80*(2), 226–238.

Quiñones, M. A. (2003). Contextual influences on training effectiveness. In M. A. Quiñones & A. Ehrenstein (Eds.), *Training for a Rapidly Changing Workplace: Applications of Psychological Research* (pp. 177–199). Washington, DC: American Psychological Association.

Rau, P. & Blanco, M. (2014). Human factors evaluation of level 2 and level 3 automated driving concepts: Concepts of operation. *Proceedings of the Transportation Research Board Annual Meeting.* Washington, DC: Transportation Research Board.

Regehr, G. & Eva, K. (2006). Self-assessment, self-direction, and the self-regulating professional. *Clinical Orthopaedics and Related Research, 449*, 34–38. doi:10.1097/01.blo.0000224027.85732.b2

Rogers, C. R. (1951). *Client-Centered Therapy: Its Current Practice, Implications, and Theory.* Oxford: Houghton Mifflin.

Romoser, M. R. E., Pollatsek, A., Fisher, D. L., & Williams, C. C. (2013). Comparing the glance patterns of older versus younger experienced drivers: Scanning for hazards while approaching and entering the intersection. *Transportation Research Part F: Traffic Psychology and Behaviour, 16*, 104–116. doi:10.1016/j.trf.2012.08.004

Rothwell, W. J., Benscoter, G. M., King, M., & King, S. B. (2016). *Mastering the Instructional Design Process: A Systematic Approach* (5th ed.). Hoboken, NJ: John Wiley.

Rudin-Brown, C. M. & Parker, H. A. (2004). Behavioural adaptation to adaptive cruise control (ACC): Implications for preventive strategies. *Transportation Research Part F, 7*, 59–76. doi:10.1016/j.trf.2004.02.001

SAE. (2016). *Recommended Practice J3114, Human Factors Definitions for Automated Driving and Related Research Topics.* Warrendale, PA: Society for Automotive Engineers.

SAE. (2018). *J3016 - Taxonomy and Definitions for Terms Related to Driving Automation Systems for On-Road Motor Vehicles.* Warrendale, PA: Society for Automotive Engineers.

Salas, E., Wilson, K. A., Priest, H. A., & Guthrie, J. W. (2006). Design, delivery, and evaluation of training systems. In G. Salvendy (Ed.), *Handbook of Human Factors and Ergonomics* (pp. 472–512). Hoboken, NJ: John Wiley & Sons.

Salinger, J. (2018). *Human Factors for Limited-Ability Autonomous Driving Systems.* Retrieved from http://onlinepubs.trb.org/onlinepubs/conferences/2012/Automation/presentations/Salinger.pdf

Sanders, M. S. & McCormick, E. J. (1993). *Human Factors in Engineering and Design.* New York: McGraw-Hill.

Sandford, R. & Williamson, B. (2005). *Games and Learning: A Handbook from Futurelab.* Bristol, UK: Futurelab.

Sarter, N. B., Woods, D. D., & Billings, C. E. (1997). Automation surprises. In G. Salvendy (Ed.), *Handbook of Human Factors & Ergonomics* (pp. 1926-1943). New York: John Wiley & Sons.

Schoettle, B. & Sivak, M. (2014). *A Survey of Public Opinion about Autonomous and Self-Driving Vehicles in the US, the UK, and Australia.* Ann Arbor, MI: University of Michigan Transportation Research Institute.

Schwartz, D. L., Lin, X., Brophy, S., & Bransford, J. D. (1999). Toward the development of flexibly adaptive instructional designs. In C. M. Reigeluth (Ed.), *Instructional-Design Theories and Models: A New Paradigm of Instructional Theory* (pp. 183–213). London: Taylor & Francis.

Shor, I. (1996). *When Students have Power: Negotiating Authority in a Critical Pedagogy.* Chicago, IL: University of Chicago Press.

Skitka, L. J., Mosier, K. L., & Burdick, M. D. (1999). Does automation bias decision-making? *International Journal of Human-Computer Studies, 51*(5), 991–1006. doi:10.1006/ijhc.1999.0252

Skitka, L. J., Mosier, K. L., & Burdick, M. D. (2000). Accountability and automation bias. *International Journal of Human-Computer Studies*, *52*, 701–717. doi:10.1006/ijhc.1999.0349

Slovic, P., Finucane, M. L., Peters, E., & Macgregor, D. G. (2006). The affect heuristic. *European Journal of Operational Research*, *177*(3), 1333–1352.

Smith-Jentsch, K. A., Jentsch, F. G., Payne, S. C., & Salas, E. (1996). Can pretraining experiences explain individual differences in learning? *Journal of Applied Psychology*, *81*(1), 110–116. doi:10.1037/0021–9010.81.1.110

Spitzer, D. R. (1984). Why training fails. *Performance & Instruction Journal & Instruction Journal*, *23*(7), 6–10. doi:10.1002/pfi.4150230704

Tannenbaum, S. I., Mathieu, J. E., Salas, E., & Cannon-Bowers, J. A. (1991). Meeting trainees' expectations: The influence of training fulfillment on the development of commitment, self-efficacy, and motivation. *Journal of Applied Psychology*, *76*(6), 759–769. doi:10.1037/0021–9010.76.6.759

Tellis, L., Engelman, G., Christensen, A., Cunningham, A., Debouk, R., Egawa, K., ... Kiger, S. (2016). *Automated Vehicle Research for Enhanced Safety* (DTNH22–05-H-01277). Washington, DC: National Highway Traffic Safety Administration.

Telsa Inc. (2018). *Telsa Model S Owner's Manual*. Retrieved from www.tesla.com/sites/default/files/model_s_owners_manual_north_america_en_us.pdf

Thorndike, E. L. (1913). *Educational Psychology, Vol 2: The Psychology of Learning*. New York: Teachers College.

Tough, A. (1971). *The Adult's Learning Projects*. Retrieved from www.infed.org/thinkers/et-knowl.htm

Trimble, T. E., Bishop, R., Morgan, J. F., & Blanco, M. (2014). *Human Factors Evaluation of Level 2 and Level 3 Automated Driving Concepts: Past Research, State of Automation Technology, and Emerging System Concepts* (DOT HS 812 043). Washington, DC: National Highway Traffic Safety Administration.

Tversky, A. & Kahneman, D. (1971). Belief in the law of small numbers. *Psychological Bulletin*, *76*(2), 105–110.

U.S. Department of Transportation. (2016). *Federal Automated Vehicles Policy*. Washington, DC: Department of Transportation.

van Merriënboer, J. J. G. & Kirschner, P. A. (2017). *Ten Steps to Complex Learning* (3rd ed.). New York: Routledge.

Victor, T. W., Tivesten, E., Gustavsson, P., Johansson, J., Sangberg, F., & Ljung Aust, M. (2018). Automation expectation mismatch: Incorrect prediction despite eyes on threat and hands on wheel. *Human Factors*, *60*(8), 1095–1116. doi:10.1177/0018720818788164

Webster, J. & Martocchio, J. J. (1993). Turning work into play: Implications for microcomputer software training. *Journal of Management*, *19*(1), 127–146.

Wickens, C. D., Hollands, J. G., Banbury, S., & Parasuraman, R. (2013). *Engineering Psychology and Human Performance* (4th ed.). New York: Psychology Press.

Wogalter, M. S., Brelsford, J. W., Desaulniers, D. R., & Laughery, K. R. (1991). Consumer product warnings: The role of hazard perception. *Journal of Safety Research*, *22*(2), 71–82. doi:10.1016/0022–4375(91)90015–N

Wright, P., Creighton, P., & Threlfall, S. (1982). Some factors determining when instructions will be read. *Ergonomics*, *25*(3), 225–237. doi:10.1080/00140138208924943

Zhang, T., Hajiseyedjavadi, F., Wang, Y., Samuel, S., & Qu, X. (2018). Training interventions are only effective on careful drivers, not careless drivers. *Transportation Research Part F: Psychology and Behaviour*, *58*, 693–707. doi:10.1016/j.trf.2018.07.004

19 Connected Vehicles in a Connected World
A Sociotechnical Systems Perspective

Ian Y. Noy
Independent

CONTENTS

Key Points 421
19.1   Introduction 422
         19.1.1 Cyber-Physical Systems 422
         19.1.2 Internet of Things 423
19.2   Benefits of CVs 423
19.3   Limitations of Current Approaches 425
19.4   The CV as an STS 428
19.5   Modeling the CV as an STS 429
19.6   Actors in AD Are STSs 432
19.7   The RTS as a System of Systems 433
19.8   CVs in a Connected World 436
19.9   Challenges and Opportunities 436
19.10 Conclusion 437
References 438

KEY POINTS

- Connected vehicles have a great potential to improve safety, mobility, and sustainability
- Current engineering efforts focus on narrow cyber-physical and communication technologies
- Important human factors and sociotechnical issues are not being adequately addressed
- The interdependencies of the various actors within a connected road transportation system (such as all connected cars, traffic control centers, service providers, original equipment manufacturers (OEMs), etc.) are critical to emergent properties such as safety and performance

- A sociotechnical system perspective requires the creation of and adherence to a shared vision and coherent architecture
- Methodologies for system-level tradeoff analyses and joint optimization are urgently needed

19.1 INTRODUCTION

We stand today at the cusp of an explosive surge of cyber-physical systems[1] (CPS), spurred by the anticipated capability of 5th generation wireless systems (5G) to deliver astonishing communication speed and bandwidth. It is projected that by 2020, there will be 30 billion connected devices (IDC, 2014). The next generation of the Internet of Things (IoT) and the evolving vision of the connected world will undoubtedly disrupt the traditional paradigm of the automobile as an independent agent, if not the entire road transportation enterprise as we know it. Indeed, the exponential advancement of robotic and communication technologies can result in a future transportation system that will be undergoing evermore frequent transformations, potentially revolutionizing the very social structure of society by altering why or how we travel.

This chapter considers connected vehicles (CV) within the broader context of the connected world[2] (aka, cyber-physical society) that is rapidly emerging. First, we briefly describe the principal technological enablers of CV, namely CPS and IoT.

19.1.1 CYBER-PHYSICAL SYSTEMS

CPSs are function-oriented, embedded, computing devices that collaborate wirelessly through the IoT to monitor and/or control physical components/processes. Advanced robotic and communication technologies are used to integrate multiple inputs to achieve desired outcomes through control-theoretic algorithms. A CV is a CPS in the sense that it comprises a collection of interconnected (to each other and to other people and objects outside the vehicle) and interdependent sensors, devices, and modules that collaborate to support intelligent computing, communication, and automated vehicle control.

Several technical communities are pursuing independent, yet overlapping research and development related to CV to achieve qualitative advances in real-time computing, machine learning, security, privacy, signal processing, and use of big data (see also, this Handbook, Chapter 23). These domains include mobile computing (MC), pervasive computing (PC), wireless sensor networks (WSNs), and CPS (Stankovic, 2014). In this context, it is interesting to note that the notion of time, as in "real time," is highly complicated, yet critical to the system design and integrity (Kitchin, 2018). Small discrepancies in computational time and synchronization can have major influences on the outcome of the system at the macrolevel.[3] While sharing

---

[1] Devices that are controlled or monitored by computer-based algorithms, tightly integrated with the Internet and its users.

[2] The connected world, or cyber-physical society, refers to the totality of prevalent cyber-physical systems.

[3] For an interesting read on the impact of microsecond-level improvement in accessing stock market exchanges by high-frequency traders readers are referred to Michael Lewis's Flash Boys (Lewis, 2017).

computations in real-time offers improved responsiveness, efficiency, optimization, and flexibility, it raises concerns around asynchronized time-based functions and their attendant socioeconomic impacts. From the point of view of ensuring system integrity and functional efficacy, sorting out system time in a common architecture may be a difficult technological challenge.

CV development is primarily focused on adaptive connectivity,[4] robust security, and privacy. They provide road safety and traffic efficiency through automated internal processes as well as by exchanging real-time information among road users, the infrastructure, and a vehicular ad hoc network[5] (VANET).

19.1.2 INTERNET OF THINGS

The cyber-physical devices all connect to the IoT that provides the communication infrastructure for efficient interaction and collaboration. The IoT is an open, digital network encompassing widespread deployment of spatially distributed smart devices. It relies on cloud-based technologies to provide data communication across a wide variety of cyber-physical devices and supports innovative services through data collection, transmission, and analytic intelligence.

A critical enabler of the IoT is the ability to measure and analyze microdata (e.g., actuator state) as well as macrodata (e.g., environmental conditions) in real time through ubiquitous sensing, and to share information across platforms (such as vehicle models, infrastructure surveillance systems, software) to form a common operating picture (COP)[6] (Gubbi, Buyya, Marusic, & Palaniswami, 2013). Moreover, as Smith (2014) points out, connectivity is broader than real-time vehicle-to-infrastructure (V2I) communication, since it will likely encompass remote update pushes from service providers and original equipment manufacturers (OEMs). For example, IoT will permit manufacturers to manage on-board CPS device algorithm updates and collect detailed information on usage patterns and performance.

19.2 BENEFITS OF CVs

A US Department of Transportation report concluded that CVs have a significant potential to improve safety, mobility, and environmental impacts, including preventing up to 575,000 intersection crashes and 5,100 fatalities per year, reducing overall delay (up to 27%), and CO_2 emissions and fuel consumption (up to 11%) (Chang et al., 2015). These estimates were based on a limited selection of V2I applications. The authors point out that the overall magnitude of benefits depends on the extent of deployment of roadside devices, on-board vehicle devices, and mobile devices. There is little question that CVs have a great potential to improve safety, mobility, and efficiency, provided they are designed in a careful and thoughtful manner through a coordinated effort on the part of industry, government, and research organizations.

---

[4] Connections that may change with conditions, technological affordances, or improvements over time.
[5] VANETs are spontaneously created wireless networks for data exchange.
[6] COP is a single, shared understanding of system operational model and current state.

We distinguish between CV, i.e., the physical platform for on-board CPS processes (including communication hardware and software), and automated driving[7] (AD), the activity of driving in which at least some aspects of the dynamic driving tasks occur without driver input. Although the USDOT recently expanded its definition of "driver" to include a computer system, the term "driver" in this chapter will refer exclusively to a human driver. In effect, AD involves the CV[8] with occupants (and driver, if there is one) going somewhere. Since AD connotes a broader involvement of entities beyond the physical vehicles, we use the acronym AD to represent the collective entities involved in carrying out the dynamic driving task unless we specifically reference the vehicle itself. The impetus for this chapter is the belief that so much more functionality and benefit can be realized if the CV of the future will interact effectively with other vehicles, both connected and traditional, as well as other road users and the road system infrastructure—in short, if it were to integrate with all other entities that it will encounter directly or remotely.

For the above to occur, it is not enough for CVs to exchange location and velocity vectors with other CVs, as seems to be the narrow focus of current automotive advances. Rather, CVs should communicate intention, adhere to harmonized decision rules, and engage other entities in negotiating strategic and tactical interactions. In addition, it seems paramount to broaden the engineering effort to take into account the myriad of ways in which the vehicle will connect to non-driving-related systems in the home, the workplace, the city and beyond. As we argue below, the problem space should be broader than the vehicle. It should even be broader than the transportation system per se. Ideally, AD should be aligned with broader societal goals because of its enormous potential to impact future society. Amaldi and Smoker (2013) identify the need to develop automation policy to address rationale, future plans and strategies with a view towards defining the impacts on socioeconomic goals such as employment, land use, and social structures. In particular, they point out that currently we lack the "methods or even methodological principles needed to gather and organize knowledge [*towards*] the construction of [*automation*] policies" (p.2, italics added).

We explore briefly what developing automation policy might entail. To do so, we are guided by systems theoretic considerations in which the system is the entire ecosystem of which the vehicle is a part. Such an approach will hopefully result in increased system resilience and safety through reduced potential for confusion, human–automation error, unwitting consequences, and foreseeable misuses. System approaches also facilitate analyses of security vulnerabilities. Given the paucity of engineering, psychological, and sociological effort being directed at integrating the CV within the broader society,[9] it seems important to draw attention to the need to consider the bigger picture.

---

[7] AD denotes the phenomenon whereby driving is accomplished with no or limited human involvement. It enables smart mobility through the use of automation technologies, comprising the universe of CVs and supporting infrastructure. Sometimes AD and CV are used interchangeably, but they are conceptually different (see this Handbook, Chapter 1).

[8] Strictly speaking, AD does not necessarily include CV if automation is achieved completely through on-board intelligence that does not entail V2V or V2X communication. However, a configuration that does not involve some level of real-time communication to cooperate or coordinate with other entities seems highly unlikely given current engineering trends.

[9] To be sure, a number of papers have discussed ethical and broader societal issues raised by AD, but they are few in number in relation to the literature devoted to advancing technical capability.

19.3   LIMITATIONS OF CURRENT APPROACHES

A great deal of the motivation and excitement behind AD is based upon unproven, or perhaps even untenable, assumptions of safety, economic, and sustainability benefits. Published estimates of the impact of ADs on safety and efficiency may overstate the benefit. Several papers in the literature have challenged the espoused benefits of AD (Bonnefon, Shariff & Rahwan, 2016; Dingus et al., 2016; Noy, Shinar, & Horrey, 2018; Sivak & Schoettle, 2015; Smith, 2012). Noy et al. (2018) identify several "blind spots" in the development of AD that raise concerns about the validity of published safety estimates. Thematic analysis of video data from a recent on-road study in the United Kingdom revealed that Tesla Model S drivers demonstrated more willingness to take higher levels of risk due to complacency and over-trust in Autopilot mode (Banks, Eriksson, O'Donoghue, & Stanton, 2018a). In the United States, on March 18, 2018, a 49-year-old woman became the first pedestrian to be killed by an autonomous vehicle. She was struck by a self-driving Uber vehicle in Tempe, Arizona. Speculations about contributing causes of the crash include fallible detection technology, an inattentive driver, and a lack of concern on the part of Uber for public safety in their rush to market as reported in the media (C. Said, San Francisco Chronicle, March 23, 2018).

Among the "blind spots" identified by Noy et al. (2018) are persistent human–system interaction concerns that relate to automation including automation surprises, lack of trust, and loss of skill, as identified in Parasuraman and Wickens (2008), Sharples (2009), as well as other chapters in this Handbook. Furthermore, Noy et al. (2018) point out that connected technologies may introduce risks that are hitherto unknown in the road transportation field. In particular, the increasing level of automation may create novel adverse interactions that occur under unplanned or unanticipated circumstances (Preuk, Stemmler, Schießl, & Jipp, 2016). The sheer number of potential component interactions, further complicated by the multiple contexts within which the interactions occur, raises the potential for system-level vulnerabilities in situations in which individual system components function as designed.

Hancock et al. (2020) point out that despite the high incidence of automobile crashes, when taking account of the magnitude of driving exposure, it turns out that human reliability is remarkably high. That human error has been identified as a contributing factor in over 90% of crashes ignores the role of human drivers in preventing crashes. Indeed, in the foreseeable future, automated technologies may have difficulty approaching human capability to prevent crashes, particularly under challenging environmental or traffic conditions. Safety, reliability, and performance metrics that match or exceed human levels might not be achievable without a qualitative improvement in machine sensing, sense-making, and communication capability. To sustain or improve on the current levels of safety in the era of AD, Hancock et al. (2020) recommend that research efforts be focused on five key areas: driver independence and mobility, failure management, driver acceptance and trust, third party testing, and political support.

Several papers also question the alleged improvement of AD on transportation efficiency and the environment. For example, Smith (2012) presented analyses that indicate that the advent of AD may exacerbate congestion due to increased mobility demand as driving becomes far more accessible. He concluded that the time-cost of travel and vehicle capacity on some highways may actually increase costs associated with congestion, emissions, and sprawl. In another study, Wadud, MacKenzie, and Leiby (2016) found that automation had a positive impact on greenhouse gas (GHG) emissions and energy consumption at low levels of automation but significantly increased levels at full automation. A related concern is the difficulty of managing traffic from origin to destination. For example, if highway throughput is increased through platooning and other means, large numbers of vehicles may converge on common destination points and create bottlenecks and significant delays that may be significantly worse than those commonly experienced today.

A comprehensive review of the relevant literature that questions the espoused benefits of AD is beyond the scope of this chapter. It is sufficient for our purpose to point out that the benefits are, in many cases, speculative, if not exaggerated, and largely based on a projection of future technology capabilities superimposed over the current landscape of the transportation system. Studies that question the assumptions underlying optimistic projections of AD benefits raise important questions that we ignore at our peril. The potential of AD to dramatically improve transportation safety and productivity is not in question. The goal of this chapter is to help prevent unintended consequences and maximize the positive benefits of AD by broadening the problem space to the sociotechnical level. Experience with economic sectors that involve complex, high-hazard operations yet remarkably few failures such as aerospace and the nuclear power industry suggests that risk management must be based on a comprehensive understanding of potential threats arising from system complexity and dynamic behavior (Leveson, Dulac, Marais, & Carroll, 2009).

It is widely accepted that the future transportation system will likely undergo frequent transformations or be in a constant state of flux. Yet, projections of future benefits are incorrectly based on current functional models of the road transportation system (RTS). Consider, for example, the USDOT report on the benefits of CVs (Chang et al., 2015), which derived its estimates from field demonstrations and analytical methods of V2I applications developed in four USDOT CV research programs. These projections are based on current deployments but ignore the potential effects of driver adaptation or altered future conditions and transportation models. This is of concern because the rules of the road, physical road infrastructure, driver behavioral norms, or even the role of transportation in society are likely to change dramatically. Yet, these topics are not receiving the attention needed to integrate CVs into the evolving intelligent transportation system. It seems clear that to exploit the benefits afforded by the new technologies, we should not constrain innovation by mimicking the existing system or functional models. But to do otherwise would require establishing a common vision of the future and a shared understanding of how to get there. We must begin by acknowledging that this is a daunting challenge, given competing interests and approaches. Yet, planning for a future transportation system characterized by uncertainty and a constantly evolving connected world is a challenge we must address. Just as one would not design a building without contextual reference to land use, available infrastructure services, soil and environmental conditions, weather, and a host of other design-critical factors, it behooves us to think more broadly about mobility in the context of the connected world. Clearly, the challenge is evermore daunting because the degrees of freedom increase dramatically with increased uncertainty and system complexity.[10] One approach to reducing complexity is to deliberately reduce the degrees of freedom by imposing operational constraints, as in the case of a railroad in which the tracks are fixed and the environment is controlled (e.g., people movers in airports). However, such an approach is antithetic to the overall goal of AD, namely increased mobility, freedom, flexibility, and connectivity.

Not only should developments in AD be more congruent with the broader context of the connected world, they should be aligned with important societal needs. For example, most of the current research and development related to AD targets healthy, relatively ambulative users. However, Hancock et al. (2020) point out that drivers who stand to benefit most include teenage and elderly drivers. Moreover, perhaps the greatest societal need may well be to extend mobility to the segment of the population that is currently not well served by automobiles (e.g., the physically disabled, visually impaired, elderly, young). The potential for AD to improve the quality of life for underserved or vulnerable populations is exciting and offers promising new value to society. Overviews of the challenges faced by disabled and older drivers have been previously identified (TRB, 2004), but to date they have not been adequately addressed. To realize the extraordinary potential of AD, however, would require that far greater effort be directed towards overcoming existing barriers to mobility, including addressing mundane practical problems such as accessibility to the vehicle in different environments, egress/ingress, and driver–vehicle interaction, which are currently not being adequately considered (Hancock et al., 2020).

AD should be considered within the broadest possible context, by which we mean a sociotechnical systems (STS) perspective. As Leveson (2011) points out, analyzing safety in complex systems using a chain of events models or reductionist approaches fails to uncover human errors arising from social and organizational factors, nor does it uncover the effects of adaptation in which systems migrate towards unsafe conditions, nor can it uncover system-level failures that are not associated with component failures. Examples of the latter abound in the safety literature. Berk (2009) provides several examples of system failures in which no single component failed (e.g., a laser targeting system and a municipal water treatment system). In these cases, the failures were attributed to components that worked as designed but interacted in ways that were not anticipated and, in the event, created unsafe conditions.

The crash of Pacific Western Airlines Flight 314 on February 11, 1978, killing 42 people on board, is another example of an STS design that failed to address plausible unsafe conditions. The pilot was attempting to abort the landing when he saw a snowplow on the runway after the plane had touched down. The plane had sufficient time to take off and avoid the plane, but the failsafe mechanism prevented retraction of the thrust reverser and it crashed moments later. The failsafe on the thrust reverse system worked precisely as designed, preventing retraction of the thrust reverse system when the gears are on the tarmac. In the ensuing accident investigation, it was determined that a calculation error on the part of air traffic control (ATC) and

---

[10] While there are more formal definitions of complexity, the analogy offered by Michael Lewis in *Flash Boys* is memorable (Lewis, 2017). "A car key is simple, a car is complicated, a car in traffic is complex."

erroneous communication with the snowplow operator were contributing causes of the crash. This example illustrates a further principle—that connected systems can adversely affect overall system resilience because a failure in one component can propagate and lead to a catastrophic system failure.

These examples suggest that only through a systems model can we begin to construct the intricate interactions and interdependencies of the various components involved, and thereby derive the maximum benefits afforded by the vast capability of emerging technologies. A reductionist approach in which system components are engineered and optimized in isolation ignores overall system-level failures that can arise when these components interact poorly or create inadvertent conflicts. Current applications of sociotechnical approaches are being implemented in a variety of industries such as gas exploration, food production, health care, military, and space systems as a result of the inadequacies of reductionist approaches to identify and prevent catastrophic failures. In the sections that follow, we outline the main components of this sociotechnical model.

19.4 THE CV AS AN STS

In this section, we adopt concepts from the STS literature to develop a model of relevant interconnected and interdependent elements of CV. It is important to highlight that our discussion applies to CV modes in which the driver has some role in the driving task, for example, as a controller or supervisor. Noy et al. (2018) propose categorizing automated vehicles along two dichotomous dimensions: (1) partial automation vs. complete automation and (2) active part of the time vs. active all of the time. This scheme is depicted in Figure 19.1. Clearly, today most vehicles fall in the category of partial automation, part of the time (e.g., intelligent cruise control that does not engage at all speeds, lane-keeping assistance that requires lane delineation, etc.). Vehicles with only partial automation but nonetheless active all of the time are making impressive progress (e.g., automatic emergency braking, electronic stability

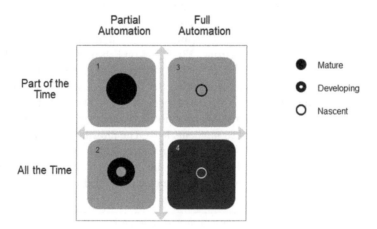

FIGURE 19.1   A scheme for categorizing levels of AD from a human-centric perspective.

control technologies, etc.; these are referred to as active safety systems earlier in the Handbook, see Chapter 2). CVs that operate under certain highway or weather conditions such as Traffic Jam Pilot will mature in time. At the current state of the art, however, it is difficult to conceive fully automated vehicle operating all of the time in the natural environment, given the complexities and uncertainties of existing road networks and use patterns.

Nevertheless, there is a widespread misunderstanding of the critical distinction between full automation that is active *part of the time* and full automation that is active *all of the time* (also see, this Handbook, Chapter 2). Most often, terms such as "autonomous vehicles" and "self-driving vehicles" conjure visions of vehicles that operate fully automatically all of the time, but this vision does not correspond to the reality of the foreseeable future. The incongruence between the idyllic future state (fully automated, all of the time) and the intermediate state (partially automated, part of the time, or fully automated, part of the time) serves to mask consideration of important risks that will inevitably arise as technology evolves towards greater automation. The distinction between *part* of the time and *all* of the time is central to the need for the application of STS approaches. In the former instance, there is a defined role for a human driver, while in the latter instance there is no need for a driver at all. The distinction is critical because the need for a driver, even as a hands-off process supervisor, raises human factors considerations around topics such as the adequacy of situational awareness (also see, this Handbook, Chapter 7), driver expectations (which are informed by a driver's mental model; this Handbook, Chapter 3), driver alertness (Chapters 6, 9), transfer of control (Chapters 7–10), and skill acquisition (Chapter 18). Thus, the presence of a human in a partially automated system raises concerns about interactions between social and technical components of dynamic, complex CVs.[11] During the transition to full automation, which may take several decades, there will be additional related complications due to the mixture of automated and traditional vehicles in the RTS. Indeed, some experts question whether there will ever be a time when it would be safe for drivers to relinquish control to CVs (Nunes, Reimer, & Coughlin, 2018).

The human element will thus be a factor for several decades to come and represent the core challenge for both existing and emerging configurations of AD. Fully autonomous vehicles in which there is no possibility whatsoever for human intervention may pose significant STS challenges that are beyond the scope of this chapter.

19.5 MODELING THE CV AS AN STS

We use the term "STS" to refer to the organization of interdependent social and technical elements that are engaged in goal-directed behavior. It extends systems theory to focus on the integration of people and technology. Figure 19.2 presents a model of STS that depicts the social sub-system and a technical sub-system. The technical sub-system represents the combination of machines, software, and data communication protocols, including cyber-physical devices that are connected via

---

[11] Of course, it makes no sense to speak of STS in the context of a fully automated vehicle that operates with no human interaction.

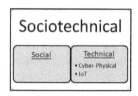

FIGURE 19.2   An STS model with social and technical sub-systems.

the IoT. Thus, the STS model depicted in Figure 19.2 comprises the universe of human and machine elements that work in concert to carry out a particular mission or set of related functions.

While the STS subsumes all elements of the system, the element that is most variable and least understood is the social sub-system (involving human–machine and human–human interactions of operators, consumers, managers, or policy makers) and its interface with the technical sub-system (comprising cyber-physical and IoT technologies). It should be noted that while the sociotechnical literature mentions other sub-systems such as organizational sub-systems and economic sub-systems, for simplicity, we regard the social sub-system to include macrolevel social influences such as organizational structure, culture, processes, and management.

For further clarity, we regard a CV that is fully automated to be essentially a CPS. If, however, there is a driver present, it becomes an STS due to influences that arise from the human element and her attendant interactions with the other system elements.

STS concepts have evolved since the early 1950s to advance our understanding of the interactions and interdependencies among people and machines in order to overcome the limitations of reductionist approaches (i.e., approaches based on delineating the system into its component parts and analyzing individual components to isolate and remove root vulnerabilities). Reductionist approaches fail to identify systemic issues that do not directly result from component failure modes. STS theory provides the hypothesis-oriented basis for identifying vulnerabilities associated with nonlinear interactions that occur across time and space among the various components, which is characteristic of complex adaptive systems. It establishes methodologies for joint optimization in which the social and technical sub-systems are considered concurrently to resolve design tradeoffs and establish the highest achievable system performance and human well-being.

Like other complex adaptive systems, AD derives three key attributes from complexity theory: (1) it is non-deterministic, (2) it cannot be easily decomposed into functional components, and (3) it has emergent properties that reflect self-organization and dynamic interactions (Pavard & Dugdale, 2006). For example, workplace safety is a system attribute that emerges from interactive and interdependent activities that occur within the sociotechnical context of the work system and that cannot be localized within the system structure (Carayon Hancock, Leveson, Noy, Sznelwar, & Van Hootegem, 2015). For a more complete overview of STS and safety, refer to Carayon et al. (2015) and Noy et al. (2015).

Although Figure 19.2 depicts an STS as comprising both social and technical sub-systems, a more analytic depiction of STS identifies the relevant hierarchical levels of the system, each level having both social and technical components, starting at the top with public policy and culminating in individual processes. Each level exerts control over the level directly below, all of which ultimately impinge on the sharp end of system design (Rasmussen & Svedung, 2000). Figure 19.3, based on the work of Rasmussen and Svedung (2000), illustrates the hierarchy of STS levels (layers) associated with CV, from government regulation to successively lower social levels and finally to the process level (the vehicle platform with embedded electronics). Information from a given layer in the hierarchy is transmitted upwards to the next level and that information informs appropriate responses in the form of direction, corrective actions, or procedures. For example, OEM experience with existing vehicles helps inform further connectivity research and development. Similarly, the CV safety experience and crash reports may be used by regulators to revise or establish new regulations. At the top layer, public opinion helps establish legislative agendas and priorities. The layers are thus connected through feedback control loops in which each loop behaves as an adaptive control element with input from the adjacent layer operating both as feedback signal and as parameter attunement (Flach, Carroll, Dainoff, & Hamilton, 2015). Information about operational experience is fed upwards and control actions, such as goals, policies, constraints, and commands are transmitted downward. Ideally, actions at each level are informed by diverse disciplines (as indicated in Figure 19.3), which contribute unique insights and

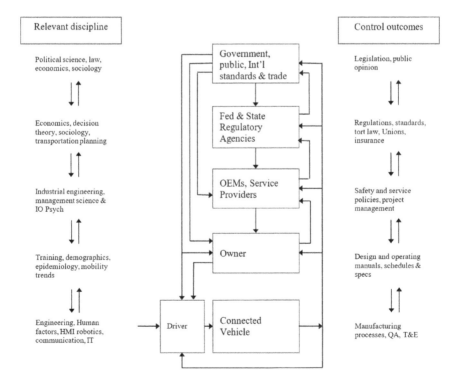

FIGURE 19.3 A simplified view of a CV as an STS. Shown are the hierarchy of layers, research disciplines that provide unique insights, and examples of control loop outcomes.

collaborative value. Taken together, feedback control loops along the hierarchy influence the overall functioning of the system. Deficiencies in the design or operation of the system can arise from suboptimal control at any point in the hierarchy that can in turn lead to unsafe behavior of individual components and interactions among the various components.

The hierarchical model of STS was further developed by Leveson (2011) for practical application in analyzing safety in system development as well as system operations. She realized that the control loops that are in play during the development of the STS differ somewhat from the control loops that exist during the operational life of the STS once it is deployed. Leveson (2013) applied systems theory to develop the "System's Theoretic Accident Model and Processes (STAMP)" as a tool for identifying system-level safety failures and its derivative hazard analysis technique, Systems-Theoretic Process Analysis (STPA). STPA is used in accident investigations to identify design-induced errors that fail to take account of human cognition, social, organizational, and management factors. Its core value is in depicting safety as a dynamic control problem rather than a component failure problem. That is, an error could arise at a given instance in time from the confluence of component conditions and performance vectors rather than from a failure of any one or more individual components. Because it identifies interdependencies among components that are amenable to further analyses and re-engineering, STPA has general application beyond accident investigation.

19.6   ACTORS IN AD ARE STSs

We have described CV as an STS and argued that safety is an emergent property that is influenced by a hierarchy of system layers, each of which comprises social and technical sub-systems that are interdependent and interact. Accordingly, it is important to understand the interdependencies of system layers and components to optimize the design of a CV. Banks, Stanton, Burnett, & Hermawati (2018b) have taken the first step towards modeling complex STSs by applying methods from distributed cognition. They demonstrate the utility of visualizing macrolevel representations using Event Analysis of Systemic Teamwork (EAST) to gain insight about how the integration of AD into the road transport system (RTS) affects overall network dynamism.

However, this view may be overly narrow since the RTS is much bigger than a single CV. It may comprise thousands of CVs[12] and other elements. The question that immediately arises in the context of systems theory is what is the appropriate unit of analysis? Is it, for example, a single CV with its embedded cyber-physical technologies and connectivity or is it the entire RTS? We posit that the unit of analysis is the smallest cohesive set of elements that work in concert towards a common mission. Accordingly, the CV, for example, is an appropriate unit of analysis. We refer to a unit of analysis as an actor in the RTS space since it has the ability to establish a unique, independent mission and act autonomously to accomplish that mission.

---

[12] Within the relevant geographical region.

An actor, such as a CV, is in effect an STS as described in previous sections. It consists of cyber-physical components, it is connected to the external environment and other actors through the IoT, and it involves a social sub-system consisting of a user such as the driver and possibly other occupants. As an STS, its behavior is shaped by a hierarchy of layers, as depicted in Figure 19.3, involving OEM policy, regulatory requirements public policy, etc.

Within the increasingly connected, multi-modal RTS, the CV interacts with a host of other actors. It is important to point out that each actor has a social element in the form of one or more people[13] who have important safety-critical roles and needs. Besides other drivers and road users, other actors may involve human operators such as traffic control operators, emergency response personnel, service providers, and infrastructure crews who need to interact in real time, providing, collecting, and sharing information. Seen from the RTS framework, AD involves more than the CV - it is a complex system in which the vehicle is but one component interacting with other actors.

The main defining characteristic of an actor is the mission it performs. A traffic control center, for example, comprises personnel as well as surveillance, analytic and communication equipment that are designed for the purpose of monitoring and intervening in prescribed ways to manage traffic flow. The CV, similarly, is a combination of embedded sensors, actuators and physical processes that are controlled or supervised by a driver to perform the driving task. Service providers, such as emergency services, have completely different missions, along with a corresponding set of equipment and processes for interacting with CVs, traffic control centers, and other systems to deliver the service for which they were designed.

Each actor can be viewed as an STS in its own right since each is subject to multiple, unique hierarchical influences. Each is a combination of social and related technical components that receive information about its environment, has goals, and has the ability to act on its environment to meet these goals. Each operator within each actor needs to trust system function and integrity as much as drivers and other users (Lee & See, 2004). Relevant questions that arise in the context of AD include "what sort of information do each of the operators involved need?," "when do they need it?," "in what form?," etc.? Equally important is how do the answers to these questions change with different levels of automation?

19.7   THE RTS AS A SYSTEM OF SYSTEMS

We have argued that the RTS consists of actors each of which is an STS in its own right. There are many possible topologies to organize the RTS. Noy et al. (2018) present the view of the system as a star network in which the various actors connect to a central node. The star network model suggested by Noy et al. (2018) is depicted in Figure 19.4, in which the central node is the RTS and the stars are the various actors.

---

[13] A traffic signal per se is not an actor since it has no independent mission and does not include a social element. It may, however, be a component in the traffic control STS if there are operators involved.

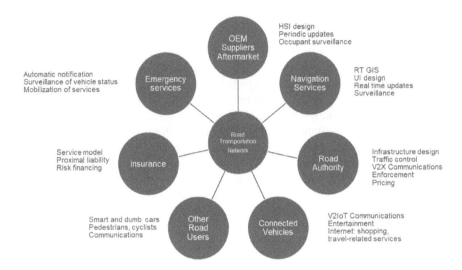

FIGURE 19.4   (See color insert.) STS Star Model of an RTS. (Adapted from Noy et al., 2018.)

In this model, the central node exercises executive power in making decisions, coordinating activities, and managing communications. For a centralized model to function properly and realize the full safety potential of AD, manufacturers, governments and research organizations need to collaborate in developing a common human-centric STS architecture, at a pre-competitive level.[14] The architecture needs to be more detailed than a topology[15] in that it should prescribe and harmonize/coordinate compatible decision algorithms, priorities, communication protocols, contingencies, etc. This means that industry and government should jointly develop an underlying strategy to ensure the consistency, reliability and functional interoperability of CVs.

Another possible topology is a decentralized network, for example a bus network, in which an actor can hook onto the common linear link without having to subscribe to a particular architecture provided it embodies a compatible connection protocol and adheres to certain principles - for example, it does not disturb any other actor or actors. This is a self-organizing system (since actors can join and alter overall system performance) in that it can adapt to changing circumstances and has sufficient flexibility to expand and transform the network to reflect new technologies. This topology too requires the establishment of rules of engagement, but there is no centralized oversight.

Figure 19.5 is an example of six actors connected to the bus network representing the RTS. In general, the RTS comprises a set of actors comprising hierarchical levels that include both social and technical sub-systems. The generic individual actors (or STS) can represent CVs, service providers, traffic control centers, and other STSs

---

[14] Pre-competitive refers to fundamental principles of operation to which all OEMs can subscribe. Competition among OEMs occurs with selection and bundling functions as well as human-machine interface features.

[15] Schematic description of how connecting lines and nodes are arranged in a network. Common networks include bus, ring, star, tree, and mesh topologies.

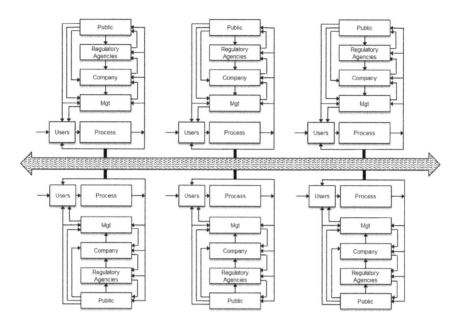

FIGURE 19.5   AD as a system of STS.

whose design and operation are shaped by technological innovations as well as a multitude of economic, social, and political forces. Hence, the RTS is effectively a system of systems. Other topologies can be easily created in the same manner. It should be noted that infrastructure elements such as roads, traffic control devices, and smart furniture can also be considered actors to the extent that they have technical components and service personnel that are subject to management oversight and public policy.

In the model depicted in Figure 19.5, each actor is considered as an STS, and the social and technical sub-systems should be optimized concurrently to address human–system integration conflicts and interactions. Whether or not a CV fails, or crashes, depends not only on the vehicle's design and embedded CPSs (that are influenced by multiple layers of control loops) but also on its interactions with other actors. In particular, cross-actor interactions are influenced by the quality of human–machine integration within each actor as well as compatibility across actors under the various operational and environmental conditions. In the end, overall system performance and safety is a function of the interactions among all components within the entire RTS. Consequently, it is important to understand the interdependence of the various actors that form the RTS to account for the effects of interactions across actors, not simply within actors.

In sum, safety and effectiveness are emergent properties of the entire RTS, not merely any particular CV. That is to say, it is not sufficient to apply joint optimization methodologies to individual actors. Rather, joint optimization should occur across the RTS as well within each actor. That requires identifying the interdependencies and interactions among CVs, traffic control centers, emergency responders, and other actors and resolving potential conflicts through tradeoff analyses.

19.8   CVs IN A CONNECTED WORLD

In the more distant future, the RTS will not exist in isolation. It will become more closely integrated with the broader connected world. Smart wearables, smart homes, and smart cities will connect to smart vehicles to exchange information concerning precise and localized services and information sharing. The interactive elements of the external connected world (i.e., outside of the RTS) can nonetheless be regarded as actors within the overall RTS model we described previously because they themselves are STSs and they interact with the CV in the same way as driving-related actors. In relation to AD, the appropriate frame of reference, however, is the RTS. For other applications, these same actors may be implicated but their behavior may be viewed differently in relation to a non-RTS frame of reference.

19.9   CHALLENGES AND OPPORTUNITIES

Given the remarkable technological transformations underway within our society, the high-level challenge we face at this juncture in the evolution of CVs is to establish the appropriate level of automation for CVs that aligns with RTS and broader societal goals, establish a suitable topology and architecture for the connected RTS, identify the interdependencies among the various actors, and define how these actors must connect to and interact with it. This can only be accomplished through a coordinated effort led by governments, both nationally and internationally, to establish a comprehensive plan for a connected RTS based on maximizing overall mobility and environmental benefits while safeguarding public well-being. Currently, the absence of government regulatory oversight frees industry to pursue fragmented and potentially unsafe systems. The lack of government leadership bodes poorly for the process of collaborative consensus building across industry, government, and academia.

The absence of a more coordinated plan for the development of AD increases the criticality of implementing STS methodologies to identify and manage system-level risks. Perhaps the most valuable aspect of identifying the interdependent actors is that a detailed, formalized framework can reveal relationships that may not otherwise be evident, and it facilitates forming empirical hypotheses for further test and development. In closely coupled systems such as CVs, the failure of one actor can have a disproportionate adverse effect on other actors, if not the entire system. This challenge is not limited to technical sub-system engineering, but as argued previously, it relates to the need to integrate the difficult human factors issues that become salient from an STS perspective. Connecting the manufacturer and other actors to the driver (or vehicle occupant) in a continuous manner raises additional social and technical concerns. From a social perspective, connectivity raises significant problems with regard to privacy, social acceptance, etc. From a technology perspective the ability of manufacturers to update algorithms on the fly/independently raises the distinct possibility of asynchronous sub-system evolution, with its attendant safety and performance issues.

Approaches and methodologies are yet to be developed to identify potential conflicts among the technical and social sub-systems and to resolve design and performance tradeoffs to achieve joint optimization within and across the various actors involved. Methodologies such as STAMP and STPA (Leveson, 2015) can help identify the key elements. Flach et al. (2015) discuss two examples (nuclear power and the limited service food industry) to illustrate how communications, defined as integration of information, and decisions, defined as controls to correct deviations from safety goals, impact the safety of STS. Both examples (one, a relatively simple system; the other, highly complex) can be described in terms of dynamical systems frameworks that provide useful approaches to assessing the fit between organizational structures and work demands. Methods for studying the interdependencies between system components might employ modeling and simulation to help identify unanticipated consequences of system design decisions, etc. (Hettinger, Kirlik, Goh, & Buckle, 2015).

Since the human element is an important source of variation in system function, a concerted effort should be directed towards the social sub-system to address traditional human–system integration considerations and interactions across multi-layered STS control loops as well as across external interactions with interconnected actors. An approach to delineate the role of humans in complex dynamic systems was created at MIT (France, 2017) to generate and analyze unsafe control actions (USC). Four categories of USC were used to demonstrate the application of the method to the Automated Park Assist system and its utility in identifying potential failure modes:

1. A control action required for safety is not provided
2. An USC is provided that leads to a hazard
3. A potentially safe control action is provided too late, too early, or out of sequence
4. A safe control action is stopped too soon or applied too long (for a continuous or non-discrete control action)

Efforts to extend sociotechnical analytic approaches are a step in the right direction, but a great deal more needs to be done. For example, DeKort (2018) advocates for the creation of a top-down scenario matrix to be used in simulation tests and algorithm development. The matrix must include every object (moving, fixed, environmental, laws/social cues) as well as degraded versions of the objects. The number of scenarios involving variations of interacting objects quickly becomes too great to test on the road. Moreover, deep learning systems need to encounter crashes, or near crashes, to learn avoidance strategies, which makes on-the-road testing inherently unethical.

While conducting exhaustive analyses on STS safety is not yet possible for lack of appropriate methodology, the perspectives presented in this chapter might at the very least provide a useful framework for considering and identifying the most problematic aspects of component integration early on so that they can be addressed as early as possible in design and/or implementation activities.

19.10 CONCLUSION

The STS approach focuses on the RTS as a whole system, not merely its parts. It assumes that important attributes of system safety and performance can only be treated adequately in their entirety, taking into account all facets relating the social to the technical aspects. These system-level properties derive from the interdependencies and interactions among the parts of systems (Leveson, 2011). While the STS approach has heretofore been primarily applied to large organizations involving many people (industry, health care, military), STS-theoretical concepts can be readily applied to AD in view of the fact that it is a complex adaptive system that can be affected by complex nonlinear dynamic interactions across time and space among a large number of components within the entirety of the system. Thus, a systems-theoretic approach is more likely to address mobility goals than the engineering design of any given component or connective technology.

The main contribution of the STS approach to AD is that it views mobility and safety as system-level outcomes, and it can establish an analytic framework for evaluating overall benefits that can inform design choices. Often, system failures occur because of incorrect assumptions on the part of one or more actors. For example, a failure of a traffic signal can cause a CV to run what might have been a red light. Understanding the decision logic and assumptions that may be implicit on the part of an actor can help create safeguards or establish common parameter definitions to avoid such misunderstanding. In well-publicized catastrophic accidents such as Deepwater Horizon or Columbia Shuttle, a common contributing failure is the lack of credence given to unsafe conditions by top decision-makers. An STS analysis can help establish information sharing pathways to avoid message filtering. Incidence reporting anywhere in the system can help guard against system creep to unsafe conditions and maintain resilience.

We posit that meaningful advances in the safety of CVs requires (1) a shift in the unit of analysis to the STS level, (2) the RTS be viewed as a system of STSs, (3) methodologies be developed to facilitate identification of important interdependencies and interactions among social and technical components within and across system actors, and (4) methodologies be developed for joint optimization to engender positive emergent properties such as system safety, resilience, and overall effectiveness. Underlying these efforts is recognition of the central role of human factors considerations in advancing the safety and utility of AD. There is much critical work that is needed to guide responsible development of CVs in the connected world.

REFERENCES

Amaldi, P. & Smoker, A. (2013). An organizational study into the concept of "automation policy" in a safety critical socio-technical system. *International Journal of Sociotechnology and Knowledge Development, 5*(2), 1.

Banks, V. A., Eriksson, A., O'Donoghue, J., & Stanton, N. A. (2018a). Is partially automated driving a bad idea? Observations from an on-road study. *Applied Ergonomics, 68*, 138–145.

Banks, V. A., Stanton, N. A., Burnett, G., & Hermawati, S. (2018b). Distributed cognition on the road: Using EAST to explore future road transportation systems. *Applied Ergonomics, 68*, 258–266.

Berk, J. (2009). *System Failure Analysis*. Materials Park, OH: ASM International.

Bonnefon, J. F., Shariff, A., & Rahwan, I. (2016). The social dilemma of autonomous vehicles. *Science, 352*(6293), 1573–1576.

Carayon, P., Hancock, P., Leveson, N., Noy, I., Sznelwar, L., & Van Hootegem, G. (2015). Advancing a sociotechnical systems approach to workplace safety–developing the conceptual framework. *Ergonomics, 58*(4), 548–564.

Chang, J., Hatcher, G., Hicks, D., Schneeberger, J., Staples, B., Sundarajan, S., … Wunderlich, K. (2015). *Estimated Benefits of Connected Vehicle Applications: Dynamic Mobility Applications, AERIS, V2I Safety, and Road Weather Management Applications* (No. FHWA-JPO-15-255). Washington, DC: US Department of Transportation.

DeKort, M. (2018). Corner or edge are not most complex or accident scenarios. *LinkedIn post.* www.linkedin.com/pulse/corner-edge-cases-most-complex-accident-scenarios-michael-dekort/

Dingus, T. A., Guo, F., Lee, S., Antin, J. F., Perez, M., Buchanan-King, M., & Hankey, J. (2016). Driver crash risk factors and prevalence evaluation using naturalistic driving data. *Proceedings of the National Academy of Sciences, 113*(10), 2636–2641.

Flach, J. M., Carroll, J. S., Dainoff, M. J., & Hamilton, W. I. (2015). Striving for safety: Communicating and deciding in sociotechnical systems. *Ergonomics, 58*(4), 615–634.

France, M. E. (2017). *Engineering for Humans: A New Extension to STPA* (Doctoral dissertation). Cambridge, MA: Massachusetts Institute of Technology.

Gubbi, J., Buyya, R., Marusic, S., & Palaniswami, M. (2013). Internet of Things (IoT): A vision, architectural elements, and future directions. *Future Generation Computer Systems, 29*(7), 1645–1660.

Hancock, P. A., Kajaks, T., Caird, J. K., Chignell, M. H., Mizobuchi, S., Burns, P. C., … Vrkljan, B. H. (2020). Challenges to human drivers in increasingly automated vehicles. *Human Factors, 62*(2), 310–328.

Hettinger, L. J., Kirlik, A., Goh, Y. M., & Buckle, P. (2015). Modelling and simulation of complex sociotechnical systems: Envisioning and analysing work environments. *Ergonomics, 58*(4), 600–614.

IDC. (2014). *Worldwide Internet of Things 2014–2020 Forecast.* Framingham, MA: International Data Corporation.

Kitchin, R. (2018). The realtimeness of smart cities. *TECNOSCIENZA: Italian Journal of Science & Technology Studies, 8*(2), 19–42.

Lee, J. D. & See, K. A. (2004). Trust in automation: Designing for appropriate reliance. *Human Factors, 46*(1), 50–80.

Leveson, N. (2015). A systems approach to risk management through leading safety indicators. *Reliability Engineering & System Safety, 136*, 17–34.

Leveson, N., Dulac, N., Marais, K., & Carroll, J. (2009). Moving beyond normal accidents and high reliability organizations: A systems approach to safety in complex systems. *Organization Studies, 30*(2–3), 227–249.

Leveson, N. G. (2011). *Engineering a Safer World: Systems Thinking Applied to Safety.* Cambridge, MA: MIT Press.

Leveson, N. G. (2013*). An STPA Primer.* Retrieved from http://psas.scripts.mit.edu/home/wp-content/uploads/2015/06/STPA-Primer-v1.pdf

Lewis, M. (2017). *Flash Boys.* New York: W.W. Norton & Company.

Noy, Y. I., Hettinger, L. J., Dainoff, M. J., Carayon, P., Leveson, N. G., Robertson, M. M., & Courtney, T. K. (2015). Emerging issues in sociotechnical systems thinking and workplace safety. *Ergonomics, 58*(4), 543–547.

Noy, Y. I., Shinar, D., & Horrey, W. J. (2018). Automated driving: Safety blind spots. *Safety Science, 102*, 68–78.

Nunes, A., Reimer, B., & Coughlin, J. (2018). People must retain control of autonomous vehicles. *Nature, 556*, 169–171.

Parasuraman, R. & Wickens, C. D. (2008). Humans: Still vital after all these years of automation. *Human Factors, 50*(3), 511–520.

Pavard, B. & Dugdale, J. (2006). The contribution of complexity theory to the study of socio-technical cooperative systems. In A.A. Minai & Y. Bar-Yam (Eds.), *Unifying Themes in Complex Systems* (pp. 39–48). Berlin: Springer.

Preuk, K., Stemmler, E., Schießl, C., & Jipp, M. (2016). Does assisted driving behavior lead to safety-critical encounters with unequipped vehicles' drivers? *Accident Analysis & Prevention, 95*, 149–156.

Rasmussen, J. & Svedung, P. (2000). *Proactive Risk Management in a Dynamic Society.* Stockholm: Swedish Rescue Services Agency.

Sharples, S. (2009). Automation and technology in 21st century work and life. In P. Bust (Ed.), *Contemporary Ergonomics* (pp. 208–217). Boca Raton, FL: CRC Press.

Sivak, M. S. & Schoettle, B. (2015). *Road Safety with Self-Driving Vehicles: General Limitations and Road Sharing with Conventional Vehicles.* Ann Arbor, MI: University of Michigan, Transportation Research Institute.

Smith, B. W. (2012). Managing autonomous transportation demand. *Santa Clara Law Review, 52*, 1401.

Smith, B. W. (2014). A legal perspective on three misconceptions in vehicle automation. In G. Meyer & S. Beiker (Eds.), *Road Vehicle Automation* (pp. 85–91). Berlin: Springer.

Stankovic, J. A. (2014). Research directions for the internet of things. *IEEE Internet of Things Journal, 1*(1), 3–9.

TRB. (2004). *Transportation in an Aging Society: A Decade of Experience.* Washington, DC: Transportation Research Board.

Wadud, Z., MacKenzie, D., & Leiby, P. (2016). Help or hindrance? The travel, energy and carbon impacts of highly automated vehicles. *Transportation Research Part A: Policy and Practice, 86*, 1–18.

20 Congestion and Carbon Emissions

Konstantinos V. Katsikopoulos
University of Southampton

Ana Paula Bortoleto
University of Campinas

CONTENTS

Key Points 441
20.1 Introduction 442
20.2 Reducing Congestion: Driver Behavior 443
    20.2.1 Effects of Vehicle Intelligence on Driver Behavior......................... 443
    20.2.2 Effects of Vehicle Connectedness on Driver Behavior 446
    20.2.3 Effects of Vehicle Automation on Driver Behavior......................... 448
20.3 Reducing Carbon Emissions: Vehicle Capacity and Driver
    Readiness to Use 449
    20.3.1 Vehicle Capacity to Reduce Carbon Emissions............................... 449
    20.3.2 Driver Readiness to Use Carbon-Emissions-Reducing Vehicles 449
    20.3.3 Eco-Driving 450
20.4 Conclusion 451
References 451

KEY POINTS

- Unless people completely surrendered driving control to ACIV, researchers must understand the effects of vehicle intelligence, connectedness, and automation on driver behavior.
- Interventions that aim at behavior modification only—such as congestion pricing or techniques of nudging—without boosting underlying processes and competencies, might fail to promote pro-environmental driving-related behaviors.
- Reasoning and decision-making, regarding one's own or others' driving, seem to be predicted by simple rules of thumb.
- The use of simple rules of thumb for parking can increase system efficiency, compared with standard game-theoretic proposals.
- Driving-related moral and social dilemmas, induced by automation, have been investigated, but more work remains to be done.

- Life cycle assessment of automated vehicles has found that they could decrease greenhouse gases emissions by 9%, but this is not taking into account possible rebound effects such as increased driving.
- Pro-environmental driving-related behaviors are tailored by personal as well as contextual factors: Being motivated towards decreasing carbon emissions is not enough to undertake a pro-environmental behavior.
- Carbon-emission information may have an impact on driver's decisions when it is provided clearly, but other criteria (e.g., fuel price, safety, time) should also be satisfied.

20.1 INTRODUCTION

According to the U.S. Bureau of Transportation Statistics (2007), the road transportation sector accounts for approximately one-third of U.S. carbon emissions from the use of energy. Previous studies have shown that congestion wastes time and money, and it also increases emissions of greenhouse gases and localized pollutants such as particulate matter (Barth & Boriboonsomsin, 2008). Can roads be de-congested, and emissions be reduced, by changing the transportation infrastructure?

One might expect that building more roads should relieve traffic congestion. Think again. Duranton and Turner (2011) conclude that increased provision of interstate highways and major urban roads is unlikely to relieve congestion. These researchers used the economic theory of supply and demand coupled with the statistics of logistic regression to estimate the elasticity of vehicle-kilometers traveled with respect to the lane kilometers in U.S. metropolitan areas between 1983 and 2003. This elasticity was estimated to be 1.03, which means that there is more driving when there is more road to drive on, with the increase in traffic being 3% in excess of the corresponding increase in road.

Decades ago, such results have been expressed by the umbrella term of a *fundamental law of road congestion* (Downs, 1962). With the benefit of hindsight, it is not very surprising that people increase the consumption of road when more road is made available to them—such behavioral adaptations are ubiquitous; haven't we all had the experience of eating more just because more was served to us? This suggests that before looking for physical interventions for decreasing congestion, one might want to look for psychological ones.

Since the publication of Thaler and Sunstein's (2008) *Nudge*, psychological interventions are often identified with behavioral ones. However, these two types of intervention are not the same. The distinction we have in mind is that in the latter, an effort is made to directly change behavior without necessarily enhancing the underlying psychological processes and their associated competencies. For example, individuals might end up eating fruits and vegetables if those are exhibited at eye level in their work cafeteria, without learning and understanding that eating fruits and vegetables (in general) enables the body's healthier function. From a human-factors perspective, this is a tricky issue as any gains from purely behavioral interventions may be transient, fail to generalize to other contexts, or could create dissonance and disappointment because it is not clear that the receivers of a nudge actually want to make the choices they are nudged toward (Sugden, 2017).

This chapter follows the *Boost* approach to making psychological interventions (Bond, 2009; Katsikopoulos, 2014; Hertwig & Grüne-Yanoff, 2017). This approach aims at first *understanding* the psychological processes—such as cognitive, motivational, and social ones—underlying behavior, and then attempting to enhance these processes so as to increase *competency* and lead to empowerement. We will not delve into conceptual discussions of these two approaches in this chapter, although we will return to the relevance of psychology in enhancing pro-environmental behavior (Bortoleto, 2014), as it relates to reducing emissions while driving, in Section 20.3.

Section 20.2 focuses on reducing congestion and, as a result, reducing carbon emissions. Unless people completely surrender driving control to automated, connected, and intelligent vehicles (ACIV), researchers must understand how driver behavior is affected by vehicle intelligence, connectedness, and automation. The next section reviews such work.

20.2   REDUCING CONGESTION: DRIVER BEHAVIOR

20.2.1   Effects of Vehicle Intelligence on Driver Behavior

Guide signs display up-to-date information useful to drivers, such as travel time on a route to a popular destination or the number of parking spots available in a busy lot. They can be located inside or outside the vehicle. Such signs can affect driver behavior (Kantowitz, Hanowski, & Kantowitz, 1997). Thus, they also represent an opportunity to alleviate traffic congestion. For instance, if one knew the percentage of drivers who, after reading a particular piece of travel time information, decided to divert to the surface streets—as opposed to staying on the highway—one could intelligently switch the sign on and off to control traffic. What do we know about the effect of information displayed on guide signs on driver decision-making?

A series of experiments run on a mid-level driving simulator (Katsikopoulos, Duse-Anthony, Fisher, & Duffy, 2000; 2002; Figure 20.1)[1] investigated the choices that highway drivers made between two routes to a common destination when information about travel time on the routes was given, as shown in Figure 20.2. The experimental instructions emphasized that I–93 was the default route in the sense that participants were currently driving on it. Thus, the choice was framed as staying on the default route or diverting to the alternative, which was Route 28.

In the first experiment, the default route always had a certain travel time (100 minutes) while the alternative route had a range of travel times (from 70 to 120 minutes; Figure 20.2). The average travel time of the alternative was taken to be the midpoint of the interval: $(70 + 120)/2 = 95$ minutes. The range of this interval is

---

[1] The University of Massachusetts at Amherst driving simulator (see Figure 20.1) was used, which, at the time, consisted of a car (Saturn 1995 SL 1) connected to an Onyx Infinite Reality Engine 2 and an Indy computer (both manufactured by Silicon Graphics, Inc.) The images on the screen subtended 60° horizontally and 30° vertically, might have been artificial or natural, and were developed using Designer's Workbench (by Centric). The movement of other cars on the road was controlled by Real Drive Scenario Builder (Monterey Technologies, Inc.). The system was assembled by Illusion Technologies.

FIGURE 20.1   Driving simulator used in the route and parking choice experiments described in the text (located at the University of Massachusetts at Amherst).

Estimated Travel Time to Downtown Boston	Route
100 min	I–93
70–120 min	Route 28

FIGURE 20.2   Sign with travel time information, to guide route choice.

120 − 70 = 50 minutes. By crossing three levels of average travel time (95, 100, 105) with five levels of range (20, 30, 40, 50, 60), 15 route choice scenarios were generated.

The results showed that a risk-averse driver is less likely to divert to the alternative route as the range increases while the average remains the same. For example, a risk-averse driver was less likely to choose Route 28 in the example in Figure 20.2 when the travel time ranged from 70 to 120 minutes than when the travel time ranged from 80 to 110 minutes (the average, in both cases, equals 95 minutes). On the other hand, a risk-seeking driver was more likely to choose the 70-to-120-minute route.

An early claim in economics was that people are risk averse. The psychological literature, however, suggests that people are risk averse when the choice belongs to the domain of *gains* but risk seeking when the choice belongs to the domain of *losses* (Kahneman & Tversky, 1979). In route choice, scenarios in which the alternative has a shorter average time than the default belong to the domain of gains. The scenario in Figure 20.2 where the alternative route ranges from 70 to 120 minutes is in the domain of gains because the average equals 95 minutes, which is less than the default of 100 minutes. If the alternative route ranged from 90 to 120 minutes, the choice

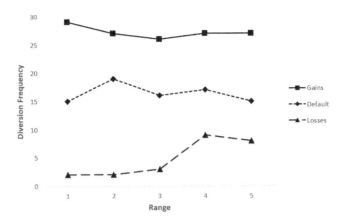

FIGURE 20.3 Number of participants diverting to the alternative route as a function of its range of travel time. On the *x*-axis, 1 means a range of 20 minutes, 2 means 30 minutes, etc. and 5 means 60 minutes.

would be in the domain of losses because the average would equal 105 minutes, and that is more than the default of 100 minutes.

Past research on route choice had only tested scenarios in the domain of gains. Katsikopoulos et al. (2000) tested route choices framed as losses. So what happened in the experiment? Figure 20.3 shows the number of participants (out of 30) that diverted as a function of the range of travel time of the alternative route (on the *x* axis, 1 means a range of 20 minutes, 2 means 30 minutes, etc. and 5 means 60 minutes). There is a decreasing trend for gains indicating risk aversion, a roughly flat line for the case where the alternative route has the default average (indicating risk neutrality), and an increasing trend for losses indicating risk seeking.

Expected utility theory (von Neumann & Morgenstern, 1947) and its modification, prospect theory (Kahneman & Tversky, 1979), are often used to model human choice in economics and psychology. According to such theories, people compute the "worth" of a decision option by summing up its possible outcomes, where each outcome is weighted by its probability. Then, people choose the option with the maximum worth. If outcomes and probabilities are taken at face value, this is the expected value theory. In expected utility theory and prospect theory, outcomes and probabilities may be translated by multi-parameter mathematical functions. A problem with such theories is that they are probably too complicated to be describing the underlying cognitive processes; rather, they are meant to be as-if representations of behavior (Katsikopoulos & Gigerenzer, 2008). In as-if theories, the claim is not that people really translate outcomes and probabilities and then sum and weight them, but rather that they make decisions "as-if" they did so.

In an alternative theory, the driver takes a *single* sample to estimate the travel time along the alternative route (where s/he perceives that the travel time of the alternative route is following a normal probability distribution over the interval of possible travel times), and chooses to divert if this estimate is less than the reference travel time. This modeling incorporates two themes of behavioral theory

(Katsikopoulos & Gigerenzer, 2008): (1) the driver is tuned to the probabilistic nature of decision-making and (2) the driver uses one piece of information to make a decision (one sample of travel time). This simple rule of thumb accounts well for a host of effects on route choice, including effects that could also be modeled by prospect theory, such as the effects of gains and losses, and effects that could not be modeled by prospect theory, such as the effects of an uncertain reference point (Katsikopoulos et al., 2002).

Such rules of thumb have been put forth for predicting parking choices as well. Hester, Fisher, and Collura (2002) ran a driving simulator study with parking scenarios where a utility model and a simple rule of thumb made different predictions. It was found that actual parking choices were more consistent with the rule of thumb than with the utility model. For instance, only 10% of the participants made all parking choices consistent with utility theory, whereas 35% of the participants made all choices consistent with the simple rule of thumb.

20.2.2 Effects of Vehicle Connectedness on Driver Behavior

Many acts of driving involve *interaction*—cooperation or competition—with other drivers. Obvious examples of interactions are when a number of drivers arrive at an intersection without traffic lights or when they are driving inside a parking lot waiting for spots to become available. Presumably, connected vehicles can support such interactions. Vehicle connectedness can, for example, utilize wireless networking technologies and mobile social applications running on smartphones, which can collect, share, and present real-time information about parking demand and availability to drivers.

The behavioral sciences, such as economics and psychology, are dominated by models of interactive situations, or *games* as they are called, that propose that people are "rational," which means that people aim at maximizing the value they receive from the game. In this sense, game theory is analogous to expected utility theory, which was discussed in the previous section. But it is even more complicated. This is so because in games, one's value depends on others' decisions, and thus "rational" decision-making needs to take into account this complexity as well. Game theory aims at discovering decisions that are mutually "optimal" for all people who play a particular game (von Neumann & Morgenstern, 1947).

Mainstream transportation theory employs game theory in order to describe and prescribe the behavioral aspects of traffic management systems, including parking management systems. Nevertheless, as noted above, it might be that parking behavior is better described by simple rules of thumb (Hester et al., 2002). More surprisingly, it might also be that such simple parking behavior in fact leads to better system performance than the purported "optimal" parking behavior suggested by game theory, as the following study found.

Karaliopoulos, Katsikopoulos, and Lambrinos (2017) consider a game where multiple drivers are competing over two types of parking resources—on-street parking and a parking lot. On-street parking is assumed to be cheaper and more easily accessible than lot parking. There is an additional excess cost for parking at the lot after searching for parking on the street (this cost expresses, among other things, additional fuel consumption). Assuming estimates of the number of drivers, the number of spots in each parking resource, and the various costs—which can be made available to ACIV drivers—Karaliopoulos et al. (2017) compared the performance of the system when drivers behave according to (1) an "optimal" equilibrium computed by game theory and (2) the following simple rule of thumb:

> *Step* 1. If the best-case costs of the two alternatives (on-street and lot parking) differ by more than a percentage of the overall worst-case cost one may incur, then search for on-street parking (because it has much smaller best-case cost);
>
> *Step* 2. Otherwise, consider the probabilities incurring the two best-case costs: If their difference exceeds a threshold, then choose the alternative with the larger probability of best-case cost;
>
> *Step* 3. Otherwise, choose the alternative with the smaller worst-case cost (independently of how small the difference between the two worst-case costs is).

To see how the simple rule works, say that there are currently two drivers on the road and there is one parking spot available on the street and three in the lot. In this case, the unit costs for on-street and lot parking are 1 and 5.5 respectively, and the excess cost is 2. Assume that for one of these drivers the percentage in Step 1 equals 10% and the threshold used in Step 2 equals 0.1. Then, the difference of the best-case costs is $5.5 - 1 = 4.5$ which is larger than 0.75 (10% of $5.5 + 2$), and thus this driver would search for on-street parking. If this driver's percentage parameter were 60%, however, Step 2 of the rule would be used. In Step 2, the difference of the probabilities— assuming that the other driver chooses one of the four parking spots randomly—of the best-case costs equals 1.0 (for the parking lot) $- 0.75$ (for on-street parking) $= 0.25$, which is larger than 0.1, and thus the driver would go to the parking lot.

The rationale for considering this particular rule of thumb is that it might be descriptive of how people choose where to park. A reason to expect so is because the rule is analogous to a rule that predicted people's majority choices better than expected utility theory and prospect theory (Katsikopoulos & Gigerenzer, 2008). Consistently, Karaliopoulos et al. (2017) provide the results of a survey of 1,120 participants, which found that the parking choices of those drivers who always park on the street or always park on a lot—19% of all participants—can be well described by this simple rule. It is not clear how to apply expected utility theory or prospect theory in this case, because it is not clear how to reliably estimate their multiple parameters.

Regarding system performance, Karaliopoulos et al. (2017) analytically derived conditions under which game-theoretic equilibrium behavior incurred larger total costs and resulted in a larger percentage of drivers competing for on-street parking than behavior consistent with the simple rule presented above. For instance, say that there are 60 drivers on the road and the number of available parking spots is 10 on the street and 15 in the lot, the unit costs for on-street and lot parking are 1 and 5.5, respectively, and the excess cost is 2. Then, it turns out that the total cost at equilibrium equals 280, whereas under the simple rule, it equals 220. And, the number of competing drivers at equilibrium is 55, whereas under the simple rule, it is 35.

In general, conditions under which the simple rule improves system performance are fulfilled for a broad range of scenarios concerning the fees charged for parking resources and their distance from the destinations of the driver's trips. This result also holds for more complicated parking games, including more than one lot. Finally, one might expect that the simple rule of thumb is more transparent than game theory to drivers, parking managers, and other stakeholders such as local authorities.

20.2.3 EFFECTS OF VEHICLE AUTOMATION ON DRIVER BEHAVIOR

As also discussed in Chapter 4, the effects of vehicle automation on driver behavior are strongly moderated by people's trust in automation. It is sometimes thought, or implied, that the more people trust automation, the better, and thus design should "invite" people to rely on automation. This, of course, cannot always be true. As Lee and See (2004) point out, the point is to design for appropriate reliance. These authors also suggest that appropriate reliance could be achieved "by making the algorithms of automation simpler or by revealing their operation more clearly" (Lee & See, 2004, p. 74). Endsley (2017) also makes the same point (Chapter 7), worrying about the opaqueness of now routine deep-learning algorithms. Such algorithms are often not even transparent to their own designers (Townsend, 2008). This point follows from the sections above and emphasizes the need for simplicity and transparency in researchers' understanding of human behavior. In particular, if the automation is going to suggest, say, following the speed limit on a smart motorway during periods of congestion, the automation needs to provide that information in a way that makes it possible for the human driver to use simple rules to come to a decision about what to do.

The design of the human–machine interface (HMI) and, in particular, how information on congestion is displayed will also influence driver's choices. Chapter 4 provides an exposition of how psychological theories of motivation, emotion, and personality, as well as the associated concepts of traits and attitudes, can inform such automation design and its customization. Additionally, Chapters 7 and 15 are also relevant, as they argue for the need for improved HMI design for automated vehicles. Endsley (2017) describes her six-month experience of driving the Tesla Model S and the challenges she faced in interacting with the automation interface. Beyond these chapters, we do not have much to add on automation here, except for the following outline of a new direction in behavioral research on automation and congestion.

Bonnefon, Shariff, and Rahwan (2016) placed Foot's (1967) classic trolley problem in the context of automated driving. The researchers presented participants in multiple surveys with questions such as the following: Should an automated-driving algorithm protect its passengers at the cost of running over pedestrians? Should such an algorithm change its course so as to run over a smaller number of pedestrians? An interesting result was that people said they want utilitarian automated vehicles (i.e., vehicles saving a larger number of people) on the road, but also seemed hesitant to buy such vehicles (and bear responsibility for their associated moral stance). Whereas it is not clear whether (1) the automated vehicles built in the future will actually have to solve dilemmas like this, and if (2) stated, hypothetical responses represent people's actual behavior (in this experiment, participants were paid only

25 cents to answer a question), such work surely needs to be followed up. It is also relevant to congestion when there is a choice between two routes and the risk of injury or fatality to other drivers and vulnerable road users is relatively high (say by a factor of four) for one route, but the travel time is shorter (say by a factor of half).

20.3   REDUCING CARBON EMISSIONS: VEHICLE CAPACITY AND DRIVER READINESS TO USE

20.3.1   VEHICLE CAPACITY TO REDUCE CARBON EMISSIONS

Today's vehicles could help reduce carbon emissions to an extent that a layperson might find surprising. For example, Berners-Lee (2011) argues that driving 10,000 miles could make a difference from 35% to 250% in emissions, depending on the type of car and how it is driven. As instances of this claim, consider the following.

About 50% of the carbon impact of driving a car comes out of the exhaust pipe, 10% comes from the fuel, and 40% is associated with the manufacturing, operating, and maintenance of the car. Small and efficient cars can save 50% of emissions compared with average cars. Accelerating and decelerating gently, and avoiding braking, save 20% of emissions under urban conditions. Driving at 60 miles per hour on highways and freeways saves 10% compared with driving 70 miles per hour. It follows from these figures that automated driving—to the extent that it is done in the right way—can make a big difference on carbon emissions.

Now, a life cycle assessment of automated vehicles showed that sub-systems could initially increase vehicle primary energy use and greenhouse gases emissions by 3%–20% due to increases in power consumption, weight, drag, and data transmission. But when potential operational effects of these vehicles are included (e.g., eco-driving, platooning, intersection connectivity), automated vehicles can lead to an overall decrease of 9% in both energy use and greenhouse gases emissions (Gawron, Keoleian, De Kleine, Wallington, & Kim, 2018).

Additionally, Igliński and Babiak (2017) argue that carbon emissions reductions will only occur after automated vehicles become very popular, and this, they say, requires developers to achieve the fifth level of automation. Of course, doing so could also lead to rebound effects such as increased driving, but this has not been estimated yet.

20.3.2   DRIVER READINESS TO USE CARBON-EMISSIONS-REDUCING VEHICLES

In a review of the literature, Stern (2000) showed that pro-environmental behavior is dependent on personal as well as contextual factors. That is, being motivated towards decreasing your carbon emissions is not enough in order to undertake a pro-environmental behavior. For instance, the behavior may be beyond your reach for a number of reasons. It may not be facilitated locally or might be costly, or it could be faced with barriers that are too difficult to overcome.

One example is the decision of buying a car. Consider a single man living in Copenhagen, Denmark. He can avoid buying a car since public transportation is comfortable enough to reach any area within the city. Besides, the city is biking-friendly.

Now, consider the same man living in Rio de Janeiro, Brazil. The overcrowded public transportation does not reach all of this city's areas, most of the roads do not even have a biking lane, and local violence should also be considered. Can the man now easily avoid buying a car in this situation? Also, what if he has three children, or lives far away from his work?

A recent survey conducted by ReportLinker in the United States (ReportLinker, 2017) found that 62% of the respondents would buy an autonomous vehicle. The following are the main reasons for doing so: using it for long-distance travel (18%), becoming able to multitask (12%), increasing the safety of roads (10%), not having to park (6%), and helping to reduce energy consumption (5%). On the other hand, 33% of the respondents said that safety was their top objection to buying an automated vehicle.

More generally, consumers have said that they would favor items with a lower carbon footprint if they were given clear information (Camilleri, Larrick, Hossain, & Patino-Echeverri, 2019). Carbon footprint labels have been suggested as a simple and clear intervention for increasing the understanding of energy use and greenhouse gases emissions for a diversity of products, thus helping to reduce environmental impacts. In Finland, 90% of consumers have stated that carbon-footprint information would have at least a little impact on their buying decisions. But this is so when other purchasing criteria (e.g., price of fuel or travel time) were satisfied (Hartikainen, Roininen, Katajajuuri, & Pulkkinen, 2014). Moreover, 86% preferred carbon labels that allowed comparisons of carbon emissions to be made across products.

20.3.3 Eco-Driving

Eco-driving is one area where the human factors issues dominate the discussion. Much more is known about these issues in the last ten years. They include everything from pre-trip eco-driving planning, to actual eco-driving during the trip, and finally to post-trip presentation of energy use (Barkenbus, 2010). A recent book has focused on one aspect of the actual driving task, in particular the presentation of in-vehicle information to the driver as the trip unfolds (Mcllroy & Stanton, 2018). The authors argue for an ecological design of the interface (e.g., Rasmussen & Vincente, 1980), one which supports the interaction of the driver with the driving task across skill-based, rule-based, and knowledge-based behaviors.

The authors of the book depart in an interesting way from the above discussion on congestion, which emphasizes the primacy of boosting underlying competencies and processes as in addition to improving actual behaviors. With eco-driving it turns out to be important to automate the decision, which is a skill-based behavior, not a knowledge-based behavior, in part because the cognitive load imposed by eco-driving needs to be minimized. The cognitive load needs to be minimized because eco-driving requires continuous input whereas route choice is engaged in sporadically and often when the driver chooses to do so. As Mcllroy and Stanton (2018) state: "the *expert* eco-driver performs the task in a way that approaches automaticity, that is, they are performing at the skill-based level of cognitive control." We refer the reader to their text for an enlightening and much more detailed discussion.

20.4 CONCLUSION

Traffic congestion increases emissions of greenhouse gases and localized pollutants such as particulate matter (Barth & Boriboonsomsin, 2008). Life cycle assessment of automated vehicles has found that automated vehicles could decrease greenhouse gases emissions by 9% (Gawron et al., 2018; although this figure does not take into account possible rebound effects). Furthermore, vehicle intelligence and connectedness are expected to bring additional efficiency gains.

This chapter reviewed research on (1) the effects of vehicle automation, intelligence, and connectedness on driver behavior and on (2) the capacity of automated vehicles to reduce carbon emissions and the readiness of drivers to use such vehicles, and (3) the strategies needed to achieve eco-driving. Related to (1), it seems that drivers tend to interface with vehicle technology by relying on the simple rules of thumb. With regards to (2), it seems that people's motivation to reduce carbon emissions is not enough for them to engage in pro-environmental behavior, but rather, informing the drivers is key. Moreover, with respect to (3) once a driver decides to purchase an environmentally friendly vehicle, it becomes important at that point to focus on the development of the actual skills required to make eco-driving a reality.

We end with a note on research methodology. Work on (1) has utilized formal analyses and driving simulation experiments, where tradeoffs encountered while driving was made explicit and controlled experimentally. In contrast, work on (2) has used engineering analyses and surveys, where tradeoffs between engaging in pro-environmental behavior and possibly giving up some convenience have not been studied in controlled laboratory settings. For example, we do not know if and to what extent people would buy a vehicle reducing carbon emissions at the expense of increased travel time (Huang, Ng, & Zhou, 2018). Finally, work on (3) has used all methods. However, the effects of eco-driving training appear to attenuate over time. How to maintain these effects is an issue of considerable interest. Investigating such questions would be a promising research direction.

REFERENCES

Barkenbus, J. (2010). Eco-driving: An overlooked climate change initiative. *Energy Policy, 38*, 762–769.

Barth, M. & Boriboonsomsin, K. (2008). Real-world carbon dioxide impacts of traffic congestion. *Transportation Research Record: Journal of the Transportation Research Board, 2058,* 163–171.

Berners-Lee, M. (2011). *How Bad Are Bananas? The Carbon Footprint of Everything.* London: Greystone Books Ltd.

Bond, M. (2009). Decision-making: risk school. *Nature News, 461*(7268), 1189–1192.

Bonnefon, J-F., Shariff, A., & Rahwan, I. (2016). The social dilemma of autonomous vehicles. *Science, 352,* 1573–1576.

Bortoleto, A. P. (2014). *Waste Prevention Policy and Behavior: New Approaches to Reducing Waste Generation and its Environmental Impacts.* Basingstoke, UK: Routledge.

Camilleri, A. R., Larrick, R. P., Hossain, S., & Patino-Echeverri, D. (2019). Consumers underestimate the emissions associated with food but are aided by labels. *Nature Climate Change, 9*(1), 53.

Downs, A. (1962). The law of peak-hour expressway congestion. *Traffic Quarterly, 16*(3), 393–409.

Duranton, G. & Turner, M. A. (2011). The fundamental law of road congestion: Evidence from US cities. *American Economic Review, 101*(6), 2616–2652.

Endsley, M. (2017). Autonomous driving systems: A preliminary naturalistic study of the Tesla Model S. *Journal of Cognitive Engineering and Decision Making, 11*, 225–238.

Foot, P. (1967). The problem of abortion and the doctrine of the double effect. *Oxford Review, 5*, 5–15.

Gawron, J. H., Keoleian, G. A., De Kleine, R. D., Wallington, T. J., & Kim, H. C. (2018). Life cycle assessment of connected and automated vehicles: Sensing and computing subsystem and vehicle level effects. *Environmental Science & Technology, 52*(5), 3249–3256.

Hartikainen, H., Roininen, T., Katajajuuri, J. M., & Pulkkinen, H. (2014). Finnish consumer perceptions of carbon footprints and carbon labelling of food products. *Journal of Cleaner Production, 73*, 285–293.

Hertwig, R. & Grüne-Yanoff, T. (2017). Nudging and boosting: Steering or empowering good decisions. *Perspectives on Psychological Science, 12*(6), 973–986.

Hester, A. E., Fisher, D. L., & Collura, J. (2002). Drivers' parking decisions: Advanced parking management systems. *Journal of Transportation Engineering, 128*(1), 49–57.

Huang, Y., Ng, E., & Zhou, J. (2018). Eco-driving technology for sustainable road transport: A review. *Renewable and Sustainable Energy Reviews, 93*, 596–609.

Igliński, H. & Babiak, M. (2017). Analysis of the potential of autonomous vehicles in reducing the emissions of greenhouse gases in road transport. *Procedia Engineering, 192*, 353–358.

Kahneman, D. & Tversky, A. (1979). Prospect theory: An analysis of decision under risk. *Econometrica, 47*, 263–291.

Kantowitz, B. H., Hanowski, R. J., & Kantowitz, S. C. (1997). Driver acceptance of unreliable traffic information in familiar and unfamiliar settings. *Human Factors, 39*(2), 164–177.

Karaliopoulos, M., Katsikopoulos, K., & Lambrinos, L. (2017). Bounded rationality can make parking search more efficient: The power of lexicographic heuristics. *Transportation Research Part B: Methodological, 101*, 28–50.

Katsikopoulos, K. V. (2014). Bounded rationality: The two cultures. *Journal of Ecnomic Methodology, 21*(4), 361–374.

Katsikopoulos, K. V., Duse-Anthony, Y., Fisher, D. L., & Duffy, S. A. (2000). The framing of drivers' route choices when travel time information is provided under varying degrees of cognitive load. *Human Factors, 42*(3), 470–481.

Katsikopoulos, K. V., Duse-Anthony, Y., Fisher, D. L., & Duffy, S. A. (2002). Risk attitude reversals in drivers' route choice when range of travel time information is provided, *Human Factors, 44*(3), 466–473.

Katsikopoulos, K. V. & Gigerenzer, G. (2008). One-reason decision-making: Modeling violations of expected utility theory. *Journal of Risk and Uncertainty, 37*(1), 35–56.

Lee, J. D. & See, K. A. (2004). Trust in automation: Designing for appropriate reliance. *Human Factors, 46*(1), 50–80.

McIlroy, R. & Stanton, M. (2018). *Eco-Driving: From Strategies to Interfaces (Transportation Human Factors)*. Boca Raton, FL: CRC Press.

Rasmussen, J. & Vincente, K. (1980). Coping with human errors through system design: Implications for ecological interface design. *International Journal of Man-Machine Studies, 31*, 517–534.

ReportLinker Insight. (2017). *Self-Driving Vehicles Navigate Twists and Turns on the Road to Adoption*. Retrieved from www.reportlinker.com/insight/self-driving-vehicles-navigate-twists-turns-road-adoption.html?mod=article_inline

Stern, P. C. (2000). New environmental theories: Toward a coherent theory of environmentally significant behavior. *Journal of Social Issues, 56*, 407–424.

Sugden, R. (2017). Do people really want to be nudged towards healthy lifestyles? *International Review of Economics*, *64*(2), 113–123.

Thaler, R. H. & Sunstein, C. R. (2008). *Nudge: Improving Decisions about Health, Wealth, and Happiness*. New Haven, CT: Yale University Press.

Townsend, J. T. (2008). Mathematical psychology: Prospects for the 21st century. *Journal of Mathematical Psychology*, *52*, 271–282.

United States Bureau of Transportation Statistics. (2007). *Transportation Statistics Annual Report*. Washington, DC: US Government Printing Office.

von Neumann, J. & Morgenstern, O. (1947). *Theory of Games and Economic Behavior* (2nd ed.). Princeton, NJ: Princeton University Press.

21 Automation Lessons from Other Domains

Christopher D. Wickens
Colorado State University

CONTENTS

Key Points 455
21.1 Introduction 456
21.2 Classic Automation Accidents 456
21.3 Features of Automation 457
    21.3.1 The Degree of Automation: What Does Automation
         Do and How Does It Do It? 457
    21.3.2 Automation Reliability 459
    21.3.3 Automation Trust and Dependence 460
    21.3.4 Out-of-the-Loop-Unfamiliarity (OOTLUF) 460
    21.3.5 Automation Modes and Complexity 461
    21.3.6 Automation Transparency 461
    21.3.7 Adaptable Versus Adaptive Automation 462
21.4 Research Findings 462
    21.4.1 Accident and Incident Data Mining: Advantages and Costs 462
    21.4.2 Experimental and Simulation Results 463
         21.4.2.1 Alerting Systems 463
         21.4.2.2 Attention Cueing 464
         21.4.2.3 Stages 2 and 3 Automation: OOTLUF 465
21.5 Solutions: Countermeasures Proposed and Implemented in
    Other Domains 465
    21.5.1 Flexible and Adaptive Automation 465
    21.5.2 Automation Transparency 467
    21.5.3 Training 467
21.6 Conclusions 468
References 468

KEY POINTS

- Automation can be defined by the stage of information processing for which it assists the human, and the level of assistance at each stage, together defining the degree of automation.
- The higher the degree of automation, the better is human–automation performance, the lower workload, and the lower situation awareness when

automation is working correctly; but the more problematic is human–automation interaction when automation fails.

- The problematic interaction results from failure to monitor what automation is doing, failure to understand its state, and degradation of skills when manual control must be exercised following the failure.
- Problems can be addressed by flexible or adaptive automation, automation transparency, and training, although benefits of adaptive automation can be difficult to realize.

21.1 INTRODUCTION

As the many other chapters in this book have made clear, automation takes many forms in vehicles. Foremost among these are the higher levels of automation and, ultimately, the total autonomy of self-driving cars. But there are numerous other examples of automation, such as headway monitors, auto-locks, a variety of alerts and warnings, anti-lock brakes and navigational systems. In designing all such systems to facilitate better interactions with the human, balancing safety versus productivity, many lessons can be drawn from accident analysis and research from other domains, particularly from aviation, which is the pioneer in the systematic investigation of human–automation (Wiener & Curry, 1980).

In this chapter, I will review the lessons that can be learned from human interaction with systems other than vehicles, including human flight in aviation (Billings, 1997; Ferris, Sarter, & Wickens, 2010; Landry, 2009), unmanned air vehicles (Cummings & Guerlain, 2007), medicine (Garg et al., 2005; Morrow, North, & Wickens, 2006), space (Li, Wickens, Sarter, & Sebok, 2014), air traffic control (Wickens, Mavor, Parasuraman, & McGee, 1998), military systems (Chen et al., 2018), consumer products (Sauer & Ruttinger, 2007), process control (Strobhar, 2012), robotics, and others.

This chapter will begin by providing a brief synopsis of three aviation tragedies directly related to breakdowns in human–automation interaction (HAI) when automation fails. Furthermore, it will describe in detail some of the key concepts in HAI that have arisen out of non-vehicle research, but are directly applicable to it. Then, I describe the empirical research bearing on several major issues in HAI. Finally, I turn to four suggested solutions to improving HAI, preventing its disasters without sacrificing the productivity that it offers, and examine the empirical evidence in support of those solutions.

21.2 CLASSIC AUTOMATION ACCIDENTS

In 1972, pilots flying on approach to Miami Airport, over the Everglades, were unable to determine if the landing gear had locked in place. All three personnel on the flight deck became engaged in the troubleshooting, and placed the aircraft autopilot on a level altitude hold. Somehow the autopilot became disengaged, but, complacent in their belief in its operation, the flight deck personnel failed to check the now slowly decreasing altitude, until it was too late, and the plane slammed into the ground.

In 1983, on KAL flight 007 over the north Pacific, pilots programmed a course into the flight management system (FMS) that was incorrect. As with the Everglades accident, pilots failed to monitor how automation was flying the plane as it flew directly into Soviet airspace and was shot down, with all lives lost (Wiener, 1988).

In 2013, an Asiana airline was on approach to San Francisco International Airport when pilots became confused regarding what the "Auto-land" system was doing. They acted in opposition to what automation was commanding the plane to do, and the plane stalled and crashed just short of the runway threshold.

All three of these tragedies—and many more (see for example Dornheim, 1995)—have identified problems in HAI encountered by highly skilled professionals. Such problems filtered through careful accident analysis and examined through flight simulation research can identify lessons learned that may be transferred to automation in ground vehicles, along with potential solutions. In the next section, I identify several key features of HAI that can be used to understand generic, cross-disciplinary applications.

21.3 FEATURES OF AUTOMATION

21.3.1 THE DEGREE OF AUTOMATION: WHAT DOES AUTOMATION DO AND HOW DOES IT DO IT?

Naturally, the most direct answer to the first part of this question is the purpose and function of automation in a particular context. For example, in aviation, the purpose of the autopilot is to stabilize and fly the plane; the purpose of an alerting system is to offload the human from continuous monitoring of some function or for some event.

Importantly, the second question—how does it do it?—can be answered at a generic level, above and beyond the specific system to which it is applied. Indeed, "how does it do it" can be defined on two generic dimensions, stages, and levels. In 1978, Sheridan and Verplank proposed the concept that automation was not a single entity but could instead be defined along a scale of levels of automation, defining the degree of authority of automation versus human in executing a task (Sheridan & Verplank, 1978). While the original scale had ten levels, these can be characterized more generally as, from lower to higher: (1) automation recommending several possible actions, (2) automation recommending only a single action, (3) automation executing that action but allowing the human to veto it, and (4) automation carrying out the action with no veto possible. The precise number of levels on any scale is less important than the change in levels, moving upward or downward to impose more, or less, automation.

Parasuraman, Sheridan, and Wickens (2000; 2008) in subsequently applying the levels to air traffic control systems, realized that the original Sheridan and Verplank (1978) scale only applied to automation support of human decisions. Parasuraman et al. (2000) postulated that there were three additional "stages" of human information processing that could benefit from the increasing level of automation assistance. These are Stage 1 (information filtering or guiding attention), Stage 2 (integrating information to support inference and situation assessment), and Stage 4 (carrying out or executing the action that was decided upon in Stage 3, action selection and decision-making). [Note: Stage 3 maps onto the automation support of human decisions as originally described by Sheridan & Verplank (1978).] For example, automation assistance for the health care practitioner may highlight particular medical problems on a patient's electronic record or alert the practitioner to a dangerous condition (Stage 1), may offer a diagnostic suggestion (Stage 2), may offer a recommended treatment (Stage 3), and a drug infusion pump may automatically administrate medicine at the appropriate time and dosage (Stage 4). As with Sheridan and Verplank's (1978) original scale, each of these stages can also be described as being implemented at varying levels of authority.

Thus, the two-dimensional structure (taxonomy) of stages and levels, shown in Figure 21.1, can be described by a higher level variable defining the degree of automation (DOA), moving from the lower left to the upper right (Onnasch, Wickens, Li, & Manzey, 2014; Wickens, 2017). When automation is implemented at the highest level of all four stages, this describes the status of full autonomy. The characterization of levels and stages of automation proposed by Parasuraman et al. (2000) can also be likened, conceptually, to the levels of automation for automated driving systems described by the Society for Automotive Engineers (SAE, 2016; this Handbook, Chapters 1 and 2).

In some key developments of the history of this research, Endsley and Kiris (1995), and Kaber and Endsley (2004; Kaber, Onal, & Endsley, 1999) carried out early research on automation and situation awareness (SA) that could be readily interpreted in the context of stages and levels of automation (i.e., DOA). More recently, Onnasch et al. (2014) carried out a meta-analysis of DOA research that examined the effect of four correlated variables that changed as DOA increased: (1) The performance of the task for which automation was designed to support increased; (2) human workload decreased; (3) humans lost SA; and as a consequence, when automation failed, (4) human failure recovery was more problematic (and sometimes disastrous).

A major reason why the taxonomy defining DOA is important in HAI is that it defines a distinction that is relevant to many automation decision support tools: Should automation advise the human user as to "what is" (diagnostic support at Stages 1 and 2) or should automation advise the human user "what to do" (decision

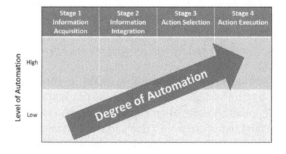

FIGURE 21.1 DOA. In this rendering, there are only two levels within each stage, but there could be several, and there does not need to be the same number of levels at each stage.

aiding at Stage 3)? Such a dichotomy exists in many areas, such as medical decision-making (Garg et al., 2005; Morrow et al., 2006), aviation conflict avoidance systems, or even statistical tools that distinguish between providing p values and confidence intervals versus advice to accept or reject the null hypothesis (i.e., decision aid; Wickens & McCarley, 2017). Given that automation is imperfect, and there may be more severe consequences if it fails at later stages of automation, this distinction needs to be considered carefully in automation design and implementation.

21.3.2 AUTOMATION RELIABILITY

The previous section alluded to the imperfections of automation: it can fail and such failures thereby characterize the concept of automation reliability. In some cases, reliability can be quantified as the ratio of correct operations to total operations—a ratio that has important meanings in certain kinds of automation, like those governing alerting systems. Research has also revealed the importance of distinguishing between the first time an automation fails (for a particular user) versus subsequent failures. The former case can often produce a much more problematic response—the so-called "first failure effect" (FFE)—than the latter (Sebok & Wickens, 2017; Yeh, Merlo, Wickens, & Brandenburg, 2003).

Beyond the quantification of automation reliability, we must also distinguish the different generic ways in which automation can fail. It can be gone, such as when the power fails, or it can be wrong, such as when an automated weather forecast is incorrect (Wickens, Clegg, Vieane, & Sebok, 2015a). In process control support, automation failures when automation is wrong appear to have a more consequential effect on human response, than when automation is gone (Eriksson & Stanton, 2017; Sauer, Chavaillaz, & Wastell, 2016; Wickens et al., 2015a).

Another way to characterize different types of automation errors is to distinguish between the following four classes of automation errors:

a. When the hardware or software truly fails: a misconnected circuit, a software bug, a hydraulic valve that is stuck open, or an errant sensor in an automatic control system.
b. When the automation is asked to perform its task in an environment in which it was not designed to perform (e.g., Global Positioning System (GPS) navigation when there is no signal).
c. When the human has an incorrect mental model of what automation should be doing, and believes that automation is wrong, even when it is not. This is not an insignificant problem when the automation is extremely complex, such as the FMS in advanced aircraft (Mumaw, 2018; Sarter, 2008).
d. When the human has programmed or set up the automation incorrectly, as in the KAL 007 disaster described above.

It is important to recognize that the latter three of these failure sources are not true "failures" from an engineering standpoint. However, from the human user's standpoint, they are perceived as automation failures, and this incorrect perception can lead to the problematic failure response, as in the case of the Asiana accident described above.

21.3.3 AUTOMATION TRUST AND DEPENDENCE

Since the classic work of Lee and Moray (1992) on process control automation, the concept of trust in automation, seen as an analogy to human trust in other humans, has become prominent in HAI research (e.g., Hoff & Bashir, 2015). This prominence was bolstered by an oft-cited article by Parasuraman and Riley (1997) that introduced concepts such as automation use, disuse, and misuse. These authors highlighted the clear link between trust, a cognitive concept, and system use (which I prefer to call dependence), a behavioral concept.

These two concepts of trust and dependence are often linked, but far from synonymous (Merritt, Lee, Unnerstall, & Huber, 2015). Thus, I can trust an agent, and not depend (partially or totally) upon it. This may be particularly true, in the case of automation, if I enjoy engaging in the processing that automation can do for me. In contrast, I can mistrust automation if it is not perfectly reliable, but still depend upon it, if I have other tasks to perform; however, in these circumstances, I should, optimally, allocate some resources to overseeing and verifying automation's activities. One can argue that the amount of such oversight (e.g., visual attention) should be directly calibrated to the reliability of automation (Moray & Inagaki, 2000).

In discussing the role of trust, it is important to distinguish between over-trust (trusting automation more than one should) and under-trust (trusting less than one should), with calibrated trust defining the balance between these. Over-trust has sometimes been referred to as "automation bias" (Mosier, Skitka, Heers, & Burdick, 1998; Parasuraman & Manzey, 2010).

21.3.4 OUT-OF-THE-LOOP-UNFAMILIARITY (OOTLUF)

A major consequence of over-trust is **complacency**, a dependence on automation that is so extensive that when automation fails, the operator, being out of the loop and hence unfamiliar with what automation is doing, will intervene slowly, inappropriately, or perhaps not at all (Endsley, 2017; Endsley & Kiris, 1995). OOTLUF can be broken down into three separate components, with the first two being associated with loss of SA at Levels 1 and 2 (Endsley, 2017; this Handbook, Chapter 7):

1. Failure to monitor the process(es) carried out by automation, the raw data that it is processing, or the performance it is producing—a failure often directly measurable by eye movements (Metzger & Parasuraman, 2005; Parasuraman & Manzey, 2010).
2. Failure to understand what automation is doing, at the time of the automation failure, and hence intervening inappropriately. A failure to understand will certainly follow from the failure to monitor but may also be associated with an absence of engagement, even when the eyes may scan the indicators of automation functioning. Such a failure is closely related to the phenomenon known as the generation effect (Slamecka & Graf, 1978), whereby we remember the state of systems better when we generate responses pertaining to those systems than when we witness other agents (e.g., automation)

carrying out identical actions. Such memory in this case translates directly into understanding of the current state of the system. The generation effect has often been applied to the concept of active learning, as a superior training technique to passive listening or reading (Dunlosky et al., 2013).

Thus, failure to monitor and understand can lead to inappropriate interventions when automation fails. To this is added the third component of OOTLUF:

3. **Deskilling** refers to the state wherein prolonged use of automation will lead to degraded operator skills in performing the task that automation is programmed to accomplish, hence further aggravating the inappropriate response to automation failure. This concern has been identified in aviation, when pilots rely too much on their automated flight systems (Casner, Geven, Recker, & Schooler, 2014).

All three of these effects: on monitoring, understanding, and manual recovery skills appear to be amplified with higher DOA, as, with this higher degree, there is less reason to become engaged in the task. Hence the more automation does during routine correct performance of automation, the more serious are the consequences when it fails. This tradeoff has been described as the "lumberjack phenomenon": the higher the tree, the harder it falls (Sebok & Wickens, 2017).

21.3.5 AUTOMATION MODES AND COMPLEXITY

More complex automation amplifies the failures to understand what automation is doing, a lesson well revealed in aviation, where the scores of functions carried out by the FMS, include many different modes of automation (Ferris et al., 2010; Landry, 2009; Mumaw, 2018; Sarter, 2008; Sebok et al., 2012). For example, there are five different ways or modes through which the FMS can lower the altitude of an aircraft. Such complexity is amplified further whenever the decision to choose or switch modes is not made by the human, but by the automation itself, given certain "triggering criteria" (e.g., crossing through a particular altitude may automatically trigger a "level off" mode, or exceeding a particular airspeed may trigger a braking mode). Thus, complexity can be often characterized jointly by the number of modes and the complexity of the decision rules used to trigger those modes. For example, "if A then do C" is less complex than "If A and B, do C." When an action is carried out by the human who is assuming that the system is in one mode, but in fact it is in another, this is defined as a mode error. Mumaw (2018) has written a compelling narrative regarding the impact of FMS mode complexity on pilot's mode errors and their challenges in understanding flight deck automation.

21.3.6 AUTOMATION TRANSPARENCY

Both the issues of complexity and OOTLUF-based failure to understand can be attributed in part to the "black box" appearance of much automation, a lesson well learned in aviation. Automation transparency (Chen et al., 2018) is a concept by which the workings of automation are made more apparent to the human.

As described in more detail below, this may be accomplished by displays of what automation is doing and why, or by verbal explanations, for example of why a decision aid came to a particular recommendation.

21.3.7 ADAPTABLE VERSUS ADAPTIVE AUTOMATION

This contrast refers to the extent to which the engagement of automation is adaptable, and can therefore be selected by the human operator (as when the pilot chooses to turn on the autopilot, or "fly by hand"), versus selected by the automation itself (adaptive automation; Dorneich, Rogers, Whitlow, & DeMers, 2016; Kaber & Kim, 2011; Kaber, Wright, Prinzel, & Clamann, 2005). In the latter case, an automated agent will itself "decide" to take automated control from the human operator or return it back to the human depending on the automation's inference about the human's momentary capacity to perform the task well.

21.4   RESEARCH FINDINGS

21.4.1 ACCIDENT AND INCIDENT DATA MINING: ADVANTAGES AND COSTS

A good deal of the understanding of problems in HAI has been gained from the aviation industry, in identifying and analyzing breakdowns that have contributed to major accidents, such as the Eastern Airlines Everglades crash described above. Such analyses appear in National Transportation and Safety Board (NTSB) report and may include HAI errors as one of the causal factors. The limitations in using such information to gain information about causality are twofold. (1) Pilots involved in crashes are often killed and there are always multiple factors involved. It is impossible to sort out, with any certainty, the extent to which HAI was the precipitating factor rather than just a contributing cause, not to mention identifying which of the many issues of HAI might have been involved. (2) Aircraft accidents are, fortunately, extremely rare. But such rarity forecloses the multiple samples that are required to draw reliable statistical inferences regarding frequency and causality. These inferences are at the foundation of the science of human factors.

A second source of data comes from incident analysis. Since 1976, the National Aeronautics and Space Agency (NASA) has collected and categorized a large volume of incident reports, contained in the Aviation Safety Reporting System (ASRS), in which pilots file, anonymously, voluntary reports of what they consider safety-compromising incidents in their flights. Such data have the advantage of large sample sizes (i.e., large N) absent in accident reports. But, being based on the recollection of pilots, who are not generally trained in human factors or psychology, the cognitive or information processing mechanisms are often not included in the narrative. Of course, since the reports are voluntary, while the numbers are large, they may be highly biased, perhaps against reporting an incident in which a pilot committed a clear violation (notwithstanding the guarantee of anonymity in such filings). Nevertheless, such a system has revealed valuable conclusions regarding HAI, recently summarized in an extensive report compiled by the Committee on Aviation

Safety (Commercial Air Safety Team, 2014) of 50 ASRS reports. Their conclusions were notable in identifying the frequency of the sorts of automation mode errors described above.

21.4.2 EXPERIMENTAL AND SIMULATION RESULTS

21.4.2.1 Alerting Systems

There has been a great deal of research on alerting systems that has been carried out in other domains besides automobiles, particularly in the medical (e.g., Seagull & Sanderson, 2001), process control (Strobhar, 2012), and aviation (Martensson, 1995; Wickens, Sebok, Walters, & McCormick, 2017) domain. One feature from these areas that differs in some respect from those in the automobile is that these other workspaces often embody hundreds of alerts and are also susceptible to "alarm flooding" in times of crisis. In contrast, the number of systems to be alerted in the automobile remains relatively limited. Nevertheless, several research findings remain relevant to all forms of alerts.

The distinction drawn in signal detection theory between alert misses and alert false alarms is vital because of their differing influence on human trust of the alerting system and response to the two different types of automation errors. Both misses and false alarms characterize the overall reliability of automation, and hence affect trust therein. Meyer (2001; 2004; Meyer & Lee, 2013) has facilitated the understanding of the distinction between human behavior in response to the two types of alerting automation errors. Reliance is the behavior relevant to automation misses, when the automation alert is "off," signaling that all is well when in fact it is not. Compliance is the behavior relevant to the alert activation. High reliance will cause the human to not notice a system failure because the automation has not been activated. High reliance is often accompanied by a failure of the human operator to monitor the automation or the system that automation is controlling. In contrast, high compliance is manifest as an immediate and consistent response to follow the guidance of the alerting system (e.g., evacuate the building upon sounding of the fire alarm), even when the alarm is false.

Both high reliance and high compliance are induced by highly reliable automation and can lead to expected consequences in the face of alerting system failure (Dixon & Wickens, 2006). High reliance can create the "double miss" (by both automation and the human). The damage done by high compliance is in the "boy who cried wolf" effect: if the alarm system activates falsely too often, the human will lose trust in it, and simply cease to adhere to its directives when future alarms occur, or even deactivate the alarm system entirely. This can lead to dire consequences when the subsequent alarm turns out to be true. Both kinds of alarm failures and their consequences are well documented in aviation (Bliss, 2003).

In all alarm systems, the balance between misses and false alarms can be adjusted by varying the sensitivity or response criterion of an imperfect alarm. This is typically adjusted in a direction that favors fewer misses at the expense of more alarm false alarms, often with a very small ratio of misses to false alarms (i.e., a very high false alarm rate). For example, Bliss (2003) has found that the rate of alert misses involved in aviation incidents is half that of alert false alarms. This adjustment is done with the understandable rationale that the consequences of the "double miss" (by both automation and the human) are very severe. But often under-appreciated are the undesirable consequences of the high false alarm rate, in terms of humans ignoring true alarms. There remains some discrepancy in research findings of the extent to which false alarms are more detrimental to overall system performance than alarm misses (Dixon et al., 2007) or the contrary (Chen & Barnes, 2012). However, the consequences to both reliance and compliance should be carefully considered by designers and human factors practitioners before an alarm system sensitivity level is chosen.

Independent of the extent to which an alert system is miss-prone or false alarm-prone, the consequences of imperfect automation alerting systems are a loss of trust and therefore dependence upon it, even when the alarm system is fairly (but not perfectly) reliable. The question then is how low can such reliability fall before the benefits of the alerting system may be abolished. Insight into this question can be gleaned from two different sources. First, the results of meta-analyses by Wickens and Dixon (2007), and Rein, Masalonis, Messina, and Willems (2013) suggest that the minimum reliability level of a system may be around 75%–80%. Above this level, such imperfection can still support performance better than the unaided human, and particularly under conditions of concurrent task load. Furthermore, this benefit appears to be observed in automation at later stages as well (Rovira, Pak, & McLaughlin, 2017; Trapsilawati, Wickens, Qu, & Chen, 2016). Below that level, the unwarranted dependence upon automation may actually produce worse detection performance than would be the case of unaided automation, not unlike grabbing onto a "concrete life preserver" in the water (Dixon & Wickens, 2006).

Second, as we have noted, the first failure experienced by an individual in his/her experience can be particularly problematic because of complacency that may have developed following experience with, up that that point in time, perfect detection (Molloy & Parasuraman, 1996; Sebok & Wickens, 2017; Yeh et al., 2003). This may be described as the FFE. As a consequence, it may be desirable to "get rid of this FFE" prior to the operator's first operational experience with the alerting system, by allowing them to experience failures during training or introduction to the system (Manzey, Reichenbach, & Onnasch, 2012; Sauer et al., 2016)—an automation failure inoculation, so to speak.

21.4.2.2 Attention Cueing

Alerting systems inform the operator that something is wrong. Automation can (and ideally should) go beyond this to inform the operator what is wrong, and/or where the dangerous condition is located. The "what" is embodied in Stage 2 diagnostic automation discussed later, but the "where" is embodied in attentional cueing systems, closely related to, but more advanced than alerting systems. Research by Yeh and her colleagues, primarily in the military domain directing a soldier's attention to the potential location of an enemy (e.g., highlighting locations or features on a map), has revealed that erroneous cueing systems (i.e., that direct attention to the wrong location) too can have serious negative consequences (Yeh & Wickens, 2001; Yeh et al., 2003; Yeh, Wickens, & Seagull, 1999). As with alerting systems, such automation errors here, are particularly problematic upon the first failure (Yeh et al., 2003).

21.4.2.3 Stages 2 and 3 Automation: OOTLUF

The concept of OOTLUF, along with the characteristics of the lumberjack effect (the more automation imposed, the more problematic is the human response to its failure) have been described above. The meta-analysis conducted by Onnasch et al. (2014) covers much of the empirical research published prior to 2013 on the general pattern of routine and failure response performance as well as the reductions in workload and SA across increasing DOA. The tradeoffs between reduced workload and loss of SA, and between routine and failure response performance have been supported by subsequent research across many domains (e.g., in military applications: Rovira et al., 2017; in medical applications: Mayo et al., 2015; in robotics: Wickens, Sebok, Li, Gacy, & Sarter, 2015b; in process control: Manzey et al., 2012).

Of particular importance in this relationship is the progressive loss of SA (understanding) as the DOA increases (Endsley & Kiris, 1995), a loss which underlies the problematic response to automation failure. As examples of this research, in process control, Manzey et al. (2012) observed that operators checked raw process variables less frequently with later stages of automation, leaving them with a degraded mental picture in the case of failure. In air traffic control, Trapsilawati, Wickens, Chen, and Qu (2017) observed that imposing Stage 3 automation of an automated conflict resolution aid reduced controller's SA of the state of conflict traffic, compared with manual conditions.

These findings lead to the valuable conclusion that assessments of SA, particularly at Endsley's level 2 (understanding), can provide a useful prediction of the problematic automation failure effect. This is because SA is more feasible to measure than is the automation failure performance (particularly the first failure), and SA is a more reliable measure given that first failure performance will only provide one data point per participant.

21.5 SOLUTIONS: COUNTERMEASURES PROPOSED AND IMPLEMENTED IN OTHER DOMAINS

The previous section has described some of the key issues and problems in HAI, as revealed by research in other, non-driving domains. There is at least an indirect hint from these problems of what might be plausible solutions. For example, to the extent that the OOTLUF problem is amplified with higher DOA, designers should probably resist the temptation to implement very high automation levels at later stages, at least in safety-critical tasks (Parasuraman et al., 2000). And if loss of SA is associated with the problematic automation failure intervention, then efforts should be made to maintain SA, as we discuss below. In this section we describe three explicit categories of solution: flexible automation, automation transparency, and training. Each of these avenues has been suggested and demonstrated in other domains to address some of the problems of OOTLUF and automation failure interventions.

21.5.1 FLEXIBLE AND ADAPTIVE AUTOMATION

If prolonged use of high DOA can induce both complacency and skill degradation, then there would seemingly be a benefit of encouraging the human user to periodically enter or reenter the control loop, to perform the task manually. Such advantages, in detecting automation failures, were indeed demonstrated by Parasuraman, Mouloua, and Hilburn (1999) who observed better failure responses during periods of automation support, when the operator was forced to engage in intermittent periods of manual performance. There are in fact two other different ways of implementing such periodic engagement, beyond the forced switching (scheduled automation) imposed by Parasuraman et al. (1999): adaptable and adaptive automation (Christensen & Estepp, 2013).

In adaptable automation, the operator is simply given the flexibility to choose when (or whether) to implement automation, although this can be encouraged by policy. In contrast, in adaptive automation, an automated agent itself makes the decision of when to implement automation, and, in turn, when to return responsibility back to the human operator. Although adaptable automation has the intuitive appeal of keeping responsibility for task switching in the operator's hands, it does impose added decision responsibilities and does not necessarily produce better performance (Christensen & Estepp, 2013; Sauer, Kao, & Wastell, 2012). However, as described below, it is considerably easier to implement adaptable automation, with a simple "on–off" mode switch at the operator's disposal. This ease is illustrated by discussing the contrasting challenges of adaptive automation.

Adaptive automation requires a set of elements, as well articulated in aviation systems by Dorneich et al. (2016). The first element is identifying what components of the task to automate (e.g., what stages and levels)—a design question equally relevant for adaptable automation. The second element concerns the aspect of human performance or cognition upon which to base the automation agent's decision to implement automation. Typically, the trigger for the automation decision to allocate is either excessive workload or poor performance or a combination of both. A third element, which we describe below, is the specific measure of human cognition/performance with which to infer that automation is needed (or no longer needed when control is returned to the human); that is, what measure triggers the decision.

This third element can be challenging because signals from the human or the human–system interface regarding both workload and performance are imperfectly reliable. It follows that multiple samples of either of these are required to draw a reliable inference about the human's capability to perform. As a consequence of this, perhaps several seconds (a non-trivial amount of time) may be required to obtain a sufficient sample in real time. The resulting delay can produce problems of instability in a dynamic environment, perhaps removing the human from the loop (because automation has inferred workload to be high) when workload has in fact been reduced (and automation is no longer needed), or more problematically, returning a task to the human at a time when workload has just spiked. To shorten this control loop further, adaptive algorithms may invite an unreliable estimate in which the automated scheduler of task allocation may infer workload to be low when it is in fact high or vice versa. It is because of such challenges that the demonstrations of successful adaptive automation with complex systems are difficult to achieve (Chen, Visser, Huf, & Loft, 2017; Sauer, Chavaillaz, & Wastell, 2017).

21.5.2 AUTOMATION TRANSPARENCY

To the extent that a problematic automation failure response may result from a loss of SA of what automation is doing, then it is reasonable to suggest that providing greater transparency of the workings of automation should support human failure intervention. Indeed, recent empirical investigations of the automation transparency concept suggest far greater success in improving HAI than is the case with adaptive automation. The results of several recent studies, outside of the automobile domain, offer unqualified support for a transparency benefit to performance, relative to a control condition with the same automation (and failure type) without a transparent automated system. Of these, eleven studies found the benefits of automation transparency to be either amplified by or specific to the conditions of automation failure (Burns et al., 2008; Chen et al., 2018; Dzindolet et al., 2003; Lai, Macmillan, Daudelin, & Kent, 2006; Mayo et al., 2015; Mercado et al., 2016; Trapsilawati et al., 2017). Hoff and Bashir (2015) offer a comprehensive review of automation transparency successes. Hergeth, Lorenz, and Krems (2017), and Seppelt and Lee (2007) described examples of transparency in support of automobile drivers.

The concept of automation transparency is broad and somewhat ill defined, but can be made more concrete by characterizing four different forms:

- Graphic representation of what automation is doing (e.g., Burns et al., 2008; Mayo et al., 2015; Mercado et al., 2016; Trapsilawati et al., 2017)
- Textual descriptions of how automation is doing it (e.g., the reasoning behind automated decision aids; Mercado et al., 2016)
- Clear presentation of the raw data being processed by automation (Trapsilawati et al., 2017)
- Estimates of automation's own degree of uncertainty in its performance (Chen et al., 2018)

The last of these is similar to the likelihood alert that signals the confidence level of an alert system that a dangerous condition exists (Wizorek & Manzey, 2014). Furthermore, in the case of textual explanations of automation functioning and reasoning, these may be offered either online, at the time a particular automation decision or diagnosis is reached, or off-line prior to the use of automation, as a form of training and instruction.

There is one important limitation of online automation transparency, and that is that it may provide a source of distraction or extra perceptual or cognitive workload that could neutralize or offset the very benefits that automation is intended to provide, particularly offsetting benefits to performance of the automation-supported task.

21.5.3 TRAINING

The last manifestation of automation transparency described above, off-line explanation of reasoning, can be thought of as a form of automation training. Like transparency in general, training has been found to be successful in buffering some of the negative effects of automation failure response, and can be offered in different forms. In the medical domain, simple instructions regarding operation of a decision aid have been observed to increase the use of decision aids (Lai et al., 2006). In process control automation, as noted above, pre-exposure to automation failures during training can buffer the negative effects of a first failure (Bahner, Hüper, & Manzey, 2008; Manzey et al., 2012; Sauer et al., 2016). Training and automation is further described in this Handbook (Chapter 18).

21.6 CONCLUSIONS

Decades of research and accident analysis have revealed that the overall benefits of automation can sometimes be mitigated by their costs, as these are often associated with imperfect reliability, leading to OOTLUF. In safety-critical environments, designers and regulators should seek solutions such as adaptive or adaptable automation and transparency to mitigate the consequences of automation errors to HAI performance. To address these consequences as much as possible, lessons learned from other domains can assist the incorporation of safe automation into the automobile; but in transferring knowledge and techniques between domains, the designer must be cognizant of the many differences between the highway driving domain and those such as aviation and process control, where the solutions described in this chapter have been identified and evaluated.

REFERENCES

Bahner, J. E., Hüper, A. D., & Manzey, D. (2008). Misuse of automated decision aids: Complacency, automation bias and the impact of training experience. *International Journal of Human-Computer Studies, 66*(9), 688–699.

Billings, C. (1997). *Aviation Automation: The Search for a Human-Centered Approach.* Englewood Cliffs, NJ: Erlbaum.

Bliss, J. (2003). Investigation of alarm-related accidents and incidents in aviation. *International Journal of Aviation Psychology, 13*, 249–268.

Burns, C. M., Skraaning, G., Jamieson, G. A., Lau, N., Kwok, J., Welch, R., & Andresen, G. (2008). Evaluation of ecological interface design for nuclear process control: Situation awareness effects. *Human Factors, 50*, 663–679.

Casner, S. M., Geven, R. W., Recker, M. P., & Schooler, J. W. (2014). The retention of manual flying skills in the automated cockpit. *Human Factors, 56*(8), 1506–1516.

Chen, J. Y. & Barnes, M. J. (2012). Supervisory control of multiple robots: Effects of imperfect automation and individual differences. *Human Factors, 54*(2), 157–174.

Chen, J. Y., Lakhman, S., Stowers, K., Sellpwotz, A., Wright, J., & Barnes, M. (2018). Situation awareness based agent transparency and human-autonomy teaming effectiveness. *Theoretical Issues in Ergonomics Science, 19*, 259–282.

Chen, S., Visser, T., Huf, S., & Loft, S. (2017). Optimizing the balance between task automation and human manual control in a simulated submarine track management. *Journal of Experimental Psychology: Applied, 23*, 240–262.

Christensen, J. C. & Estepp, J. R. (2013). Coadaptive aiding and automation enhance operator performance. *Human Factors, 55*(5), 965–975.

Commercial Air Safety Team. (2014). *Airplane State Awareness.* JSAT Final Report.

Cummings, M. L. & Guerlain, S. (2007). Developing operator capacity estimates for supervisory control of autonomous vehicles. *Human Factors, 49*, 1–15.

Dixon, S. R. & Wickens, C. D. (2006). Automation reliability in unmanned aerial vehicle flight control: A reliance-compliance model of automation dependence in high workload. *Human Factors, 48*, 474–486.

Dixon, S. R., Wickens, C. D., & McCarley, J. M. (2007). On the independence of reliance and compliance: Are false alarms worse than misses? *Human Factors, 49*, 564–572.

Dorneich, M. C., Rogers, W., Whitlow, S. D., & DeMers, R. (2016). Human performance risks and benefits of adaptive systems on the flight deck. *International Journal of Aviation Psychology, 26*, 15–35.

Dornheim, M. A. (1995). Dramatic incidents highlight mode problems in cockpits. *Aviation Week and Space Technology, 142*(5), 57–59.

Dunlosky, J., Rawson, K. A., Marsh, E. J., Nathan, M. J., & Willingham, D. T. (2013). Improving students' learning with effective learning techniques: Promising directions from cognitive and educational psychology. *Psychological Science in the Public Interest, 14*(1), 4–58.

Dzindolet, M., Peterson, S., Pomranky, R., Pierce, L., & Beck, H. (2003). The role of trust in automation reliance. *International Journal of Human Computer Studies, 58*, 697–718.

Endsley, M. R. (2017). From here to autonomy: Lessons learned from human-automation research. *Human Factors, 59*, 5–27.

Endsley, M. R. & Kiris, E. O. (1995). The out-of-the-loop performance problem and level of control in automation. *Human Factors, 37*, 381–394.

Eriksson, A. & Stanton, N. A. (2017). Takeover time in highly automated vehicles: Noncritical transitions to and from manual control. *Human Factors, 59*(4), 689–705.

Ferris, T., Sarter, N., & Wickens, C. (2010). Cockpit automation: Still struggling to keep up. In E. Salas & D. Maurino (Eds.), *Human Factors in Automation* (2nd ed.). Amsterdam, The Netherlands: Elsevier.

Garg, A. X., Adhikari, N. K., McDonald, H., Rosas-Arellano, M. P., Devereaux, P., & Beyene, J. (2005). Effects of computerized clinical decision support systems on practitioner performance and patient outcomes. *Journal of the American Medical Association, 293*, 1223–1238.

Hergeth, S., Lorenz, L., & Krems, J. F. (2017). Prior familiarization with takeover requests affects drivers' takeover performance and automation trust. *Human Factors, 59*(3), 457–470.

Hoff, K. & Bashir, M. (2015). Trust in automation integrating empirical evidence on factors that influence trust. *Human Factors, 57*, 407–434.

Kaber, D. B. & Endsley, M. (2004). The effects of level of automation and adaptive automation on human performance, situation awareness and workload in a dynamic control task. *Theoretical Issues in Ergonomics Science, 5*, 113–153.

Kaber, D. B. & Kim, S. H. (2011). Understanding cognitive strategy with adaptive automation in dual-task performance using computational cognitive models. *Journal of Cognitive Engineering and Decision Making, 5*, 309–331.

Kaber, D. B., Onal, E., & Endsley, M. R. (1999). Level of automation effects on telerobots performance and human operator situation awareness and subjective workload. In M. W. Scerbo & M. Mouloua (Eds.), *Automation Technology and Human Performance: Current Research and Trends* (pp. 165–170). Mahwah, NJ: Erlbaum.

Kaber, D. B., Wright, M. C., Prinzel, L. J., & Clamann, M. P. (2005). Adaptive automation of human-machine system information-processing functions. *Human Factors, 47*, 730–741.

Lai, F., Macmillan, J., Daudelin, D. H., & Kent, D. M. (2006). The potential of training to increase acceptance and use of computerized decision support systems for medical diagnosis. *Human Factors, 48*(1), 95–108.

Landry, S. (2009) Flight deck automation. In S. Nof (Ed.), *Handbook of Automation* (pp. 1215–1239). Berlin: Springer.

Lee, J. D. & Moray, N. (1992). Trust, control strategies and allocation of function in human-machine systems. *Ergonomics, 35*, 1243–1270.

Li, H., Wickens, C. D., Sarter, N., & Sebok, A. (2014) Stages and levels of automation in support of space teleoperations. *Human Factors, 56*(6), 1050–1061.

Manzey, D., Reichenbach, J., & Onnasch, L. (2012). Human performance consequences of automated decision aids: The impact of degree of automation and system experience. *Journal of Cognitive Engineering and Decision Making, 6*, 1–31.

Martensson, L. (1995). The airplane crash at Gottrora: Experiences of the cockpit crew. *International Journal of Aviation Psychology, 5*, 305–326.

Mayo, M., Kowalczyk, N., Liston, B., Sanders, E., White, S., & Patterson, E. (2015) Comparing the effectiveness of alerts and dynamically annotated visualizations (DAVs) in improving clinical decision making. *Human Factors, 57*, 1002–1014.

Mercado, J. E., Rupp, M. A., Chen, J. Y., Barnes, M. J., Barber, D., & Procci, K. (2016). Intelligent agent transparency in human–agent teaming for Multi-UxV management. *Human Factors, 58*(3), 401–415.

Merritt, S., Lee, D., Unnerstall, J., & Huber, K. (2015). Are well calibrated users effective users? Association between calibration of trust and performance on an automated aiding task. *Human Factors, 57*, 34–47.

Metzger, U. & Parasuraman, R. (2005). Automation in future air traffic management: Effects of decision aid reliability on controller performance and mental workload. *Human Factors, 47*, 35–49.

Meyer, J. (2001). Effects of warning validity and proximity on responses to warnings. *Human Factors, 43*, 563–572.

Meyer, J. (2004). Conceptual issues in the study of dynamic hazard warnings. *Human Factors, 46*, 196–204.

Meyer, J. & Lee, J. D. (2013). Trust, reliance, and compliance. In J. D. Lee & A. Kirlik (Eds.), *The Oxford Handbook of Cognitive Engineering* (pp. 109-124). New York: Oxford University Press.

Molloy, R. & Parasuraman, R. (1996). Monitoring an automated system for a single failure: Vigilance and task complexity effects. *Human Factors, 38*, 311–322.

Moray, N. & Inagaki, T. (2000). Attention and complacency. *Theoretical Issues in Ergonomics Science, 1*, 354–365.

Morrow, D., North, R., & Wickens, C. D. (2006). Reducing and mitigating human error in medicine. *Reviews of Human Factors and Ergonomics, 1*, 254–296.

Mosier, K. L., Skitka, L. J., Heers, S., & Burdick, M. (1998). Automation bias: Decision-making and performance in high-tech cockpits. *International Journal of Aviation Psychology, 8*, 47–63.

Mumaw, R. J. (2018). *Addressing Mode Confusion Using an Interpreter Display* (Contractor Technical Report). Moffett Field, CA: San Jose State University Research Foundation. doi:10.13140/RG.2.2.27980.92801

Onnasch, L., Wickens, C. D., Li, H., & Manzey, D. (2014). Human performance consequences of stages and levels of automation: An integrated meta-analysis. *Human Factors, 56*(3), 476–488.

Parasuraman, R. & Manzey, D. (2010). Complacency and bias in human use of automation: An attentional integration. *Human Factors, 52*, 381–410.

Parasuraman, R., Mouloua, M., & Hilburn, B. (1999). Adaptive aiding and adaptive task allocation enhance human-machine interaction. In M. W. Scerbo & M. Mouloua (Eds.), *Automation Technology and Human Performance: Current Research and Trends* (pp. 119–123). Mahwah, NJ: Lawrence Erlbaum.

Parasuraman, R. & Riley, V. (1997). Humans and automation: Use, misuse, disuse, abuse. *Human Factors, 39*(2), 230–253.

Parasuraman, R., Sheridan, T. B., & Wickens, C. D. (2000). A model for types and levels of human interaction with automation. *IEEE Transactions on Systems, Man, & Cybernetics: Part A: Systems and Humans, 30*(3), 286–297

Parasuraman, R., Sheridan, T. B., & Wickens, C. D. (2008). Situation awareness, mental workload, and trust in automation: Viable, empirically supported cognitive engineering constructs. *Journal of Cognitive Engineering and Decision Making, 2*(2), 140–160.

Rein, J. R., Masalonis, A. J., Messina, J., & Willems, B. (2013). Meta-analysis of the effect of imperfect alert automation on system performance. *Proceedings of the Human Factors and Ergonomics Society Annual Meeting* (Vol. 57, No. 1, pp. 280–284). Los Angeles, CA: SAGE Publications.

Rovira, E., Pak, R., & McLaughlin, A. (2017). Effects of individual differences in working memory on performance and trust with various degrees of automation. *Theoretical Issues in Ergonomics Science, 18*(6), 573–591.

SAE. (2016). Taxonomy and definitions for terms related to driving automation systems for on road motor vehicles (SAE Standard J3016 201609). Warrendale, PA: SAE International.

Sarter, N. B. (2008). Investigating mode errors on automated flight decks: Illustrating the problem-driven, cumulative, and interdisciplinary nature of human factors research. *Human Factors, 50,* 506–510.

Sauer, J., Chavaillaz, A., & Wastell, D. (2016). Experience of automation failures in training: Effects on trust, automation bias, complacency and performance. *Ergonomics, 59*(6), 767–780.

Sauer, J., Chavaillaz, A., & Wastell, D. (2017). On the effectiveness of performance-based adaptive automation. *Theoretical Issues in Ergonomics Science, 18*(3), 279–297.

Sauer, J., Kao, C., & Wastell, D. (2012). A comparison of adaptive and adaptable automation under different levels of environmental stress. *Ergonomics, 55,* 840–853.

Sauer, J. & Ruttinger, B. (2007). Automation and decision support in interactive computer products. *Ergonomics, 50,* 902–909.

Seagull, F. J. & Sanderson, P. M. (2001). Anesthesiology alarms in context: An observational study. *Human Factors, 43,* 66–78.

Sebok, A. & Wickens, C. D., (2017). Implementing lumberjacks and black swans into model-based tools to support human-automation interaction. *Human Factors, 59,* 189–202.

Sebok, A., Wickens, C. D., Sarter, N., Quesada, S., Socash, C., & Anthony, B. (2012). The automation design advisor tool (ADAT): Development and validation of a model-based tool to support flight deck automation design for NextGen operations. *Human Factors and Ergonomics in Manufacturing and Service Industries, 22,* 378–394.

Seppelt, B. D. & Lee, J. D. (2007). Making adaptive cruise control (ACC) limits visible. *International Journal of Human-Computer Studies, 65*(3), 192–205.

Sheridan, T. B., & Verplank, W. L. (1978). *Human and Computer Control of Undersea Teleoperators.* (Technical Report, Man-Machine Systems Laboratory, Department of Mechanical Engineering). Cambridge, MA: MIT Press.

Slamecka, N. J. & Graf, P. (1978). The generation effect: Delineation of a phenomenon. *Journal of Experimental Psychology: Human Learning, Memory, and Cognition, 4,* 592–604.

Strobhar, D. (2012). *Human Factors in Process Plant Operation.* New York: Monument Press.

Trapsilawati, F., Wickens, C. D., Chen, C.-H., & Qu, X. (2017) Transparency and conflict resolution automation reliability in Air Traffic Control. In P. Tsang, M. Vidulich & J. Flach (Eds.), *Proceedings of the 2017 International Symposium on Aviation Psychology* (pp. 419–424). Dayton, OH: ISAP.

Trapsilawati, F., Wickens, C. D., Qu, X., & Chen, C.-H. (2016). Benefits of imperfect conflict resolution advisory aids for future air traffic control. *Human Factors, 58,* 1007–1019.

Wickens, C. D. (2017). Stages and levels of automation: 20 years after. *Cognitive Engineering and Decision Making, 12*(1), 35–41.

Wickens, C. D., Clegg, B. A., Vieane, A. Z., & Sebok, A. L. (2015a). Complacency and automation bias in the use of imperfect automation. *Human Factors, 57*(5), 728–739.

Wickens, C. D. & Dixon, S. R. (2007). The benefits of imperfect diagnostic automation: A synthesis of the literature. *Theoretical Issues in Ergonomics Science, 8*(3), 201–212.

Wickens, C. D., Mavor, A., Parasuraman, R., & McGee, J. (1998). *The Future of Air Traffic Control: Human Operators and Automation.* Washington, DC: National Academy Press.

Wickens, C. D. & McCarley, J. (2017). Commonsense statistics in aviation safety research. In P. Tsang, M. Vidulich, & J. Flach (Eds.), *Advances in Aviation Psychology: Volume 2* (pp. 98-110). Dorset, UK: Dorset Press.

Wickens, C. D., Sebok, A., Li, H., Gacy, A., & Sarter, N. (2015b). Using modeling and simulation to predict operator performance and automation-induced complacency with robotic automation: A case study and empirical validation. *Human Factors, 57*, 959–975.

Wickens, C. D., Sebok, A., Walters, B., & McCormick, P. (2017). *Alert Design Considerations* (FAA Technical Report). Boulder, CO: Alion Science & Technology.

Wiener, E. L. (1988). Cockpit automation. In E. L. Wiener & D. C. Nagel (Eds.), *Human Factors in Aviation* (pp. 433–461). San Diego, CA: Academic Press.

Wiener, E. L. & Curry, R. E. (1980). Flight deck automation: Promises and problems. *Ergonomics, 23*, 995–1012.

Wizorek, R. & Manzey, D. (2014). Supporting attention allocation in multi task environments: Effects of likelihood alarm systems on trust behavior and performance. *Human Factors, 56*, 1209–1221.

Yeh, M., Merlo, J. L., Wickens, C. D., & Brandenburg, D. L. (2003). Head up versus head down: The costs of imprecision, unreliability, and visual clutter on cue effectiveness for display signaling. *Human Factors, 45*, 390–407.

Yeh, M. & Wickens, C. D. (2001). Attentional filtering in the design of electronic map displays: A comparison of color coding, intensity coding, and decluttering techniques. *Human Factors, 43*, 543–562.

Yeh, M., Wickens, C. D., & Seagull, F. J. (1999). Target cuing in visual search: The effects of conformality and display location on the allocation of visual attention. *Human Factors, 41*, 524–542.

22 HF Considerations When Testing and Evaluating ACIVs

Sheldon Russell and Kevin Grove
Virginia Tech Transportation Institute Center
for Automated Vehicle Systems

CONTENTS

Key Points 474
22.1 Introduction 474
    22.1.1 Testing Approach: Avoiding Significant But Not Meaningful 477
    22.1.2 Driving Automation Characterization 477
        22.1.2.1 Automated Features 478
        22.1.2.2 Request to Intervene 479
        22.1.2.3 Other Alerts 479
        22.1.2.4 Connected Vehicle Capabilities 479
    22.1.3 Participant Training 480
    22.1.4 Training and Improper Use 481
22.2 Commercial Vehicle Testing 481
    22.2.1 Drayage 482
    22.2.2 Platooning 482
22.3 Testing Early in Development: Data Analysis and Driving Simulation 483
    22.3.1 Naturalistic Driving Data for Scenario Development 484
    22.3.2 Driving Simulator Testing 486
22.4 Mid-Late Testing: On-Road Experiments 487
    22.4.1 Scenario Selection 487
    22.4.2 WO Approaches 488
22.5 Late Stage Testing: NDSs 488
    22.5.1 Participant Selection 490
    22.5.2 Data Sampling & Reduction 490
        22.5.2.1 Data Sampling 490
        22.5.2.2 Data Reduction 491
22.6 Conclusion 492
Acknowledgments 493
References 494

KEY POINTS

- Accurately characterizing the automated driving system(s) present in a platform is critical for testing, allowing participant training materials to be developed that accurately inform participants about system function, and aids in the selection of testing scenarios throughout the development process.
- Testing of commercial vehicles involves unique operational domains as well and specialized automation such as platooning.
- Methods for testing in early development are expected to be iterative, and to build a foundation for later stages of prototype testing.
- Analysis of existing naturalistic driving data will allow testing of models of automated features to undergo bench testing, and also aid in selection of testing scenarios.
- Simulator testing provides a method of iterative testing of early feature designs with a large degree of experimental control, but reduced external validity.
- Mid-development testing approaches should include on-road testing, using a prototype or Wizard of Oz approach to increase external validity of testing, at the expense of iterative testing.
- Late stage testing (including post-development) can include naturalistic driving studies in which large datasets of drivers actively using automation are collected.

22.1   INTRODUCTION

The purpose of this chapter is to provide an overview for feature development and human subjects' testing of various aspects of automated driving systems. Considerations for driving automation when testing heavy commercial vehicles (e.g., tractor trailers, busses) are also included. Although they can exist as stand-alone systems, connected vehicle features are not covered separately in this chapter; for the purposes of testing they are described here as a feature of an automated driving system. Other Level 1 features (e.g., automated emergency braking, blind spot warning, etc.) may also be part of the driving system, but are referred to generally here as Advanced Driver Assistance Systems (ADAS).

The terminology used in this chapter is intended to be generally consistent with SAE J3016 in referring to the driving automation system and/or features of said system (rather than a vehicle) (SAE International, 2016). However, some distinctions in classification are noted. For the purposes of this chapter driving automation is considered to be any sustained automation of both lateral and longitudinal functions (i.e., Level 2 and above). Furthermore, SAE J3016 defines a specific type of alert unique to automated driving systems, a Request to Intervene (RTI). An RTI is an alert or notification from the automation system to the driver that an intervention is needed. Per SAE J3016, RTIs are defined in the context of Level 3 (or above) systems; however, one could consider "hands-on wheel type alerts" designed to keep the driver engaged in the driving task as RTIs as well (insofar as an intervention is requested from the driver; Russell et al., 2018).

Previous chapters have covered issues that are fundamental to the human driver, driving automation, driving in general, as well as potential problems associated with driving automation and connected vehicle systems. Potential solutions to these problems such as driver training (this Handbook, Chapter 18), driver monitoring (this Handbook, Chapter 11), and human–machine interface (HMI) designs have also been presented (this Handbook, Chapters 15, 16). No matter how principled, any potential solution is just that until it is tested and evaluated. This chapter will include high-level overviews of steps critical to testing of driving automation systems. These steps include system characterization, commercial vehicle considerations, and testing methodologies (e.g., driving simulator, test track or live road experiments, and naturalistic driving).

SAE standard J3018 provides some guidance for testing at automation Levels 3 and above (SAE International, 2015a). The standard advocates a graduated approach where the expertise required for testing scenarios decreases over the course of testing, while the complexity of the testing scenario increases over the course of development. The standard provides definitions for expert test drivers, experienced test drivers, and novice test drivers. Expert test drivers are typically engineers who are designing the automated features themselves and can interact with the systems at the mechanical and software level; experienced drivers are trained on the systems but are unable to interact at the software level; and finally novice drivers have received only cursory training (if any). Testing locations and road scenarios should also be graduated in complexity; potential testing variables listed in the standard include

- Location
  - Test track
  - Closed campus operations (e.g., military base, corporate or university campus)
  - Public roads
- Roadway Type
  - Limited access freeway
  - Highway (single or multi-lane),
  - Arterial roads
  - Residential streets
  - Driveway, parking lot, or structure
- Traffic Environment
  - Traffic density
  - Vehicles
  - Pedestrians
  - Signage
  - Irregular—construction, crash scenes, road detours, flooding
  - Complex intersections, merges
  - Regional variations in road design
  - Traffic control devices (signals, signs, curbs, guardrails, etc.)
- Time of Day
  - Lighting conditions (day vs. night)
- Seasonal
  - Weather conditions

While providing some guidance, SAE J3018 does not provide any specific methodology for testing during phases. Furthermore, a system and associated features will undoubtedly go through many different tests throughout their development cycle. Assuming that a system is being developed from start to finish, the methodologies described herein are intended to build upon one another across the development cycle (e.g., data modeling in early development leads to scenarios for driving simulator testing, production systems are then tested via naturalistic driving studies (NDSs)). The scope of graduated testing is broad; it includes tests that may not consider the driver (i.e., engineering evaluations). While these tests are critical to the development cycle, the scope of this chapter is focused toward human subjects testing. It is assumed that engineering evaluations of the systems and components have been completed, and the features themselves are operable.

Testing is not limited to systems early in the development cycle. Testing of post-development cycle (i.e., commercially available) driving automation will always be necessary. Post-development testing can be categorized into two primary focuses of post-release monitoring and testing of broader safety benefits. Post-release monitoring refers to maintaining system reliability and performance once features are out in the world, with a focus on "black box" or vehicle data monitoring by an original equipment manufacturer (OEM) or other researcher. Analysis of this data may lead to over-the-air software updates (e.g., Tesla autopilot software changes) or otherwise inform future system designs. Furthermore, automated systems that have been deployed on public roads may allow for testing safety benefits within the larger transportation system, for example, crash rate comparisons between vehicles equipped with driving automation and non-equipped vehicles, which may then lead to rulemaking and/or policy considerations. Although distinct, these two approaches are complimentary. Post-release monitoring and safety benefit testing results may lead to insights into the overall capabilities and limitations at each level of automation and understanding limitations of current platforms should inform the design of future iterations of driving automation. Figure 22.1 provides a summary of testing throughout the development cycle, beginning with heuristic evaluation of the concept, usability testing of a prototype, user studies of pre-production models, and in-service monitoring of the released product.

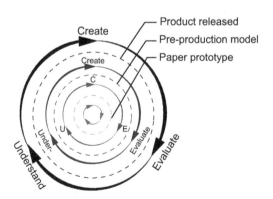

FIGURE 22.1 Overview of the development cycle. (From Lee, Wickens, Liu, & Boyle, 2017.)

22.1.1 TESTING APPROACH: AVOIDING SIGNIFICANT BUT NOT MEANINGFUL

Avoiding non-meaningful results is an unstated goal that is key to scientific discovery in general. Still it bears repeating, that when testing driving automation systems, it is critical that the researchers make their best effort to avoid focusing on statistical significance at the expense of practical significance, even if the result is a non-significant statistical test. The testing methodology, the driving scenario tested, and the participant characteristics will all determine the nature of results obtained and will all be critical for validity of testing. Additionally, there is a clear distinction in testing approaches; attempting to troubleshoot or find failure modes will lead to a different testing procedure than testing theory using inferential statistics (Lee et al., 2017). It will be up to the researchers designing and conducting the tests to select an appropriate series of tests as part of feature development and/or evaluation. Not an easy task to be sure. Methodologies should be layered, and testing scenarios should be designed that cover a broad spectrum of situations and use cases. Scenarios may be orchestrated in order to test a known limitation or edge case of the system. Alternatively, scenarios may be exploratory, in that they are designed to uncover edge cases rather than test them. The following sections include discussions of the basic underpinnings of system characterization and participant training which are relevant to nearly any test methodology.

22.1.2 DRIVING AUTOMATION CHARACTERIZATION

At the risk of stating the obvious, it is of utmost importance before any tests are designed that the intended function of the driving automation system is clearly understood by the researchers, a process referred to here as characterization. Whether the system of interest is in the conceptual phase or already in production, an understanding of the system functions should be developed by the researcher(s). Characterization supports multiple aspects of testing and evaluation, at multiple phases of development. In particular, it serves as a heuristic evaluation of a system, helps guide research question development and testing procedures, serves as a foundation for training participants on system function, and finally guides data sampling and reduction in naturalistic studies. Characterization includes identifying the expected performance envelope(s) (e.g., speed ranges, steering torque, etc.), the timing, modality, and type(s) of the RTIs, the presence of ADAS (e.g., collision mitigation braking), alerts associated with ADAS features (e.g., forward collision warning—FCW, blind spot warning), and so on. Characterization needs will vary and evolve depending on the stages of development and testing (e.g., early, mid, late) as capabilities are refined.

Any particular make, model, or brand of feature will have qualities that will alter driver interactions with the systems, and therefore inform scenario design. For example, a lateral automation feature that requires longitudinal automation to activate may have different use patterns than a lateral feature that can activate alone. HMI design considerations, such as RTI characteristics (e.g., multi-modal), cluster display, and type (e.g., screen size, head-up vs. head down, etc.) are likely to differ across different system designs (see e.g., this Handbook, Chapter 15). Different monitoring methods may allow for a sequence of activities that "defeat" the monitoring system or provide an avenue for improper use of the automation (e.g., steering torque detection vs. driver gaze detection; see also, this Handbook, Chapter 11).

Although a critical step, there is no one right way to go about characterizing the driving system; the approach to characterization will vary based on the overall intent of the tests. Essentially, individuals evaluating a system should let their research questions guide the level and depth of characterization needed. The approach laid out in this chapter should be considered only as a starting point. Guidelines have been put forward for characterizing interface designs for driving automation (Campbell et al., 2018), which were based on the output from early studies of Level 2 and 3 systems (Blanco et al., 2015).

22.1.2.1 Automated Features

First and foremost, in terms of characterization, is the type of driving automation features that are present in a particular platform (also see this Handbook, Chapter 2). As described in SAE J3016, a vehicle may have multiple features that operate at different levels of automation in different combinations of activation or different operational driving domains (ODDs). There are any number of other items that may be of interest to a particular test or research question; a short example characterization is included in Table 22.1. The type of feature, speed ranges of activation, methods of activation, and whether or not the feature can activate alone are all important details for characterization. Although included in the table as Level 3, low-speed traffic jam assist features may or may not be classified as Level 3 by the manufacturer, which could even vary based on location.

For commercially available systems, a great deal of information may be contained in manufacturer sources (e.g., owner's manuals, manufacturer's website); this may include the specified level of automation for a system. As part of characterization, experimenters should operate systems and verify that the published specifications

---

TABLE 22.1

Example Characterization of Features for Level 1, Level 2, and Limited Level 3 Capability

Feature	Lateral/ Longitudinal	Speed	Activation Method	Can Be Activated Alone	SAE Level	Alerts
Adaptive cruise control	Continuous longitudinal support		Steering wheel button	Yes	1	FCW
Lane centering	Continuous lateral support	Above 40 mph	Automatic when speed is crossed; system setting to disable	No—Requires ACC	2	RTI
Traffic jam pilot	Continuous lateral and longitudinal support	Below 35 mph	Steering wheel button; HMI notifies when available	No	3	RTI

for the system are accurate and if there are affordances from the system, such as extended periods of hands-off driving or feature activation on improper road types, which are not within the intended use but are nonetheless available to the driver. This type of characterization helps to determine the most likely avenues of misuse or other improper use cases that may be observed in naturalistic settings or otherwise require testing.

22.1.2.2 Request to Intervene

RTIs are notifications to the driver that the driving automation requires an intervention. To the best of the researchers' abilities, the source or trigger for the RTI should be characterized (e.g., hands-off wheel detection, driver gaze tracking, and/or external conditions). It may be determined that these alerts may be generated by vehicle software based on combinations of different variables. The timing, modality, and the identified trigger for the RTI should be noted for the testing platform, an example is included in Table 22.2.

22.1.2.3 Other Alerts

In some cases, alerts or notifications and driver responses to these alerts will be the primary focus of study. Alerts for novel or new applications, such as connected vehicle alerts should be characterized insofar as the alert triggering conditions, modality, etc. of the alert are known to the researchers. Alert specifications not only allow for interpretation of the data and scenario development but also provide information that may be explained to participants as part of training.

Alerts may not be of primary interest to a testing scenario. However, the nature and presence of alerts should be noted (e.g., FCW, Blind Spot Warnings, Lane Departure Warnings). Again, the triggering conditions and operational ranges, should be understood by the research team in order to provide information to participants as necessary. Particularly for a naturalistic research study, alerts are likely to be encountered by a participant driver.

22.1.2.4 Connected Vehicle Capabilities

In addition to characterizing automated features, the presence and characterization of any connected vehicle technology should be included as part of the system characterization. Connected vehicle technology enables communication between vehicles (vehicle to vehicle, V2V) and between vehicles and infrastructure

TABLE 22.2
Example RTI Characterizations

	Source (Alert Trigger)	Number of Stages	Stage Duration (seconds)	Total Duration (seconds)	Consequence
Level 2 RTI	Steering wheel torque	2	15	30	Lane centering is disabled for duration of trip
Level 3 RTI	External conditions	3	15	45	Vehicle slows to a stop in lane

(vehicle to infrastructure, V2I) using short-range radio frequencies. This communication can be used by various in-vehicle systems and/or driving automation to provide information to the driver/automated system, such as crash warning system (CWS) alerts. V2I features may provide navigation, variable road signage, or other infrastructure-based information to the driver and/or driving automation system (also see Chapters 2 and 19).

Although existing radar and camera base systems can provide information on FCW situations, V2V messaging may allow for other types of CWS alerts, such as a left turn across path warning or an alert that a vehicle is crossing the intersection. This type of alert is likely novel to most drivers; as such aspects of the HMI should be noted. Among the aspects of the HMI to consider are the modalities (visual, auditory, and tactile) of the interface that are used to notify the driver of an impending crash conflict (also see Chapter 15).

22.1.3 PARTICIPANT TRAINING

No matter how participants are selected (paid volunteers from the surrounding community, university students seeking course credit), it's likely that at least some of them have heard the term "self-driving car" or "autonomous vehicle" used in news reports, YouTube videos, or other media, but have little practical experience with any driving automation. Participants may have already decided that they do not trust driving automation or have a preconceived notion of how well it will work (see also, this Handbook, Chapters 4, 5). A recent survey found 71% of participants were afraid to ride in a self-driving car (AAA Public Affairs, 2019). Generally speaking, participant training should be designed to provide information for a driver to safely and effectively operate the system(s) of interest, and not have participants rely only on these existing notions to guide their use of the automated systems. For example, Russell et al. (2018) created training for naturalistic driving participants to mimic an ideal dealership experience. It should go without saying that system characterization must inform training; providing inaccurate information may bias the results of testing or otherwise lead to improper use of driving automation (see Section 22.1.4).

The goals of the desired test must also be considered; different situations will have different instructional requirements. The level of detail provided in participant instructions or training should be tuned to the expected use case for the systems being tested (e.g., this Handbook, Chapter 18). For example, if the desired scenario is for an unfamiliar driver experiencing automation that unexpectedly activates (e.g., a rental car), providing little or no training to a participant may be necessary. If the test case is not intended to represent a completely unaware driver, providing no information to a participant will probably lead to situations where the participant is confused about how to operate the system or respond appropriately to system alerts or messages. Participants who receive too much (or overly technical) training may not accurately reflect the knowledge of the "typical" user. However, these results may generalize better to commercial vehicle drivers or other highly experienced populations. Likely training should be somewhere in between in most cases: a general description of the features, what they do, how they activate, and what alerts or messages may be displayed to the driver. A test drive where a participant can ask questions and experience the features for the first time is a second step that should help reinforce instructions.

Training considerations may not be limited to the use and activation of automation systems, it may be necessary to provide training for participants on any other unfamiliar or novel aspect of a test, such as a specific non-driving task. Even if the tasks are not novel, the results could be biased if the application of the task is confusing or misunderstood by the participant. Finally, if the goal of the test is to develop a training system or to compare different training methods, designing tests to compare the methods will be required. For example, a written or practical evaluation to determine what information was understood and/or retained from the training material by the participants.

22.1.4 Training and Improper Use

A key question for the rollout of driving automation will be how drivers use and adopt the systems; automation that is improperly used may lead to crashes. As noted previously, there may be scenarios where a system affords the driver behaviors that are outside the design intent, but otherwise possible (e.g., hands-off wheel, performing a non-driving task). Testing these affordances will require deliberate creation of improper use cases, such as asking a driver to perform a non-driving task during periods of automation use in a controlled setting. For example, Blanco et al. (2015) asked participants to respond to emails using an experimenter-provided tablet during Level 2 operation, and asked participants to watch videos during Level 3 operation. Russell, Atwood, and McLaughlin (in press) provided training to participants that described lateral automation capabilities above or below the actual capability (i.e., described a highly capable lane centering system when in reality the capability was more akin to a lane departure prevention system).

22.2 COMMERCIAL VEHICLE TESTING

The potential economic benefits of automated driving systems for commercial vehicles may lead commercial fleets to become early adopters of driving automation. These potential benefits include flexibility in driver's hours of service, reductions in labor costs, reduced liability for crashes, and fuel economy improvements (Poorsartep & Stephens, 2015). Commercial vehicle drivers may not have a choice when adopting driving automation; they may be assigned a tractor trailer that is already equipped with ADAS or other driver assistance technology. Existing research on commercial applications of ADAS systems (e.g., ACC, FCW, and Lane Departure Warning Systems) suggests that the systems still produce some false activations and extraneous low-level feedback which may frustrate drivers or reduce their trust in the systems (Grove, Atwood, Blanco, Krum, & Hanowski, 2017). It was also found that these drivers were willing to use low-level automation such as ACC in adverse weather conditions against manufacturer recommendations (Grove, et al., 2015) and that there was a small, measurable difference in their visual behavior while using ACC and following a lead vehicle (eyes-off-road time was slightly higher; Grove, Soccolich, Engstrom, & Hanowski, 2019). Recently the commercial vehicle industry has begun to recognize issues with driver confusion or frustration towards FCW and lane departure alerts due to variations between brands of systems, generations of system, and integration approaches (Technology & Maintenance Council, 2018). The best practices from these efforts may inform designs for higher levels of automation for commercial vehicle ADAS or commercial AVS (Technology & Maintenance Council, 2019).

In addition to issues with driver acceptance, commercial vehicles also introduce additional challenges for testing and development. Testing of commercial vehicles presents specific challenges in terms of ODDs as well as specialized automation that may be different than driving automation for passenger vehicles. For example, trucking operations that swap trailers may be limited to placing sensors only on the truck/tractor itself, creating visibility challenges unique to commercial vehicles. Commercial vehicles are also subject to roadside inspections, and automated systems may need to interact with state personnel who perform these inspections or their infrastructure. Commercial vehicles may also operate on private properties or non-public roadways as part of their operation where road markings are limited. Considering these issues during testing will be critical to uncovering the edge cases for driving automation for heavy vehicle applications. Additional examples of these challenges specific to heavy commercial trucking are described in the following sections.

22.2.1 DRAYAGE

One example of a non-public roadway usage for commercial vehicles is drayage. Drayage trucks typically work to transfer cargo between mixed modes of transportation, such as offloading cargo from a ship to then be transported by heavy truck or train. These operations are often conducted in a space with controlled access, such as a port of entry or rail yard facility. These characteristics make the domain attractive for heavy vehicle automation; the operation space is confined, there is little to no mixed traffic, there are relatively fixed transit routes, and there are lower speeds of operation (see Smith, Harder, Huynh, Hutson, & Harrison, 2012, for an overview of drayage operations and facilities). Drayage may also offer unique challenges to automated commercial vehicles as roadway infrastructure could differ significantly from public roads. Additionally, while most driving may take place in areas closed to public traffic, there may be some driving on public roads in order to get the cargo to a nearby destination for further transit. These transitions between private and public space may provide test cases of transfer of control between automated driving and manual driving.

22.2.2 PLATOONING

Another application of automation that may be unique to commercial vehicles is platooning. Platooning involves a series of vehicles following each other at relatively short headways for extended periods of time. Platooning is typically accomplished by V2V communication in order to synchronize throttle or steering and assist a driver, but future implementations could also include higher levels of automation which could operate without a driver present. Platooning offers the opportunity for the following vehicles to reduce airflow drag and improve fuel economy at higher speeds, which presents an economic opportunity in commercial vehicle operations that drive long distances at highway speeds. However, the following distances necessary to achieve optimal gains are expected to be relatively short, as low as 10 feet (Tsugawa, Jeschke, & Shladover, 2016) and as high as 50 feet (Lammert, Duran, Diez, & Burton, 2014).

The short headways necessary for platooning reduce the time in which the following vehicles can react and create blind spots in front of the following vehicles in which sensors cannot see beyond the leading vehicle. This can be overcome with communication between leading vehicles and following vehicles, but the lead vehicle may need to consider whether there is a platoon following it in choosing how to react to potential conflicts. Additionally, the points at which a human driver or the automation needs to engage/disengage a platoon are another area of potential testing unique to commercial vehicles. Depending on how platoons are designed to form or dissolve, they may involve a human that is in control at what would typically be considered an unsafe headway or the automation must transition over time from an independent state at a longer headway to a synchronized state at a shorter headway (or vice versa). These handover or transition points could lead to edge cases and testing requirements that are unique to commercial vehicles.

A final concern for platooning testing is how the systems should react to vehicles around a platoon. Light vehicles may attempt to cut between platooning trucks (depending on the following headway required for a platoon), and platoons may impact how surrounding vehicles attempt to enter, exit, or change lanes on a highway. Anticipating these behaviors around platoons and including them in testing scenarios will be critical for ensuring safe deployment in the future.

22.3 TESTING EARLY IN DEVELOPMENT: DATA ANALYSIS AND DRIVING SIMULATION

For the purposes of this chapter, early stages of testing are those prior to development of a working prototype in a moving vehicle. Data analysis and modeling along with simulator testing are reviewed briefly, as they relate to participant testing. These two methods can be implemented in an iterative fashion to improve feature performance before a prototype system is built. For example, consider a traffic scenario that may be familiar to the reader; a slowed vehicle is revealed to a driver. As outlined in Figure 22.2, three vehicles are traveling down a roadway (Panel 1), when the lead vehicle brakes. In certain circumstances, this slow/stopped vehicle cannot be seen by the driver of the third vehicle. When the middle vehicle changes lanes to avoid the slowed lead vehicle (Panel 2), the slow/stopped vehicle is revealed to the third driver who must then respond, potentially with an evasive maneuver (e.g., hard braking; panels 3 & 4).

As part of the development of a longitudinal automation feature, examples of this type of interaction can be selected from existing naturalistic driving datasets (e.g., the Second Strategic Highway Research Program (SHRP2); Dingus et al., 2014). Using the video, acceleration, global positioning system (GPS), and other data collected as part of the study, the various actors within the scenario can be modeled in a computer-based simulation (e.g., CarSim, MATLAB, or other software program).

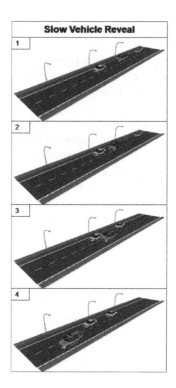

FIGURE 22.2   Slowed vehicle reveal scenario.

Automated features or other ADAS systems can then be modeled (for example, implementing a connected vehicle application that detects the lead vehicle slowing in the model). These modeled features can then be implemented in a driving simulator to test the features with a naïve driver; the driver response data can then be used to further refine model parameters, leading to additional tests.

22.3.1   Naturalistic Driving Data for Scenario Development

Testing of automated driving systems will require millions if not billions of miles to determine overall reliability and safety which would take tens or hundreds of years to achieve based on a recent RAND report (Kalra & Paddock, 2016). Analysis of the millions of miles of existing naturalistic driving data may speed up the testing process. The following chapter will cover the analysis of large datasets in further detail (this Handbook, Chapter 23); however, data analysis and modeling techniques are included here for their contributions to early stage testing and development of automated features. Mining existing datasets, determining points of comparison, and developing models and simulations of features can provide a valuable and easily repeatable method for testing features in early stages of development.

Although the end goal of driving automation systems may be to replace the human driver (or at least to augment the driver), automated systems will first be compared to the human driver they intend to replace. A reasonable place to start is to use existing data on typical drivers, namely naturalistic driving data, such as that found in the SHRP 2 dataset (Dingus et al., 2014). Most approaches to naturalistic data have been to study rates of safety-critical events (SCEs; crashes and near crashes) along with the factors present during each event.[1] The remaining data are used to select baseline cases without crashes for computing odds ratios and other analyses. Crashes are rare events overall, and the majority of the data are uneventful in the conventional sense. However, this "leftover" data provide a multitude of driving scenarios that could be informative for automated system development and testing. Essentially, system function can be compared to data from actual drivers in the scenario; as noted previously these scenarios can then be modeled in a computer-based simulation and system parameters tuned to desired performance specifications prior to testing with a human driver. This is an emerging approach for driving automation system development, and most work is being conducted in the private realm but is conceptually similar to hardware in the loop simulation (see e.g., Murray-Smith, 2012).

To what standard should a system be tested? Should a system be developed to be similar to the mean following distance? What about following closer? Why not choose one of the outliers for an added margin of safety? There are no right answers to any of these questions; but as a researcher, decisions will need to be made about how the system should be tested, and this decision will have an impact on the outcome of the test. There are a wide variety of driving styles; and driving style is going to change depending on the surrounding traffic, weather and environmental factors, and so on. This makes even what is on the face a simple question, such as how closely should a longitudinal feature follow lead traffic, more challenging. At some point in development, it has to be determined what type of driving the automated system is going to achieve. Figure 22.3 shows boxplots calculated from following distance

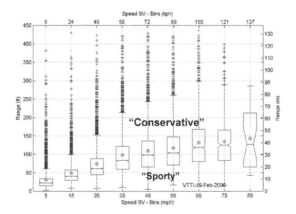

FIGURE 22.3 Boxplots of mean following distance data for 10,000 trips from the 100-car NDS with range (*Y* axis) and speed (*X* axis) in metric and standard units.

---

[1] Note that some controversy surrounds the use of near crashes to understand the factors that influence crashes (Guo, Klauer, McGill, & Dingus, 2010; Knipling, 2015)

observed during the 100-car NDS (Dingus et al., 2006). Data are shown for 10,000 trips, binned across the speed ranges on the *x*-axis. Based on the following distance classification, those on the higher ranges could be classified as "conservative" with the shorter following distances considered "sporty."

Consider the example of a slowed vehicle reveal scenario (Figure 22.2) and designing vehicle automation that can resolve the scenario safely. Using this type of driving data, iterative models of different following distance envelopes can be used to narrow down design considerations (e.g., following at a safe distance to resolve a vehicle reveal). Again, analysis methods for large datasets are covered in further detail in other chapters (this Handbook, Chapter 23). For the purpose of iterative testing, upon narrowing the feature parameters, testing with human operators in a driving simulator is likely appropriate.

22.3.2 DRIVING SIMULATOR TESTING

The majority of published tests using human subjects and vehicle automation utilize a driving simulator of some type. Driving simulator options run the gamut from simple monitors on a tabletop computer, large screens with a typical vehicle interior setup, to immersive driving simulators with actual movement (instead of visual-only motion cues) (see Weir, 2010, for an overview driving simulator applications in relation to HMI development). Testing using a driving simulator has some advantages, namely cost and logistics, when compared with testing on live roads. Re-programming of simulations allows a researcher to collect data for a wide variety of different conditions, features, scenarios, and so on.

Aside from issues of motion sickness (see Brooks et al., 2010, for an overview of simulator sickness and potential detection and mitigation), simulators pose little safety threat to a participant. However, this can be both a blessing and a curse for testing. On one hand, a crash imminent scenario can be implemented without any real danger; but without any consequence for an incorrect decision a driver's response may not be representative of what might be observed on real roads (Ranney, 2011). The key then is making sure the test of interest is matched with the simulator. Specifically, the fidelity of the simulator and the accuracy at which automated features are replicated need to be matched. Even if there is some non-visual motion cue, it is likely that it does not have the same acceleration profile as a vehicle traveling on an actual road surface. This lack of a realistic lateral motion cue may delay driver responses to lateral automation problems such as lane drifts (Kaptein, Theeuwes, & Van Der Horst, 1996).

Since the test environment is programmed at the direction of the experimenter, it certainly allows for more overall experimenter control than naturalistic data collection. The scenarios can play out exactly the same way for different participants. This may be particularly useful in early stages of system development or testing, in which individual aspects of a system are being tested iteratively and rapidly (e.g., alert timings, tones, icons). Physical configurations and layout changes can also be implemented iteratively in a simulator where mistakes made have little consequence. Still, at some point, testing must move onto live road testing using vehicle-based driving automation. These early steps should establish design guidance for on-road system development.

22.4 MID-LATE TESTING: ON-ROAD EXPERIMENTS

Early stages of testing are essential for development and refinement of system parameters, physical layout of controls and displays, HMI messages and timing, and so on. Early and iterative testing approaches should lead to a better initial design of a testable prototype system. Even based on the optimum modeling and simulator tests, a system will require further testing on live roadways (including both test track and public roads). There may be additional mechanical system limitations that arise only when the automated systems are implemented in a prototype vehicle, such as steering system torque limitations, machine vision limitations when reading lane markings, and so on.

As noted above, testing at this stage will involve on-road testing using driving automation. The focus of this section is on scenario selection and on one particular method of on-road testing, the Wizard of Oz (WO) approach. Similar to analysis of existing naturalistic driving data and simulator testing, scenario selection is critical for on-road testing. Although on-road testing can be conducted with commercially available vehicles (Endsley, 2017; Banks, Eriksson, O'Donoghue, & Stanton, 2018), the WO approach is included as it will be a critical path for testing automated driving technologies that are not commercially developed (e.g., Levels 3 and 4 driving automation).

22.4.1 SCENARIO SELECTION

Some on-road scenarios may have already been selected from previous data analysis and/or extensions of driving simulator tests (e.g., the slowed vehicle reveal example, Figure 22.2). At this stage of testing, cost and logistics will limit a researcher's ability to iteratively revise and re-test many scenarios. As such, testing on live roadways (either test track or public roads) requires the researcher to narrow down the types of scenarios to those of most interest.

Tests of driver interactions with prototype or WO automated systems may be conducted on public roads, depending on the safety and the legality at the state and local level. Essentially, a participant operates an automated system on public roads while one or more researchers provide instruction, analyze behaviors, record participant preferences, and so on. Manipulations may include changes to system characteristics, such as HMI displays and/or messaging strategies (see Blanco et al., 2015), and/or automated system capabilities (see Russell et al., in press). This approach will provide data for driver interaction with automated systems in typical settings, with traffic conditions that will vary over the course of testing. This approach is akin to a quasi-naturalistic study; an experimenter is present, but traffic scenarios develop as they would occur normally in the testing environment.

It may not be an effective use of time or resources to "wait and see" what scenarios develop; a researcher may need to create or orchestrate a specific scenario. Multiple traffic scenarios, environmental variables, and other sources of scenario complexity were noted in the introductory section to this chapter, per SAE J3018. Additional aspects of the automated systems (e.g., lateral features, alert severity, etc.) add another layer of complexity. Given the sheer number of possibilities, it will certainly be helpful to narrow down options. One potential scenario classification method is based on how quickly a response is needed (i.e., urgency), whether the scenario is expected or not (i.e., predictability), and the consequences of not responding or intervening (i.e., criticality) (Gold, Naujoks, Radlmayr, Bellem, & Jarosch, 2017).

Again, there is no single test or even set of tests that should always be conducted; testing will vary based on the feature of interest. However, the highest return on investment may be a "worst-case scenario" approach. Using the classification above, this would be a scenario with high urgency, low predictability, high criticality, assuming it can be implemented safely. These scenarios are dependent on the capabilities of the system or feature of interest, but some examples may include an unresponsive driver (asleep, impaired), major system malfunction (improper steering, sensor failure), or a driver performing a non-driving task during an RTI. This method is dependent on the researcher's ability to perform these tests safely but can yield valuable data on driver responses to these critical situations.

22.4.2   WO Approaches

Although likely familiar to most readers, WO testing refers to testing a system that operates as though it were automated but is not. WO "automation" is achieved through mechanical intervention (e.g., secondary controls), using only pre-determined routes, GPS guidance, or similar methods. This approach may be necessary for a variety of reasons; it is possible that the researchers are interested in testing a general category of automation (e.g., Level 2) rather than a specific platform or system and would therefore avoid using a stock testing platform. Alternately, the state of development or other testing limitations may be such that the system of interest is not ready for deployment but may work on a test track with a pre-programmed path. Alternatively, a lateral control feature may be classified as Level 2, and is capable of maintaining lane position but requires the driver to maintain steering input; adding a mechanical steering system from a confederate driver in a rear seat may allow for testing of Level 3 scenarios on live roads.

As a final note, the WO testing is a flexible testing approach not limited to driver testing. For example, the approach can be used to test interactions between Level 4 driving automation and other road users. Rather than augmenting vehicle technology, a driver may be hidden from view of anyone outside the vehicle. The "ghost driver" (Rothenbucher, Li, Sirkin, Mok, & Ju, 2015) method can give the appearance a vehicle is autonomous, allowing testing for general reactions from other road users and pedestrians, such as responses to signal lights (Ford Motor Company, 2017). Responses can be gathered via post-exposure interviews or video-based reduction of observer response.

22.5   LATE STAGE TESTING: NDSS

For a moment, assume the best-case scenario: A feature was designed using existing naturalistic driving data; models of the feature were tested in computer-based simulation, initial concepts of this feature were then refined iteratively in a driving simulator, and finally a functioning prototype of the feature was subsequently tested in a series of on-road experiments. As discussed, many interesting scenarios were likely not tested for logistical and budgetary reasons. There may be unintended consequences or other unforeseen edge cases that may only be observed when deployed in real-world conditions. While not guaranteed to reveal new edge cases, testing new technology via an NDS may reveal new scenarios of interest as well as provide a test bed for existing assumptions about feature use and adoption. Finally, NDSs may provide insight into changes in driving automation use over time; a long experiment may last several hours, and use of a system for days, months, or even years is likely necessary to test how automation use patterns change over time (see also this Handbook, Chapter 12).

Methods and results for two different NDSs using commercially available Level 2 systems have been recently published (Russell et al., 2018; Fridman et al., 2019). Russell et al. (2018) deployed five different makes and models equipped with Level 2 features to 120 participants, two of each model were listed below:

- 2017 Audi Q7 Premium Plus 3.0 TFSI Quattro with Driver Assistance Package
- 2015 Infiniti Q50 3.7 AWD Premium with Technology, Navigation, and Deluxe Touring Package
- 2016 Mercedes-Benz E350 Sedan with Premium Package, Driver Assistance Package
- 2015 Tesla Model S P90D AWD with Autopilot Convenience (software version 8.0)
- 2016 Volvo XC90 T6 AWD R with Design and Convenience Packages

Vehicles were instrumented with a data acquisition system including camera views of the driver and forward roadway, and accelerometer and GPS data. Each participant drove the vehicles for four weeks. Participants received an introduction to the driving automation, including a test drive before their participation period. A total of 216,585 miles were driven, with 70,384 miles driven with both lateral and longitudinal features active. Fridman et al. (2019) recruited drivers who were already owners of Level 2 capable systems (specifically Tesla drivers), with analyses reported for 21 vehicles and 323,384 total miles (112,427 miles driven with Level 2 active). Cameras were used to record driver behavior and the forward roadway.

The intent here is not to provide a "how to" for the logistics and data collection procedures for conducting an NDS, but some primary issues will be reviewed briefly. The focus in this section is on the issues relating to the participant and data collection, data sampling, and data reduction for testing of automated driving systems. Researchers will need to plan for what types of systems and participants will be studied (e.g., owners of candidate systems or participants to which systems will be loaned). Researchers will also need to decide on what instrumentation (e.g., aspects of the data acquisition system: cameras, accelerometers, GPS, vehicle network information, storage capacity) should be deployed. Likely dedicated staff will be needed to monitor the data collection, to replace and/or re-align cameras, and to manage data as data storage fills up during the data collection period. For a more detailed review of methods for field operational tests and NDSs, see the overview published by the FESTA Consortium (2017).

22.5.1 PARTICIPANT SELECTION

Participant selection for NDSs has a few distinct differences from a test track study. The SHRP 2, the largest NDS as of 2019 (32 million miles, ~3,500 drivers), recruited volunteers willing to have their own personal vehicles instrumented with cameras and other data collection equipment (Dingus et al., 2014). This approach to recruiting is dependent on volunteers that already own a vehicle with automated driving capabilities and are then willing to have it instrumented with cameras and other data collection equipment. As such, using this method for testing may be limited due to lack of market penetration and/or finding willing volunteers.

As an alternative, vehicles that include automated driving systems may need to be procured and loaned to participants for some length of time, as was done for the Level 2 NDS (L2NDS; Russell et al., 2018). For either approach, limited budget and funds will call for tradeoffs in the study length and the number of participants selected. Depending on the specific needs of the test, it may be better to conduct a study with longer exposure per driver with fewer participants, or it may be better to collect data for a larger number of participants for a shorter time. For example, longer exposures may be more likely to reveal driver adaptations and habituation to using driving automation (this Handbook, Chapter 12). If the approach is to loan vehicles to participants, then training methods for participants (as described in the introduction section) should be considered before their data collection period begins.

22.5.2 DATA SAMPLING & REDUCTION

The hallmark of an NDS is that data are recorded continuously any time that the vehicle is in operation. This amounts to an enormous volume of data for even a single drive taken by a single participant, let alone across an entire dataset that consists of trips taken over months or years. The typical approach is to sample the continuously recorded data for subsequent reduction and analysis. The number of samples, the duration of each sample, and types of samples (i.e., the sampling strategy) will vary based on the size of the dataset and the goals of the analysis. The events to be sampled will be tailored to the needs of the particular test but will likely include specific alerts or other transitions points of the driving automation (e.g., RTI alerts issued by the automated features, or automation activation or deactivation). It is noted that alternative methods that use machine vision and other neural net analytics (e.g., Fridman et al., 2017) to analyze data continuously (as opposed to discrete samples) are under development.

22.5.2.1 Data Sampling

Sampling strategies may require comparisons between levels of automation, particularly for driving automation that can operate at different levels (depending on which features are activated). Baseline selection may be somewhat tricky, and random sampling of "non-eventful" driving may lead to results that are not representative of the full range of capabilities a system may have. Aside from excluding alerts or other situations of interest, a sampling plan should be based on information from the characterization process, in order to select samples that include all levels of

TABLE 22.3
SCE Descriptions as Used in SHRP 2

Severity Level	Description
Most severe	Any crash that results in any injury requiring medical attention, or one that includes an airbag deployment or requires vehicle towing
Police reportable crash	A crash that does not meet the requirements for a Level I crash, but does include sufficient property damage that warrants being reportable to the police
Minor crash	A crash that does not meet the requirements for a Level II crash, but does result in minimal damage
Low risk crash	Tire/curb strike
Near crash	Any circumstance that requires a rapid evasive maneuver by the subject vehicle, or any other vehicle, pedestrian, cyclist, or animal, to avoid a crash

automation for which a system is capable and appropriate ODDs for different levels of automation. For comparisons between automated driving and non-automated (or non-assisted) driving, samples can be taken from the same ODDs with and without features activated. If the automated driving system has one or more of those features that can be activated separately, samples should be taken from each "level" of activation. For example, naturalistic studies of vehicles equipped with both lateral and longitudinal automated features sampled Level 0 (no features active), Level 1 (one feature active), and Level 2 (both features active) during similar driving scenarios (e.g., highway driving above 40 mph; Russell et al., 2018).

SCEs consisting of crashes and near crashes will certainly be of interest in most if not all test cases. If not reported to the research team directly (by participant drivers themselves or by law enforcement) SCEs can be detected by kinematic data and subsequently confirmed by reviewing video data (see data reduction section). Table 22.3 shows the definitions of SCEs used in the SHRP 2 study (Antin et al., in press).

As a final note, driver behaviors themselves do not automatically elevate a sample to an SCE, even if the behavior is egregious (e.g., visibly intoxicated, sleeping). This also applies to the state of automation in relation to driver behavior (e.g., texting while driving with the automated feature).

22.5.2.2 Data Reduction

Data reduction, sometimes referred to as data annotation, is a step prior to analysis in which each sample epoch is reviewed and the relevant driver, vehicle, and environmental factors are classified by a researcher following a specific protocol. Typically, the time-synchronized vehicle data (e.g., automation state, driver inputs, etc.), sensor data (accelerometer, GPS, etc.), and video data are reviewed frame by frame as necessary (based on the recording rate of the video). Relevant driver factors include presence of non-driving tasks, driver gaze patterns, hands-on-wheel behaviors, pedal behaviors (e.g., accelerator release, brake presses, etc.), and any visible signs of impairment, etc. Driver behaviors that occur with and without driving automation, including the presence of non-driving tasks will likely be of importance to any naturalistic study.

Describing the driving scenario is an intricate process. In addition to the driver factors, environmental factors should also be annotated. For example, the researcher should describe the scenario in relation to the bulleted list of driving scenarios and domain descriptions listed in the introduction of this chapter (a modified list follows).

- Roadway Type
  - Limited access freeway
  - Highway (single or multi-lane),
  - Arterial roads
  - Residential streets
  - Driveway, parking lot, or structure
- Traffic Environment
  - Level of service
  - Vehicles present (e.g., leading and adjacent vehicles)
  - Pedestrians
  - Animals
  - Relation to intersection
  - Traffic control devices (signals, signs, curbs, guardrails, etc.)
- Time of Day & Lighting Conditions
  - Day
  - Night (unlit)
  - Night (lit)
  - Dusk
  - Dawn
- Inclement weather present
  - Rain
  - Snow
  - Fog

Specific operation definitions used for the analysis should be compiled into a "data dictionary" for each study. The data dictionary includes operational definitions used to reduce each scenario that can be referenced by any reductionist at any time and can be used to train new reductionists. An example data dictionary can be consulted to guide protocol development and descriptions of the vehicle, driver, and environmental factors present in the scenario (Virginia Tech Transportation Institute, 2015). Other sources for operational definitions and calculating driver performance metrics include Green (2012) and SAE International (2015b).

22.6 CONCLUSION

We sit at a critical transition point for driving automation deployment and testing. Testing of Level 4 and higher automated driving systems is currently underway by multiple companies (e.g., Waymo, Cruise); however, the many engineering challenges for autonomous driving leave timelines for deployment are still uncertain.

Characterization of these Level 4 systems, including their expected ODDs as well HMIs, will be critical for testing these systems. ADAS (e.g., Level 1) and partial driving automation (e.g., Level 2) are becoming available in more affordable makes and models. Early test track studies of these systems conducted with prototype vehicles (Blanco et al., 2015) have provided understandings and underpinnings of human factors guidance for driving automation development (Campbell et al., 2018; see also, this Handbook, Chapter 15). Additional research is likely underway in the private realm, as OEM and Tier 1 suppliers test and develop systems as part of system and product development.

As driving technology continues to improve, a question remains as to the degree to which drivers will be monitoring systems as opposed to actively driving with automation. As noted above, two naturalistic studies of Level 2 driving automation use have recently been published. The results for these studies indicate that drivers using Level 2 automation were using driving automation in a largely active fashion; there was no difference in non-driving task prevalence observed by Russell et al. (2018) nor were more dangerous tasks more likely while Level 2 was active (e.g., texting or browsing on a smartphone) nor any relationship between Level 2 use and SCEs. Fridman et al. (2019) specifically looked at periods of de-activation of Level 2 states, finding 90.6% of deactivations categorized as anticipatory in nature. It will be of great interest to see if similar results are observed with Level 3 and higher systems.

As automated driving technology continues to become more prevalent, testing and development will be critical to the safe deployment of the technology. Again, the goal for testing should be of practical significance; when evaluating driving automation technology, the end result may be one when "nothing happens." Reaction times or distractions observed in early stages should be tested in naturalistic settings to determine if there is an overall safety risk associated. As on-road and naturalistic testing with vehicle automation continue, we should not be surprised as results emerge that seem counterintuitive when compared with findings from simulator or on-road experimental findings.

Finally, the testing approaches in this chapter were largely focused on system development and evaluation with a targeted focus on driver testing. As automated driving of all levels becomes more widespread, testing must continue to interrogate broader effects of automated driving at the "transportation system" level (this Handbook, Chapters 13, 19). Interactions with other "non-automated" driving in a mixed fleet will likely lead to novel cases that cannot be easily predicted until they emerge. Similarly, interactions with pedestrians, cyclists, and other vulnerable road users at a broader scale should be optimized as informed by testing and evaluation.

ACKNOWLEDGMENTS

This chapter was heavily informed by the authors conducting funded studies from the National Highway Traffic and Safety Administration as well as other proprietary sponsors. A special thanks to the editors and other reviewers for valuable feedback on early drafts of this chapter.

REFERENCES

AAA Public Affairs. (2019). *Three in Four Americans Remain Afraid of Fully Self-Driving Vehicles.* Retrieved May 12, 2019, from https://newsroom.aaa.com/tag/autonomous-vehicles/

Antin, J. F., Lee, S., Perez, M. A., Dingus, T. A., Hankey, J. M., & Brach, A. (in press). Second strategic highway research program naturalistic driving study methods. *Safety Science, 119,* 2-10.

Banks, V. A., Eriksson, A., O'Donoghue, J., & Stanton, N. A. (2018). Is partially automated driving a bad idea? Observations from an on-road study. *Applied Ergonomics, 68,* 138–145.

Blanco, M., Atwood, J., Vazquez, H. M., Trimble, T. E., Fitchett, V. L., Radlbeck, J. ... Morgan, J. F. (2015). *Human Factors Evaluation of Level 2 and Level 3 Automated Driving Concepts* (DOT HS 812 182). Washington, DC: National Highway Traffic Safety Administration.

Brooks, J. O., Goodenough, R. R., Crisler, M. C., Klein, N. D., Alley, R. L., Koon, B. L., ... Wills, R. F. (2010). Simulator sickness during driving simulation studies. *Accident Analysis & Prevention, 42*(3), 788–796.

Campbell, J. L., Brown, J. L., Graving, J. S., Richard, C. M., Lichty, M. G., Bacon, L. P., ... Sanquist, T. (2018). *Human Factors Design Guidance for Level 2 and Level 3 Automated Driving Concepts* (DOT HS 812 555). Washington, DC: National Highway Traffic Safety Administration.

Dingus, T. A., Hankey, J. M., Antin, J. F., Lee, S. E., Eichelberger, L., Stulce, K., ... Stowe, L. (2014). *Naturalistic Driving Study: Technical Coordination and Quality Control* (SHRP 2 Rep. S2-S06-RW-1). Washington, DC: National Academies.

Dingus, T. A., Klauer, S. G., Neale, V. L., Petersen, A., Lee, S. E., Sudweeks, J., & Knipling, R. R. (2006). *The 100-Car Naturalistic Driving Study, Phase II Results of the 100-Car Field Experiment* (DOT HS 810 593). Washington, DC: National Highway Traffic Safety Administration.

Endsley, M. (2017). Autonomous driving systems: A preliminary naturalistic study of the Tesla Model S. *Journal of Cognitive Engineering and Decision Making, 11,* 225–238.

FESTA Consortium. (2017). *FESTA Handbook Version 7.* Retrieved from https://fot-net.eu/Documents/festa-handbook-version–7/

Ford Motor Company. (2017). *Ford, Virginia Tech Go Undercover to Develop Signals That Enable Autonomous Vehicles to Communicate with People.* Retrieved from https://media.ford.com/content/fordmedia/fna/us/en/news/2017/09/13/ford-virginia-tech-autonomous-vehicle-human-testing.html

Fridman, L., Brown, D. E., Glazer, M., Angell, W., Dodd, S., Jenik, B., ... Abraham, H. (2017). MIT autonomous vehicle technology study: Large-scale deep learning based analysis of driver behavior and interaction with automation. *arXiv:1711.06976.* doi: 10.1109/ACCESS.2019.2926040

Fridman, L., Brown, D., Kindelsberger, J., Angell, L., Mehler, B., & Reimer, B. (2019). *Human Side of Tesla Autopilot: Exploration of Functional Vigilance in Real-World Human-Machine Collaboration.* Cambridge, MA: MIT. Retrieved from https://hcai.mit.edu/tesla-autopilot-human-side.pdf

Gold, C., Naujoks, F., Radlmayr, J., Bellem, H., & Jarosch, O. (2017). Testing scenarios for human factors research in level 3 automated vehicles. *International Conference on Applied Human Factors and Ergonomics* (pp. 551–559). Berlin, Springer.

Green, P. (2012). Standard definitions for driving measures and statistics: Overview and status of recommended practice J2944. *Proceedings of the 5th International Conference on Automotive User Interfaces and Interactive Vehicular Applications* (pp. 28–30). Eindhoven, The Netherlands.

Grove, K., Atwood, J., Blanco, M., Krum, A., & Hanowski, R. (2017). Field study of heavy vehicle crash avoidance system performance. *SAE International Journal of Transportation Safety, 5*(1), 1–12.

Grove, K., Atwood, J., Hill, P., Fitch, G., DiFonzo, A., Marchese, M., & Blanco, M. (2015). Commercial motor vehicle driver performance with adaptive cruise control in adverse weather. *Procedia Manufacturing, 3*, 2777–2783.

Grove, K., Soccolich, S., Engstrom, J., & Hanowski, R. (2019). Driver visual behavior while using adaptive cruise control on commercial vehicles. *Transportation Research Part F: Traffic Psychology and Behavior, 60*, 343–352.

Guo, F., Klauer, S., McGill, M., & Dingus, T. (2010). *Evaluating the Relationship Between Near-Crashes and Crashes: Can Near Crashes Serve as a Surrogate Safety Metric for Crashes?* (DOT HS 811 382). Washington, DC: National Highway Traffic Safety Administration.

Kalra, N. & Paddock, S. M. (2016). Driving to safety: How many miles of driving would it take to demonstrate autonomous vehicle reliability? *Transportation Research Part A: Policy and Practice, 94*, 182–193.

Kaptein, N. A., Theeuwes, J., & Van Der Horst, R. (1996). Driving simulator validity: Some considerations. *Transportation Research Record, 1550*, 30–36.

Knipling, R. (2015). Naturalistic driving events: No harm, no foul, no validity. *Proceedings of the Eighth International Driving Symposium on Human Factors in Driver Assessment, Training and Vehicle Design.* Iowa City: University of Iowa.

Lammert, M., Duran, A., Diez, J., & Burton, K. (2014). Effect of platooning on fuel consumption of Class 8 vehicles over a range of speeds, following distances, and mass. *SAE International Journal of Commercial Vehicles, 7*(2), 626–639.

Lee, J. D., Wickens, C. D., Liu, Y., & Boyle, L. N. (2017). *Designing for People: An Introduction to Human Factors Engineering.* Charleston, SC: CreateSpace.

Murray-Smith, D. (2012). *Modelling and Simulation of Integrated Systems in Engineering.* Philadelphia, PA: Woodhead Publishing.

Poorsartep, M. & Stephens, T. (2015). Truck automation opportunities. In G. Meyer, & S. Beiker (Eds.), *Road Vehicle Automation 2. Lecture Notes in Mobility.* Berlin: Springer.

Ranney, T. (2011). Psychological fidelity: Perception of risk. In D. Fisher, M. Rizzo, J. Caird, & J. Lee (Eds.), *Handbook of Driving Simulation for Engineering, Medicine and Psychology.* Boca Raton, FL: CRC Press.

Rothenbucher, D., Li, J., Sirkin, D., Mok, B., & Ju, W. (2015). Ghost driver: A platform for investigating interactions between pedestrians and driverless vehicles. *Proceedings of the 7th International Conference on Automotive User Interfaces and Interactive Vehicular Applications.* ACM, New York, NY, USA, pp. 44–49.

Russell, S. M., Atwood, J., & McLaughlin, S. (in press). *Driver Expectations for System Control Errors, Engagement, and Crash Avoidance During Level 2 Driving Automation.* Washington, DC: National Highway Traffic Safety Administration.

Russell, S. M., Blanco, M., Atwood, J., Schaudt, W. A., Fitchett, V. L., & Tidwell, S. (2018). *Naturalistic Study of Level 2 Driving Automation Functions* (DOT HS 812 642). Washington, DC: National Highway Traffic Safety Administration.

SAE International. (2015a). *Guidelines for Safe On-Road Testing of SAE Level 3, 4, and 5 Prototype Automated Driving Systems.* Warrendale, PA: Society for Automotive Engineers.

SAE International. (2015b). *Operational Definitions of Driving Performance Measures and Statistics.* Retrieved from https://saemobilus.sae.org/content/J2944_201506/

SAE International. (2016). *Surface Vehicle Recommended Pratice J3016: Taxonomy and Definitions for Terms Related to driving Automation Systems for On-Road Motor Vehicles.* Warrendale, PA: Society for Automotive Engineers.

Smith, D., Harder, F., Huynh, N., Hutson, N., & Harrison, R. (2012). Analysis of current and emerging drayage practices. *Transportation Research Record, 2273*, 69–78.

Technology & Maintenance Council. (2018). Technological advances in next generation collision warning driver interfaces. *The Trailblazer: The Technical Journal of TMC's 2018 Annual Meeting.* ATA, Arlington, VA, USA,. 40–45.

Technology & Maintenance Council. (2019). RP 430 update. *The Trailblazer: The Technical Journal of TMC's 2019 Annual Meeting.* ATA, Arlington, VA, USA, 39–40.

Tsugawa, S., Jeschke, S., & Shladover, S. E. (2016). A review of truck platooning projects for energy savings. *IEE Transactions on Intelligent Vehicles, 1*(1), 68–77.

Virginia Tech Transportation Institute. (2015). *SHRP 2 Researcher Dictionary for Video Reduction Data Version 4.1.* Blacksburg, VA: VTTI.

Weir, D. H. (2010). Application of a driving simulator to the development of in-vehicle human-machine-interfaces. *IATSS Research, 34*, 16–21.

23 Techniques for Making Sense of Behavior in Complex Datasets

Linda Ng Boyle
University of Washington

CONTENTS

Key Points 497
23.1 Introduction 498
23.2 Data Collection Tools 499
    23.2.1 On-Road Settings...499
    23.2.2 Driving Simulators 501
    23.2.3 Other Data Collection Tools 503
23.3 Data..504
    23.3.1 Data Cleaning, Extraction, and Preparation 504
    23.3.2 Accounting for Context..506
    23.3.3 Accounting for Outliers 508
    23.3.4 Data Visualization 508
23.4 Understanding the Controller 509
    23.4.1 Human Behavior..510
    23.4.2 System Behavior 511
    23.4.3 Human–System Interaction 511
23.5 Analytical Tools..512
    23.5.1 Exploratory Models 512
    23.5.2 Inferential Models..512
    23.5.3 Predictive Models 514
23.6 Conclusion 515
References...515

KEY POINTS

- Given advances in sensors and technology, a great deal of transportation data can be gathered, and these data can help answer questions related to road user behavior and help understand the behavior of the vehicle. However, the challenge is being able to extract the appropriate data for the research question of interest and being able to use the data to model the anticipated outcomes.

- The data needed to understand behavioral changes in an automated or connected vehicle can be gathered from various sources, including simulated and on-road environments, among others.
- Databases from simulator and on-road studies can contain terabytes to petabytes of data, making it a challenge to parse the data into something meaningful. In addition to data cleaning, extraction, and preparation, data visualization provides a way to better identify the outcomes to examine and the explanatory variables to include.
- The controller is the entity that is responsible for the performance of the vehicle system. Depending on the level of vehicle automation that is in operation, the controller can be the human or the system. Hence, it is important to understand the behavior of both controllers.
- There are many analytical models that can be used to examine the dependent variables, and the models can be generally grouped into three categories: exploratory, inferential, and predictive.

23.1 INTRODUCTION

Human factors researchers are often asked to provide insights on road user's behavior. In the context of automated, connected, and intelligent vehicles, the research questions relate to the human–system interaction and can be tailored toward the human side or the system side. Given advances in sensors and technology, we have amassed a great deal of transportation data, and these data can help answer questions related to road user behavior and help predict the behavior of the vehicle. However, the challenge is being able to extract the appropriate data for the research question of interest and being able to use the data to model the anticipated outcomes. As we move forward with more advanced vehicle systems, the goal becomes one of understanding how user behavior changes as they interact with these systems over time and whether the changes will impact safety (also see this Handbook, Chapter 12).

Oftentimes, researchers begin with an analytical model rather than a theoretical model. That is, researchers tend to begin integrating data into sophisticated predictions or inferential models without any knowledge or understanding of the data. Theory provides a framework for exploring the research problem. Its purpose is to clarify the concepts and proposed relationships among the variables of interest. To gain a better understanding of the theoretical side, researchers should take time to explore and understand the data before moving toward inferences and predictions. For that reason, this chapter focuses more on how best to parse and process the data and how to visualize the data (and less on the analytical models). That said, the reader is provided some high-level examples of tools that go beyond regression models to provide the reader a good starting point for classifying, inferring, and predicting the outcomes of interest. In general, safe driving depends on the ability of the controller to devote attention to the roadway demand (Lee, Young, & Regan, 2009). For the vehicle systems discussed in the context of this book, the controller can be the human or the system. This chapter considers research questions that can be addressed using data related to automated, connected, and intelligent vehicles.

The goal of this chapter is to provide techniques that can help the analyst make sense of complex datasets. This chapter is organized into four topics:

- *Understanding the data collection tools*: Data collection tools used to make sense of the behavior given the context, user, and the technology itself.
- *Understanding the data*: Types of data often used for examining behavior and ways to manipulate, visualize, and explore your data.
- *Understanding the controller*: Types of research questions that may be of interest from the human and system controller perspective, and what human–system interactions can be modeled.
- *Understanding the analytical methods*: Types of models that can be developed that will account for changes in behavior over time and location.

23.2 DATA COLLECTION TOOLS

The data needed to understand behavioral changes in an automated or connected vehicle can be gathered from various sources. What makes the data complex is that the observed behavior is changing over time, location, and environment. Hence, we need to capture the controller's performance (speed, braking, steering) over these changes and be able to make sense of the data, especially when the performance appears to be a significant change from the operator's normal behavior. For the human operator, we also want to be able to capture the changes in travel motivation and perceived risks.

There are many data collection tools that can be used to capture these changes. This chapter focuses on data collected from simulated environments and on-road studies using instrumented cars. These simulator and on-road studies capture changes in performance of the road user in a given context. For on-road studies, the context can be determined from audio and video recordings, roadway sensors, and land surveys. For simulated road conditions (driving simulators or test tracks), the context can be manipulated by the researcher. In both sets of studies, data related to the operator's reaction time, speed, acceleration, and lateral deviations for any given moment in time and location are possible with sensors placed on the pedal and steering wheels.

23.2.1 ON-ROAD SETTINGS

On-road data collected from instrumented vehicles can be categorized into controlled observational studies, field operational tests (FOTs) and naturalistic studies (Carsten, Kircher, & Jamson, 2013). These real-world settings are designed to observe subjects in their natural environment. The controlled observational studies and FOTs introduce an intervention or a semi-controlled environment, but participants are allowed to drive as they normally would. Naturalistic studies do not make any attempts at interventions or control on the part of the researcher.

These on-road datasets are complex for many reasons. The vehicles are instrumented with many sensors to capture driver speed, steering, and location. GPS systems as well as accelerometers are also used to capture the spatial locations of the road user. These sensors need to be mapped to the driver's actions and the roadway conditions. The driver's action is often captured using audio or video equipment. The integration of all these sensors often requires some manual coding; this is true even if one has access to video reduction software. The reason for this is that the event of interest needs to be initially identified by the researcher/analyst and that may take extensive time to complete.

While complex, such datasets are incredibly useful in providing insights on driver-initiated activities and any adaptive behavior associated with existing technology equipped in the driver's vehicle. The datasets can also provide insights on user behavior during safety-critical events that cannot be observed otherwise. They can also capture surprising issues that the researcher may not have deemed relevant. This includes unexpected interactions with passengers, pedestrians, and other road users, changes due to road conditions, and trip purpose. For example, most drivers do not interact with their mobile device continuously while driving. Rather, they moderate their use based on the road, environment, and traffic (Oviedo-Trespalacios, Haque, King, & Washington, 2019). In fact, studies show that many drivers commonly perform secondary tasks while stopped at a red light (e.g., Kidd, Tison, Chaudhary, McCartt, & Casanova-Powell, 2016). And, while using a mobile device at an intersection may be observed by other road users as more of an annoyance than a safety concern, studies do show that a driver's continual need to engage with their mobile device may be an indication of more harmful and negative behavior (Billieux, Maurage, Lopez-Fernandez, Kuss, & Griffiths, 2015), which can impact overall driver safety. Data collected from the real world can provide insights on these subtle differences as well as help the researcher understand the sensitive issues that participants are not as willing to reveal in surveys or controlled studies.

On-road data can also provide insights on the appropriateness of crash surrogate measures. For example, are lane departures or the distance to a leading vehicle really good indicators of "crashes"? On-road data can be used to map out the trajectory of the vehicle in the context of other road users. In doing so, one may see that the sharp deviation in lane position may have actually been necessary to help the driver avoid an otherwise more dangerous situation.

As of this writing, several of these datasets are available to users from the organization that collected the data. Several are housed on data.gov such as:

- University of Michigan Transportation Research Institute (UMTRI): Safety Pilot Model Deployment Data (SPMD)[1] with a one-day sample available[2]
- Washington State Department of Transportation (DOT) data from a connected vehicle study to demonstrate Intelligent Network Flow Optimization (INFLO) Prototype System[3]

---

[1] SPMD: https://catalog.data.gov/dataset/safety-pilot-model-deployment-data.
[2] SPMD for one day: https://catalog.data.gov/dataset/intelligent-transportation-systems-research-data-exchange-safety-pilot-model-deployment-on-f77e5.
[3] INFLO: https://catalog.data.gov/dataset/intelligent-network-flow-optimization-prototype-infrastructure-traffic-sensor-system-data.

Other data are housed at the institution that collected the data and often require a data sharing agreement. Examples include:

- Waymo https://waymo.com/open/
- VTTI 100 Car study: https://dataverse.vtti.vt.edu/dataset.xhtml?persistentId= doi:10.15787/VTT1/CEU6RB

Due to protection of human subjects, other datasets (e.g., Strategic Highway Research Program 2 (SHRP 2) data) cannot be downloaded directly from the website and a data use license is required to access the data. SHRP 2: https://dataverse.vtti.vt.edu/dataverse/shrp2nds.

These real-world datasets can capture differences given the driver's natural habitat, but analysis of such data can be a challenge. For example, it would be rare to capture a crash that occurred in the same environment, weather, and traffic condition for more than one study participant. While the behavior that is related to a crash or safety-critical event may be interesting to explore in the real world, it would be a challenge to identify all the factors that may have caused the event. Oftentimes, the "abnormal" behaviors are those that warrant further investigation. But how can we best infer policies, education, or engineering changes from the few observations observed with behavioral characteristics. That is the value of data collected from driving simulators.

23.2.2 Driving Simulators

Trends or patterns observed in naturalistic studies can be further examined using driving simulators. These tools allow researchers to test and re-test situations that may pose risks to drivers in the real world. Driving simulators provide researchers the ability to test proposed systems using well defined tasks. There are many types of simulators ranging from desktops to full motion based. Readers interested in more details on the use, fidelity, and validation of simulators are referred to Fisher, Rizzo, Caird, and Lee (2011).

In general, high-fidelity driver simulators are increasingly affordable and provide realistic vehicle dynamics that can render highly representative driving scenes (Boyle & Lee, 2010). They have the capability to generate repeatable outcomes with more equal sample sizes (than possible in on-road studies) for each independent variable of interest. While data from the real world afford more opportunities to look at behavior over time, there are also opportunities to capture some time variations in driving simulators (Ranney, Simmons, & Masalonis, 1999; Miller & Boyle, 2019).

With respect to automated vehicles, driving simulators are an excellent tool to examine the ability of drivers to take over control from a highly automated vehicle given a particular traffic density, level of secondary task engagement, and even given individual differences. Likewise, we can manipulate the system in a simulated environment to capture how quickly drivers can respond when the system hands back control. That is, researchers can use driving simulators to:

- control the context in which a safety-critical situation occurs;
- simulate a system hand off;
- assess the ability of a driver to take over when the automation is not working as expected;

- simulate a situation where the system is not behaving as expected by the operator (e.g., slowing down or speeding up for no obvious reason); and
- assess how well the driver can monitor the system if they are engaged in a secondary task.

Thus, driving simulators are useful to provide practical feedback before system deployment and help guide the design of automated systems. Driving simulator studies can be used to observe the effects of safety-critical events for varying circumstances. For example, a study by Peng, Boyle, Ghazizadeh, and Lee (2013) used a driving simulator and eye glance data to extract the level of risk associated with entering and reading text while driving (Figure 23.1a). Given that the participants were exposed to the exact same secondary task condition, the study was very useful to examine individual differences (Figure 23.1b). More specifically, the study showed that there was greater variation between participants for entering text when compared with reading text. And drivers that had long glances away from the road when typing 12 characters also tended to have long glances when typing four characters.

Because driving simulator studies are not the real world, there are always issues regarding the validity of the outcomes and how well can they really generalize to

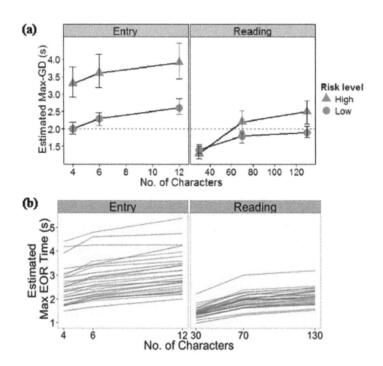

FIGURE 23.1   (See color insert.) Data from a study on secondary task engagement. The data in (a) was reported in Peng et al (2013). A review of the data afterwards showed that the spread of maximum eyes-off-road time was greater in the text entry task when compared with the text reading task (b).

the real world. There is also the possibility of simulator sickness, which can create sampling biases. That is, the types of people who can participate are skewed toward those that do not get simulator sickness.

23.2.3 OTHER DATA COLLECTION TOOLS

Most of the studies that examine the interactions between the driver and the advanced vehicle systems are based on data collected from sensors in real cars and in driving simulators. These tools are often supplemented with other types of data collection tools such as eye trackers, heart rate monitors, electroencephalogram (EEG), and other physiological measures. There are advantages and disadvantages of using these supplemental data collection tools in the real world and in a simulated environment. For example, eye movements are easier to interpret in a simulator (e.g., reaction time from a stimulus onset), but movements are restrained to what is depicted on the screen. There are more motions (including head and body) observed in real-world settings, but this creates greater challenges for measurement, modeling, and analysis (Lappi, 2015).

Data from crash reports can also be used to supplement the data collected from the real world and simulators. Even in the real world, it is rare to have a participant involved in a crash during the data collection period if the sampled data came from a representative sample of drivers. It is even more rare to have two or more participants involved in a crash with the same traffic, road, and environmental conditions.

Crash reports provide additional insights on the likelihood of observing a crash on specific roadway, traffic, vehicle, and environmental conditions. Crash reports can also provide details on the crash history of an individual. Crash data are not often considered to be "complex" in the same way that data collected on-road and in simulators are complex. The reason is that crash data do not provide insights on a road user's behavior over time. However, these data are still part of a large set of relational databases that takes time to process. For example, in the United States, the analytical user's manual for the 2016 crash database is 275 pages long.[4] And while these datasets alone cannot provide insights on behavior, they can provide reasonable cause. For example, a crash that occurred on a snowy day where the road was wet may help explain why a driver was unable to slow down in time and collided with another vehicle.

Crash data can also be mapped to specific areas where on-road data are also collected. For example, vehicle–pedestrian and vehicle–bicyclist crashes have been associated with the built environment (Dai & Jaworski, 2016), land use (Kim, Ulfarsson, Shankar, & Mannering, 2010), and local-area socioeconomic characteristics [e.g. population density and percentage of elderly residents (Kim, Kho, & Kim, 2017)]. While these insights can be captured from crash data, the number of pedestrian and bicyclist crashes is small in comparison with the number of vehicle–vehicle crashes. Given the sample size and the large variation in the data, traditional statistical models may not find any significant differences in speed, weather, or traffic conditions with crash data only. On-road data may help provide additional insights on why the driver was speeding, why the driver did not see the pedestrian/bicyclist, and why the driver felt comfortable engaging in secondary task at a crosswalk.

---

[4] https://crashstats.nhtsa.dot.gov/Api/Public/ViewPublication/812510.

Likewise, survey or questionnaire data are not often viewed as complex datasets. But these studies can include longitudinal travel diaries that can be time consuming to review. For example, the Puget Sound Regional Travel Study had some participants filling out a multi-page travel diary for seven days (RSG, 2018).[5] A review of these diaries shows that they contain thousands of records over separate datafiles for the household, person, vehicle, day, and trip.

23.3 DATA

Extracting meaningful information from terabytes to petabytes of data can be costly and time consuming. The data also need to be prepped before data analysis. This often requires review of several sources of data: video, vehicle kinematics, and physiological data. You also need to link the questionnaires and other subjective measures to the objective measures.

As part of data prepping, it is important to consider whether there are any underlying subgroups. In human factors research, we often group the data by age and gender, but there may be other more meaningful subgroups such as those related to risk taking (e.g., Figure 23.1a). However, placing people into groups may obscure other individual differences which may impact their use of new technology. In summary, how you chop, slice, dice, or julienne the data will matter. And it is important to take the time to explore the data in different ways to understand the driver's behavior while they operate automated, connected, and intelligent vehicles.

23.3.1 DATA CLEANING, EXTRACTION, AND PREPARATION

The data you collect from on-road and simulator studies are rarely used in their raw form. Preparing the data for use in analytical models can take considerable time involving manual coding, filtering algorithms, and creation of statistical codes to help parse, clean, and extract the data. Manual coding is often required for video and audio data. The time needed to manually process data will depend on the data collection tool, the amount of data you have collected, and the amount of noise or artifact expected. For example, in driving simulator data, information from traffic signals is already coded prior to the data collection and is automatically mapped to the driver's performance. However, in the real world, such roadway data may need to be observed through video recorders and manually coded.

Let's say we are interested in examining driver behavior at intersections. In a naturalistic study, drivers may travel through many different intersections and we cannot examine every single one. One way to approach this is to first identify intersections with high numbers of reported crashes (using historical crash data[6]). This subset of intersections can then be matched to intersections that did not have any

---

[5] www.psrc.org/household-travel-survey-program.
[6] Many states post their historical crash data on data.gov. Federal crash data on a sample of crashes across the country is available at: BTS.gov.

reported crashes but did have similar road and lane configurations. In this way, you control for the roadway factor while focusing on the human or system behavior.

The ability to observe traffic patterns in the real world is where on-road data provide great value. As an example, data from a signalized intersection in Ann Arbor, MI, as part of the SPMD study, provide data on drivers as they traversed through an intersection at Washtenaw Ave. and S. Huron Pkwy in Ann Arbor, Michigan. Between 2013 and 2017, there was an average of 43 crashes per year at this intersection (SEMCOG, 2019) making it an intersection of great concern for transportation engineers and the public.

In understanding the reasons crashes may be high at this intersection, a first step is to identify the buffer area of interest and how best to extract out behavior you may not have observed in crash or self-reported data. The time and distance to the traffic light are ways we can buffer the data. The distance to the traffic signal has the advantage of ensuring that every participant's data begin at the same location in which they should have been able to view the traffic signal.

The SPMD data show that the majority of movements through this intersection were drivers that were going straight in the eastbound ($n = 856$) and westbound ($n = 1191$) direction (Figure 23.2a). For those that are traveling straight in the eastbound direction, we can examine traces for all traffic light conditions (red, yellow, and green) (Figure 23.2c). As observed with many complex datasets, when all the traces of data are mapped on a 2D plot, it becomes difficult to extract useful information. One can try and color code the data based on the color of the signal as in Figure 23.2c, but the traces for signals that were encountered often (red signal = 332 samples, green

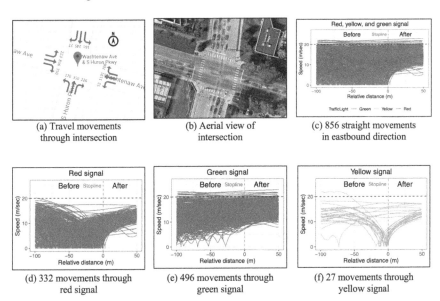

| (a) Travel movements through intersection | (b) Aerial view of intersection | (c) 856 straight movements in eastbound direction |

| (d) 332 movements through red signal | (e) 496 movements through green signal | (f) 27 movements through yellow signal |

FIGURE 23.2 **(See color insert.)** Washtenaw Ave & S Huron Pkwy, Ann Arbor, MI 48104 traffic patterns. (a) Travel movements through intersection, (b) aerial view of intersection, (c) all movements in eastbound direction, and number of movements by (d) red signal (e) green signal, and (f) yellow signal.

signal = 496) will overshadow those with less driver exposure (yellow signal = 27). However, decades of studies show that the yellow light provides the most insight on the decision processes and risk taking of the driver (Konecni, Ebbeson, & Konecni, 1976; Edwards, Creaser, Caird, Lamsdale, & Chisholm, 2003; Papaioannou, 2007). From Figure 23.2f, we can see two major driving patterns as drivers approach a yellow light: (1) those that continue through without changing speed and (2) those that slow down and wait for the signal to change to red.

Upon further investigation of the movements for those that encountered the yellow light (Figure 23.2f), we see that one driver was above the speed limit on the approach (top line at $x = -100$ m) while another is actually speeding up (bottom line at $x = -100$ m). These individuals are not going through this intersection in the same manner as the others. This can be viewed by other drivers as "unexpected" and create a safety-critical situation for those that are making a left turn in the westbound direction.

The traces of vehicle movements shown in Figure 23.2 took many months to code. The analyst needed to identify the start and end point for data extraction, review the video data for 856 movements, and manually code the color of the traffic signal using the forward video data. This data cleaning (or data wrangling) needs to be done before extracting out the speed, acceleration, and deceleration of each movement.

While video or face detection algorithms can help automate some of the data processing, the accuracy of the detection greatly depends on the technology available at the time of data collection and the quality of the video. In the examination of this intersection, a video detection algorithm was created to automate the process of traffic signal color detection. Unfortunately, it was only 50% reliable given the quality and resolution of the data. While clearly time consuming, it is still very important to review the data before any analytical models are created. Things that need to be checked and cleaned in raw data include:

- Data entry errors: Depending on the statistical package used, upper and lowercase letters are viewed as different units. This is the case for those who use the R statistical package. Are spikes in the data true outliers that warrant further investigation or are they due to data entry errors (e.g., the majority of individuals report driving 100 miles per week ± 20 miles, but a few entries are recorded with over 10,000 miles per week).
- Notation for missing data: Depending on how many people have entered data, missing data can be listed in many ways (e.g., period, NA, na, blank).
- Numbers vs. characters: Check to see if some numbers are stored as text. Creating a table of summary statistics can show which entries are not included in numerical summaries (e.g., means, sd).
- Removing extra spaces: Some entries include extra spaces, which may also cause some numbers to be read as characters.

23.3.2 Accounting for Context

Parsing out differences observed in complex datasets may be challenging given variations in people, vehicles, roads, weather, and environments. In fact, the variation even within one variable of interest may not be consistently observed for all drivers.

There are also varying sample sizes which can impact the capability to state something is significantly different.

An example of why context matters is related to studies on driver distraction. It appears to be clear from years of research that driver distraction can negatively impact safety (e.g. Strayer et al., 2013). However, some outcomes show secondary tasks can actually have a protective effect, or surprisingly help you be more attentive (Young, 2013).

The impact of driver distraction on crash risk (in the real world) depends on many things including: the type of distraction (eating, playing music, talking on the phone), type of mobile device (phone, tablet, laptop), physical location of device (lap, purse, car stand), type of interaction with the device (dialing, texting), as well as the context in which the distraction occurs (highway, heavy traffic, snowy conditions).

One measure of safety associated with driver distraction is brake reaction time (Consiglio, Driscoll, Witte, & Berg, 2003; Salvucci, 2002), where studies show that distracted drivers tend to have higher brake reaction times (Strayer & Drews, 2004) or longer reaction time (Lee, Caven, Haake, & Brown, 2001). The majority of these findings were observed in a simulated environment. However, in a naturalistic setting, the engagement in distractions is not controlled. The value of data from the real world is the ability to capture these subtle differences, but it is greatly dependent on how the analyst portrays the findings.

For example, let's say we collected a sample of 1,200 events in the wild that captures the driver's brake reaction time while making a call on a mobile device. The general conclusion could be that brake reaction time was actually quicker while using the phone (see Figure 23.3). However, if we separated out the events into the type of action used when making a call, we may see a difference in brake reaction time. For those drivers that have to physically touch the screen to enter a phone number, a much slower response time is observed when compared with those who used their voice while holding a hand-held device, and this may still be slower than those who used Bluetooth dialing. Given that holding your device while driving is illegal in many U.S. states, we may see that there are very few drivers that dial using the touch screen or hand-held voice.

In fact, if we were to examine the data more closely, we may observe that the 150 observations for touch screen dialing is based on 75 unique participants (or 1–3 events

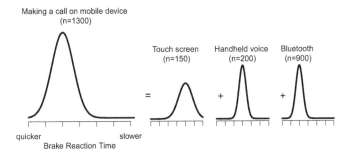

FIGURE 23.3 A hypothetical example of differences in brake reaction time for making a call on a mobile device given the type of interaction.

per person), while the 900 observations for Bluetooth dialing is based on 225 unique participants (or 2–6 events per person). In summary, the differences in sample size, variation, and type of device can impact the conclusions and need to be accounted for.

23.3.3 Accounting for Outliers

It is important to take time to explore the outliers in your data. The two individuals in Figure 23.2 that were speeding as they drove through a yellow light and speeding up toward the yellow would be considered outliers as their behavior does not match those of the other traces or movements. While some may want to exclude these two individuals from the analysis or group them under an "Other" category, it is important to recognize that crashes do not occur every day for each person. They are rare events and often fall in the outskirts of the data. For example, looking at interactions with the lowest number of touches to the in-vehicle display may reveal elements in the interface that the designer failed to account for as the user is interacting with the system. There was clearly a reason why a button or feature was included. But it may be difficult for the user to view, interact, or find the feature while driving. Alternatively, the ones that are used most often and without any crash impacts may not be the best place to focus system improvements. Some outliers may never be explained but that is not to say it should not be worth at least a few minutes' review.

23.3.4 Data Visualization

Visualization provides a way to better understand the outcome to examine and the explanatory variables to include. While summary statistics such as means, medians, and standard deviations are important point estimates, exploration of complex datasets benefit from visualizing relationships, trends, and distributions. These visualizations can help you find interesting features of the data that would not be observed otherwise. For example, you can visualize where an autonomous vehicle has traveled over the course of the day and use color coding based on travel speed to reveal how fast the car is going in specific areas. This data can then help researchers, planners, and engineers identify commute patterns and the best locations for infrastructure changes.

Visualization is also a first step in identifying the best analytical model. For example, distribution plots will show whether the data follows a normal distribution or not. It can also provide insights on whether the sampled population comes from one group (unimodal distribution), two (bimodal) groups, or more. Visualizing data from complex datasets can help identify differences among individuals in terms of the spread of data. One can observe the differences in variation within and between subjects and over time. This was observed in Figure 23.2. In this figure, we can see that there was greater variation in the text entry task when compared with the text reading task, but the variation within participants was not as large. In other words, those drivers that looked away for long periods when entering 4 characters would also look away for long periods when entering 6, or 12 characters (Figure 23.2b). This would not have been as easily observed in an on-road study given the lack of control in the number of characters that may be entered, the manner in which text is entered, and even the type of device used to enter text.

Time can be segmented in many ways: by minutes, hours, days, weeks, months, etc. You can also consider information by each sequential trip. The time scale selected will greatly impact the amount of variation you see in the data as well. However, that is the actual benefit of looking at your data over time, i.e., that you are able to get a sense of the variation in the data and thereby gain insights on the spread in the data. Of course, these graphs should not replace rigorous confidence interval calculation, but they can help you quickly see large changes and help explain any statistically significant differences.

A 2-D correlation plot of data from complex datasets is not often meaningful given the number of data points on the graph; the graph tends to look like a blob or ink splatter. However, a correlation matrix table that shows the correlation coefficients between variables in one table may be very useful. While these matrices may seem like a fishing expedition, it can help the analyst focus on only those relationships that may be of interest, especially when there are hundreds or thousands of variables. These correlations can then be placed in order from largest to smallest. It will then be up to the analyst to decide how reasonable, surprising, or unrealistic are the correlations.

One way to overcome overplotting is to use heat maps. Heat maps provide a way to visualize the volume of events or locations. They are called heat maps as they are used to show the "hot" areas in an otherwise meaningless blob. Bubble charts are also great for showing relationships among discrete outcomes. They are essentially a correlation plot in which each data point is replaced by a bubble. The size of the bubble represents the amount of data that has the same x and y values. Bubble charts provide another way to view three dimensions of information in a 2D space.

While not as intuitive, cluster analysis provides a way to group a set of people together based on how similar their responses are across several variables. Nielsen and Haustein (2018) used cluster analysis to group drivers based on their expectations of self-driving cars (skeptics, indifferent, enthusiasts). These groupings in the data may not have been as apparent otherwise. Clustering drivers based on their driving profiles, trust level, or driving experience are ways that can help us better design advanced vehicle systems based on the individual needs and preferences.

23.4 UNDERSTANDING THE CONTROLLER

The controller is the entity that is responsible for the performance of the vehicle system. Depending on the level of vehicle automation that is in operation, the controller can be the human or the system. Hence, it is important to understand the behavior of both controllers. For example, in SAE Level 2 the human driver has a high level of responsibility in terms of monitoring the system performance and acting appropriately. Hence, understanding the shared responsibilities between the human and the system becomes increasingly important (see also, this Handbook, Chapter 8). These complex datasets from on-road and simulated environments can provide insights on this behavior.

23.4.1 HUMAN BEHAVIOR

A model that includes the human as the controller should relate the driver's action to their perception of safe driving, driving experience, and preferences over a wide range of possible traffic situations (Donmez, Boyle, & Lee, 2009; Xiong & Boyle, 2012; Prokop, 2001). It is easy to assume that in a crash, the human did something wrong. As a society, we often fault the driver for being inconsiderate, aggressive, unaware, or otherwise not paying attention. Behavior has a large impact on performance and vice versa. That is, a driver who has just run a red light (performance) may have done so for several reasons that include a general propensity to take risks, not being able to stop in time, or just not paying attention given the context. In reality, the driver may be compensating for changes given the driving environment. Alternatively, some drivers are just better at assessing the situation. Relating back to the previous section, a cluster analysis can help identify these subgroups of drivers.

The understanding of individual differences will help us understand how to better design the system given the human's limitation. Each time the human operator uses the system, he or she becomes more familiar with the operational limitations of the system. That said, there are also situations that the system has not encountered. Humans often extrapolate the ability of the system to a broader context than what has been experienced in the past from a specific context. The system worked in the rain, so it should also work for all adverse weather conditions (also see Chapter 18).

Humans use their sensory perception and expert knowledge to predict the vehicle's future behavior for the next timeframe (Prokop, 2001). As drivers become more familiar with the use of the system, their driving performance may alter accordingly. It is possible that the human's belief in the mode of operation may not match the actual level of automation in operation at that moment in time. And while a vehicle may be operating at Level 2 at the operational level, it is not necessarily operating at that same level strategically and tactically. Hence, the probability of a mismatch is impacted by the amount of exposure the road user has in that level of automation for a particular road, environmental and traffic situation. Of interest is the amount of time that a human operator has to make a good decision once they are aware that a decision has to be made. These complex datasets can help us observe the planning involved in curve negotiations, left turn decisions, and route selection for various situations. They can help us answer research questions related to the human operator's engagement and performance in the system over time and space:

- How do drivers respond to transitions in automation states?
- Will the human operator know when the automation fails?
- Will the human operator have sufficient time to take over when the automation fails?
- How quickly can drivers resume control from a highly automated vehicle, and does this resumption change with greater system exposure?
- What are the negative or positive effects of automation on driver's performance over time?

Some example research related to these questions can be found in Merat, Jamson, Lai, Daly, and Carsten (2014) and in many chapters in this Handbook; however, clearly more research is needed. For example, while Merat et al. (2013) showed that it can take up to 15 seconds for drivers to resume control, other studies show takeover times from 4 to 6 seconds (Eriksson & Stanton, 2017; Gold, Happee, & Bengler, 2018) depending on other road, traffic, and driver conditions. As noted earlier, the examination of takeover time is often examined in a driving simulator, and these studies used precisely that data collection tool with some studies supplemented with eye trackers.

23.4.2 SYSTEM BEHAVIOR

A model that includes the system behavior should consider the capability of the system to adjust and maintain path tracking, heading, and stability of motion while effectively varying environments given internal and external disturbances. This requires information over time and for various locations. While most systems are based on algorithms that try to predict the next action given past actions, not all future actions can be predicted. The algorithm is only as good as the data that is provided. Hence, the research questions that can be examined with complex datasets include:

- In what context will the automation fail? When will the automation fail?
- How can the system be designed to prevent these failures?
- Will the vehicle be able to stop in time when the car in front is closing in?
- Will the vehicle be able to detect when a pedestrian will cross in front of it?
- How did the combined lateral and longitudinal control system operate?
- What operational conditions influenced the availability of the automated functions?

23.4.3 HUMAN–SYSTEM INTERACTION

The previous research questions in this section are from the driver's or vehicle's perspective. However, there are also research questions that need to be addressed given the interactions between the human and system. For vehicle–pedestrian interactions, such questions can include the following: "Will the pedestrian trust that the vehicle will stop as he or she is crossing the street?." On-road data may include videos of pedestrian and vehicle interactions, and these videos can be studied to identify the scenarios that might be considered dangerous for pedestrians and also likely to result in potential vehicle–pedestrian conflicts (Tian et al., 2014). Other questions related to the interaction between the human and system include:

- Were the combined braking and steering controls functionally compatible with driver expectations for the automation?
- Were driver expectations of the advanced vehicle system consistent across various types of roadways, driving conditions, speeds, etc.?

23.5 ANALYTICAL TOOLS

The overarching goal of this chapter was to provide readers ways to make sense of behavior in complex datasets. The focus has been on understanding the data, the user, and the context. This understanding will help researchers better define the outcomes of interest as well as the analytical method that would be best to address the research questions. As noted earlier, the on-road and simulator studies are often supplemented with eye trackers, heart rate monitors, EEGs, and even tools that collect body posture (Lotz & Weissenberger, 2018). Once these datasets are integrated together, the analyst then decides which variables should be included in the analytical model.

The value of data collected from on-road and simulator studies is the ability to capture changes over time and location. Some examples of dependent measures (or outcomes) of interest include takeover time, takeover probability, time-to-collision, and crash probability (Gold et al., 2018). There are many analytical models that can be used to examine these dependent variables, and while there are many perspectives (see Donoho, 2015), the models can be generally grouped into three categories: exploratory, inferential, and predictive.

23.5.1 EXPLORATORY MODELS

There are various models to help you segment the operator's behavior into subgroups. These models are used to group data together based on some measure of inherent similarity (i.e., taxonomies). In machine learning terminology, these exploratory models are considered unsupervised because they are centered on the tools that assume no pre-defined groups or classes. The most common include principal components analysis (PCA), cluster analysis, and factor analysis (FA).

PCA is a dimension reduction tool. It converts a set of observations of possibly correlated variables into a set of linear uncorrelated variables. FA is also used to convert a set of observations to a set of uncorrelated variables. However, unlike PCA, the goal of FA is to find hidden (or latent) factors in the data. Cluster analysis was described earlier. It is a method used to discover groups in the data. It is often used as a preprocessing step for other algorithms.

For each of these tools, there are many variations on how data is reduced, variables are uncovered, and groups are discovered. There are also time- and space-varying versions of these analytical methods (e.g., functional PCA).

23.5.2 INFERENTIAL MODELS

Inferential models are used to identify a set of variables that can help explain whether an observed pattern (e.g., takeover when automation fails) will likely be observed. This is the most common statistical analysis tool and is used with a formal research hypothesis. The goal is to assess whether sampled data can be used to infer something about the larger (or entire) population. Analysis of variances constructed from controlled experiments would be placed in this category.

The examination of data collected to examine autonomous vehicles often suggests that there are associations between some outcome and the various explanatory variables. The datasets described in this chapter often include multiple observations per participants. Hence, there is a need to consider the presence of serial correlation (measures in one time period are related to the measure in the next time period) and spatial autocorrelation (correlations exist between geographical locations). In terms of visualizing the autocorrelation, the use of a variogram (Gringarten & Deutsch, 2001) is a useful tool to get a sense of the degree and range of autocorrelation in your data. On-road data contain such time separated data and widely used models for accounting for the serial correlations including auto-regressive moving averages (ARMA) and auto-regressive integrated moving average (ARIMA).

Because multiple observations are collected for each participant, the covariance structure is an important consideration. Models often used to account for subject-specific effects are random coefficient models. Depending on the research domain, these modeling tools are also called linear mixed-effects models, multi-level models, or hierarchical linear models (Verbeke, Molenberghs, & Rizopoulos, 2010). Analysis of variance with repeated measures can also be considered in this category. In general, these models provide a way to account for both population average (fixed effects) and subject-specific effects (random effects).

In the driving domain, these models have been used to compare the effects of different levels of automation while drivers are engaged in secondary tasks (Naujoks, Purucker, & Neukum, 2016), and for evaluating the safety effects of bicycle tracks (or dedicated bike lanes) at intersections (Zangenehpour, Strauss, Miranda-Moreno, & Saunier, 2016). In both these examples, each person has their own random intercept and slope. The random intercept and slope from a random coefficient model allow one to predict the responses for a new subject who is not in the study sample but from the same subject population, given the coefficient estimates on a "similar" subject. It can also provide information on the variability of the responses among subjects over time. The model would also include an estimated intercept and slope for the population (fixed effect).

Another model that has been used in the transportation field for longitudinal data is the general estimating equation (GEE). This model is used to estimate the parameters of a generalized linear model when the correlation between outcomes is unknown. That said, some suitable correlation matrix method is needed to provide a general data structure for the model (Ziegler & Vens, 2010). GEE has been used to examine the driver's likelihood to intervene when adaptive cruise control was activated (Xiong & Boyle, 2012) and to the effect of system information on a driver's decision to takeover while driving a partially automated car (Boelhouwer, van den Beukel, van der Voort, & Martens, 2019).

Both modeling techniques can capture specific variance–covariance structures. The difference is that the GEE is often preferred when one is interested in uncovering population average effects while random coefficient models are preferred when uncovering effects specific to individuals (or subjects), especially if the individual differences are large.

23.5.3 PREDICTIVE MODELS

Predictive models are models that can help you predict one measurement (outcome) from a set of measurements (features). This modeling technique uses a combination of data mining and probability to forecast outcomes. The predictor variables are those that are likely to influence future results. Many inferential and classification models can be used as predictive models. These include logistic regression, linear regression, and confirmatory FA. There are some subtle differences. The most important difference is whether the analyst is interested in inferring something about the human or system (e.g., will the human operator trust the system) or predicting whether the system will work in some context (e.g., the system will work in snowy conditions). Predictive models do not necessarily explain why a predictor variable is a good selection, just that it can help improve prediction. Another difference is in how the sampled data is used. In inferential models, the entire sampled data are used to create the model. In predictive models, the sampled data are often separated into training and testing sets. And only the training set is used to create (and train) the model. This is important as the analyst needs to be able to test the predictive ability of the model on a different dataset. The testing involves assessing the accuracy, precision, and recall. Accuracy provides an overall measure of how many events were correctly predicted. Precision is a measure of exactness or quality. It identifies the proportion of events that was correctly classified into any predicted level. And recall is a measure of completeness or quantity (i.e., the proportion of correctly predicted events to an assigned level). There is also a fourth measure, F1, used for when you want to examine the balance between precision and recall.

There are actually many predictive models. For complex datasets, more sophisticated models are often used, such as Artificial Neural Networks (ANN), Classification and Regression Tree (CART), Random Forest, Support Vector Machine, and XGBoost. It is important to test out different models to find the best fit for the data you are working with.

Another modeling framework that has been used for predictive modeling of multidimensional time-series is hidden Markov models (HMMs). This modeling framework has been considered for predicting the human's ability to transition from an automated to a manual state (Janssen, Boyle, Kun, Ju, & Chuang, 2019). It can be used to formalize the beliefs of the human and the expected mode in which a semi-automated vehicle is operating (Janssen et al., 2019). The reason it can do this is based on the Markov process assumption that the "future is independent of the past given the present." In other words, assuming we know our present state, we do not need additional historical information to predict the future state. This process assumes that there are unobservable states with different state transition probabilities. In a Markov process, each state only depends on the previous state and not on any other prior states.

Because HMMs are probabilistic models, they allow the joint probability of a set of hidden (or latent) states to be computed given a set of observed states. And once we know the joint probability of a sequence of hidden states, we can then determine the best possible sequence and better predict the human's ability to take over when needed.

23.6 CONCLUSION

This chapter describes four things needed to make sense of behavior in complex datasets. First, know your data collection tools and what type of variables can be collected from these tools. Second, know your data and the context that the data was collected. While data cleaning can be time consuming, it is an important first step after you collect data. There are many ways to look at your data. The simplest is to do point estimations (means, medians, standard deviation), but complex data often require data visualization and that requires time to explore. As Yanir Seroussi notes,[7] good data scientists "… are ready to get their hands dirty by writing code that cleans the data…" Third, know the research questions to ask to create good, thoughtful controllers. This is achieved by understanding the goals of the controller as well as their limitations. Lastly, know the analytical methods and the goal of the analysis; are you interested in data exploration, inferring the likelihood of a system to be used correctly, or predicting when the system will fail?

One of my graduate students gave the following advice at a panel session on naturalistic data, "Don't be scared." Alternatively, I have colleagues who believe graduate students should, "be scared, be very scared." Depending on the day, I think that both sentiments can be true. Complex data merit extreme respect because they contain many pitfalls and we need to proceed carefully and with humility.

Finding the event of interest may be a challenge as it is often rare, and some events, such as a crash on a freeway, may not be observed even after extensive data collection efforts. However, capturing the differences in behavior can help ensure that automated systems can enhance traffic flow, that drivers can be safer, and that our transportation network accommodates a wide range of road users with comfort.

REFERENCES

Billieux, J., Maurage, P., Lopez-Fernandez, O., Kuss, D. J., & Griffiths, M. D. (2015). Can disordered mobile phone use be considered a behavioral addiction? An update on current evidence and a comprehensive model for future research. *Current Addiction Reports*, 2(2), 156–162.

Boelhouwer, A., van den Beukel, A. P., van der Voort, M. C., & Martens, M. H. (2019). Should I take over? Does system knowledge help drivers in making take-over decisions while driving a partially automated car? *Transportation Research Part F: Traffic Psychology and Behaviour*, 60, 669–684

Boyle, L.N. & Lee, J.D. (2010). Using driving simulators to assess driving safety. *Accident Analysis & Prevention, 42*(3), 785–787.

Carsten, O., Kircher, K., & Jamson, S. (2013). Vehicle-based studies of driving in the real world: The hard truth? *Accident Analysis & Prevention, 58*, 162–174.

Consiglio, W., Driscoll, P., Witte, M., & Berg, W. P. (2003). Effect of cellular telephone conversations and other potential interference on reaction time in a braking response. *Accident Analysis & Prevention, 35*(4), 495–500.

Dai, D. J. & Jaworski, D. (2016). Influence of built environment on pedestrian crashes: A network-based GIS analysis. *Applied Geography, 73*, 53–61. doi:10.1016/j.apgeog.2016.06.005

---

[7] https://yanirseroussi.com/2014/10/23/what-is-data-science/.

Donmez, B., Boyle, L., & Lee, J. D. (2009). Designing feedback to mitigate distraction. In
    M. A. Regan, J. D. Lee, & K. Young (Eds.), *Driver Distraction: Theory, Effects, and
    Mitigation*. Boca Raton, FL: CRC Press.

Donoho, D. (2015, September 18). 50 years of Data Science. *Based on a Presentation at the
    Tukey Centennial Workshop*, Princeton, NJ. Retrieved from http://courses.csail.mit.
    edu/18.337/2015/docs/50YearsDataScience.pdf

Edwards, C. J., Creaser, J. I., Caird, J. K., Lamsdale, A. M., & Chisholm, S. L. (2003).
    Older and younger driver performance at complex intersections: Implications for
    using perception-response time and driving simulation. *Proceedings of the Second
    International Driving Symposium on Human Factors in Driver Assessment, Training
    and Vehicle Design* (pp. 33–38). Iowa City, IA: Public Policy Center.

Eriksson, A. & Stanton, N. A. (2017). Takeover time in highly automated vehicles: Noncritical
    transitions to and from manual control. *Human Factors, 59*(4), 689–705.

Fisher, D. L., Rizzo, M., Caird, J., & Lee, J. D. (2011). *Handbook of Driving Simulation for
    Engineering, Medicine, and Psychology*. Boca Raton, FL: CRC Press.

Gold, C., Happee, R., & Bengler, K. (2018). Modeling take-over performance in level 3 con-
    ditionally automated vehicles. *Accident Analysis & Prevention, 116*, 3–13

Gringarten, E. & Deutsch, C. V. (2001). Teacher's aide variogram interpretation and model-
    ing. *Mathematical Geology, 33*(4), 507–534.

Janssen, C. P., Boyle, L. N., Kun, A. L., Ju, W., & Chuang, L. L. (2019). A hidden Markov
    framework to capture human–machine interaction in automated vehicles. *International
    Journal of Human–Computer Interaction, 35*(11), 947–955.

Kidd, D. G., Tison, J., Chaudhary, N. K., McCartt, A. T., & Casanova-Powell, T. D. (2016).
    The influence of roadway situation, other contextual factors, and driver characteris-
    tics on the prevalence of driver secondary behaviors. *Transportation Research Part F:
    Traffic Psychology and Behaviour, 41*, 1–9.

Kim, J. K., Ulfarsson, G. F., Shankar, V. N., & Mannering, F. L. (2010). A note on model-
    ing pedestrian-injury severity in motor-vehicle crashes with the mixed logit model.
    Accident Analysis & Prevention, 42(6), 1751–1758.

Kim, M., Kho, S. Y., & Kim, D. K. (2017). Hierarchical ordered model for injury severity of
    pedestrian crashes in South Korea. *Journal of Safety Research, 61*, 33–40. doi:10.1016/j.
    jsr.2017.02.011

Konecni, V., Ebbeson, E. B., & Konecni, D. K. (1976). Decision processes and risk taking
    in traffic: Driver response to the onset of yellow light. *Journal of Applied Psychology,
    61*(3), 359.

Lappi, O. (2015). Eye tracking in the wild: The good, the bad and the ugly. *Journal of Eye
    Movement Research, 8*(5), 1–21.

Lee, J. D., Caven, B., Haake, S., & Brown, T. L. (2001). Speech-based interaction with in-
    vehicle computers: The effect of speech-based e-mail on drivers' attention to the road-
    way. *Human Factors, 43*(4), 631–640.

Lee, J. D., Young, K.L., & Regan, M.A. (2009). Defining driver distraction. In M. A. Regan,
    J. D. Lee, & K. L. Young (Eds.), *Driver Distraction: Theory, Effects, and Mitigation*
    (pp. 31-40). Boca Raton, FL: CRC Press.

Lotz, A. & Weissenberger, S. (2018). Predicting take-over times of truck drivers in condi-
    tional autonomous driving. *International Conference on Applied Human Factors and
    Ergonomics* (pp. 329–338). Berlin: Springer.

Merat, N., Jamson, A. H., Lai, F. C., Daly, M., & Carsten, O. M. (2014). Transition to
    manual: Driver behaviour when resuming control from a highly automated vehicle.
    Transportation Research Part F: Traffic Psychology and Behaviour, 27, 274–282.

Miller, E. E. & Boyle, L. N. (2019). Adaptations in attention allocation: Implications for take-
    over in an automated vehicle. *Transportation Research Part F: Traffic Psychology and
    Behaviour, 66*, 101–110.

Naujoks, F., Purucker, C., & Neukum, A. (2016). Secondary task engagement and vehicle automation–Comparing the effects of different automation levels in an on-road experiment. *Transportation Research Part F: Traffic Psychology and Behaviour, 38*, 67–82.

Nielsen, T. A. S. & Haustein, S. (2018). On sceptics and enthusiasts: What are the expectations towards self-driving cars? *Transport Policy, 66*, 49–55.

Oviedo-Trespalacios, O., Haque, M. M., King, M., & Washington, S. (2019). "Mate! I'm running 10 min late": An investigation into the self-regulation of mobile phone tasks while driving. *Accident Analysis & Prevention, 122*, 134–142.

Papaioannou, P. (2007). Driver behaviour, dilemma zone and safety effects at urban signalised intersections in Greece. *Accident Analysis & Prevention, 39*(1), 147–158.

Peng, Y., Boyle, L.N., Ghazizadeh, M., & Lee, J. D. (2013). Factors affecting glance behavior when interacting with in- vehicle devices: Implications from a simulator study. *Proceedings of the Seventh International Driving Symposium on Human Factors in Driver Assessment, Training and Vehicle Design* (pp. 474–480). Iowa City, IA: University of Iowa. doi:10.17077/drivingassessment.1529

Prokop, G. (2001). Modeling human vehicle driving by model predictive online optimization. *Vehicle System Dynamics, 35*(1), 19–53.

Ranney, T. A., Simmons, L. A., & Masalonis, A. J. (1999). Prolonged exposure to glare and driving time: Effects on performance in a driving simulator. *Accident Analysis & Prevention, 31*(6), 601–610.

RSG. (2018). *Draft Final Report: 2017 Puget Sound Regional Travel Study.* Retrieved from www.psrc.org/sites/default/files/psrc2017-final-report.pdf

Salvucci, D. D. (2002). Modeling driver distraction from cognitive tasks. *Proceedings of the Annual Meeting of the Cognitive Science Society, 24*(24), 792-797.

SEMCOG. (2019). *High-Frequency Crash Locations: Washtenaw County: Washtenaw Ave - Huron Pkwy S Detail Crash List.* Retrieved from https://semcog.org/high-frequency-crash-locations/point_id/81012727/view/individualcrashreport

Strayer, D. L., Cooper, J. M., Turrill, J., Coleman, J., Medeiros-Ward, N., & Biondi, F. (2013). *Measuring Cognitive Distraction in the Automobile.* Washington, DC: AAA Foundation for Traffic Safety. Retrieved from https://aaafoundation.org/wp-content/uploads/2018/01/MeasuringCognitiveDistractionsReport.pdf

Strayer, D. L. & Drews, F. A. (2004). Profiles in driver distraction: Effects of cell phone conversations on younger and older drivers. *Human Factors, 46*, 640–649

Tian, R., Li, L., Yang, K., Chien, S., Chen, Y., & Sherony, R. (2014). Estimation of the vehicle-pedestrian encounter/conflict risk on the road based on TASI 110-car naturalistic driving data collection. *2014 IEEE Intelligent Vehicles Symposium Proceedings* (pp. 623–629). Piscataway, NJ: IEEE.

Verbeke, G., Molenberghs, G., & Rizopoulos, D. (2010). Random effects models for longitudinal data. In K. van Montfort, J. H. L. Oud, & A. Satorra (Eds.), *Longitudinal Research with Latent Variables* (pp. 37–96). Berlin: Springer.

Xiong, H. & Boyle, L. N. (2012). Drivers' adaptation to adaptive cruise control: Examination of automatic and manual braking. *IEEE Transactions on Intelligent Transportation Systems, 13*(3), 1468–1473.

Young, R. A. (2013). Naturalistic studies of driver distraction: Effects of analysis methods on odds ratios and population attributable risk. *Proceedings of the 7th International Driving Symposium on Human Factors in Driver Assessment Training and Vehicle Design* (pp. 509–515). Iowa City, IA: Public Policy Center.

Zangenehpour, S., Strauss, J., Miranda-Moreno, L. F., & Saunier, N. (2016). Are signalized intersections with cycle tracks safer? A case–control study based on automated surrogate safety analysis using video data. *Accident Analysis & Prevention, 86*, 161–172.

Ziegler, A. & Vens, M. (2010). Generalized estimating equations. *Methods of Information in Medicine, 49*(5), 421–425.

24 Future Research Needs and Conclusions

Donald L. Fisher
Volpe National Transportation Systems Center

William J. Horrey
AAA Foundation for Traffic Safety

John D. Lee
University of Wisconsin-Madison

Michael A. Regan
University of New South Wales

CONTENTS

Key Points 520
24.1 Introduction 520
24.2 The State of the Art: ACIVs 521
24.3 Issues in the Deployment of ACIVs (Problems) 522
    24.3.1 Drivers' Mental Models of Vehicle Automation (Chapter 3) 522
    24.3.2 Driver Trust in ACIVs (Chapter 4) 523
    24.3.3 Public Opinion about ACIVs (Chapter 5) 523
    24.3.4 Workload, Distraction, and Automation (Chapter 6) 524
    24.3.5 Situation Awareness in Driving (Chapter 7) 525
    24.3.6 Allocation of Function to Humans and Automation and the
           Transfer of Control (Chapter 8) 525
    24.3.7 Driver Fitness in the Resumption of Control (Chapter 9) 526
    24.3.8 Driver Capabilities in the Resumption of Control (Chapter 10) 527
    24.3.9 Driver State Monitoring for Decreased Fitness to Drive
           (Chapter 11) 527
    24.3.10 Behavioral Adaptation and ACIVs (Chapter 12) 528
    24.3.11 Distributed Situation Awareness (Chapter 13) 528
    24.3.12 Human Factors Issues in the Regulation of Deployment
           (Chapter 14) 529
24.4 Human-Centered Design of ACIVs (Solutions) 529
    24.4.1 HMI Design for ACIVs (Chapter 15) 529
    24.4.2 HMI Design for Fitness-Impaired Populations (Chapter 16) 530

24.4.3 Automated Vehicle Design for People with Disabilities (Chapter 17)... 530
24.4.4 Importance of Training for ACIVs (Chapter 18) 531
24.5 Special Topics 531
24.5.1 Connected Vehicles in a Connected World: A Sociotechnical
Systems Perspective (Chapter 19) 532
24.5.2 Congestion and Carbon Emissions (Chapter 20) 532
24.5.3 Automation Lessons from Other Domains (Chapter 21) 532
24.6 Evaluation of ACIVs 533
24.6.1 HF Considerations in Testing and Evaluating ACIVs (Chapter 22) 533
24.6.2 Techniques for Making Sense of Behavior in Complex Datasets
(Chapter 23) 533
24.7 Conclusions 534
Acknowledgements 534
References 534

KEY POINTS

- The exact point at which a given vehicle technology will be present in the majority of the vehicle fleet is very difficult to predict, but no one expects Level 4 or 5 technologies to be a majority of the fleet in the next ten years.
- Given that the majority of the vehicle fleet for the next ten years will be Level 0–3 vehicles, along with vehicles which have active safety systems, the overwhelming majority of the chapters in the Handbook are relevant to today's human factors concerns and those concerns for at least the next decade.
- The best summaries of the future research needs are in the chapters themselves.
- This chapter focuses on the various research needs in each topical area which the editors believe are most critical, based on their broad reading of all of the chapters.

24.1   INTRODUCTION

Our collective experience professionally, as editors of this Handbook, and as avid readers of the chapters contained herein, provides us with a perspective on the issues surrounding the human factors concerns that are raised by the advance of automated, connected, and intelligent vehicles (ACIVs) that we would like to share with our readers. This experience may make us blind to the real concerns, but we hope not. And in so far as it does, you should turn, as we do, to the chapters themselves where the real expertise and wisdom lies.

So, what we want to do in closing is give readers the editors' reflections on the chapters and topics, which includes in addition to summaries of the chapters our best sense of the most critical human factors research, development, practice, planning, and/or policy needs as relevant to the chapter being discussed. We encourage you to disagree, to push the limits of what is possible, in whatever area is of most interest to you. We certainly do not have privileged access to what will unfold in the future. With that as background, we hope you find at least one or two nuggets in what follows.

24.2 THE STATE OF THE ART: ACIVS

The future of ACIVs has always been just around the corner, at least since the late 1930s (Chapter 2). But it has stalled several times since. The question facing us today is whether we are at another inflexion point. The author of Chapter 2 supplies example after example of the steady march forward of ACIVs. The progress truly is stunning and we encourage readers to go back to this chapter whenever they doubt the steady progress of technology. In fact, when we, the editors, started this Handbook we were not sure it would be relevant because industry leaders predicted that driverless cars would be on our streets by 2018 (Yadron, 2016). Yet a recent article in the *New York Times* in July of 2019 (Boudette, 2019) suggests that driverless cars are far into the future. And why?

The answer is telling: "the delay [is due] to something as obvious as it is stubborn: human behavior (Boudette, 2019)." This is much like blaming the victim. The human driver may be stubborn; but he or she is also very good at dealing with *ordinary* driving, which can involve complex negotiations between drivers in unexpected circumstances (as opposed to situations where, say, automatic emergency braking can be a real life saver). It is our advanced technologies that cannot yet do in many cases what the human driver does more or less faultlessly, human drivers having only 1.25 fatal crashes per 100 million miles. Yet we are now asking Level 1–3 automobiles to perform the functions that humans now perform so well: ordinary driving. But having Level 1–3 technologies substitute for what the human driver does can create unexpected problems.

The various chapter authors have argued throughout the Handbook why it is that Level 1–3 vehicles create human factors challenges. As has been pointed out so eloquently by the authors, drivers of current vehicles at these levels of automation can easily become disengaged, lose situation awareness (SA), and fail to take over control when necessary or do so too slowly to avoid a crash. A number of vehicle manufacturers are trying to change the image the public has of Level 1–3 vehicles by rebranding them. For example, at the 2019 Consumer Electronics Show, Toyota differentiated between its two new automated vehicle lines, Toyota Guardian (Levels 1–3) and Toyota Chauffeur (Levels 4–5). The CEO of the Toyota Research Center says of the Guardian mode: "With Guardian, the driver is in control of the car at all times except in those cases where Guardian anticipates a pending incident, alerts the driver and decides to employ a corrective response in coordination with driver input…. In this way, Guardian combines and coordinates the skills and strengths of the human and the machine (Global Toyota, Feature Innovation Automated Driving Technology Region U.S, 2019)." This concept (the car as Guardian rather than Chauffeur) was referred to as directed automation in Chapter 4 and its importance discussed there.

This appears to be a real shift from an emphasis on "driverless" cars to cars which serve as a backup for the human driver when that driver is no longer able to control the vehicle. The research question that really needs to be asked here is whether a vehicle that serves as a guardian can actually be developed, a vehicle in which the driver does have full control and, critically, does not disengage, until he or she is no longer capable of avoiding a crash. It simply is not known how much control the driver needs to maintain over the vehicle in order to keep engaged. Some vehicle manufacturers now allow drivers to keep their hands off the wheel as long as their eyes are on the road for some period of time in a given interval. Other vehicle manufacturers allow drivers to do almost anything as long as their hands are on the wheel for some period of time in a given interval. However, no one knows what exactly a driver must do to remain safely engaged at all points in time during a trip, knowing that the car he or she is driving will serve as a guardian if the situation warrants it. Research is needed into the most basic of questions.

24.3 ISSUES IN THE DEPLOYMENT OF ACIVS (PROBLEMS)

Before deploying ACIVs, it will be critical to understand issues centering on the driver—alone in the vehicle or embedded within a larger system, on the public's acceptance of such vehicles, and on the regulation of these vehicles. The key questions in our opinion that need to be addressed in each of the topics covered in the section on deployment (Chapters 3–14) are discussed briefly below.

24.3.1 DRIVERS' MENTAL MODELS OF VEHICLE AUTOMATION (CHAPTER 3)

Mental models might be one of the most critical considerations when dealing with the intersection between humans and automated systems. They inform almost every aspect of the driver's interaction with the system: when and how to use the system and what they do when the system is engaged, to name only a few. And, some of these interactions will have cascading effects on other aspects of behavior and performance (e.g., workload, situation awareness (SA), etc.). It is also easy to see connections between a driver's understanding of a system and their trust and acceptance of the system. In spite of their importance in this space, there is still much we do not know about driver's mental models of automated systems. We do know that drivers show significant deficits in their understanding of automated systems, including knowledge of their functional limitations (e.g., McDonald et al., 2018). We know less about how these gaps in understanding map directly onto driver decisions and behavioral and safety outcomes. More work is needed and, fortunately, many enterprising organizations are working to provide data points on this front. At the same time, we need to further develop our understanding of how mental models are formed in the first place and what types of information, branding, experiences, and so on might help to improve user's mental models. This can have important implications for automobile manufacturers as they consider the development of owner's manuals and related resources and for dealerships as well as marketing folks. Already, as part of this, we are seeing some initiatives to promote a common nomenclature for the systems comprising market-ready automation (e.g., AAA, 2019). This also will have an impact on our approaches to driver training and education. Ultimately, we would like to know the best way of tailoring a driver's mental model to the system that is present in their own vehicle. The list of specific research needs goes well beyond the space afforded in this chapter—suffice it to say that there is much work to do.

24.3.2 Driver Trust in ACIVs (Chapter 4)

Like mental models, trust has a pervasive influence on how people rely on and accept technology. Inadequate trust leads people to reject potentially useful technology and too much trust leads people to rely on technology when they shouldn't. Trust becomes an increasingly powerful influence as the complexity of automation makes it difficult to form a complete mental model. The advances in automotive technology confront drivers with a more diverse and complicated set of devices, and so trust is likely to play an increasingly important role in how drivers interact with vehicles.

Two general areas of research merit particular attention. The first concerns overtrust and the automation paradox. Efforts to improve the performance of automation makes it less likely that drivers will confront edge cases where they will need to take control. The less often a driver needs to take control the fewer opportunities drivers have to learn where the automation might fail, leaving the driver less able to build up enough exposure to systems to accurately calibrate their trust. How might feedback from the automation be accentuated to enhance exposure to systems? How might drivers be able to interact with automated systems and to generate feedback? Such design concepts require evaluation, which prompts the question: what techniques can be used in experimental settings to evaluate trust (often in absentia of long-term exposure)? The second general research area concerns the role of trust in driverless vehicles where the people take on the role of passengers. Such situations introduce a broad range of trust relationships that must be supported: between the passenger and the vehicle, between and among passengers in shared-ride situations, and between passengers and the companies operating the vehicles. People need to trust a vehicle to safely transport them to their destinations, but driverless vehicles will likely leave passengers with little opportunity to influence the vehicle, which might lead to feelings of dread risk that can undermine trust. What interaction architectures and interfaces can mitigate dread risk and promote trust? Because driverless vehicles are unlikely to be owned by individuals, an important research question is how to promote trust and trustworthy behavior in those who might share the vehicle. Likewise, companies operating vehicles will control people in new ways, raising the question of how to design transport services to operate in a trustworthy manner?

24.3.3 Public Opinion about ACIVs (Chapter 5)

As vehicles equipped with automated driving systems start entering the market, it is important to gauge public opinion about them—even if the public at large has had little or no direct exposure to them. This point was underscored by the authors in Chapter 5 and elsewhere (Cunningham, Regan, Horberry, & Dixit, 2019). As noted by these authors, an understanding of public opinion about automated driving systems can benefit different stakeholders in society in different ways: benefit governments, for example, in making future planning and investment decisions; benefit industry, for example, in helping technology developers design and refine their products in response to the perceived needs of end users; and benefit research organizations, in identifying new directions for research. Ultimately, as argued in Chapter 5, the predicted benefits of automated vehicles may never materialize unless there is societal acceptance of them. If not, people may refuse to purchase them, refuse to travel in them, or interact with them in ways unintended by designers.

The following are some research needs in this area, as distilled from Chapter 5 and elsewhere (Cunningham, Regan, Horberry, & Dixit, 2019). First, most studies of public opinion of automated vehicles have, to date, been cross-sectional in design, precluding formulation of any conclusions about causality between the constructs of interest. Future studies would benefit from the use of longitudinal designs to track changes in public opinion over time and discern the factors that underlie identified changes. Second, highly and fully automated passenger vehicles (SAE Levels 3–5) are not yet commercially available in large numbers. Consequently, measurement of public acceptability in most studies to date has been based on people having to imagine how such vehicles might operate in the future. An important area for future research is research on acceptance of such vehicles, after people have had direct exposure to them, and to compare the findings with those obtained prior to exposure—to determine to what extent measures of acceptability are predictive of measures of acceptance. Third, there is evidence that, while there is some commonality in opinion about automated vehicles across countries and cultures, there is also some divergence of opinion. Further research is needed to understand these differences across a wider range of countries and cultures, to inform local needs and to inform those who seek to market vehicles and systems in other countries and cultures. Finally, there are many population demographic and other variables (e.g., age, gender, willingness to pay) that can be used to predict societal acceptability and acceptance of highly automated driving systems. Further research is needed to understand which of these demographic and other variables, individually and collectively, account for most of the variance in public acceptability and acceptance of these technologies.

24.3.4 WORKLOAD, DISTRACTION, AND AUTOMATION (CHAPTER 6)

The introduction of automation into the vehicle will have an impact on the driver's roles and responsibilities. By extension, their experience of cognitive, visual, and manual workload will change, leading possibly to underload or to increased engagement in other tasks. These concerns play into other topics, such as SA (Chapter 7) and driver fitness to resume control (Chapter 9). Although research is now documenting the impact of vehicle automation on driver workload, arousal, and in-vehicle activities (e.g., potentially distracting tasks), there are many questions as of yet unanswered. Where the interface between driver workload and driver distraction is concerned, a number of issues warrant further research. For example, to what extent do prolonged periods of automated driving (low workload) encourage driver involvement in potentially distracting activities? To what extent does driver distraction lead to automation surprises, and associated spikes in workload, and how? To what extent do drivers self-regulate their workload (monitoring of the forward roadway) when distracted to prepare for takeover when required? In the context of a mixed fleet, will drivers of vehicles not capable of operating autonomously have workload spikes and be distracted by the behavior of other vehicles operating autonomously (in some or all operational design domains)—in much the same way that we are distracted now by drivers who drive erratically in traffic? In addition to the many specific research questions, there are also more general conceptual questions regarding workload, distraction, and automation. For example, will workload, distraction, and inattention, more generally, remain as issues in vehicles equipped with SAE Level 4 and 5 technologies that are operating autonomously (Cunningham & Regan, 2018)? For these levels, is it possible for such self-driving vehicles themselves to be distracted or overloaded (i.e., can the complexity of a situation exceed the processing power of the hardware and software to take the appropriate countermeasures)? We might call this "vehicle distraction" or "vehicle overload." Here, again, the frame of reference from which to conceptualize distraction and workload will change. But what competing activities, if any, could divert a vehicle's "attention" (or computational resources), more generally, away from activities critical for safe driving? In fact, what might it mean for a vehicle driving autonomously to be inattentive or be overloaded; and if it was, what might be the mechanisms of inattention and resource constraints?

24.3.5   SITUATION AWARENESS IN DRIVING (CHAPTER 7)

Situation awareness (SA) characterizes the driver's understanding of the state of the vehicle and the environment that informs decision-making in the complex, dynamic traffic setting. SA involves perceiving and attending to information in the environment, integrating and making sense of this information, and projecting future status based on this information. It follows that SA implicates driver attention, information processing, and various cognitive constructs, such as working memory. As automation changes the driver's roles (Chapter 8), the manner in which drivers regulate their attention in support of SA will likewise change (e.g., Chapter 6, 9). The importance of complete and accurate SA is most salient when considering takeover situations where the driver needs to resume control of the vehicle. Many research questions beg answers (and these are highly intertwined with topics covered in other chapters). For example, what is the best way to keep the driver engaged (i.e., with good SA) in the driving task when the automation is operating? What factors degrade driver SA in advance of, during, and after takeover situations and by how much? While some of the solutions described below might offer possible benefits, there is a definite need for more data.

24.3.6   ALLOCATION OF FUNCTION TO HUMANS AND AUTOMATION
AND THE TRANSFER OF CONTROL (CHAPTER 8)

The allocation of different driving functions, that is, the assignment of different responsibilities to either the driver or the automated system, has been the topic of much debate. Indeed, the decisions regarding this assignment have significant impacts on many of the topics discussed throughout this Handbook (e.g., workload, Chapter 6; SA, Chapter 7, driver impairments, Chapter 9, etc.). Functional allocation also factors into many of the discussions regarding design solutions (e.g., adaptive automation, Chapter 16). Discussions of allocation of function date back to the 1950s, where common guidance was based on who did a particular task better: humans or machines? If humans were superior at a task, then they should be responsible for that task (and vice versa). If one thing should be clear from the Handbook and Chapter 8 in particular, it is that the situation is not so straightforward. Humans are very good at driving and its subtasks; the crash rate per mile driven is very low (while acknowledging that every crash is a tragic event to be avoided). Automation is very good at performing many driving tasks as well. Therein lies one challenge: the degree of overlap in those tasks that both drivers and automation are good at. Who should be assigned what task? Another challenge relates to those tasks and subtasks at which drivers and automation are not very adept (e.g., driving in adverse conditions). As the authors note, this is especially critical when the human–automation system needs to negotiate control transfers. The appropriate allocation of functions and authority continues to be an important area for research, including questions of can a function be safely automated, should it be, and what happens when situational factors change? Unfortunately, or perhaps ironically, some of the decisions regarding the allocation of function are products of a design philosophy and occur far removed from the eventual moment by moment interactions between the driver, system, and the traffic environment. Automation does not simply replace the person in performing certain functions, but creates new functions associated with coordinating the automation, an important research question remains: How to anticipate and support that coordination?

24.3.7 Driver Fitness in the Resumption of Control (Chapter 9)

If the functions have been allocated appropriately to the human and automation, it is still the case that problems can arise in the transition of control from the vehicle back to the driver. As discussed by the authors in Chapter 9, the driver may not be fit to take back control for several different reasons, including distraction, fatigue, impairment, and simulator sickness. One question looms large. In particular, consider the general issue of impairment. The issue around impairment becomes more central every day as more and more states legalize marijuana, as the number of prescription and over-the-counter (OTC) drugs increases, especially among the elderly, and as the number of illegal drugs spread rapidly throughout the population. Recent estimates in the United States suggest that prescription, OTC, and illegal drugs are present in some 13% of the driving population (Kelley-Baker, Berning, Ramirez, Lacey, & Compton, 2017). But prevalence is not the only problem. Testing for the presence of marijuana or any of the myriad prescription, OTC and illegal drugs in blood or breath specimens drawn at the roadside can be very difficult for a wide variety of reasons. A number of countermeasures have been proposed (Smith, Turturici, & Camden, 2018). But none have been suggested that use real-time driver and vehicle monitoring algorithms to identify behavioral measures of impairment due to prescription, OTC, and illegal drugs [the Driver Alcohol Detection System for Safety is already being explored for alcohol before the driver actually starts driving (NHTSA, 2016)]. If a real-time behavioral test were used, it would not matter what substance was on-board, if any. This is largely a scientific question at this point. Actual application of such a behavioral evaluation of impairment based on real-time monitoring would require legal and other hurdles to be overcome. But, until one asks whether such a system has a high predictive validity and very few false alarms, one simply cannot take the next logical step towards implementation.

24.3.8 DRIVER CAPABILITIES IN THE RESUMPTION OF CONTROL (CHAPTER 10)

Drivers may not be ready to take back control at any given moment during a trip because of issues with their fitness to drive, issues which are transient. But, at a more fundamental level, they may not either currently be able to drive or are at an especially high risk when driving because of limitations in their basic capabilities. Drivers with such limitations include those who are medically at risk, those with visual impairments, and those with neurological and neurodegenerative impairments. For this class of drivers, the authors of Chapter 10 focus on the core clinical characteristics, the associated functional performance deficits, the effect of these deficits on driving behaviors, and the potential of automated vehicle technologies (SAE Level 0, 1, and 2) to mitigate the effects of the functional performance deficits and to enable the driver to resume driving. Although the empirical literature supports most of the sections in the chapter, the authors note that the last section (i.e., the potential of automated technologies to reduce performance deficits) represents an attempt on their part to map between the particular performance deficits and the advanced vehicle technologies that might address these deficits, an attempt which is grounded solely in the persuasiveness of the argument. Clearly, there is a pressing need to determine whether the advanced vehicle technologies can indeed mitigate the potential deficits of those drivers who are medically at risk, who have visual impairments, or who have neurological of neurodegenerative disorders. As just one example, consider older drivers. The authors point to intersection assistance as one technology that could decrease crashes among older drivers at intersections. If this were the case, it could have a large impact on the number of crashes of older drivers given that intersection crashes are among the riskiest and most prevalent for this group of drivers (Mayhew, Simpson, & Ferguson, 2006). Moreover, as pointed out in Chapter 10, a feature such as intersection assistant is also potentially useful for drivers with glaucoma, Parkinson's disease, and younger drivers with attention-deficit hyperactivity disorder. The question of real significance here is now largely a practical one: Will intersection assistant actually reduce the crash rates at intersections as intended?

24.3.9 DRIVER STATE MONITORING FOR DECREASED FITNESS TO DRIVE (CHAPTER 11)

Knowing that the driver is not fit to take back control or capable of more basic control is central to implementing the countermeasures that were described above. Driver state monitoring sensors and algorithms have taken a huge leap forward as discussed by the authors of Chapter 11. Remote sensing of the position of the eyes and the hands has now become standard. For example, the position of the eyes is now used to monitor both visual and cognitive distractions (Seppelt et al., 2017). However, much still remains to be done to sense and monitor drowsiness in real time, especially the occurrence of microsleeps. As noted in a previous chapter, in the United States sleepiness is estimated to be a factor in 100,000 police-reported crashes each year, including over 71,000 injuries and 1,550 fatalities (National Safety Council, 2019). Recent algorithms have been developed which can predict the second microsleep, but not the first (Watson & Zhou, 2016). Improving the algorithms so that they could detect the first microsleep is critical. Moreover, these algorithms still rely on electrocardiogram sensors which must be worn as a vest. Less cumbersome sensors need to be developed to further advance the application of the microsleep prediction algorithms.

24.3.10 BEHAVIORAL ADAPTATION AND ACIVs (CHAPTER 12)

We have spoken above as if drivers' behavior vis-à-vis the different features of an automated vehicle is static for the most part. However, as is made clear by the authors in Chapter 12, the issue of behavioral adaptation to automated vehicles and their various features is a complex one and absolutely critical to understanding the safety benefits of said vehicles (Sagberg, Fosser, & Saetermo, 1997). Ideally, it would be possible to predict with some assurance whether a given feature was going to lead to adaptation, either positive or negative. This issue of behavioral adaptation is made still more complex by the fact that the features of an automated vehicle are constantly changing as new upgrades are made to the software. Behavioral adaptation as it has been traditionally used refers to changes in behavior which occur over time as a function of exposure to a given feature. With automated vehicles this definition still holds. But now, the definition is broadened to include whatever behavioral changes the driver might incorporate into his or her behavior when changing from a manual vehicle to an automated vehicle in order to adapt his or her behavior to the assumed feature or features of the automated vehicle. Regardless of which definition is chosen, it seems that researchers will always be playing catch up unless a model can be developed which will predict with some confidence how drivers will adapt to a given feature, either over time, when changing from a manual to an automated vehicle, or when going from one version of the software to an updated version. Perhaps the development and evaluation of such a general model is too much of a stretch, but it stands as paramount to preventing unexpected effects of automation.

24.3.11 DISTRIBUTED SITUATION AWARENESS (CHAPTER 13)

SA is, as noted previously, an important consideration in the design of automated vehicles and the road systems in which they will operate. The authors of Chapter 13 suggested that distributed situation awareness (DSA) is the most useful construct when considering the design and analysis of automated vehicle systems, primarily because it considers the SA needs of both human and non-human agents as well as the required interactions between them. To demonstrate this approach, they presented an overview of a DSA model and an analysis of the recent Uber–Volvo fatal collision in Tempe, Arizona. In going forward, the authors of Chapter 13 present a framework to support DSA-based design in road transport, which provides an overview of the kinds of analyses required to ensure that DSA can be understood and catered to during autonomous vehicle design life cycles. The framework includes the use of on-road naturalistic studies to examine DSA and road user behavior in existing road environments, the use of systems analysis methods such as Cognitive Work Analysis to identify key design requirements, the use of sociotechnical systems theory (STS) and an STS-Design Toolkit to generate new design concepts, and, finally, the use of various evaluation approaches to evaluate design concepts. This framework may be usefully applied in future to the design, testing, and implementation of automated vehicles.

24.3.12 HUMAN FACTORS ISSUES IN THE REGULATION OF DEPLOYMENT (CHAPTER 14)

In Chapter 14, the author sets out to answer the question "What Human Factors issues need to be considered in preparing policy and regulation for the deployment of automated vehicles?" The focus of the chapter was on how much policy makers need to be involved in attempting to influence: (1) how companies design systems for human users and (2) how humans interact with those systems. As noted by the author, this is difficult in areas of new and evolving technology, where the risks may not yet be clear, let alone optimal solutions to address those risks. Consequently, the author argued that policy makers may need to consider moving towards less prescriptive and more outcomes-, or principles-based, approaches to policy and regulation to ensure that risks are managed and companies and individuals can be held responsible, while allowing for system innovation and evolution, but with less standardization, which can create its own risks. The author underscored in Chapter 14 the need for governments to continually update policy as human behavior changes and the understanding of human risks surrounding automated vehicles evolves, and identified two main areas for future research. First, there is the need for governments to better monitor road safety, due to the data that automated vehicles will provide. Second, there is a strong need for research by governments to identify new risks, to understand their likelihood and severity, and to react accordingly—in relation to the human–machine interface (HMI) between the vehicle and the passengers/driver/fallback-ready user inside the vehicle, how other road users interact with these vehicles, and how their behavior evolves over time, and what this might mean for safety.

24.4 HUMAN-CENTERED DESIGN OF ACIVS (SOLUTIONS)

As was made clear throughout the earlier chapters, the design of the HMI is central to the driver maintaining SA and trust, to the driver understanding the operational design domain, and to helping the driver maintain the right level of workload. The HMI should be designed to ensure that changes in drivers' fitness or capabilities are considered and the appropriate responses are taken. Where possible, the HMI should also actively serve to train the driver how to operate the vehicle, given the usual ways of informing the driver (manuals, dealerships) about the features of automation do not seem to be working and given that the features of the vehicle are constantly changing through real-time software upgrades. As has been previously noted in the Handbook, there is much more to the human-centered design of ACIVs other than just a consideration of the HMI. Some of these considerations are discussed in the chapter on capabilities, but the issues are much larger than that and could be a Handbook in itself.

24.4.1 HMI DESIGN FOR ACIVS (CHAPTER 15)

Design guidelines for interfaces that remain fixed and for vehicle hardware and software that remain fixed were discussed at length by the authors in Chapter 15. As highlighted there, the design guidelines will need to be updated and modified for automated vehicles. The opportunities for teaming the human and the automated vehicle through the use of artificial intelligence (AI) are huge here (Kamar, 2016), not only in the original conception of the software but also for updates to the software. In such a teamed system, the driver will take actions based on recommendations from the AI partner. The combined system (human, automation, and AI) can arguably make better decisions than either component alone. What remains particularly problematic, and has been identified in a recent report as such (Bansal, Nushi, Kamar, Lasecki, & Horvitz, 2019), is the fact that updates to the software are often incompatible with a driver's previous mental model of the software. Thus, the driver ends up making a decision that is incorrect with the updated software that would have been correct with the original software. Even though the updates improve the AI performance they might not improve the combined AI–human system. So, ideally there needs to be backward compatibility of the software updates with the existing mental models. But it is not always the case that such can happen. In such cases one could potentially retrain the driver or share the AI's confidence in a prediction. However, these two approaches come with real drawbacks (Bansal, Nushi, Kamar, Lasecki, & Horvitz, 2019). Research is needed to deal with the backward compatibility problem and all of its ramifications for HMI design.

24.4.2   HMI Design for Fitness-Impaired Populations (Chapter 16)

With automation, the role of the driver will change in the context of an automated system (Chapter 8), and this in turn will impact driver workload and in-vehicle driver behaviors (Chapters 6 and 9). It follows that an understanding of the state of the driver, including their current behaviors, will be an integral part of future vehicle design. This chapter examined the intersection between driver state monitoring (Chapter 11) and adaptive automation, with a particular focus on drivers who are fitness impaired (e.g., distracted, drowsy, under influence of drugs and alcohol). The framework for adaptive automation offers the potential to improve the safety of the systems, or perhaps more accurately, the safety of the human–automation interactions. However, as the author rightly notes, such systems will only be useful if they can effectively leverage the capabilities of the automated system to complement or compensate for the momentary change in driver capacity. As with other potential design approaches (i.e., "solutions"), there is a dire need for data to corroborate the overall effectiveness of adaptive automation in terms of overall safety and performance as well as driver decision-making, in-vehicle behaviors, system use, trust, and acceptance. As it relates to impaired drivers, there are also many other related questions regarding human behavior. For example, how will the existence of (and, by extension, the use and reliance on) such systems impact drug and alcohol usage? As adaptive automation reacts to the momentary needs and capacities of an individual driver, will there be cascading effects throughout the entire system (e.g., Chapter 19)?

24.4.3   Automated Vehicle Design for People with Disabilities (Chapter 17)

Chapter 17 highlights some of the design challenges for people with disabilities. This includes a more deliberate consideration of accessibility. The social model of disabilities and many of the tenets of universal design provide some inroads for addressing the needs of these very different driving populations. They also reinforce the notion of the whole trip; for these and many other individuals, getting to and from the vehicle and getting into the vehicle are as important as what the automated vehicle does while enroute. For them it is not the first mile/last mile which is the only problem; rather, it is the first 100 and last 100 ft. Neglecting research that addresses the whole trip can amplify rather than reduce the inequities of those who are transportation disadvantaged. While many of the design principles discussed can improve driver's experiences, there is a general need for research that documents and quantifies these improvements. Given the breadth of drivers with disabilities, it will be important to understand how different driver characteristics interact with system effectiveness. Moreover, as the technology creates the potential for new drivers who were previously unable to drive, there will need to be a more profound understanding of how automation impacts their safety and mobility.

24.4.4 Importance of Training for ACIVs (Chapter 18)

Not all of the difficulties that face drivers can necessarily be overcome with good HMI design. The authors in Chapter 18 point to the importance that training can play in the successful deployment of various advanced technologies, addressing both the wide variation in driver's mental models and capabilities and the real challenges of learning the complex features of the advanced technologies. A number of recommendations were made given the current state of the art. First, training needs to be adaptable, changing in content with the constantly changing capabilities of the automated vehicles on the road. Second, training programs (however delivered) need to be evaluated. Such is generally not the case now. It is not clear just which types of training might eventually prove most efficient and effective. Third, driver state monitoring systems may well contain elements that are important to training programs and of which training programs could take advantage. For example, imagine a program that attempts to train the driver to remain engaged. It would be important for the driver to know when he or she became disengaged, something which can be difficult to self-monitor. A driver state monitoring system could provide drivers with information on their performance which might not otherwise be available. Fourth, guidelines need to be developed by state transportation officials and other stakeholders which detail just what essential elements need to be covered in a training program. Although such guidelines now exist in at least one state (California), they apply only to individuals who are field testing automated vehicles, not to the general population of drivers.

24.5 SPECIAL TOPICS

The special topics that were discussed in this section are the ones which pushed the envelope of our understanding of and proposed use for ACIVs, either by looking towards the future or by taking a step backwards and considering what has been learned in other domains. We now look back on these chapters and single out those issues that we see in most need of attention among the many that were mentioned.

24.5.1 CONNECTED VEHICLES IN A CONNECTED WORLD: A SOCIOTECHNICAL SYSTEMS PERSPECTIVE (CHAPTER 19)

It is fitting that this section starts with a discussion of the need to consider the overall sociotechnical system in which drivers are embedded, and that only by inspecting and acting upon the human factors concerns that arise out of this consideration can one truly experience the full promise of ACIVs (Chapter 19). At one level, this chapter can be considered a broadening of the importance of DSA that was discussed Chapter 13. The focus is on the overall system because it is only by doing this that emergent properties such as system safety, resilience, and overall effectiveness can be identified and promoted. We agree and, along with the author of Chapter 19, end up where he does. In particular, methodologies for system analysis and joint optimization are urgently needed to create the foundations that promote positive emergent properties such as the above.

24.5.2 CONGESTION AND CARBON EMISSIONS (CHAPTER 20)

Just as a consideration of systems in Chapter 19 is a broadening of the focus of the majority of chapters in the Handbook on individual vehicles to system safety, so too is a consideration of congestion and carbon emissions in Chapter 20, a broadening of the focus of the Handbook from the individual vehicles to system efficiency. Reductions in congestion and carbon emissions are critical, and the authors discuss ways to do so that rely on understanding psychological decision making. The authors start with a discussion of congestion. They describe how interventions that aim at behavior modification only—such as congestion pricing or techniques of nudging—without boosting underlying processes and competencies, might fail to promote pro-environmental driving-related behaviors, that reasoning and decision-making about one's own or others' driving seems to be predicted by simple rules of thumb, and that these simple rules of thumb, at least for parking, can increase system efficiency, compared to standard game-theoretic proposals. They suggest that research is needed on the design of the HMI that displays information to the driver in a way that influences their simple rules of thumb. Second, the authors go on to discuss carbon emissions. The editors have themselves taken note of the renewed research interest in eco-driving (McIlroy & Stanton, 2018). There is clear evidence that eco-driving can reduce carbon emissions by up to 45% with relatively low implementation costs (Sivak & Schoettle, 2012). However, the authors show that being motivated is not enough to undertake a pro-environmental behavior. Rather, providing drivers with clear information on their eco-driving performance appears to be the key. So, they recommend again a focus in future research on the design of the HMI that displays this information.

24.5.3 AUTOMATION LESSONS FROM OTHER DOMAINS (CHAPTER 21)

Humans have been interacting with automated systems in other domains for decades. In spite of the differences in operating environments, tasks, and user populations (to name only a few), many of the same challenges exist in these other domains as are now being discussed in the context of surface vehicles. While many of the relevant issues and human factors concepts focused on automation are discussed in other parts of the

Handbook, this chapter reminds us that there is important knowledge to be gleaned from other applications—aviation, air traffic control, military, process control—especially concerning some of the solutions that have been tried and tested. While the translation to driving might not be perfect in every case, it is worthwhile to reflect on this adjacent body of knowledge, especially given its sheer magnitude relative to the driving context. An open question remains: How to identify what lessons from other domains translate to what elements of vehicle automation?

24.6   EVALUATION OF ACIVs

Ultimately, one will want to evaluate ACIVs, not only on simulators but also in the field of both controlled and naturalistic studies. Many countermeasures to the human factors problems that might arise for drivers of such vehicles were discussed in the above chapters. However, no matter how principled, any solution is only a potential one until it is tested and evaluated. The remaining two chapters focus on evaluation.

24.6.1   HF CONSIDERATIONS IN TESTING AND EVALUATING ACIVs (CHAPTER 22)

The authors in Chapter 22 discuss in detail the many considerations which go into the design of evaluations of ACIVs, including iterative evaluations first in driving simulators, then in field demonstrations, and finally naturalistic studies. The authors emphasize the need in such evaluations to characterize the automation accurately for participants in order to develop training materials appropriate to the automation. In the absence of the 100s of millions of miles of observations that are needed to get meaningful information on crashes, we the editors believe the most critical determination still needs to be made of just what dependent variables should be used to evaluate the safety of the various systems. As just one example, at the macroscopic level there is still considerable debate on whether the dissection of near crashes can provide information relevant to the understanding of actual crashes (Knipling, 2015). How can we ever hope to evaluate the safety of automated vehicles when we have relatively little information on exposure if we cannot come to some agreement about what dependent variables best predict safety outcomes when such limitations arise? At the microscopic level, we are now relying on the distribution of especially long glances inside the vehicle to determine how risky are glances away from the forward roadway (Klauer et al., 2014). Yet, increasingly, information suggests that sequences of especially short glances on the forward roadway are also very risky (Samuel & Fisher, 2015). In short, we believe that more research is needed to isolate the dependent variables that best predict crashes and the behaviors that lead to crashes when exposure data are limited. The limited exposure data are a particularly acute challenge with increasingly automated vehicles because the automation and the automation–driver system have different failure mechanisms than drivers of conventional vehicles.

24.6.2   TECHNIQUES FOR MAKING SENSE OF BEHAVIOR IN COMPLEX DATASETS (CHAPTER 23)

The author in Chapter 23 provides abundant examples of tools that can be used to inspect complex datasets and tools that can be used, once a theoretical model is available, to analyze the complex dataset that are appropriate to the questions being asked (appropriate to the theoretical model). These complex datasets can help provide much more complete answers to research questions than were previously available, questions related to the human operator's engagement and performance in a vehicle over time and space. The research questions that can now be addressed (and need to be addressed) include: How do drivers respond to transitions in automation states? Will the human operator know when the automation fails? Will the human operator have sufficient time to take over when the automation fails? How quickly can drivers resume control from a highly automated vehicle, and does this resumption change with greater system exposure? What are the negative or positive effects of automation on driver's performance over time?

24.7 CONCLUSIONS

We want to conclude by thanking once again all of the authors who contributed to this Handbook. The creation, development, and completion of the Handbook have been a journey for many years and have enriched us as well as we hope the contributors and ultimately our readers. We have already probably gone on too long. So we leave you with one theme which has played centrally throughout the entire Handbook.

ACIVs are one key to reducing some of society's greatest problems. The sheer magnitude of the 1.4 million annual vehicle fatalities around the world is astronomical in terms of economic costs and psychological burden. On a more personal level, tragically so many of us know only too well someone who has fallen victim to a distracted driver, a driver who has fallen asleep, or perhaps a driver who has simply accelerated when he or she had meant to decelerate. The economic costs and psychological burden of carbon emissions are all becoming too clear. Congestion is reaching levels that were almost unbelievable ten years ago. More generally, commuting opportunities are the key factors in social mobility, even more so than factors related to crime and education (Bouchard, 2015). Individuals with mobility impairments could for the first time be freed from the very real constraints on transportation needs that come with those impairments. But the promise of ACIVs cannot be fully realized until the human factors concerns are addressed; in fact, the promise could be delayed considerably unless these concerns are addressed before they become an issue ("dread risk," Chapter 4).

ACKNOWLEDGEMENTS

Donald Fisher would like to acknowledge the support of the Volpe National Transportation Systems Center for portions of the preparation of this Handbook. The opinions, findings, and conclusions expressed in this publication are those of the authors and not necessarily those of the Department of Transportation, the John A. Volpe National Transportation Systems Center, the AAA Foundation for Traffic Safety, or the University of New South Wales.

REFERENCES

AAA (2019). *Advanced Driver Assistance Technology names: AAA's recommendation for common naming of advanced safety systems.* Retrieved from https://www.aaa.com/AAA/common/AAR/files/ADAS-Technology-Names-Research-Report.pdf

Bansal, G., Nushi, B., Kamar, E. W., Lasecki, W., & Horvitz, E. (2019, August 17). *A Case for Backward Compatibility for Human-AI Teams*. Retrieved from Computer Science. Human Computer Interaction, https://arxiv.org/pdf/1906.01148.pdf

Bouchard, M. (2015, May 7). Transportation Emerges as Crucial to Escaping Poverty. *The New York Times*. Retrieved September 28, 2019, from www.nytimes.com/2015/05/07/upshot/transportation-emerges-as-crucial-to-escaping-poverty.html

Boudette, N. (2019, July 17). Despite high hopes, self-driving cars are 'way in the future'. *The New York Times*. Retrieved from https://www.nytimes.com/2019/07/17/business/self-driving-autonomous-cars.html

Cunningham, M. & Regan, M. (2018). Driver distraction and inattention in the realm of automated driving. *IET Intelligent Transport Systems, 12*, 407–413.

Cunningham, M., Regan, M., Horberry, W., & Dixit, V. (2019). Public opinion about automated vehilces in Australia: Results from a large-scale national survey. *Transportation Research Part A, 129*, 1–18.

Global Toyota, Feature Innovation Automated Driving Technology Region U.S. (2019, January 8). *Dr. Gill Pratt, CEO, Toyota Research Institute CES 2019 Remarks*. Retrieved from Global Toyota, https://global.toyota/en/newsroom/corporate/26085202.html?_ga=2.245291527.341054311.1565554409–1615031356.1565554409

Kamar, E. (2016). Directions in hybrid intelligence: Complementing AI systems with human intelligence. *International Joint Conference on Artificial Intelligence*. Retrieved August 17, 2019, from https://pdfs.semanticscholar.org/38b5/fec2730e4e3224d97469-4d4b3522c0778e45.pdf?_ga=2.268500209.1071098841.1566058726–1666421389.1566058726

Kelley-Baker, T., Berning, A., Ramirez, A., Lacey, J. C., & Compton, R. (2017). *2013-2014 National Roadside Study of Alcohol and Drug Use by Drivers: Drug Results*. Washington, DC: National Highway Traffic Safety Administration. Retrieved August 17, 2019, from www.nhtsa.gov/sites/nhtsa.dot.gov/files/documents/13013-nrs_drug-053117-v3-tag_0.pdf

Klauer, S., Guo, F., Simons-Morton, B., Ouimet, M., Lee, S., & Dingus, T. (2014). Distracted driving and risk of road crashes among novice and experienced drivers. *New England Journal of Medicine, 370*, 54–59.

Knipling, R. (2015). Naturalistic driving events: No harm, no foul, no validity. *Proceedings of the Eighth International Driving Symposium on Human Factors in Driver Assessment, Training and Vehicle* Design. Iowa City: University of Iowa.

Litman, T. (2019). *Autonomous Vehicle Implementation Predictions: Implications for Transport Planning*. Victoria, British Columbia: Victoria Transport Policy Institute. Retrieved August 10, 2019, from www.vtpi.org/avip.pdf

Mayhew, D., Simpson, H., & Ferguson, S. (2006). Collisions involving senior drivers: High risk conditions and locations. *Traffic Injury Prevention, 7*, 117–124.

McDonald, A., Carney, C. & McGehee, D. V. (2018). Vehicle Owners' Experiences with and Reactions to Advanced Driver Assistance Systems. Washington, D.C.: AAA Foundation for Traffic Safety.

McIlroy, R. & Stanton, M. (2018). *Eco-driving: From Strategies to Interfaces (Transportation Human Factors)*. Boca Raton, FL: CRC Press.

National Safety Council. (2019). *Drowsy Driving Is Impaired Driving*. Itasca, IL: National Safety Council. Retrieved August 17, 2019, from www.nsc.org/road-safety/safety-topics/fatigued-driving

NHTSA. (2016, September 24). *Driver Alcohol Detection System for Safety*. Retrieved from NHTSA. Vehicle Safety, www.nhtsa.gov/Vehicle+Safety/DADSS

Sagberg, F., Fosser, S., & Saetermo, I.-A. F. (1997). An investigation of behavioural adaptation to airbags and antilock brakes among taxi drivers. *Accident Analysis and Prevention, 29*, 293–302.

Samuel, S. & Fisher, D. (2015). Evaluation of the minimum forward roadway glance duration critical to latent hazard anticipation. *Transportation Research Record, 2518*, 9–17.

Seppelt, B., Seaman, S., Lee, J., Angell, L., Mehler, B., & Reimer, B. (2017). Glass half-full: On-road glance metrics differentiate crashes from near-crashes in the 100-Car data. *Accident Analysis and Prevention, 107*, 48–62.

Sivak, M. & Schoettle, B. (2012). Eco-driving: Strategic, tactical, and operational decisions of the driver that influence vehicle fuel economy. *Transport Policy, 22*, 96–99.

Smith, R., Turturici, M., & Camden, M. (2018). *Countermeasures Against Prescription and Over-the-Counter Drug-Impaired Driving.* Washington, DC: AAA Foundationn for Traffic Safety. Retrieved August 17, 2019, from https://aaafoundation.org/wp-content/uploads/2018/10/VTTI_Rx_OTC_FinalReport_VTTI-FINAL-complete-9.20.pdf

Watson, A. & Zhou, G. (2016). Microsleep prediction using an EKG capable heart rate monitor. *IEEE First International Conference on Connected Health: Applications, Systems.* Retrieved February 28, 2019, from https://ieeexplore.ieee.org/document/7545850

Yadron, D. (2016, June 2). Two years until self-driving cars are on the road – is Elon Musk right? *The Guardian.* Retrieved August 11, 2019, from www.theguardian.com/technology/2016/jun/02/self-driving-car-elon-musk-tech-predictions-tesla-google